ECO-REGIONAL APPROACHES FOR SUSTAINABLE LAND USE AND FOOD PRODUCTION

Systems Approaches for Sustainable Agricultural Development

VOLUME 4

Aims and Scope

The book series *Systems Approaches for Sustainable Agricultural Development* is intended for readers ranging from advanced students and research leaders to research scientists in developed and developing countries. It will contribute to the development of sustainable and productive systems in the tropics, subtropics and temperate regions, consistent with changes in population, environment, technology and economic structure.

The series will bring together and integrate disciplines related to systems approaches for sustainable agricultural development, in particular from the technical and the socio-economic sciences, and presents new developments in these areas.

Furthermore, the series will generalize the integrated views, results and experiences to new geographical areas and will present alternative options for sustained agricultural development for specific situations.

The volumes to be published in the series will be, generally, multi-authored and result from multi-disciplinary projects, symposiums, or workshops, or are invited. All books will meet the highest possible scientific quality standards and will be up-to-date. The series aims to publish approximately three books per year, with a maximum of 500 pages each.

The titles published in this series are listed at the end of this volume.

Eco-regional approaches for sustainable land use and food production

Proceedings of a symposium on eco-regional approaches in agricultural research, 12–16 December 1994, ISNAR, The Hague

Edited by

J. BOUMA
Department of Soil Science and Geology, and C. T. de Wit Graduate School for Production Ecology, Wageningen Agricultural University, Wageningen, The Netherlands

A. KUYVENHOVEN
Department of Development Economics, Wageningen Agricultural University, Wageningen, and Netherlands Economic Institute, Rotterdam, The Netherlands

B. A. M. BOUMAN, J. C. LUYTEN
DLO-Research Institute for Agrobiology and Soil Fertility, Wageningen, The Netherlands

and

H. G. ZANDSTRA
International Potato Center, Lima, Peru

SPRINGER-SCIENCE+BUSINESS MEDIA, B.V.

Library of Congress Cataloging-in-Publication Data

Symposium on Eco-regional Approaches in Agricultural Research (1994 : ISNAR)
Eco-regional approaches for sustainable land use and food production : proceedings of a Symposium on Eco-regional Approaches in Agricultural Research, 12-16 December 1994, ISNAR, The Hague / edited by J. Bouma ... [et al.]
p. cm. -- (System approaches for sustainable agricultural development ; v. 4)
Includes index.
ISBN 978-94-010-4058-7 ISBN 978-94-011-0121-9 (eBook)
DOI 10.1007/978-94-011-0121-9
1. Agricultural ecology--Congresses. 2. Food crops--Ecology--Congresses. 3. Regional planning--Environmental aspects--Congresses. 4. Land use, Rural--Environmental aspects--Congresses. 5. Sustainable development--Environmental aspects--Congresses. I. Bouma, Janneke J. II. Title. III. Series.
S589.7.S94 1994
338.1'4--dc20 95-21138

ISBN 978-94-010-4058-7

Printed on acid-free paper

Contents

Preface

In the next four decades, the globe will need to triple its food and feed production, and almost all of the increase is needed for developing countries. This observation has a socio-economic and a biophysical side.

On the biophysical side, increasing productivity is a necessity. However, many intensive crop and animal production systems do not use natural resources (land, water, biodiversity) in ways that are physically and economically sustainable. Much of global agriculture overuses the scarce suitable land, and extends into fragile marginal lands. As a result, a spiral of unsustainability occurs there with disastrous results. At the other end of the spectrum, excessive use of inputs, too narrow crop rotations, and loss of biodiversity prevail. Unsustainability in eco-technical terms is threatening food production and food security. These trends are becoming more and more apparent. Achieving sustainable use of natural resources is therefore an indispensable component for increasing food production.

On the socio-economic side, increasing production requires broad-based economic development, as poverty must be overcome for the landless to buy food and for farmers to purchase external inputs to produce sufficient food. Economic development takes place in communities and regions.

Three inseparable sides of the development triangle are therefore: a regional approach to reduction of poverty, efficient and sustainable use of natural and human resources in the region, and increased food production. raditional component and commodity research addresses these issues too narrowly and at too small a scale. Rural development needs an eco-regional approach that integrates biophysical and socio-economic understanding of cropping systems, livestock, the environment, and natural resources.

As natural resources management becomes a more important part of the CGIAR's agenda, systems analysis and modelling capabilities are more needed in CG-centers and national agricultural research systems. The integration of insights in the biophysical and socio-economic aspects of commodity technology, cropping systems, livestock and natural resource management will require these new capabilities. Experimentation and hypothesis testing will more and more be based on the results of systems analysis and modelling work, as will extrapolation of results of location specific observations.

The Directorate General for International Cooperation (Special Program 'Research') of the Netherlands Ministry of Foreign Affairs initiated and supported generously the organization of this meeting. ISNAR and the Nether-

J. Bouma et al. (eds.), Eco-regional approaches for sustainable land use and food production, ix–x.

lands Ministry of Agriculture, Fisheries and Nature Management also provided welcome support. The symposium emphasized that the rural development triangle requires major changes in focus of international agricultural research, and allowed discussion of some case studies in which eco-regional approaches are practiced. The meeting stimulated collaboration among participants and wider application of natural resources research techniques discussed.

We hope that this book will inspire many other scientists and research leaders.

F.W.T. Penning de Vries (AB-DLO)
R. Rabbinge (TPE-WAU)
H.G. Zandstra (CIP)

Introduction

The symposium "Eco-regional approaches for sustainable land use and food production" was initiated by a request from the Directorate General for International Cooperation (DGIS) of the Ministry of Foreign Affairs in The Netherlands. The eco-regional approach to research is increasingly being embraced by CGIAR Institutes and NARCs. This calls for periodic meetings where methodology is discussed and improved, while results obtained are subjected to discussions within interdisciplinary teams which are well versed in systems analysis. This book presents invited papers, results of group discussions and some general conclusions and recommendations on eco-regional approaches.

Eco-regional studies emphasize the use of interdisciplinary approaches to characterize relations between different agro-ecological production systems and limited natural and socio-economic resources of the region. Each region has specific characteristics to the extent that it can be considered as a relatively homogeneous unit of analysis. As is made clear in the papers being presented, the region is only one level to be considered, be it a level that is attractive from a land use planning point of view. Data at farm level, however, form the backbone for defining land use options, even at regional level, while the regional level feeds analyses at higher levels ranging from a province or state to the international level.

One element of the discussions during this symposium has been raised many times before and deals with the most appropriate focus of eco-regional studies in general. During the conference, emphasis was placed in several papers on exploratory studies of the spatial effects of different agro-ecological land-use scenario's in terms of defining realistic options for land use. Whether or not such options can be realized depends on the socio-economic environment: services and infrastructure determine which enterprises can be supported and markets determine whether and to what extent it is worthwhile to produce certain crops, lifestock or forestry products. Defining options that are feasible from an agro-ecological point of view can certainly be illuminating, but lack of further research in terms of supporting the ultimate selection process where many other factors have to be considered as well, can make agro-ecological options by themselves rather academic.

Exploratory studies based on agro-ecology provide important "building blocks" to be used or discarded when final decisions about land use are derived by farmers and planners. For them, agro-ecology is only one element

J. Bouma et al. (eds.), Eco-regional approaches for sustainable land use and food production, xi–xiii.

to be considered. Clearly, current land use patterns in the world are not a primary result of agro-ecological possibilities, but, rather, of more restrictive socio-economic conditions. Also, land units often do not correspond with areas that are characterized by homogeneous socio-economic conditions. So, truly scenario-type rather than exploratory studies are more difficult and comprehensive, and these types of studies have not been well represented at this conference. Aside from recommendations made during this conference, the Editorial Board stresses the need for more scenario studies in which the implementation process of agro-ecological options is being explored. Information technology has to play a crucial role in this process as is illustrated for some exploratory studies. In fact, our desire to realize exploratory and scenario studies on land use may be realistic only because we are only recently in a position to collect, manipulate and visualize immense volumes of data from different disciplines.

The ample availability of data sets on eco-regional issues also adds a new dimension to the problem of effective multi-disciplinary co-operation. Whereas the need for a multi-disciplinary approach to agro-ecological questions is widely recognized, the different and often conflicting perceptions held by participating scientists regarding their own discipline and paradigms versus those of others often leads to prolonged debates on the interpretation of data, concepts and models. For such debates to be effective, a common perspective on agro-ecology is essential. The Editorial Board is confident that papers and discussions during the symposium have contributed towards reaching such a common perspective. Scientists from biophysical and socio-economic disciplines have not merely commented on each other's work, but in an encouraging number of contributions shown how co-operation in eco-regional research can be achieved.

As various micro-level studies in the different sessions illustrate, multi-disciplinary co-operation appears at present most succesful at the farm household level. Agro-technical options for the various relevant enterprises are manageable, the socio-economic environment in terms of markets, institutions and infrastructure is largely exogenous to the farm household, and the actor's preferences (and constraints) translate fairly well into a simulation of land use through crop, livestock and forestry selection mechanisms. As a result, the impact of the farmer's allocation of resources of e.g. the development of new agro-technical options, the removal of institutional constraints, changes in socio-economic incentives or the provision of improved infrastructure can be estimated within reasonable margins of confidence.

The focus on farm households has, however, limitations as research findings are seldom directly applicable at higher levels of aggregation. In most cases agro-technical improvements require simultaneous changes in the socio-economic environment, an area which traditionally remains outside the scope of farming systems analysis. A focus on agriculture from an eco-regional perspective opens the possibility of linking findings at the farm household level with policy advise regarding the incentive function of the socio-economic environment, while maintaining a meaningful integration of bio-physical and socio-economic contributions. A regional perspective further enables the incorporation of linkages between agriculture and other sectors of the economy. This adds to the comprehensiveness of the analysis as a considerable proportion of the rural incomes are derived from other than agricultural activities. Findings at the regional level can in turn be related to the national level, facilitating policy analysis which reflects broad goals of development such as income and employment generation, fairness in distribution and environmental concerns. Among the micro building blocks, farm households are of central importance in such a regional approach, and the common ground for exchange among the participating disciplines remains unimpaired.

Higher levels of analysis often lack such an immediate common disciplinary ground because at these scales agro-ecological as well as socio-economic circumstances show substantial internal variation, and require a different conceptualisation and categorization for meaningful analysis. Aggregative exploratory studies reflecting agro-technical outer boundaries pose, as explained above, such dilemmas of conceptualisation. Being supply-oriented, the neglect of socio-economic demand and distributional considerations may blind the eye for the consequences of continued relative price decreases for regional agricultural development: discouragement of investment in agricultural and environmentally-related activities, slower technical advances in research and development, and, as a result, an increased socio-economic relevance of physical and ecological constraints on the future expansion of food production and agricultural activities in general. In this case the nature of the constraints is not necessarily agro-ecological. The problem takes on a socio-economic nature: incentives to mobilize the resources to conserve the environment and to develop new knowledge and better infrastructure may simply be lacking.

The editors

SECTION A

Eco-regional approaches

Eco-regional approaches, why, what and how

R. RABBINGE

Netherlands Scientific Council for Government Policy, P.O. Box 20004, 2500 EA The Hague, and Department of Theoretical Production Ecology, Wageningen Agricultural University, P.O. Box 430, 6700 AK Wageningen, and C.T. De Wit Graduate School for Production Ecology, Wageningen, The Netherlands

Key words: aggregation level, objectives, systems boundaries

Abstract. Eco-regional studies and approaches are more and more demanded by policy makers. They help to make strategic decisions and to explore options for development. The possibilities for such studies have increased as appropriate research tools became available, information could be collected and integration of disciplinary knowledge at various aggregation levels could take place through systems approaches.

The combination of detailed biophysical information, with farming systems research and with socio-economic studies on macro and meso levels enabled the development of explorative, predictive and analytic studies at regional level.

Thus, eco-regional research helps in priority setting at lower integration levels and may give an answer to questions formulated at the regional level. A solid scientific basis of these studies could be given by systems approaches at all integration levels that integrate detailed biophysical studies and socio-economic research.

Introduction

Eco-regional approaches may become the buzzword for the research agenda of the nineties and could easily be considered as old wines in new bottles. That would be a pity as eco-regional approaches may help to answer questions at the regional level and fulfil the promise of integration of disciplinary approaches which took place in the biophysical and socio-economic sciences. They may integrate studies at various integration levels, bridge the gaps between basic sciences and applied sciences, form an unifying problem oriented concept for the disciplinary contributions and create new rewarding research questions. Their possibilities as vehicle in understanding, gaining insight and exploring options for development are increasing as many new tools are becoming available.

A description of eco-regional approaches is given in this contribution, why they are so important, and how they can take place.

J. Bouma et al. (eds.), Eco-regional approaches for sustainable land use and food production, 3–11.

Eco-regional approaches, why and what

The need to understand better how regional development and changes affect the possibilities and limitations of individuals in a region – farmers, other land users and other economic actors – is increasing. Studies at lower levels of integration than farm level cannot be generalised since they do not take into account changes in the environment at regional level that dictate the possibilities at lower integration levels. Land use studies and productivity studies at field or farm level often neglect the change in het biophysical and socio-economic environment in terms of constraints or new chances. Therefore the assumption that the environment is considered as an independent forcing variable is not true.

These changes may result in new constraints for example water availability and nutrient shortage or in new chances due to better possibilities to buy external inputs.

Eco-regions may be considered as systems with well defined boundaries within which farms and other elements and their interaction take place. The eco-regional approach enables the systematic study of (changes in) land use, and the study of agricultural systems within these systems. Biophysical and socio-economic knowledge and insight is further developed and integrated or synthesized to investigate and understand better how these systems function and how agricultural systems operate within these regions. Those studies are combined in eco-regional studies that explore possibilities at the regional level. Various objectives and constraints may be distinguished at the regional level that are completely different from the same characteristics at the farm level.

In the eco-regional approach a region is identified by its natural, administrative or socio-economic boundaries, within which the main rural and land developmental issues are made explicit. All important relations between agro-ecological systems and other forms of land use are specified, in particular those related to limited natural resources and limited socio-economic resources. In that way the eco-regional approach emphasizes the specific characteristics of a region. These specific characteristics help to identify the possibilities and constraints of these regions. Those constraints may be biophysical or socio-economic. The first may be dictated by climatic conditions and limited natural resources such as water and nutrient availability or external inputs or by lack of technological insight and knowledge. The latter by (skilled) labour, capital or insight in and availability of markets.

Biophysical and socio-economic objectives may play in all situations an important role. The need of studies at an eco-regional level has increased considerably as policy objectives and constraints are often defined at that level and determine the possibilities of farming systems.

Eco-regional studies are at the intermediate level between the farming systems approaches and the macro economic analysis that more and more consider the world as a global village. During the last decade many tools have become available that enable eco-regional approaches of a high quality. Geographical information systems (GIS) are used as appropriate tools for describing, analyzing and characterising eco-regional systems. Crop models may be used as integrating tools and explorative instruments to determine the potential and attainable yields. Many types of scenario approaches such as Interactive Multiple Goal Linear Programming (IMGLP) are becoming available with which various scenarios may be developed. Both help to derive options and to make general intentions and aims such as sustainable development operational.

Eco-regional approaches, how

The aim of the eco-regional approach dictates the methodology and the tools that should be used. Explorative studies aiming at the exploration of various possibilities are different from predictive studies that investigate what may be expected in the near future: which are again different from descriptive analytic studies that aim at understanding the interactions within the eco-regional system, and their response to changes in the environment.

In predictive and explorative studies a time horizon is needed. Predictive studies using the analysis of the present and contemporous development as main elements need a time horizon that is not too far away (less than 10 years). Trends and developments of the recent past dictate in those studies the possibilities and expectations for the near future The predictive studies are like weather forecasts. The closer the time horizon the more reliable the outcomes of such studies. Long term predictions are dangerous as they presume absence of discontinuities in trends. They may accept some variability in outcomes but always around a general trend. In predictive studies, trend analysis of certain and uncertain characteristics of systems play an important role. The reliability of forecasts (predictions) is determined by the accuracy of the input relations in connection with the environment. Plausibilities in those studies are measured as reliability of the prediction. Therefore probability is a good quality criterium for such studies.

There are many examples of such studies. The majority of eco-economic models for regions or for countries such as those developed by LEI-DLO, SC-DLO, Worldbank and many other institutions are used for these predictive analysis. The study of the Dutch Central Economic Planning Agency designs three scenarios, 'global shift, European renaissance and Balance Growth'. In each of these scenarios present developments are considered and amplified or cut back. The spectrum of possibilities is wide and shows the biotechnical, agro-technological or technical economic negative feedback are implicit in those studies and based on judgements of the authors.

Explorative studies are of a completely different nature. They explore the future on basis of technical knowledge and normative and technical objectives and constraints. Technical knowledge is in most cases based on insight of the way biophysical systems operate. In agro-ecosystems such technical limits may be found in the amount of radiation arriving from the sun, the temperature and conservative characteristics of the systems such as the physiological traits that describe photosynthesis. Potential, attainable and actual production levels may be distinguished that are different in different production situations (Rabbinge 1993).

It is the task of biophysical research to identify how growth and yield defining factors can be used at field but also at higher integration levels and how they can be affected in such a way that in technical terms optimal crop and cropping systems are developed. In technical terms means in this case that the efficiency of the external inputs per unit of product is maximized or that the emission of nutrients per unit of product is minimized. It may be desirable that for ecological reasons the emission of nutrients and immission of pesticides per unit of area should be minimized and that would mean in good production situations a suboptimal production technique in terms of efficiency per unit of product for the other inputs.

Thus the biophysical sciences enable an explicit discussion on what preference should be given to various objectives. It shows also that some ecological aims are within the ecological domain conflicting and not conflicting with socio-economic aims. It is the contribution of biophysical sciences to eco-regional studies to design, production technologies, to identify and quantify the various aims and the possibilities to reach those objectives, to gain better insight in agro-ecosystems and to use that in the improvement of those systems.

Examples of eco-regional approaches

Demonstrations of explorative studies in this book are given by Penning de Vries *et al.* (1995), Van Latesteijn (1995), and Aggarwal *et al.* (1995).

The study of Van Latesteijn (1995) is explorative and shows how various objectives (socio-economic, eco-logical and agri-cultural) can be achieved by optimizing land use. Various scenarios of land use depending on the relative weight of various objectives are developed. The study of Penning de Vries *et al.* (1995) is also explorative but of a different nature. The biophysical limits to feed the world population by various production systems are investigated under potential and water limited circumstances. That type of study was also done some twenty years ago (Buringh *et al.* 1975). However, the explorative studies of Penning de Vries *et al.* (1995) take water availability, production technologies and required food in terms of different diets into account. Moreover, a more detailed land evaluation using grids and dominating biophysical soil and climate characteristics is used.

The results show that there are wide margins on a world scale and considerable differences between various regions. In some regions 'anything goes', in other regions all means are necessary to attain sufficient food for the population in those areas. The possibilities to feed a world population with different technologies and diets vary between $11 \cdot 10^9$ to $44 \cdot 10^9$ capita. In the latter case, still only 30% of the world's area is in use for agriculture that is maintained in a sustainable way.

The danger of these explorative studies to overestimate the real possibilities to achieve the potentials can not be denied. However, studies of a predictive nature tend to overestimate negative developments such as soil degradation. Studies of the Worldwatch Institute predict that $5 \cdot 10^9$ humans are too much for the carrying capacity of the world of Brown *et al.* (1993). These studies indicate that $2 \cdot 10^9$ human beings are more or less the maximum carrying capacity of mother earth.

The considerable differences between predictive and explorative studies may help to gain understanding in how to intervene. For that reason, analytic knowledge at various integration levels is necessary. Institutional constraints may prevent a proper use of the possibilities and limit the use of adequate policy instruments. To identify such constraints and to enable interventions that may help to reach the desired explored perspectives, good knowledge of farm systems and the way they function in regions is needed.

In that way explorative, predictive and descriptive/analytical studies are complementary. They may stimulate each other and through synergism help to stimulate regional development.

Examples of eco-regional approaches that aim at prediction are found in many programs of the CGIAR institutes. The eco-regional initiatives in West Africa and the Andean highlands are at present dominated by predictive analysis of the changes in land use. They show the devastating effects of a continuation of such developments. It is for that reason that there is an increasing demand for explorative studies that complement these discouraging explorative analysis. A detailed and appropriate analysis of the way the various actors function in eco-regional systems may help to design efficient and effective interventions that bring explorative possibilities more in line with predictive dystopia.

Linkages between various integration levels

Tools such as crop modelling, GIS, and IMGLP techniques have become well developed and mature instruments that enable problem oriented eco-regional approaches.

Systems approaches have already been used to identify knowledge-based management strategies to improve resource use management at the crop level with a minimum negative environmental impact of Rabbinge (1993). Tools such as crop simulation models, that can predict the yields attainable in a range of environments, have been developed to generate technologies targeted at specific local environments by Van Latesteijn (1995).

Constraints for growth in productivity stem from biophysical and socio-economic characteristics at regional levels including water resources. Systems approaches should therefore be extended to regional levels in order to identify the relevance of these constraints and to develop solutions. Options for land use at regional levels that meet region-specific goals must incorporate spatial information on environmental and socio-economic constraints. Several tools, such as GIS and land evaluation systems are used, and new methodologies of Penning de Vries *et al.* (1995) have been developed to integrate agro-ecological and socio-economic approaches at regional levels. Although spatial information is often incomplete, an eco-regional approach is meant to integrate that information, determine lacks of knowledge and to make the knowledge and insight operational. The eco-regional approach to problems in agriculture is schematically presented in figure 1.

The relationships between components of production systems and the environment must be studied to understand better the behaviour of the systems. Knowledge from studies at the level of genome, plants and crops of water and nutrient use, of consequences of pests, and of effects of alternative management techniques are integrated and engaged to define production technologies tailored to specific needs and opportunities. Implications of different scenarios for environmental changes and improved production systems, can be made visible by using optimization techniques. Socio-economic objectives, constraints and analyses are an important part of such optimizations. Depending on the priority given to various objectives such as cost minimization, employment, or use of natural resources, the results of the scenarios are very different.

The eco-regional approach makes explicit the choices for various ways of agricultural land use that can be made, and the unavoidable trade offs between objectives. The new insights can also be used to prioritize crop research and to identify needs for additional biophysical and socio-economic information. The various links are presented in figure 2.

These studies are still characterized by explorative analysis and not so much by realistic designs for future developments. That may only be possible when a synthesis between the various approaches is operational. Another advantage of this approach is that detailed biophysical research at crop or cropping systems level is focused and integrated to higher aggregation levels.

The detailed research projects and questions at the lower level are derived from the higher aggregation level. By an interative way of working priority

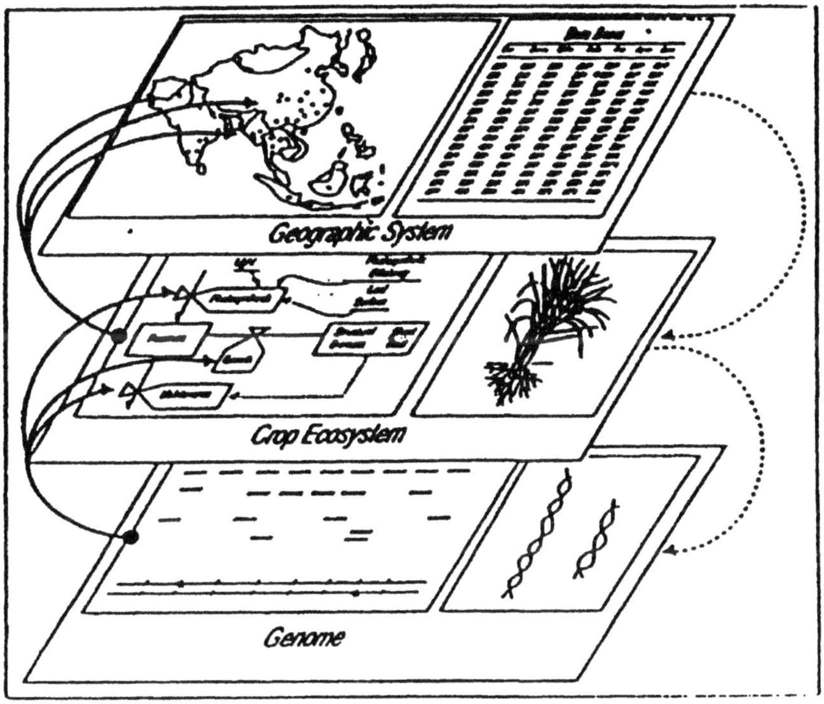

Fig. 1. Systems approaches to link agricultural research activities at different scales (source: IRRI 1994).

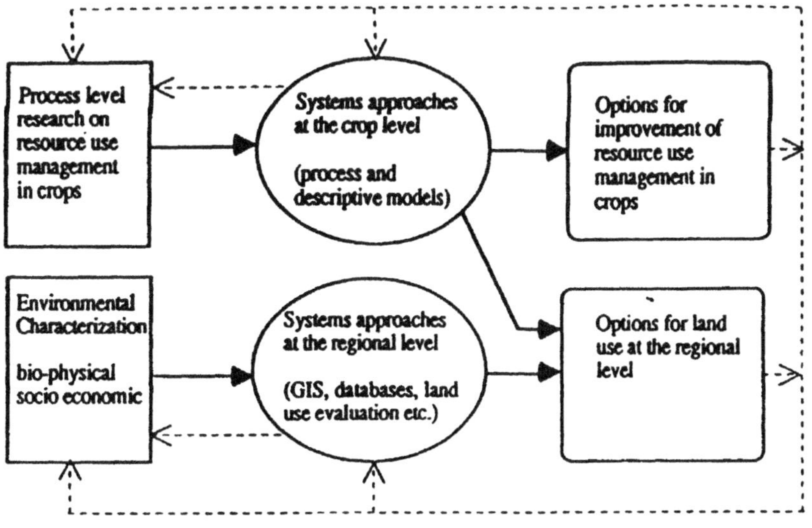

Fig. 2. Research of structure eco regional networks. Dotted lines indicate feed-back information.

setting in detailed process research is determined by the problem solving synthesizing research activities at higher levels. The model activities at the regional level are in this example of an explorative nature.

Perspectives for eco-regional approaches

The perspectives of eco-regional approaches are bright. They may fulfil various expectations such as:

- problem oriented research;
- take into account the biophysical and socio-economic characteristics of regions;
- integrate contributions of various disciplines;
- are wide without loosing deepness;
- enable the operationalisation of vague intentions such as sustainable agriculture;
- demonstrate how resource use efficiency may be optimized at regional levels.

There are, however, also some dangers in the eco-regional approach:

- they may be restricted to methodologies;
- they may lead to too much tool orientation losing the problem orientation;
- they may lead to delay of interventions and may be used as an alibi for not doing things.

It is the task of the CG-institutes, the NARS and advanced systems groups to use the chances and to prevent a development away from problem-oriented research. In this book the various tools and institutional requirements will be discussed. Ultimately that should lead to an improved collaboration and to problem solving at regional level. Eco-regional studies are no aim in themselves, they should be used for a better guidance of regional development within the context of a global rapidly changing environment.

Acronyms

CGIAR	Consultative Group on International Agricultural Research
GIS	Geographical Information System
IMGLP	Interactive Multiple Goal Linear Programming
NARS	National Agricultural Research System

References

Aggarwal P K, Kalra N, Bandyyopadhyay S K, Selvarajan S (1995) A systems approach to analyze production options for wheat in India (pages 166–186 of this volume).

Brown, L R, Durning A, Falvin C, *et al.* (1993) State of the World. W.W. Norton Company, New York, USA.

Buringh, P, Van Heemst H D J, Staring G J (1975) Computation of the absolute maximum food production of the world. Dept. of Tropical Science, Wageningen Agricultural University, Wageningen, The Netherlands.

IRRI (1994) Mid term plan. International Rice Research Institute (IRRI), Los Baños, Philippines.

Van Latesteijn H C (1995) Scenarios for land use in Europe: agro-ecological options within socioeconomic boundaries (pages 43–63 of this volume).

Penning de Vries F W T, Van Keulen H, Rabbinge R (1995) Natural resources and limits of food production in 2040 (pages 65–87 of this volume).

Rabbinge R (1993) The ecological background of food production. Pages 2–29 in Crop protection and sustainable agriculture (Ciba Foundation Symposium 177), Wiley, Chichester, UK.

Facing the challenge of the Andean Zone: the role of modeling in developing sustainable management of natural resources

R.A. QUIROZ, RUBÉN DARÍO ESTRADA, C.U. LEÓN-VELARDE and H.G. ZANDSTRA

Respectively, Livestock Production Systems Specialist, Agro-Economist, Livestock Systems Specialist, CONDESAN/CIP, and Director General CIP, International Potato Center, Apartado 5969, Lima, Peru

Key words: Andes, eco-region, modeling, simulation

Abstract. The Andean eco-region faces problems of population, agricultural productivity, pressure for land, and overexploitation of natural resources. Actual knowledge of the Andean eco-region is vast, but results from research and development have been site-specific and have been seldom successfully integrated. CONDESAN, the Consortium for the Sustainable Development of the Andean eco-region – through associate institutions is developing a methodology to evaluate different scenarios to improve land and water management for a sustainable and more productive agriculture.

The methodology integrates the available knowledge of crop, livestock and soil loss simulation models. Technical coefficients obtained from research at different sites are used as inputs in animal (alpaca, sheep, cattle) production simulation models. These models are linked to the EPIC model to evaluate the impact of alternative management practices on long term soil erosion and crop production. A regional economic impact model combines the outputs of the previous models to evaluate productivity, loss of soil fertility, adoption of land use alternatives, and marginal benefits accrued by farmers and by society over a time span of 30 years.

The paper addresses different land use scenarios at both farm and regional level, simulating the most commonly encountered constraints in the Andean farming systems. Alternatives that minimize natural resource erosion and increase households' profitability were selected and analyzed. The IRR ranged from 5 to 56% with and overall value of 9.23%. Benefits accrued roughly equally to farm households and to society as a whole.

Introduction

The Andean eco-region is increasingly receiving national and international attention. The reasons include high population density, low productivity of agricultural systems, large agro-ecological variability, threats to biodiversity, increasing pressure on land use, and over-exploitation of natural resources, among others.

Several research and development (R&D) projects, from different funding sources, have attempted to improve the well-being of Andean households. But results have been site-specific and seldom successful. This implies that

J. Bouma et al. (eds.), Eco-regional approaches for sustainable land use and food production, 13–31.

traditional ways of designing R&D projects for the Andes must change in order to allocate resources more efficiently and therefore alleviate widespread poverty.

Current knowledge of the Andean eco-region is vast, but it has not been integrated; therefore, it cannot be efficiently used to design and implement R&D projects (CIP 1993).

Reasons for unsatisfactory implementation of many agricultural projects in the Andes do not differ from the ones given for projects in other less developed regions: inappropriate technologies, inadequate support systems and infrastructure, the failure to understand and appreciate the social environment, project administration problems, and general policy environment (Anaman 1993). Poor design and appraisal also affect project performance.

How to improve the possibilities of success for agricultural and land and water resources management in the Andes is a key question that must be addressed by people involved in projects.

CONDESAN (the Consortium for the Sustainable Development of the Andean Eco-region), via its associate institutions throughout the Andes, is currently involved in developing a methodology that could be used to improve the efficiency of natural resources management for agricultural projects in the Andes. Several strategies have been or are being implemented. They emphasize integration of available knowledge at different hierarchical levels, efficient communication between people and institutions within and outside the eco-region, field studies that evaluate the performance of land use and commodity systems, and policy studies.

Simulation models have been widely used to study biophysical relationships in both plants and animals. Farm models are also used to analyze the interaction between the biophysical and socio-economic components. Models that integrate information from different levels in a hierarchy are much less frequently used.

This paper describes the experience developed in CONDESAN in the use of simulation models that operate at different hierarchical levels to design R&D projects in the Andes.

Issues

New trends in sustainable agriculture have raised questions that require a closer look by researchers and developers. Among the key issues we find:

- **Micro/macro relationships**. Can research results developed at a micro level be linearly extrapolated to a region? Is it realistic to assume massive adoption of technologies just because they proved to be better at a micro level? What mechanisms should we use to increase the amount of grass-roots information considered in policy making?

- **The trade-off between poverty and degradation of natural resources.** Should we expect alienated, poor peasant families to adopt conservation practices irrespective of economic outcomes? Who benefits the most from conservation practices: rural households, urban society, present or future generations?
- **Sustainable technological alternatives.** Are families using sustainable practices? What changes are required? What is the expected long-term impact of improved technologies? Are technologies that are sustainable at a micro level also sustainable at a regional level?

Some of these questions can be addressed by simulating different scenarios at both the farm and regional level.

Methodology

The proposed methodology integrates the experience accumulated by R&D projects and simulation models. The first step is the exhaustive review of R&D efforts in a target area. Validated technologies are analyzed and conditions and technical coefficients are summarized, and then used to simulate different scenarios and select the most promising ones for improving profitability and reducing erosion of the natural resources involved.

Technological alternatives are selected by using linear programming models, by generating a response surface with a central composite rotable design (León-Velarde and Quiroz 1994), or both. The temporal frame of the evaluation is 30 years or more. An optimal response is always designed, based on profitability and erosion of natural resources, and only the alternatives giving the required responses, at a farm level, are used for the next steps in the analysis. All the others are rejected.

Selected alternatives are evaluated at the regional level. This evaluation includes variables such as soil erosion, productivity, and the adoption rate of the alternatives. Soil erosion can be evaluated by two different approaches. For shallow soils, we can use the decline in productivity. But this does not work in deep soils, because no relevant changes in productivity are detected. Here, we use international prices for the most important nutrients.

Two adoption curves are generated. The first one refers to natural adoption of alternatives. This implies that changes are expected even when R&D projects are absent. The second curve describes the adoption expected with the application of the project. The difference between the curves is used to estimate the benefits for the economic analysis of the project.

Total changes in practices of target farmers can seldom be attributed to a particular project. Previous projects in the area and research results from institutions outside the project make important contributions to the outcomes. Therefore, just part of the outcome (the difference between the two curves) should be used for the economic analysis. The proportion of the outcome directly attributed to the project is different in every case.

Within a regional perspective, we estimate the marginal benefit for each hectare for which a particular technology is adopted. The benefit is divided into internal (accruing to the farmer) and external (accruing to the society). Society benefits from decreased erosion, the added value of specific products, and decreased silt accumulation in dams for hydroelectric plants and potable water. We also calculate a marginal benefit for the whole target region.

1. Simulation models

Several models are used, according to specific objectives. Farm and regional models are always used together. Farm models include soil erosion, crop productivity, and animal production. The results from the farm models are inputs for the regional impact model.

2. The EPIC model

The erosion/productivity impact calculator (EPIC) is a model designed to predict the impact of different crop management practices on long-term soil erosion and crop production. It has been described in detail by Sharpley and Williams (1990). Basic data for this model include: physical-chemical characteristics of the soil, slopes, monthly rainfall and its deviation, skewness coefficient (optional), wind speed and directions, crops and crop rotation, and tillage. Predicted soil erosion did not differ from the values found by Felipe-Morales (1994).

3. Animal production

Three simulation models, developed within CONDESAN, are used to assess the bioeconomic outcomes of different technological scenarios at the farm level. All three models have been validated with data collected over several years. Model predictions did not vary from actual values ($P > 0.05$).

a. Alpaca

This model was described by Arce *et al.* (1994). It is energy driven. Its most important subroutines are: pasture growth, voluntary intake, energy metabolism, gestation, lactation, and offspring. The voluntary intake subroutine has a random normal function. All the other subroutines are deterministic. The model simulates the performance of all the categories of animals in a flock, grazing up to three different types of pastures.

b. Sheep
The sheep model is similar to the alpaca model. Functional relationships for the metabolic processes in sheep were adapted for the smaller animals encountered in the Andes (Cañas *et al.* 1991).

c. Beef/dairy cattle
This is a single animal model. Besides energy metabolism, the model considers protein, calcium, and phosphorus requirements. Animal production is controlled by the most limiting nutrient. In addition to the random function for voluntary intake, the model has a random sampler for minimum temperature and wind speed. The energy required to maintain the animals within a thermal comfort zone is estimated. A linear program is built in to complement the model, so the user can optimize the ration to be used before running the simulation model. Once an optimal ration is obtained, the user can vary any desired feedstuff or supplement (Quiroz *et al.* 1990).

4. Regional economic impact

This model was developed in collaboration with the International Center for Tropical Agriculture (CIAT), the International Potato Center (CIP), and CONDESAN. It was originally developed to analyze the impact of crop management practices on natural resources conservation within a watershed. It has been modified to fit the needs of different CONDESAN projects throughout the Andes, including animal production.

This unpublished model has several features:

- Several crops or animal production systems can be simulated simultaneously for the number of years defined by the user.
- The most important inputs are: crop rotations, productivity, soil erosion, prices, and costs.
- The economic impact of technological alternatives is based on logistic adoption curves (see León-Velarde and Quiroz 1994). We use two curves: the first refers to a natural adoption, which assumes technological changes without the presence of the project. The second is the adoption due to the project's efforts to change natural resources management in the target population. The difference between the two curves is defined as the impact of the project.
- Total profitability considering the time value of money (net present value – NPV) is estimated, at arbitrary discount rates, for reasonable time frames (usually 35 to 50 years).
- Total investment and total profitability are used to estimate the project's internal rate of return (IRR).

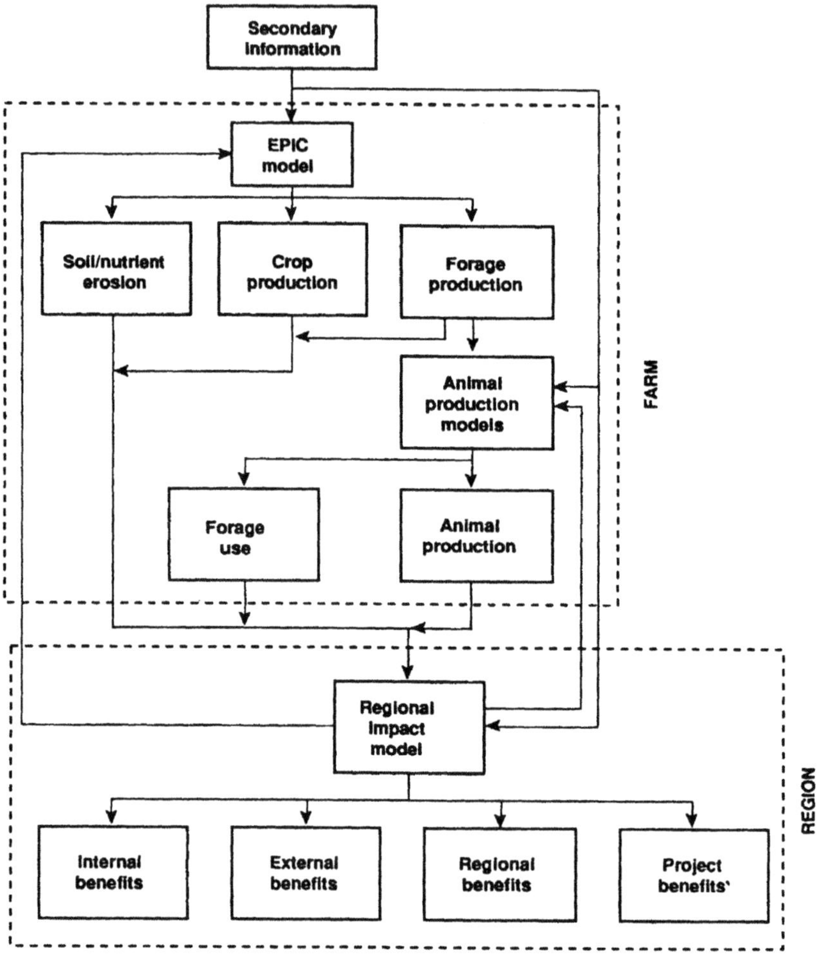

Fig. 1. Information flow among simulation models.

- Two types of benefits are estimated: those accruing to the household (internal to the farm) and those accruing to society (external to the farm).

5. *Integration*

The EPIC model is used to predict soil erosion and productivity of the different crops involved (figure 1). In addition, this model estimates forage production. The amount of soil eroded and the nutrients lost (converted to the equivalent amount of money that would be required to replace the nutrients in deep soils, or changes in productivity in shallow soils), and the level of production for each crop, are inputs for the regional economic impact model.

Forage production, as growth rate (kg DM day^{-1}), estimated with EPIC is an input for the animal models. An estimated value of the monthly digestibility of the forages is introduced into the model. Flock composition and grazing management are defined for each case. The models calculate the animal performance for all the chosen simulation conditions. Optimal alternatives for a single type of farm are selected using a central composite rotable design. For instance, if we want to evaluate three different treatments with three levels, 20 treatment combinations are run. Only the central treatment of the response surface is replicated six times (six runs). We assume that the variability of the central treatment is common to the axial treatments. Every 11th run is taken as a replication, to have a good spreading of the outcome of the random generation function.

Production values (meat, fiber, milk, manure) from the animal models are inputs for the regional economic impact model. The data generated with all the farm models are complemented with average market prices for each product. Labor and inputs to the farms for each activity considered are given a value and used in the calculation. Expected project expenses are introduced as costs to calculate the benefit.

Only the alternatives selected by the farm models as viable are evaluated with the regional economic impact model.

Case study

The following example shows how the methodology works. All the models described were used to formulate a project for the southern Sierra of Peru. The target area includes eroded areas above 2000 m a.s.l. of four departments: Apurímac, Ayacucho, Cusco, and Puno. The project aims to improve soil and water management in 800 peasant communities with an innovative training and extension methodology that relies on contests among households and communities (Van Immerzeel and Núñez del Prado 1994).

The most important farming systems encountered can be grouped in three types:

a. Alpaca production systems

These systems are located above 4000 m a.s.l. Frost risk is present throughout the year. Alpaca is the predominant animal species. Llamas and, on a minor scale, sheep are found. Most of the farmers have access to more than 80 ha, of which about 90% are dry rangelands. The rest of the area has natural irrigation, with water coming from high snowy peaks. The irrigated areas are called 'bofedales'. Forage species growing in this area have excellent nutritional quality and produce a relatively large amount of biomass.

Rotation patterns include grazing rangelands during the rainy season (December through March) and grazing the rest of the year in the 'bofedales'.

b. Mixed farming systems
A vast majority of the farmers manage mixed crop/livestock farming systems. Their productivity varies according to the altitude, which is associated with the weather. The crop subsystem usually consists of a rotation of four years: potato, quinoa, or barley during the second or third years and a legume in year four. If the farmer owns a small piece of land, fallow periods are seldom encountered. As farm size increases, fallow periods can reach six years. No soil cover is planted during the fallow period, allowing soil erosion on hillsides.

It is common to find farmers who own two to six cows, which are used for milk and meat production and for traction and fuel (manure). The best forage produced is given to the cow, which is a highly valued animal.

Sheep are common in most of these farming systems. They are used as reserves for household needs and for producing manure for crops. They must gather their feed by grazing natural rangelands and only during rainy years, when forages are produced abundantly, may sheep grazing be supplemented from 'bofedales'.

c. Inter-Andean valleys
These mesothermic valleys are highly productive, because they have good soils and weather. Both crop and animal production are profitable. Most of them have access to water for irrigation. Unfortunately, the areas with these conditions are small.

6. Scenarios used to evaluate soil erosion

Several ground covers were simulated in order to evaluate the impact of management on soil erosion and productivity (table 1). Evaluations considered a range of slopes from 0 to 50%, with increments of 10%. Soil parameters used as inputs for the model are shown in table 2. These are average values for the zone. A specific evaluation must use specific soil characteristics in order to improve predictability. A continuous crop was considered as the control (traditional). The alternatives included sowing plots with either natural pastures or alfalfa after cropping periods.

The weather generator option of EPIC was used, with average monthly rainfall and the corresponding standard deviation of 23 years of data from 'Granja Cayra', Cusco, as seed values (table 3).

a. Crop production scenarios
Two crop scenarios were selected (between 2000 and 3500 m): improved fallow and improved irrigation. Improved fallow implies the seeding of either

Table 1. Soil covers evaluated by EPIC for erosion impact in the southern Sierra of Peru.

Cover	Y1	Y2	Y3	Y4	Y5	Y6	Y7	Y8	Y9	Y10
Continuous crop	POT	BAR	BAR	WPE	POT	BAR	BAR	WPE	POT	BAR
Crop/natural pastures	POT	BAR	BAR	WPE	NPA	NPA	NPA	NPA	NPA	NPA
Crop/alfalfa	POT	BAR	BAR	WPE	ALF	ALF	ALF	ALF	ALF	ALF
Range	NPA	NPA	NPA	NPA	NPA	NPA	NPA	NPA	NPA	NPA
Alfalfa	ALF	ALF	ALF	ALF	ALF	ALF	ALF	ALF	ALF	ALF

Y1 = year 1, POT = potato, BAR = barley, WPE = winter peas, NPA = natural pastures, ALF = alfalfa

Table 2. Physical characteristics of the soil input to EPIC.

Characteristic	Unit	Value
Bulk density	$t\ m^{-3}$	1.25
Sand	%	20
Silt	%	40
Clay	%	40
Organic nitrogen	$g\ t^{-1}$	1190
Nitrates	$g\ t^{-1}$	75
Available phosphorus	$g\ t^{-1}$	25

Table 3. Average monthly rainfall at 'Granja Cayra', Cusco (over 23 years).

Month	Rainfall (mm)	Standard deviation
January	138.3	25.0
February	113.1	25.9
March	99.3	27.6
April	46.5	20.6
May	7.1	6.5
June	2.7	3.6
July	4.2	6.4
August	7.2	8.8
September	21.7	14.1
October	50.4	27.5
November	70.4	23.7
December	105.9	30.7

Table 4. Main characteristics of the alpaca scenarios evaluated with simulation models (100 ha).

Scenario	Dryland pasture ha	Bofedal pasture ha	Stocking rate* AU ha^{-1}	Maximum DM-prod. dryland kg ha^{-1}	Maximum DM-prod. bofedal kg ha^{-1}
Traditional	90	10	0.07	1000	1600
Alternative 1	90	10	0.07	1000	1500
Alternative 2	90	10	0.14	2000	1500
Alternative 3	80	20	0.21	2000	3000

* AU = animal unit (400 kg bodyweight).

natural or improved pastures at the beginning of year five in the rotation system.

Cropping areas with access to irrigation are limited in the Sierra. Wherever irrigation is found, water management is inappropriate. One alternative deals with improving water management and including a second crop within the agricultural year. A fourfold increase in irrigation efficiency is estimated.

b. Animal production scenarios

The scenarios described refer to those selected through optimization. Biologically, all of them are viable for the target farm conditions:

- alpaca production (above 4000 m)
- sheep production (above 2500 m)
- dairy production (between 2000 and 3500 m)

Alpacas. Table 4 shows the main characteristics of the scenarios evaluated. Alternative 1 implies adequate rotational grazing management and reclamation of rangelands by re-seeding the areas with natural species with high nutritional value. In addition to the improvements in alternatives 2, alternative 3 involves an improvement in water management for naturally irrigated grazed areas called 'bofedales'.

Sheep. Sheep are traditionally managed in overgrazed rangeland. Natural rangelands can be improved by re-seeding them with indigenous species of proven good quality (table 5), thus increasing carrying capacity.

Dairy. Milk production in the Andes is limited because of current feeding conditions. Fodder quality and availability can be improved and milk production incremented (table 6). Both annual and perennial pastures can be grown in the Andes. Productivity is highly enhanced when farmers have access to irrigation.

Table 5. Main characteristics of the sheep scenarios evaluated with simulation models (100 ha).

Scenario	Maximum DM production kg ha^{-1}	Stocking rate* AU ha^{-1}
Traditional	2000	0.07
Alternative 1	3000	0.14

* AU = animal unit (400 kg bodyweight).

Table 6. Main characteristics of the dairy scenarios evaluated with simulation models.

Scenario	Potential milk produc. kg $lact.^{-1}$	Birthrate %	Lactation length days	Rangeland ratio %	Annual improved forage %	Permanent improved forage %
Traditional	550	50	180	70	30	0
Alternative 1	800	60	210	60	20	20
Alternative 2	1200	60	210	0	0	100
Alternative 3	3500	70	305	0	0	100

Overall, ten different technical alternatives proved most promising: two for alpaca production (alpaca 1 & 2), one for sheep production (sheep), two for crop rotation systems (rota 1 & 2), three for dairy production (milk 1, 2, & 3), and two for areas under irrigation (irrig 1 & 2).

Results and Discussion

Optimization

The first step in the methodology is to find optimal production conditions. This is illustrated by the case of alpaca production. Optimum resource management at the farm level is obtained when the stocking rate is 0.14 AU ha^{-1}, pasture growth rates are 0.6 and 8 kg DM ha^{-1} day^{-1} for dryland and bofedal pastures, respectively, and digestibility of the forage selected by the animal is 54 and 67 for the two pastures (figure 2). These conditions are met by the two selected alternatives.

Erosion

Current management of most cultivable land in the Andes threatens the sustainability of the agricultural systems and thus jeopardizes the future of the

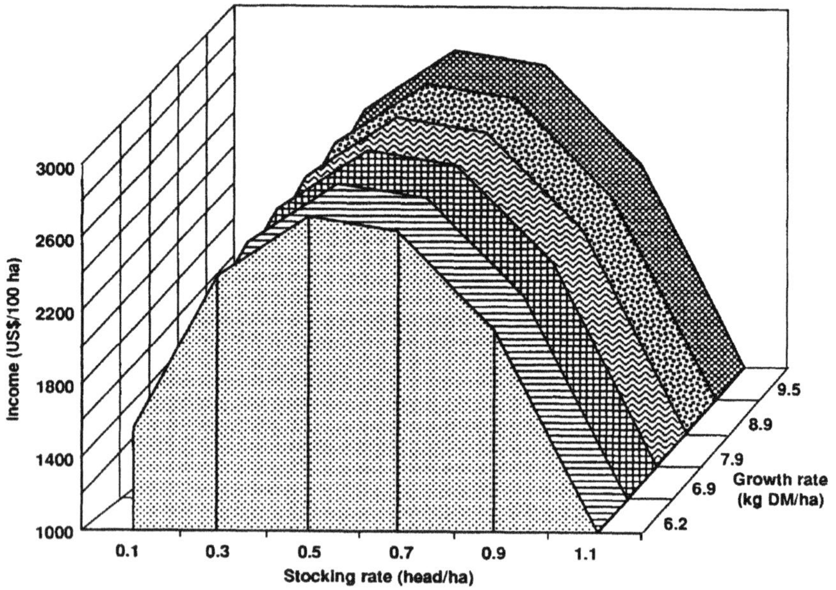

Fig. 2. Optimum pasture management in an alpaca production system in the southern Sierra of Peru.

region's inhabitants, because of excessive soil erosion. Under these circumstances, only animal production is suitable. Some erosion exists, even in extremely steep terrains with well-managed rangelands.

Considering a threshold value of soil lost for the Sierra of 5 t ha^{-1} (Felipe-Morales 1994), even at minimum slopes erosion could occur. If no conservation practices are used, only flat areas would be suited for crop production (table 7).

Because slopes must be used for crop production, due to increasing local demand for food, soil conservation practices are mandatory. If we look at the most important nutrients only, the amounts of N and P added to the soil, mainly through fertilization with sheep manure, are far below the amounts lost in runoff water with the predominant management.

Existing soil and water management practices are not adequate. The erosion produced can, however, be decreased while crop and animal production are increased (table 8). Some of the alternatives may increase erosion when compared to using natural pastures as soil cover. This is true when natural pastures are replaced by alfalfa. For the system to be sustainable, the erosion caused with the alternative must be below or equal to the threshold value for soil regeneration. This is an example of a trade-off that must be considered in order to stretch profitability for the households to the highest value possible.

Table 7. Soil and nutrients lost in runoff water with different soil covers and slopes, in the southern Sierra of Peru.

	Soil cover				
Slope %	Range	Alfalfa	Crop/ range	Crop/ alfalfa	Crop w/out fallow
		Soil lost, t ha^{-1}			
0	0.00	0.01	0.04	0.05	0.08
10	0.20	1.31	6.44	8.39	14.02
20	0.75	5.41	24.00	31.16	52.42
30	1.68	12.97	53.72	69.22	117.05
40	3.00	24.57	96.44	123.58	209.30
50	4.75	40.75	155.59	199.72	330.71
		N lost, kg ha^{-1}			
0	0.04	0.20	1.94	2.19	4.58
10	0.39	2.15	8.36	10.41	21.00
20	1.14	7.00	21.60	26.99	52.21
30	3.28	14.85	39.15	48.25	90.47
40	3.82	25.60	56.68	68.84	125.57
50	5.73	38.87	75.33	89.47	155.43
		P lost, kg ha^{-1}			
0	0.01	0.03	0.30	0.34	0.72
10	0.07	0.35	1.33	1.66	3.30
20	0.20	1.14	3.47	4.33	8.25
30	0.40	2.42	6.35	7.83	14.41
40	0.68	4.17	9.28	11.26	20.20
50	1.01	6.35	12.41	14.74	25.27

These results show that current management practices are not sustainable, regarding soil erosion, and that changes must be incorporated if we expect to achieve sustainable agriculture in the Sierra.

Table 8. Cummulative (35 years) erosion control and marginal agricultural production for each selected alternative (1000 metric tons).

Alternative	Decreased erosion	Increased erosion	Animal fiber	Meat	Milk	Potato	Barley	Hortic
Alpaca 1	6215	0	0.54	0.73	–	–	–	–
Alpaca 2	6330	0	1.31	0.83	–	–	–	–
Sheep	12430	0	1.14	1.90	–	–	–	–
Rota 1	7368	3688	–	–	–	74.04	25.38	–
Rota 2	16210	0	–	–	–	162.90	55.85	–
Milk 1	0	1565	–	2.02	20.79	–	–	–
Milk 2	0	302	–	2.17	24.89	–	–	–
Milk 3	0	0	–	0.5	119.40	–	–	–
Irrig 1	1951	334	–	–	–	58.33	10.00	33.40
Irrig 2	0	0	–	–	–	0	0	402.00

Productivity

Productivity for a farm adopting the proposed management practices will begin to increase after the second year, as in the alpaca case (figure 3). By year 20, a threefold increase in productivity can be attained. This holds for all the scenarios presented in table 9.

Although productivity is increased, if we look at the magnitude of the increment, it is evident that it is very low. This implies that these farming systems, in which a great deal of their natural resources are in the process of erosion, cannot afford external inputs. Profitability is so low that it is difficult to expect conservation of the natural resources for future productivity when current food security is jeopardized.

On the other hand, present and future societies benefit from the conservation of natural resources. Therefore, it is important for the society to invest in the conservation of these resources in the southern Sierra of Peru.

Some of the alternatives showed a high return (i.e., milk 3 and irrig 1 & 2). All three are alternatives for the inter-Andean valleys, where the size of the area limits the possible economic impact at the regional level.

The time frame for improving the well-being of peasant households is long (figure 4). We cannot expect farmers with a low marginal income to wait 20 or 30 years to obtain benefits from conserving natural resources. This is another factor to consider when natural resource management projects are designed and implemented.

It is evident that society may benefit as much as or more than farmers from farmers' effort? to preserve natural resources. Benefits for society include

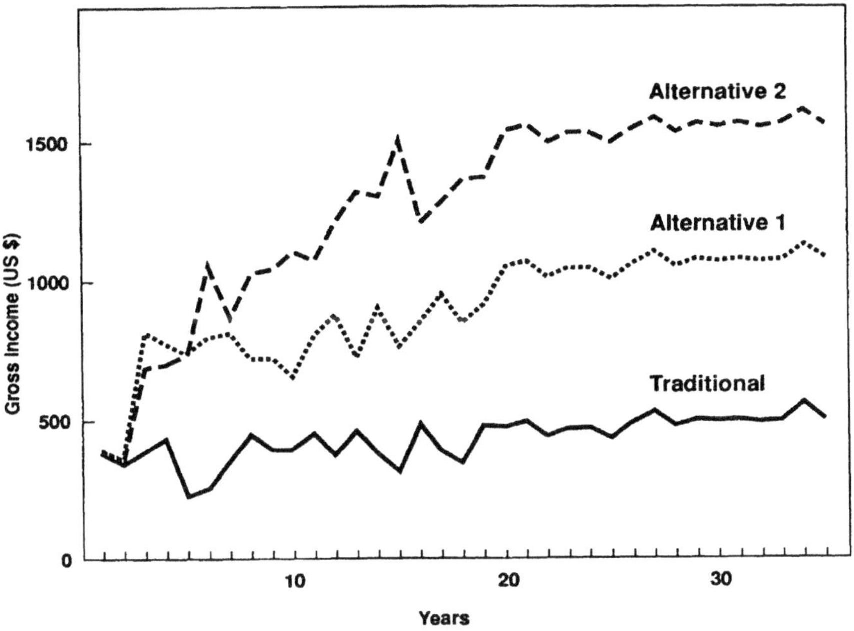

Fig. 3. Effect of technological alternatives on the gross income of alpaca production systems.

Table 9. Marginal benefits* accruing to farmers and society for each alternative selected for the southern Sierra of Peru (US$ ha^{-1}).

	Household			Society		
Alternative	Year 10	Year 20	Year 30	Year 10	Year 20	Year 30
Alpaca 1	0.37	0.65	0.61	0.66	1.47	1.59
Alpaca 2	0.16	1.21	1.61	0.37	2.58	3.37
Sheep	0.17	0.43	0.46	0.27	1.04	1.09
Rota 1	19.94	41.06	45.66	21.60	42.60	47.26
Rota 2	8.07	18.26	22.78	9.80	19.86	24.38
Milk 1	2.36	11.92	12.73	1.74	11.30	12.17
Milk 2	3.63	34.06	58.62	3.31	33.69	58.30
Milk 3	52.61	308.60	497.30	52.61	308.60	497.30
Irrig 1	15.64	49.12	77.60	17.89	51.38	79.30
Irrig 2	84.00	84.00	84.00	84.00	84.00	84.00

* Benefits for society include those accruing to farmers as well as consumers in cities.

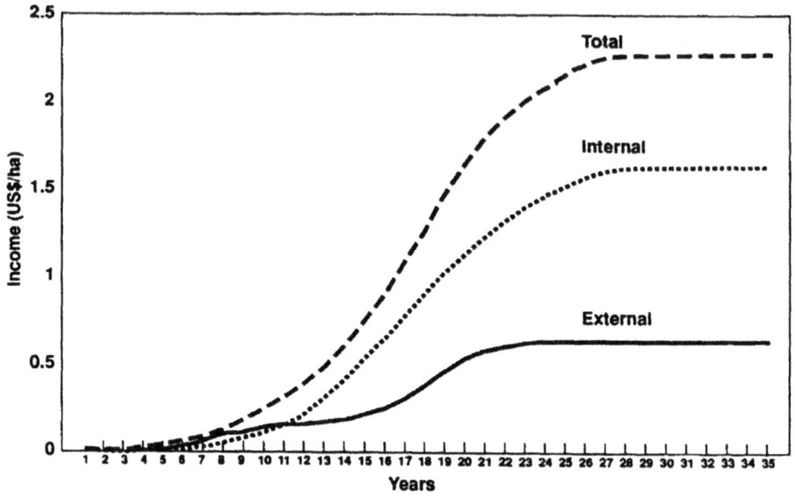

Fig. 4. Marginal benefits accruing to farmers and society in the southern Sierra of Peru (US$ ha^{-1}) when improved alpaca management practices are adopted.

those derived from productivity increases, reduced soil erosion, and added value. For instance, for every kg of alpaca fiber produced, US$3.50 accrues to the peasant farmer and the society may obtain up to US$25 in added value.

Society can invest in natural resource management and conservation projects. These projects should be designed in such a way that positive returns can be obtained, for both farmers and society.

The present case study shows sequentially every step for an ex ante evaluation of all the possible scenarios for a natural resource management project. Results from the exercise should show the profitability of the project, within a reasonable time frame.

When all the technological alternatives in the example are combined, including the cost of implementing all the expected changes in soil and water management in the Sierra, a positive internal rate of return is obtained (table 10). It is important to note that, on the average, only 30% of the benefits in this economic analysis have been attributed to the project. The analysis assumes that other institutions and people have contributed significantly to achieving the goal, including research institutes, training institutions, and peasant families.

Considering the internal rate of return, for every selected scenario and its respective target area, we can see that 35 years of use of these alternatives not only decreases the erosion of natural resources in the Sierra but also produces an economic payoff.

The project is viable if the technologies are adopted at the rate used for the ex ante evaluation. This implies that the adoption curve can be used for monitoring and evaluation. Any deviation from the curve observed in the

Table 10. Economic evaluation of technical alternatives (in million US$): regional income, net present value – NPV, internal rate of return (IRR), and target area.

		Technological alternative									
Item	Total	Alpaca1	Alpaca2	Sheep	Rota1	Rota2	Milk1	Milk2	Milk3	Irrig1	Irrig2
Area (000ha)	901	200	200	400	25	55	3.50	3.5	2	8	4
Productivity	49.26	1.05	1.74	1.27	4.09	8.13	2.94	4.05	9.84	4.30	11.85
Nutrients	6.47	0.77	0.81	1.53	0.66	2.74	–0.24	–0.05	0.00	0.25	0.00
Aggr.value	33.89	0.57	1.14	0.15	17.80	13.45	0.00	0.00	0.00	0.78	0.00
Subtotal	89.62	2.39	3.69	2.95	22.55	24.32	2.70	4.00	9.84	5.33	11.85
Due to project	26.45	0.95	1.48	1.18	5.08	8.63	0.81	1.20	1.96	1.61	3.55
Investment	18.01	0.52	0.52	1.29	3.24	6.80	0.66	0.64	1.39	0.65	2.30
NPV (5%)	11.11	0.50	1.02	0.04	2.23	2.92	0.223	0.63	1.05	1.04	1.45
IRR	9.23	9.97	11.91	5.21	8.80	7.33	6.89	8.89	9.22	11.55	55.60

monitoring process must be analyzed and corrected. If the analysis indicates that the parameters used for the ex ante analysis were wrong, they can be corrected and a more appropriate adoption curve can be generated for future monitoring.

Conclusions

CONDESAN's multidisciplinary team is now quantifying many years of diagnostic and research experience in the Andes in simulation models. As these models become successfully validated in different conditions of the Andean eco-region, they will constitute an important tool for designing and implementing R&D projects for improved productivity and sustainability of land use systems.

In most areas of the Andean eco-region, the technical coefficients required for model operation are available to researchers and developers from previous experiments in crop and livestock production. Past experiences incorporated into simulation models can be used for designing and implementing R&D projects in natural resource management, as in the case study presented. In cases where the required data do not exist, the models can be used for sensitivity analysis. The most sensitive variables may require appropriate research for a better understanding of their effects on the specific production system, thus giving the model another important application: resource allocation for research.

The integration of simulation models describing systems at different hierarchical levels is possible and useful for understanding relationships between the micro, meso, and macro level. Technologies proven good at the farm level must go through adoption processes that in most cases require a longer time than expected. In order to avoid disappointment, the use of scenarios with regional models brings us closer to reality.

In the southern Sierra of Peru, conservation of natural resources is not attractive per se, if peasant families have to pay the bill alone. Society must invest in the future. In the case study presented, the current government is aware of the issue and is willing to invest in the project.

CONDESAN has begun work to develop the know-how required for integrating the results obtained by modeling with geo-referenced data and traditional statistical data on production, prices, and input and product flows. This will greatly improve our capability for better assessing the long-term impact of natural resources management policies on the productivity and sustainability of land use in the Andean mountain ecology.

Acronyms

CIAT Centro Internacional de Agricultura Tropical (International Center for

	Tropical Agriculture)
CIP	Centro International de la Papa (International Potato Center)
CONDESAN	Consorcio para el Desarollo Sostenible de la Ecorregion Andina (Consortium for the Sustainable Development of the Andean Eco-region)
EPIC	Erosion/Productivity Impact Calculator

Acknowledgments

This work was carried out in the framework of the Sustainable Highland Agriculture project supported by the International Development Research Centre (IDRC) of Canada. Information from the Highland Agricultural System Research Project (PISA, the Spanish acronym) carried out by the National Institute of Agricultural Research (INIA) of Peru, and the National Agrarian University, La Molina, is greatly appreciated. The authors acknowledge the most helpful comments from Dr. B.A.M. Bouman.

References

Anaman K A (1993) Risk analysis in the economic appraisal of a smallholder agricultural development project in Papua New Guinea. Agricultural Systems 43:397–414.

Arce B A, Aguilar C, Cañas R, Quiroz R A (1994) A simulation model of an alpaca system in the dry puna of the Andes. Agricultural Systems 46:205–225.

Cañas R, Aguilar C, Edwards G (1991) Modelo de simulación de producción ovina para la zona de Puno-Perú. Pages 89–137 in Arguelles L, Estrada R D (Eds.) Perspectiva de la Investigación Agropecuaria para el Altiplano. PISA. Lima, Peru.

CIP (1993) Planificación y Priorización de Actividades de Investigación y Acción para el Desarrollo Sostenible de la Ecorregión Andina, PPO. Lima, Peru.

Felipe-Morales C (1994) Pérdida de agua, suelo y nutrientes bajo diversos sistemas de cultivo y prácticas de conservación de suelo en zonas áridas, subhúmedas y muy húmedas en el Perú. Pages 637–656, Perú: El problema agrario en debate. SEPIA V. Lima, Peru.

Van Immerzeel W H M, Núñez del Prado J V (1994) Pacha Mama Raymi: Un Sistema de Capacitación para el Desarrollo de Comunidades. Tercera Edición. EUROCONSULT. Cusco, Peru. 169 pp.

León-Velarde C U, Quiroz R A (1994) Análisis de Sistemas Agropecuarios: Uso de Métodos Bio-matemáticos. CIRMA/CONDESAN. Edi-Graf, La Paz, Bolivia. 238 pp.

Quiroz R A, Arce B A, Cañas C, Aguilar C (1990) Desarrollo y uso de modelos de simulación en la investigación de sistemas de producción animal. Pages 335–370 in Ruiz M, Ruiz A (Eds.) Memorias de la reunión general de RISPAL (9, 1990, Zacatecas, México) San José de Costa Rica, IICA – RISPAL.

Sharpley A N, Williams J R (Eds.) (1990) EPIC – Erosion/Productivity Impact Calculator: 1. Model Documentation. U.S. Department of Agriculture. Technical Bulletin No. 1768. 235 pp.

Eco-regional approaches and Agenda 21: a policy view

R.D. VAN DEN BERG*

Directorate General for International Cooperation, Netherlands Ministry of Foreign Affairs, P.O. Box 20061, 2500 EB, The Hague, The Netherlands

Key words: aid effectiveness, agricultural research, development funding, impact analysis, sustainability

The paradox of awareness

Having been asked to present a policy view on eco-regional approaches in the light of Agenda 21, the first question that needs to be answered is whether there is still any light remaining in Agenda 21. Is Agenda 21 still an issue in international relations? It seems to slowly fade away. Almost no additional funding has emerged for Agenda 21, apart from a valiant effort by Norway. The Netherlands had promised an annual increment of 0.1 percent of the Gross National Product for activities in the framework of Agenda 21, but on the condition that a sufficient number of other countries would do the same. And unfortunately, Norway is not a sufficient number of other countries.

Agenda 21 was an impossible document anyway. I never met anybody who claimed to have read more than a few paragraphs. In the research community most people probably tried to read through the section on research, but that did not make much sense either. So, after a few years we can conclude that UNCED was a failure and Agenda 21 a waste of paper. However, at the same time we need to recognize that whatever we conclude about UNCED or the Agenda 21 document, has no bearing on the urgency of problems of the environment and sustainability. I sometimes hear people say that because governments are not willing to spend money on Agenda 21, the environment is not a crucial issue anymore. This is like saying that because the CVSE Conference was not able to come to any decision on the war in Bosnia, this apparently is not such a bad war at all. It is of course the other way around: not only is the war in Bosnia horrifying, it is also horrifying that the CVSE was not able to do anything to stop this war. It augments rather than diminishes the crisis. The conclusion that UNCED was a failure, and that Agenda 21 can be thrown in a wastepaper basket, add to the urgency of the problem of sustainability.

* This paper expresses the personal viewpoints of Mr. Van den Berg, and should not be read as a policy statement by the Netherlands Ministry of Foreign Affairs.

J. Bouma et al. (eds.), Eco-regional approaches for sustainable land use and food production, 33–40.

Yet it needs to be recognized that on the highest political level environmental issues are no longer of the first importance. At the same time, it appears that the highest political level is at the moment rather impotent, to put it mildly. Issues which are on the highest political agenda, such as Bosnia, are not getting solved either.

If we step down to a level slightly lower: that of the highest civil servants and diplomats, the circus of international negotiations, we find that there environmental and sustainability issues are very much part of the agenda. To name but a few: the Montreal Convention on Global Climatic Change, the Desertification Convention, the Global Environmental Facility, negotiations on trade in tropical timbers, and so on. There is a general sense of urgency in the negotiations on these problems, yet there is also a general lack of money and political will coming from the highest level to implement what is being agreed upon.

If we would step down to ground level, it is my feeling that we would encounter a paradox. On the one hand it would appear that the urgency of a solution to sustainability issues becomes clearer. It always strikes me that people everywhere, both in the South as well as in the North, are so very much aware of the deterioration of their environment. We know that in the Netherlands the air is not clean, that our ground water is not as clean as it should be, that the soil itself cannot be trusted blindly, that the amount of traffic is unhealthy. In the South I often hear that erosion is sharply perceived, that problems of salinity, lowering levels of groundwater, overgrazing, and so on, are becoming more urgent everyday. Yet at the same time, most of these problems cannot be solved by the people themselves.

On the other hand, it would appear that especially in the North the willingness is eroding to spend money on the solution of these problems. And to be more specific: the willingness is eroding to spend public money on the solution of these problems. Development aid is slowly losing its ground. Budgets are being lowered in almost all countries. When this trend was first perceived some five years ago, it was thought that the integration of development issues with environmental issues would help to stop that process, would provide development budgets with a broader and more solid basis. This did not materialize. The 'greening' of aid resulted in some new compartimentalizations, some new budget categories, some new money, but not in a general revitalization of development aid. As a result I sometimes hear people argue in favour of returning to a more narrowly focused development aid. In the case of the CGIAR, one sometimes hears arguments in favour of returning to the old core of agricultural research: crop improvements: that is what has been proven to work, and that is what one should focus on. I do not believe that that is the direction to take. Here again it should be stated that the fact that no additional money is forthcoming, does not diminish in any way the urgency of the problems. Returning to a more narrow agenda is both a misreading of the context, and of the problems to be solved. I will deal with the problems later on, let me first return to the context.

Crisis in public funding

The paradox I put forward was that on the one hand, people are more than ever aware of developmental and sustainability problems, yet on the other hand they are less and less willing to spend any money on the solution of these problems. There is a pervasive mood in Northern countries that enough tax has been paid, and that budget deficits cannot go on indefinitely. At the height of the financial crisis in Southern countries in the mid-eighties, many development economists pointed out that the budget deficit in a country like the United States of America was relatively higher than in many a developing country. Yet no World Bank or International Monetary Fund was ever able to dictate government spending in the USA. But what the World Bank and IMF were not able to do, is now being accomplished by the American voter. And voters in other Northern countries will, most likely, repeat this pattern. What we could be facing is democratically guided structural adjustment in the North. And just like the structural adjustment in the South the question is going to be whether it will be adjustment with a human face. The chances are that it will not have such a face. The message in the USA is: more death penalties, more prisons, a strong army, less government, no complete health care coverage, less spending on soft and social issues such as welfare, the arts, or development aid. Jesse Helms, the envisaged new chairman of the powerful senate committee on Foreign Affairs, has already stated his firm intention to cut on development aid.

The reader may think that this picture is too harsh. In the Netherlands, you may say, things are looking much better. The Dutch still give more than the required 0.7% of their Gross National Product, and Minister Jan Pronk defended his budget for 1995 in Parliament without any problems. That may be the case, or on the other hand, it may be the quiet before the storm, as we say in the Netherlands. During the election campaign in the beginning of 1994 no mention was made whatsoever of the outside world, except on the one issue of migrants. The international issues highlight of the campaign was when one of the party leaders let himself be interviewed at the border with Belgium. He pointed to all the cars rushing over the border as the metaphor of the outside world getting into the Netherlands without any proper controls.

Perhaps the development community made a mistake in the last decade, when we tried to argue for development out of 'enlightened self-interest'. The argument was put forward most clearly by Jan Tinbergen, who time and again said that if we would not provide the South with the means to develop, the South would come to us to get it. It may very well be that the answer to this argument will be that Europe will close its borders, will strengthen its armies, and will stop caring about the rest of the world. To some extent, we see this already happening. Frits Bolkestein, the leader of the Dutch conservative party, recently said that he wanted to spend government money only on humanitarian issues; no longer on traditional development aid.* This may seem very caring, very humanitarian. On the other hand it shows also a lack of caring about the underdevelopment of the world. Bolkestein seems to say that if you quickly starve of hunger, you will get aid, yet if you do it slowly, no one will come. Furthermore, he will give the hungry fishes, instead of learning them how to fish, to use that ancient Chinese proverb once again. Actually, with the depletion of fish stocks in the world seas, it is not an appropriate proverb anymore. As an aside, it should be noted that the proper metaphor now should read: "learn a person how to *farm* fishes, and he or she will never starve". This again shows that truly the problems have changed – it is not a question of fashion, but a question of analysis. But let me return to the public spending on development aid.

Jan Pronk once analyzed the opposition to the spending of government money on development aid along the following lines. He said that opponents in the North usually take their reasoning from four subsequent arguments against aid. These arguments are:

1) The North does not have any money anymore;
2) Even if the North would have the money, it should not be spend in the South, but on more urgent problems in the North;
3) Even if the problems in the South are more pressing, the North still should not spend money on them, because aid is not helping anyway;
4) Even if aid would help, the people in the North still do not want to pay tax for it.

The last argument can only be answered by an appeal to humanity. The third argument, that aid is not helping at all, is the crucial one for the development cooperation community, since it deals with issues of impact, effectiveness and efficiency. Does aid help? Is it being spend in the right way, on the right subjects and themes, or are we missing opportunities? In the second half of my speech I will deal with these issues. The second argument, that local Northern problems are more urgent, has to be answered by showing the interconnectedness of problems, by showing that solving only our problems is simply not possible: we cannot solve our problems in isolation. The first argument, that we simply do not have the money anymore to provide aid, is manifestly untrue. In one of the poorest countries of the European Community, Ireland, some of the largest non-governmental aid agencies are active, and the Irish people are giving generously. In fact they are the most generous of

* Bolkestein later developed his argument in an article in De Volkskrant on Monday February 6, 1995: "Development aid is ready for a moratorium on new spending" (Ontwikkelingshulp is toe aan een moratorium nieuwe uitgaven).

the European Community. At the same time, the Irish government spends a very low amount of the Irish Gross National Product on official development aid. This may be the future of development aid in Europe: a shift towards non-governmental sources. What may happen is that Western governments in the next decade or so will indeed reduce official development aid, because most of the budget will go towards reducing the governmental debt.

Getting the aid Jumbo of the ground

A crucial issue for the future of official development aid is in my view whether or not we are able to show impact of the money we spend on development. And here development aid is in trouble. The general idea, often expressed in debates about aid effectiveness, is that approximately a third of development projects could be considered a success. Another third did not achieve much, but did not do much damage either. The last third was a failure and actually worsened the situation instead of improving it. But even if we look at the successful third, the overall picture remains bleak. It was found that even if a country has a lot of successful development projects, this does not necessarily mean that it will have successful development. The growth rates that were achieved in the project environment, were not reflected in the growth rates of the economy. This is called the micro/macro paradox: successful efforts on the micro level somehow are not translated to the macro level.*

One of the answers to this problem has been to enlarge the development project to that macro level. The process can roughly be described by the following parallel. A development project in the traditional sense has been likened by an expert from the World Bank to building a Boeing 747 Jumbo. Lots and lots of different nuts and bolts, aluminium, electronic gadgets and so on, are put together by experts. Why do so many projects fail? The World Bank expert likened these projects to a Boeing 747 without wings. Just a major design failure. If it would be that easy, development projects would by now all be very successful. Yet they are not. What is the problem?

To enlarge the metaphor, it would appear that even if you build a perfect Boeing 747, the plane still will not take off, because there is no runway. OK, so we include the runway in our project. Even now, the plane does not take off. We need fuel supply, air control, or to conclude: we need the infrastructure of an airport. Now the plane is taking off alright, only suddenly we find that it has flown straight to Switzerland.

This in short is a comic history of what can go wrong with development projects, and our efforts to rectify our mistakes. We can fit in some of the

* The general ideas and reasoning can be found in overall aid evaluations such as Cassen (1986) and the World Development Reports of the World Bank; see for example the 1990 report on Poverty. The Cassen report is actually more generous on the effectiveness of aid: it concludes that *"the majority of aid is successful in terms of its own objectives"* (Ch. 10.1, p. 294).

episodes of the recent past: for example the role of women in agriculture, especially in Africa. Only very recently, about ten years ago, many an agricultural extension project in Sub-Saharan Africa failed, because only males were trained in extension, and they visited male farmers. Women farmers, responsible for more than 80% of the agriculture, were overlooked. This is indeed in the order of magnitude of forgetting the wing of an airplane. Similarly, the environmental impact of project interventions was often neglected. Irrigation leading to problems of salinity is just one example. Socio-economic factors were also quite often overlooked. When I started as a junior civil servant in the Dutch Ministry of Foreign Affairs, I was involved in a promising fisheries project in an African country. In a certain village pools would be build, and fish would be nursed, grown and sold to the profit of the village. Technically, this was no problem at all. The pools could be build locally, there was enough water, maintenance was not that demanding, food for the fish was cheap and in some cases even free, fish hatchings were available. Yet in half a year the project was abandoned a total failure. The crucial issue was maintenance. In an evaluation of the project is was established that the work force of the village was already overburdened, because approximately half the village had migrated to other countries to earn money. Of the remaining work force, nobody had any time left for the maintenance tasks for the fish pools.

The challenge for the eco-regional approach

What is the relevance of these anecdotes for agricultural research? Many in the CGIAR believe that the current crisis in development aid should not touch upon the CGIAR, because it is in fact a success machine, with enormous impact. In their view, it is a sorry mistake of the general public to mix the efforts of international agricultural research with the general kind of development project that indeed often fails miserably. If only the public could be informed a bit better, then the money would return manifold to the valiant fighting men and women of the CGIAR Centers. Impact in this context is something like the green revolution: the CGIAR provides the farmer with a new variety of a crop, the farmer adopts this, yields rise significantly, and the resulting gains in food availability, income for the farmers, and so on, can be attributed to the original introduction of the new variety.

I have objected in many a CGIAR meeting to this rather simplistic picture, because I feel that it does not guarantee any success in the future. The green revolution happened because a lot of circumstances were just right for it. It happened because farmers were willing to invest in the new varieties. It did not happen in Africa, where the circumstances were different. It did not happen in the Punjab province of Pakistan, although the circumstances there were almost the same as those in the neighbouring areas in India. Why not? Studies suggest that the level of education in the Indian Punjab was much higher than in the Pakistan Punjab. So, to follow simplistic reasoning once again, we could conclude that the green revolution shows the impact of education!

What we know about development and about progress in general, seems to be on the one hand more and more, and on the other hand less and less. Our technical expertise has grown enormously in the past decades. Yet we still do not know what the crucial factors are that decided that for example Singapore grew from a quiet, relaxed Asian port some twenty five years ago, to a nation which has in many ways developed beyond the traditionally 'developed' countries. And the success of Singapore is quite a different success from that of Korea, or more recently, that of Malaysia. When talking about impact, we have to recognize that we simply do not know the actual chains of causes and effects which shape society. To some extent, these processes are a bit similar to the weather forecast. We know the ingredients: high or low pressure areas, the effects of land or sea on the climate, and so on, but we do not know when and where the wing of the butterfly is going to start the storm on the other side of the ocean. Mathematical theories of complexity prove that such chains of action and reaction can start from very humble and very crucial beginning.

Again I seem to digress from agricultural research, and I have not mentioned the eco-regional approach yet. In my view the role of research in development is to find and provide for the very humble and crucial beginnings that I mentioned in reference to the weather. The CGIAR was founded on the basis of the analysis of a crucial problem in the sixties. The conclusion was that improved crop varieties could make a difference.

The world we are facing now will not allow for such a simple end-conclusion. What we need now is solutions for specific situations: solutions for concrete geographical locations. It may be that drought resistance is the crucial factor in a certain area, yet of less importance elsewhere. Erosion could be the crucial problem in another region.

These may be open doors, yet they may provide us with an answer to the growing criticism on development aid in the North. One of the elements in that answer may be that we have to keep the lines between funding and actual improvements implemented as short as possible. The traditional CGIAR philosophy may well be too long-term in nature: strategic research, which leads to applied research, which leads to adaptive research, which leads to implementation in a farming system. Most of this chain from invention to implementation is out of the grasp of the CGIAR, but in the hands of NARS, extension services, and most importantly the farmers themselves. To shorten the chain, and to provide the Northern tax payer with the guarantee that his or her money will be spend on actual improvements in the South, the eco-regional approach is one very promising avenue of cooperation with the other partners involved.

This partnership is also important from another viewpoint. In a lot of Southern countries the NARS, farmers organizations, extension services, agricultural faculties of universities have become quite strong. This has to be recognized. Secondly, local knowledge has so often been rejected as unscientific, unmethodological. Indeed, it is unscientific, for it is local knowledge: by definition specific rather than universal. The other side of the coin is that much of scientific knowledge cannot be applied locally, because it is not specific enough for the situation at hand. An anthropological study discovered that 'scientific experts' coming from the outside into a certain problem area were well recognized as 'experts' by the local people. They were indeed astounded by the wealth of knowledge that these experts had available. Yet at the same time the local people all agreed that these scientists were complete fools when it came to the application of this knowledge to the local situation (Cohen 1993).

Eco-regional research should have a mechanism to cope with all of these factors: true partnership, sustainability questions, productivity concerns, socio-economic aspects, cultural aspects, local knowledge. It is starting to resemble a Boeing 747 in its complexity. Do we know whether it will work? There the optimistic message is that yes, there are methodologies to make it work, just as there is an airplane industry which knows how to build airplanes. One of the most promising methodologies in my view is the one developed at the Wageningen Agricultural University. The Netherlands' government wants to make this methodology available if needed for eco-regional programmes. For that reason it sponsored this Conference. An important part of these methodologies is that in order for the research to start, one first needs to make a careful analysis on whether we perceive the problems in the right way, and whether we can tackle the crucial factors. Consequently my motto for this Conference is: sometimes we need more research before we can do the research which will make a difference. I wish you success in this endeavour.

Acronyms

CGIAR	Consultative Group on International Agricultural Research
IMF	International Monetary Fund
NARS	National Agricultural Research System
UNCED	United Nations Conference on the Environment and Development

References

Cassen R (1986) Does Aid Work? Clarendon Press, Oxford, UK.

Cohen A P (1993) Segmentary knowledge: a Whalsay sketch. Pages 31–42 in Hobart M (Ed.) An Anthropological Critique of Development: the growth of ignorance. Routledge, London, UK.

SECTION B

Resources availability and demand studies at a global scale

Scenarios for land use in Europe: agro-ecological options within socio-economic boundaries

H.C. VAN LATESTEIJN

Netherlands Scientific Council for Government Policy, P.O. Box 20004, 2500 EA The Hague, The Netherlands

Key words: agricultural production, European Union, future studies, scenarios

Abstract. An eco-regional approach to research is called for to address emerging problems related to regional food supply. In an eco-regional analysis information can be gathered on constraints to and possibilities for agriculture in its physical and socio-economic environment. This information can reveal ways to increase agricultural production while at the same time conserving the natural resource base.

In this paper attention is given to the methodology of exploring future possibilities for land based agriculture in the European Union. To that end future technical possibilities of primary production are confronted with a number of different objectives that are put to land based agriculture. The technical possibilities are derived from regional production potentials of different indicator crops that have been assessed with the aid of a crop growth simulation model. In an Interactive Multiple Goal Linear Programming (IMGLP) procedure four different scenarios of optimal land use are calculated, considering four distinct political priority settings.

The results of the study have been published in a report to the Government by the Netherlands Scientific Council for Government Policy, an independent advisory body to the Dutch Government. In a public debate following the publication of the report, several different groups have discussed the usefulness of this type of analysis. The initial reaction to the study was somewhat reserved, because this type of exploration of the future gives rather different results than the more usual future studies that rely on extrapolated trends. However, in the discussions that followed the complementarity of both approaches has been accepted.

Introduction

Throughout the world agriculture is faced with emerging problems that are related to the interaction of different agricultural activities with their environments. In developed countries environmental degradation occurs as a result of overuse of arable land and intensive livestock farming. In developing countries environmental degradation is also apparent, but here mainly as a result of soil degradation and erosion due to farming practises on marginal land. In both developed and developing countries structural changes in agriculture will be necessary to overcome these problems. However, these changes must take place in a physical and socio-economic environment that puts a number of constraints to the possible directions of change. In Europe and other western countries agricultural policies in the past have led to a situation of over production for some agricultural products. There is also a growing concern

J. Bouma et al. (eds.), Eco-regional approaches for sustainable land use and food production, 43–63.

for the breakdown of social structures in rural areas and environmental problems have led to a strong call for major adaptations in ways of production to safeguard the environment. In most developing countries population growth calls for a major increase in food production and rapid urban and industrial development claim an increasing part of the available land area. This implies that growth in agricultural production has to be achieved on a declining area available for agriculture.

Although the problems are quite different, the approach to tackle some of the questions involved appears to be the same. Information on constraints to and possibilities for agricultural production is needed in all regions to address the emerging problems. In its document 'The Eco-regional approach to research in the CGIAR' the TAC recognised the need for this type of research (TAC 1993). The main focus, however, is on the relation between increasing agricultural production and the conservation of natural resources. To achieve these contrasting goals the TAC encourages research to identify ways to relieve the constraints at the field, farm, community, land-use system and policy level. To that end research has to be extended from the more traditional level of individual plants and crops to cropping systems and regions. Moreover, if options for land use must be identified, explorations into the future will be necessary.

So, both at the spatial and at the temporal scale research activities have to be initiated to come up with information that may identify constraints and possibilities for developments in agriculture. Moreover, the expansion of research activities to the regional level can also help to focus research activities at crop level, because information from the regional level can for example be used to identify promising developments in improving characteristics of crops. Another benefit may be the identification of minimum data requirements for the characterisation of the relevant (bio)physical and socio-economic environment.

Examples of application of this approach to agricultural research are still scarce. In this paper one example will be demonstrated that deals with the identification of options for land use. In 1988 the Netherlands Scientific Council for Government Policy (WRR) initiated a policy-oriented survey of land based agriculture at the level of the European Union (EU) that resulted in a report to the Dutch Government entitled 'Ground for choices, four perspectives for the rural areas in the European Community' (WRR 1992). In this study much attention was paid to the development of apt methodologies to assess future options for land use. This paper focuses on some of the aspects of that methodology, especially on the problem of how to incorporate the future in an analytical study. The choice for a certain method is closely linked to the exact formulation of the questions that need an answer. For that reason the background of the study is also presented in some detail. The results of the study are summarised and some information is given on the use of the results.

Problems with current and future agricultural policies in the European Union

Improving production conditions, increasing knowledge on cultivation techniques and high-yielding varieties has led to a period of growth in agricultural productivity whose end is not yet in sight. In the EU growth in productivity is further stimulated by a set of policy measures in the Common Agricultural Policy (CAP). However, to date the success of the CAP seems to cause new problems, because it resulted in a situation of growing surplus production for several products. With a considerable amount of subsidies these surpluses are sold on the world market. Every year the EU budget had to rise to keep up this policy, which led to political strain in the Union. Member states argue about the maximum level of support that should be observed, there is conflict with important trading partners over the subsidised dumping of EU surpluses on the world market, the market is distorted – mainly to the detriment of developing countries – and there are increasing environmental problems resulting from current production methods.

A general feeling arose that the problems would become intractable if we were to carry on this way. Production and farmers' income should no longer be the primary objectives of policies, but concern for our common social en physical environment ought to have an influence on developments in agriculture also. However, most of these objectives call for formerly unexplored pathways of developments. This forms a policy dilemma. A successful policy like the CAP calls for new reactions to mitigate the adverse effects that have become apparent. Information on how to change policies best is commonly lacking, and simultaneously the benefits of present policies for some stakeholders are evident. Any policy proposal that aims to enhance any of the new public objectives in agriculture is likely to find massive opposition on its way.

The different policy goals that are attributed to the agricultural sector can have very different consequences for the developments in agriculture. Some of these objectives are even in sharp conflict with each other. Obviously, striving for free-market conditions (prices at world market level, no trade barriers) will be difficult to combine with a policy goal of maintaining regional labour. Still, these goals are all taken very seriously in the policy reforms and the statement that all goals are of equal importance can be heard regularly in the policy arena. In this setting it will be very difficult to come up with initiatives that are in line with all the needs and wishes expressed in the policy debate. To break this impasse, information must be made available that illustrates the consequences of different policy aims given the potentials of the agricultural sector.

Dilemmas can only be overcome if all parties can be convinced that for the benefit of all a revision of positions for all parties is necessary. Such a revision requires at least information on the trade-offs between the various objectives that are at stake. Scientific information might illuminate these trade-offs and shed some light on the potential conflicts that linger beneath the surface of the current discussion on policy changes.

The possibility to attain some or all of these different policy goals will depend on the performance of the agriculture sector in the future. A starting point to measure this performance is soil productivity, because this figure relates directly to the type and location of agricultural land use, and land use is very relevant to all policy goals involved. Type and intensity of land use determine the agricultural production at a given location, but also the possible impact on the environment and the economic performance of the agricultural sector. So, data on changes in soil productivity can be used to infer information on a number of other relevant criteria. All developments in agriculture will have an effect on the type and location of land use throughout the EU. This will be reflected in changes in regional soil productivity over time.

A historic evaluation of wheat yields reveals a remarkable similarity in the development of soil productivity in different parts of the world (De Wit *et al.* 1987). For most western countries a sharp increase in productivity growth occurs shortly after the Second World War. This shift can explained by the introduction of modern farming techniques such as improved nitrogen application, the use of herbicides and new forms of mechanisation. Together these changes resulted in a 'green revolution' with overwhelming results. Productivity growth boomed from on average less than 5 kg ha^{-1} $year^{-1}$ to over 50 kg ha^{-1} $year^{-1}$. The possibilities for a further growth in productivity will be crucial to the possibilities of achieving certain specified policy goals. So, the question is whether this rise in productivity will be levelling off in the foreseeable future. Intuition tells us that a growth like this cannot go on eternally. At a given point in time technical limitations will prevent a further rise in production per hectare. The first assignment for an exploration of future options in agriculture is to assess the 'ceilings' in productivity growth.

Ultimately, these limits to the rise in productivity will define the possibilities of the agricultural system in the future. So, first these potentials must be assessed, or in other words the potential levels of a further rise in productivity must be estimated. Once information is available on these potentials, questions regarding the different policy goals may be assessed. Does the system allow a continuation in agricultural employment? Does the system allow a certain level of protection of the environment? Does the system allow a sustained generation in agricultural income? These questions should be answered if an informed debate on the aims of agricultural policies is pursued. Therefore, exploring the limits of the productivity growth in agriculture should be the first step in a procedure to assess policy-driven options for the future.

The exploration of the future must therefore be aimed at opening up all possibilities that are present in the agricultural system. This implies that an exploration of this nature should not consider the possible impacts of relatively inertial structures in the system. For example, several development theories start from the assumption that agri-business complexes or current levels of investment in infrastructure and management skills put a restriction to future developments. Several of these constraints might be very important to short-range explorations, but an exploration of long-term possibilities should not be limited by incorporating these *a priori* restrictions to the future. For this type of exploration one should primarily consider the technical and/or logical constraints so that the full range of possible options for the future will come to sight.

Methodological considerations of exploring future options

Two research questions result from this line of thought:

1. What are the upper bounds of productivity rise in land-based agriculture?
2. What are the consequences of different policy goals for future developments in land-based agriculture?

The questions do not include a reference to the time dimension. The exploration is aimed at identifying the boundaries of the system. The question whether these boundaries will be faced on short notice is secondary. Both questions must be addressed at the regional level. Current differences in productivity between countries and between regions is evident. Most probably the potentials of different regions will also show considerable differences. Regions where the scope for a further rise in productivity is limited will have less options for future developments than regions that still have a long way to go. Once regional potentials are known, the possibilities for attaining policy goals related to land use can be assessed. These possibilities can illustrate the consequences of opting for specific policy goals. If the exploration indicates that the consequences are not very favourable, some numbers can even be put to the 'price' of the different goals. This type of policy-oriented future research informs the policy makers of the trade-offs that are present in the policy arena on land use issues. Especially this type of information is lacking in the current debate so most discussions on policy reforms are either noncommittal or the discussion shifts away from the aims to the means by focusing on policy measures.

Most experience in exploring future developments has been collected in economics. Policy-oriented future studies are mostly focused on the poten-

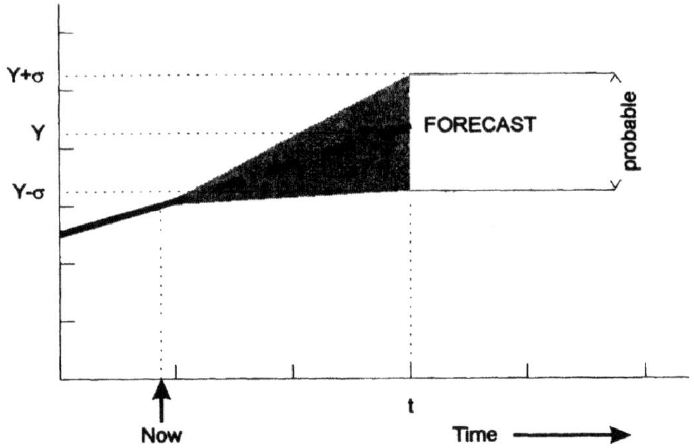

Fig. 1. Extrapolation of historical data and the accompanying confidence interval. Y is the calculated value for time t. $Y+$ and $Y-$ denote the boundaries of the distribution of possible outcomes of the forecast due to statistical errors.

tials of economic developments and the consequences thereof for the rest of society. Generally these future studies are based on the assumption that gathering information can facilitate policy-making by narrowing down the uncertainties that are inherent to policy decisions. So, a future study generates information about the future with a certain degree of accuracy. If one takes all the information that is available on a certain topic and feeds this information into some kind of extrapolating technique, the future will reveal some of its secrets to us. If we can predict our future in this sense, some of the uncertainties of the future are eliminated. For policy makers this means that the chances of making a wrong decision are narrowed down a little.

This expectation of the results of future research can be very appealing to politicians who are reluctant to take full responsibility for the decisions that have to be made. As a result, strong pressure has been put to the research community to come up with this type of analyses, which resulted in the mainstream of future research aiming to produce forecasts with a certain degree of accuracy. Current facts and knowledge are extrapolated one way or another into the future, leading to an estimate with an accompanying interval of confidence. A generally accepted method to assess this interval is calculating the standard deviation of the forecast and consider all values between plus and minus. Although the methodologies can differ substantially, the outcome of this procedure will take the form of the graph given in figure 1.

Forecasts aim at plausibility

A forecast can give information on the probable development of an observed phenomenon over a limited period of time in the future. By definition, forecasts are based on information gathered in the past and the present. Additional information on causality between the observed input and output of the system is used to calculate future values. Because there is no way that we can assess the relation of future developments with those in the past and present other than the notion of continuity, there will always remain a great deal of uncertainty to this type of future research.

As with all model studies, a forecast generated with the aid of an explicit model will always be based on a part of a system. This implies that for the rest of the system assumptions have to be made, so all models incorporate a number of exogenous variables that describe the 'environment' of the forecast. Exogenous variables can account for a number of differences between the results of future studies on the same topic. Cole and Miles (1978) conclude that the differences in exogenous variables are critical to the debate and can be traced back to differences of opinion in:

1. resources and technical change;
2. political objectives and norms;
3. social and political processes that change society.

In a survey of future studies the WRR (1988) concluded that the role of these exogenous variables must not be underestimated. It can even be demonstrated that these assumptions are dominant for the results obtained. In a number of cases the outcome of the future research has more to do with the selection of 'relevant exogenous variable' than with the specification of the model itself. These exogenous variables generally comprise a forecast of relevant variables. For example: a study of future educational needs may use the developments in international trade as one of its exogenous inputs. However, there are several conflicting theories based on other types of future research on the developments in international trade that might be expected. The selection of one of these conflicting forecasts of future international trade can have a decisive impact on the results of the study of future educational needs. In conclusion: the difference between 'objective' information from the forecast and the 'subjective' information put into the assumptions regarding the exogenous variables is hard to assess.

The discussion on the accuracy of the estimated parameters has always troubled the users and critiques of these models. As an example: the operation of the economic process and the decisive factors in its dynamics are generally acknowledged to be improperly understood. This means that caution is needed when it comes to the use of models in which the behaviour of the various economic actors is specified in some detail. In a review article several economists stated that "no economic theory tells you exactly what the equations should look like" resulting in 'fiddle and fit' operations until the model seems to be working well on data from the past (Kolata 1986).

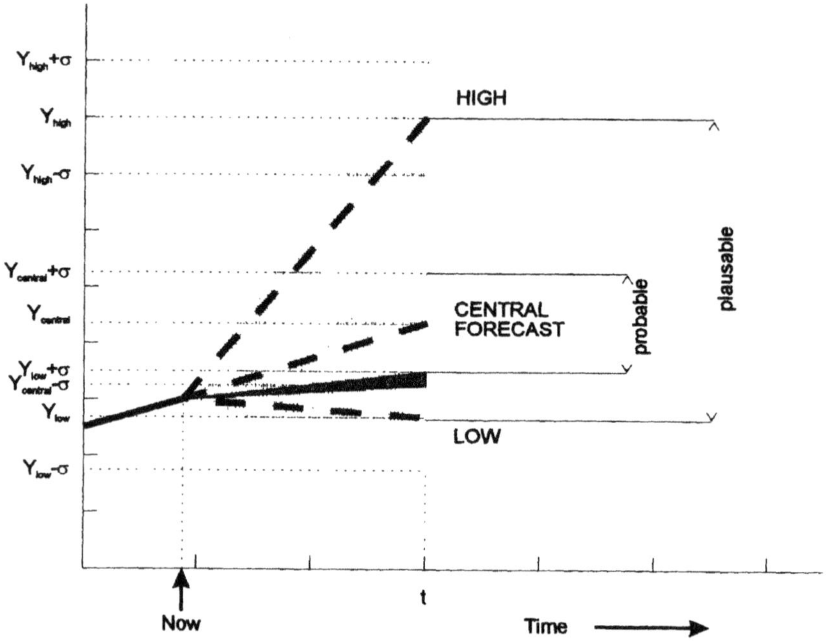

Fig. 2. Alternate forecasts (central, high and low) resulting from different assumptions regarding the specification of exogenous variables of the forecasting model.

One way to get round the problem of decreasing reliability of model specifications is to specify several alternative *conditional* forecasts. In this way the same model can generate different forecasts depending on the exact specification of the model parameters and/or the exogenous variables. This is illustrated in figure 2. Based on different estimations of the model parameters or other estimates of exogenous variables a high and a low forecast are calculated next to the original or central forecast. Because the high and low forecasts imply different assumptions regarding the model parameters and/or the exogenous variables, the three varying forecasts are generally indicated as *scenarios*. So, a scenario presents a forecast of future development with a certain accuracy, given a number of assumptions regarding the model that is used to obtain the forecast. Each of the scenarios (central, high and low) is still accompanied by an interval that denotes the accuracy of the forecast. The different scenarios are believed to be plausible forecasts of future developments that result from varying assumptions regarding the input of the model. Each of these inputs by itself is considered equally plausible, so no ranking can be made in the plausibility of the scenarios.

If a forecast is calculated for a longer period of time (mid-term or long-term forecasts) the confidence intervals of the different scenarios will tend to overlap. Of course this overlap complicates the interpretation of the results, especially since each of the scenarios must be regarded equally plausible. Still, most of the contemporary policy-oriented forecasts fall into this category (United Nations demographic forecasts, FAO world food scenarios, OECD economic outlooks, etc.) because users of future studies are generally more interested in findings stated in term of plausibility and probability than in term of possibilities.

Explorations aim at feasibility

When it comes to research into possible developments over the somewhat longer term, the danger arises that alternative developments and policy options may be prematurely lost to sight. Similarly, detailed behavioural models are less suitable for evaluating the consequences of policies expressly aimed at altering behavioural patterns. If one's concern is to survey long-term prospects and if breaks in the trend are not to be ruled out in advance, conditional forecasts might not render the desired information.

An exploration of future possibilities can give this type of information, but does not allow any conclusions with respect to probability and plausibility. Such an exploration is based on properties of the system, not on observed realisations of a process in a historical time-series. There is no extrapolation involved of any time-series, so there need not be an underlying assumption of continuity and breaks in the observed trends are perfectly sensible within an exploration. What is assumed, however, is that properties of the system under investigation will not change – or will reveal only minor changes – over time. This implies that any future development should be within the limits that are prescribed by the properties of the system. If an exploration is properly conducted it will show us the scope of possibilities that the future has in store for us.

As the methods of exploring and forecasting show considerable differences the information contents are also very different. An exploration of the future does not render a specified path in time, but it shows us the boundaries of any future development at any given point in time. This is illustrated in figure 3.

The focus of the research is no longer on developments that *might be expected*, but on developments that *might be feasible*. So, an exploration does not lead to the identification of probable developments. This may seem a step back in evolution. However, the usefulness of a well-founded finding relating to future possibilities might be more relevant to policy making than a disputable assessment of probability of a particular development.

The aim of an exploration of the future is not to reduce uncertainty, but to open up possibilities. Exploring here means to search for the boundaries of future developments. Not to reduce the uncertainty for those who have to decide on policy issues, but to show to them the realm of possibilities that the future has in store. So the ultimate goal of exploring the future is to influence

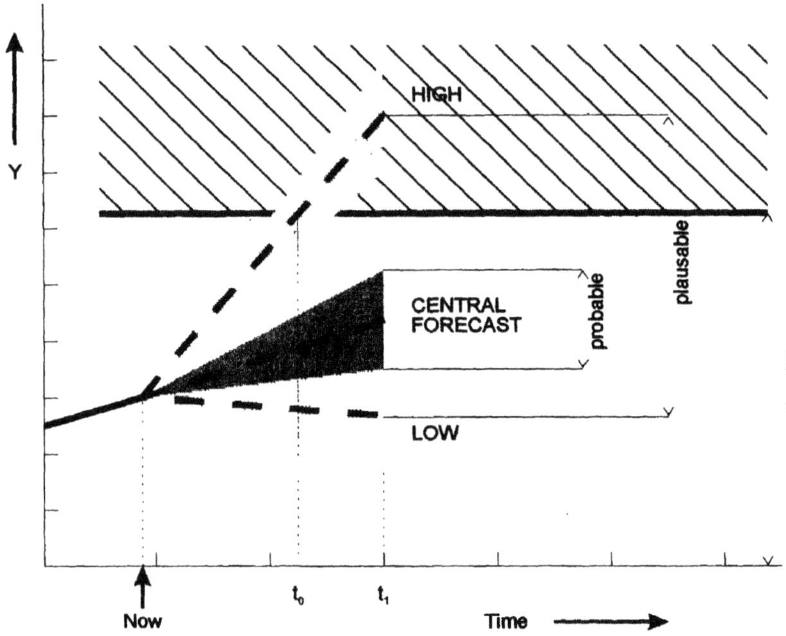

Fig. 3. Forecasts compared to an exploration: the forecast indicates three different scenarios with their corresponding confidence interval. The exploration indicates the feasible scope for future developments. It can be concluded that the high forecast will not be feasible. After t_1 the forecast is out of feasibility bounds.

future developments by showing what is possible. In this way future research cannot be used to relieve a policy maker of his or her responsibility. Rather, explorations show to an executive what exactly can be the consequences of a decision, thereby clarifying the responsibility.

An exploration uses the identified properties of the system to assess the boundaries of any future development of the system. Next, these boundaries can be compared with the projected developments from a partial forecast, or they can be used to assess the possibilities of changes that are expressed in explicit or implicit goals for the future. With an exploration, future possibilities can be mapped out. Because the present does not act as a reference also possibilities are included that do not follow directly from current practices. In an exploration the future is not seen as a function of the present, rather future possibilities are restricted only by limiting factors of the system that we can define.

Exploring future options in agriculture

To come up with information on future possibilities of the land-based agricultural production system in the EU it is necessary to identify the constraints on future developments and confront them with goals that relate to the desired performance of the system. Using available models and data it is possible to take the different policy goals to their logical conclusions. In the case of agricultural production the technical limitations of the system are well-known.

At the level of individual plants and crops, agronomic research has brought understanding of the processes involved in the productivity rise. Much of the agricultural research in the last decades has been focused on improving the productivity in agriculture by improving management, cultivars, machinery, inputs, pest control etcetera. This gave rise to an extensive body of scientific research on basic processes in individual plants and crops, including the limitations of those individual plants and crops. However, policy decisions are taken at much higher levels of aggregation. Policy is formulated at EU-level or at national level. Policy problems are perceived at regional level. The scientific basis of these policy issues at higher levels of scale is generally constrained to economic analysis. So, the available and needed information are not readily available at the same level of scale. If we want to benefit from both the agronomists knowledge at the lower levels and the economists knowledge at the higher levels we have to bridge the gap using some type of methodology. A systems approach turns out to be the answer to that question (Rabbinge and Van Latesteijn 1992).

Using engineers knowledge it is possible to construct a model-representation of agriculture. Using economists knowledge it is possible to translate policy goals into quantified objective functions and integrate them in the model. With a model based on the physical limitations of the agricultural system the flexibility of the agricultural system as a whole can be assessed, given the fact that various goals are to be fulfilled within this single system. This gives us information on the possibilities within the agricultural system based on the properties of the system itself. Again, it is explicitly not the intention to come up with more or less reliable predictions for the future of agriculture within the EU, but rather to explore the possibilities of the agricultural system.

So, information on physiological processes at the level of individual plants and crops is used to assess the properties of the system. A stepwise translation of the available information is needed to arrive at a model description of the land based agricultural system at the level of the EU. Next, this model can be used to explore possible future developments on the direction we want the system to evolve in the guise of several scenarios.

To construct the scenarios a Linear Programming (LP) model is used in an Interactive Multiple Goal Programming (IMGP) procedure. A LP-model is generally used to optimise a single objective function. The IMGP proce-

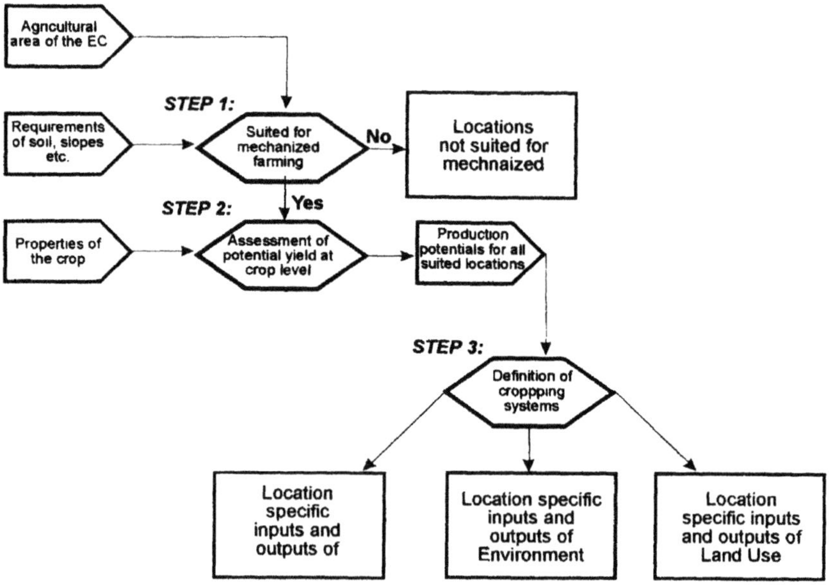

Fig. 4. Three steps to assess the necessary data on farming systems based on the assumption of Best Technical Means.

dure enables the optimisation of a set of objective functions in an iterative process. This reveals the trade-offs between different goals that are modelled by the objective functions. The core of the procedure is formed by the GOAL (General Optimal Allocation of Land Use) model of (agricultural) land use in the EU. The model can choose from a limitative set of types of land use to meet an exogenously defined demand for agricultural and forestry products. A number of policy aims are coupled to types of land use in the form of objective functions. These aims indicate the variety of notions that are considered to be essential for future land use.

Future demand for agricultural produce is assessed by using demographic data and information on the possible changes in diet related to a rising income. Assumptions are made regarding the share of import and export. The combination of these figures yields the needed yearly agricultural production within the EU. The LP-model must calculate a solution that meets this demand by choosing between different production techniques that will be available in the near future. For that reason current agricultural practices should not be used as a reference. Instead, types of land use that can be envisaged over a longer period of time (about 25 years) are defined. Moreover, these types of land use are available in all regions of the EU in the model, hence regional differences in skills and infrastructure are assumed to disappear within this 25 year period. The data on future farming systems are gathered in three steps, which are illustrated in figure 4 (Van Latesteijn 1993).

In the first two steps a mixed qualitative/quantitative approach is used to assess the production potential of the EU for a number of indicator crops. This mixed approach considerably reduces the number of model calculations (Van Lanen *et al.* 1992a). In the first step the ALES (Automated Land Evaluation System) is used to identify areas without any production potential for (mechanised) farming by looking at factors like steepness, salinity, and stoniness of the soil. The evaluation is executed at the level of Land Evaluation Units (LEUs), a combination of soil and climate conditions that is assumed to be homogeneous. For the EU some 22.000 units are necessary to cover the total area. A Geographical Information System (GIS) is used to identify the LEUs and process the resulting data. The first step results in a number of maps that state the suitability for mechanised farming of grass, cereals and root crops as well as the suitability for rough grazings and perennial crops (including forestry) (Van Lanen *et al.* 1992b).

In the second step production potentials for suited locations are calculated by means of a simulation model. This can be denoted as quantitative land evaluation. The quantitative land evaluation is accomplished through the use of the WOFOST crop growth simulation model (Van Keulen and Wolf 1986). The simulation model uses as its inputs: technical information on regional soil (such as water holding capacity) and climate properties and relevant properties of the crop (such as phenological development, light interception, assimilation, respiration, partitioning of dry-matter increase over plant organs and transpiration). Using this input information the attainable yields of winter wheat, maize, sugarbeet, potato, and grass in both rainfed and irrigated situations, are assessed. In the rainfed situation maximum yields can be limited by the availability of water at any point during the growing season. In that case the model simulation gives an indication of the attainable yields when no irrigation is applied. In the irrigated situation there are no limitations to crop growth other than those by climate and soil conditions and properties of the crop. In that case the model simulation indicates the maximum attainable (or potential) yield at a given location.

In the third step, the calculated potential yields of indicator crops are translated into cropping systems. Cropping systems involve various sets of crop rotations and various production techniques. The variety of production techniques found at present in various regions of the EU is tremendous. To eliminate historical differences between regions and inefficient or ineffective current production techniques, the concept of Best Technical Means (BTM) was introduced. Crops and cropping systems are produced using the best technological insight and procedures. This does not necessarily mean that only one way of producing is possible. Several different combinations of crop rotations and combinations of inputs are distinguished. Inevitable losses are accepted in for example fertiliser use and pesticides, but efficiencies are as high as possible by using the synergistic effects of combination of inputs. These forerunners are used as a reference for future developments.

Because in the model these techniques may be adapted and adopted in all regions, the results of the model calculations are not biased by the current differences in (agricultural) regional development. BTM can be regarded as sustainable agriculture put into practice, because emissions to the environment are minimised. By definition the efficiency of inputs being transformed into outputs can never be 100%, which implies that losses are inevitable. However, BTM will maximise the efficiency of the inputs, leading to minimal losses of inputs per unit of output. Cropping systems comprise a rotation scheme, strategic management decisions and well defined inputs. Results of field experiments and expert knowledge are used to set up a limited set of rotation schemes, consequently deducing the input and output coefficients (De Koning and Jansen 1992). For a given level of production (the rainfed and irrigated production levels) the minimal levels of inputs are assessed. The theoretical foundation of this optimisation is dealt with by De Wit (1992). He describes this optimum as the situation where each variable production resource is minimised to such a level that all other production resources are used to their maximum.

To arrive at feasible production techniques, expert knowledge is used to define cropping systems that are acceptable from both an economic and agronomic point of view. If only economic and agronomic efficiency is taken into account a set of rotation schemes results which we call Yield Oriented Agriculture (YOA). In this definition objective information on yield potential is combined with subjective information on required labour and capital goods. Another set of techniques emerges if environmental hazards are taken into account. This implies that less environmentally hazardous inputs (such as pesticides and fertilisers) per hectare are used, even if this means a slight decrease in yield (and thus technically sub-optimal). This set of systems we call Environment Oriented Agriculture (EOA). A third set is based on land use concerns. Under all circumstances it can be foreseen that the agricultural area within the EU will diminish. This can be detrimental to the maintenance of the countryside in some regions. A relatively low production per hectare is characteristic for this set which we call Land use Oriented Agriculture (LOA). These technically sub-optimal techniques are only feasible for grass and cereals. Root crops and perennial crops do not allow such techniques because for these crops they result in low efficiencies, high losses and low yields. In total the GOAL comprises some 6400 definitions of different BTM production techniques (YOA, EOA and LOA differing in rotation, input and outputs on a regional basis). All these different techniques are available for further analysis.

Next, information on land use related policy goals is incorporated in the model. The goals selected for this purpose are:

1. maximisation of yield per hectare;
2. maximisation of total labour;

3. maximisation of regional labour;
4. minimisation of total pesticide use;
5. minimisation of pesticide use per hectare;
6. minimisation of total N-fertiliser use;
7. minimisation of N-fertiliser use per hectare; and
8. minimisation of total costs.

Contrasting political *philosophies* on land use in the EU are fed into the model by assigning different preferences to these goals. This is done interactively in an IMGP-procedure by alternately restricting the objective functions to a certain domain while minimising or maximising another objective. For example: total labour is not allowed to drop below 6 million manpower units (MPUs) while minimising total costs. In this way scenarios can be constructed that show the effects of policy priorities and the trade-offs between different goals. For example: to maintain the labour force the model will have to select types of land use with a relatively high input of labour, thus resulting in relatively higher costs.

Four contrasting scenarios have been developed to illustrate the contrasting political philosophies that constitute the main movements in the current debate on agriculture. These are extreme philosophies, in which the ideas put forward in the debate are taken to their logical conclusions.

(a) *Scenario Free market and Free trade (FF)*: under this scenario agriculture is treated as any other economic activity. Production is as low-cost as possible. A free international market for agricultural products has been assumed, with a minimum of restrictions in the interests of social provisions and environment. The GATT is based on this philosophy.
(b) *Scenario Regional Development (RD)*: this scenario accords priority to regional development of employment within the EU, which creates income in the agricultural sector. This philosophy can be regarded as a continuation and extension of current EU policy.
(c) *Scenario Nature and Landscape (NL)*: under this scenario the greatest possible effort is made to conserve natural habitats, creating zones separating them from agricultural areas. Besides protected nature reserves, areas would also be set aside for human activity. Nature conservation groups are exponents of this philosophy.
(d) *Scenario Environmental Protection (EP)*: the primary policy aim under this scenario is to prevent potential hazardous substances from entering the environment. In contrast to scenario NL, the main aim is not to preserve or stimulate certain plant and animal species, but to protect soil, water and air. Natural and agricultural areas are therefore not physically separated but integrated. Farming may take place everywhere, but subject to strict environmental restrictions. This philosophy is in line with the concept of integrated agriculture as developed during the last decade.

The results of the four scenarios indicate that agricultural land use can be considerably lower in future than at present. The highest production per hectare is achieved in scenario NL, where the area of agricultural land is smallest. The discrepancy between the area of land currently in use and the area technically necessary for food production shows that the rise in productivity can still continue. Even if techniques are considered that show a relative low production per hectare (LOA) the land needed for production is still much less than at present. These results indicate that the present set-aside schemes in the EU to adverse the surplus production can only be the very beginning. If technical progress continues there seem to be little scope for a policy aiming at maintaining all current agricultural land in use.

Another striking results is that all scenarios show much lower figures for agricultural employment. Even in scenario RD with the objective to keep as many people as possible employed in land-based agriculture, employment declines, i.e. from 6 million MPUs (1988/89) to 2.8. These results indicate that preserving the current level of employment boils down to maintaining hidden unemployment (in some regions up to 50%) at high costs. In all scenarios considerable effort is required to accommodate the decrease of labour in agriculture.

In figure 5 the trade-offs between the eight policy goals are illustrated by showing the normalised values of each of the goals in the four scenarios. The value that can be reached for each of the goals if no other restriction is put to any of the other goals is set to 1. The graphs therefore show the deviation of the maximum values that can be attained for each of the goals. The graphs therefore summarise the very complex issues in assessing the costs and benefits of different policy options. If only these goals are valued the scenario RED would seem to be the most sensible policy choice. However, one must be aware that the spatial results of the scenarios do also show large differences and a final conclusion on the priority of each of the goals involved should also take into consideration that information.

Implications for policy making and the political debate

The GOAL-model is used to develop different to attain a set of well-founded objectives within the technical possibilities of the land-based agricultural sector. Policy instruments, such as price changes and assumptions on the behaviour of actors as well as institutional obstacles, are excluded. For tactical policy decisions concerning the use of instruments this will not be adequate, but for strategic policy planning purposes this type of analysis gives rise to valuable information. Hence, this is not a study of the effects of possible amendments to the CAP, although its results indicate the technical limitations to such changes. In many other policy areas such an assessment of technical limitations would pose considerable problems. For example: when should a

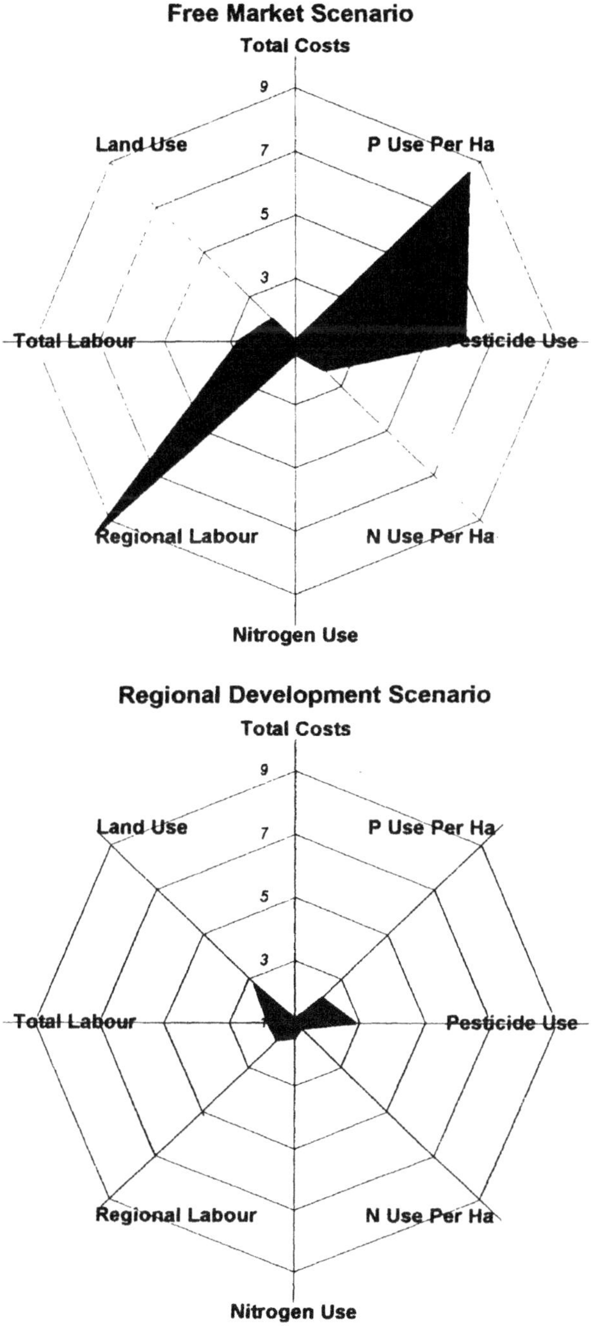

Fig. 5a, b. The standardized results of the policy goals in each of the four scenarios. The results are made proportional to the best value each goals variable can attain if no other restriction is put to any of the other goals. A most ideal solution would therefore be a unit circle.

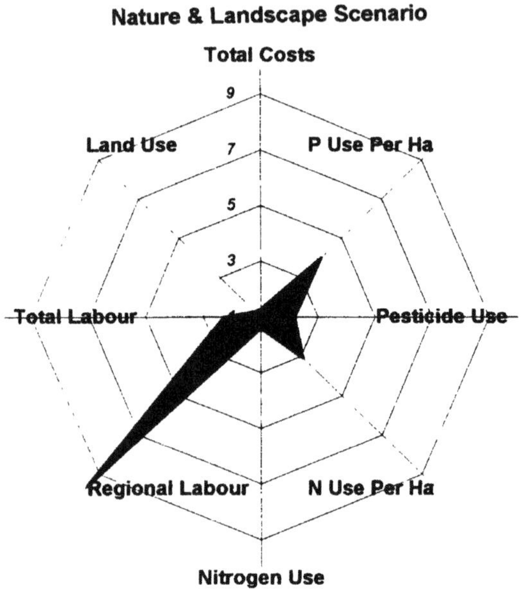
Nature & Landscape Scenario
Total Costs
9
7
5
3
Land Use
P Use Per Ha
Total Labour
Pesticide Use
Regional Labour
N Use Per Ha
Nitrogen Use

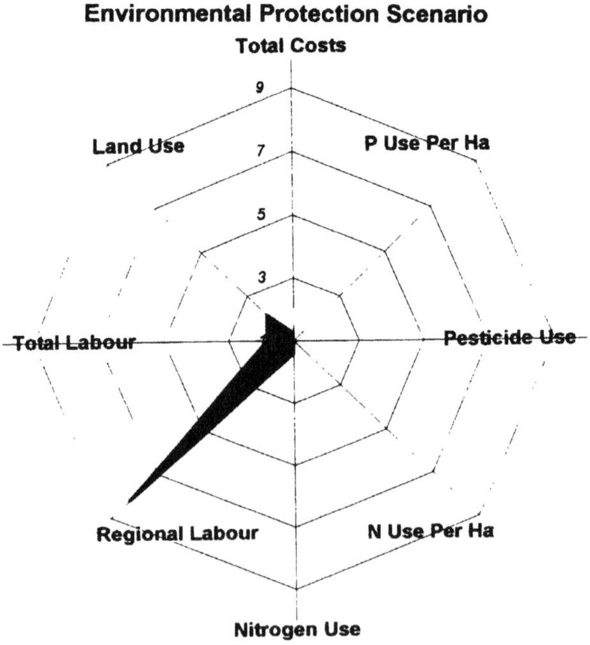
Environmental Protection Scenario
Total Costs
9
7
5
3
Land Use
P Use Per Ha
Total Labour
Pesticide Use
Regional Labour
N Use Per Ha
Nitrogen Use

Fig. 5c, d.

country be considered 'full', or what level of prosperity is 'enough'? This is possible for land-based agriculture in the EU, though, because the limitations can be deduced from biological and physical laws, supported by ample quantitative data (demand for agricultural products, technologies, possible use of land, etcetera).

Since its publication the report 'Ground for choices' has been discussed in various circles. Initially, several agricultural economists – or in general social-scientists – were rather reluctant to accept the results and methodology of the study. In their view this methodology was of not much use in view of the agricultural problems in the real world that ask for a solution. Generally the discussions revealed that much of this opposition can be traced back to a certain degree of misunderstanding or misinterpretation. There is a rather strong culture of executing future studies for the agricultural sector, generally based on econometric modelling. The results of these forecasts (viz. predictions) were compared to the outcome of the four scenarios and the observed differences led to a rejection of the whole study, including the methodology. Only after some time the complementarity of the two approaches came to the fore. Hopefully it will only be a matter of time before a real integration of the two approaches is institutionalised. Scenario studies along the lines of the GOAL-model can then be used to rephrase the questions and formulate the boundary conditions for further research with 'econometric' models. The question then becomes how a technically feasible and politically desired situation might be attained, given the present situation and information on behavioural constraints of all parties involved.

Among agronomists discussion arose on the scope of the farming systems defined in the GOAL-model. Especially advocates of ecological and/or organic farming stated that these types of farming were not sufficiently represented by the three orientations (YOA, EOA and LOA) in the model. By itself, of course, can be a legitimate observation, but the real challenge lies in the formulation of 'other' farming systems in usable model specifications. If the critics of the study would take up that challenge, the discussion on the alleged advantages of a specific type of agriculture might be rationalised considerably with the aid of a series of model calculations. Consensus building on these issues would benefit greatly from such an endeavour.

A separate discussion that runs through all other debates is on the usefulness of the concept of Best Technical Means. In other words: is it realistic to assume that at any given location high skilled farmers will be able to put optimal management techniques into practice and realise optimal efficiencies. For a number of critics this 'Utopia' is too far fetched. However, the problem can be reversed. Are there any sound arguments to assume that certain farmers, or certain regions, will be structurally less successful in the future? Will the northwestern European farmer be superior to his mediterranean colleague at all times? An explorative study that would postulate this kind of structural constraints would be very difficult to defend indeed. The same hold true for more general 'inevitable efficiency losses'. If one would adopt this line of thought, how much less than optimal is 'realistically' feasible? At this point in time there is no clear answer to these questions, so for the moment a well-argued definition of BTM seems the only sensible approach.

Policy makers discussed the outcome. They are investigating the possibilities to use the information on options and ask to what extent current policy can cope with the major developments generated in the scenarios. The discussion on the reform of the CAP did not end with the policy reforms effectuated in 1992. The budgetary problems are still there and the shift from price-support to direct income sharply revealed to a larger public the enormous amounts of money that are involved. In the near future an estimate should therefore be made of the effort required to achieve objectives in agricultural policy, depending on whether we will have to 'go against the tide' or simply go with it. The results of the scenarios can serve as guidelines for these future policies.

One obvious source of conflict might be that in all four scenarios agricultural land use is much lower than the 127 million hectares currently in use in the EU. At present great effort is given to maintain the situation that all this land is used for agricultural production. The results in the different scenarios demonstrate that this policy will be disastrous. The costs of the CAP will not decrease, environmental objectives will not be achieved, surpluses will increase and socio-economic goals will be jeopardised. Therefore, a more drastic adaptation is needed. The scenarios may help in defining that change, and it might well be that, all things considered, other goals should be given preference. Such questions arise from simply defining technical possibilities. The generation of information on what is possible is a valuable contribution of scientific research to the political debate. The answer to the question what we want must be formulated in a political decision process; this study is instrumental to that end.

Acronyms

BTM	Best Technical Means
CAP	Common Agricultural Policy
CGIAR	Consultative Group on International Agricultural Research
EU	European Union
GOAL	General Optimal Allocation of Land Use
IMGLP	Interactive Multiple Goal Linear Programming
LEU	Land Evaluation Unit
LP	Linear Programming
TAC	Technical Advisory Committee
WRR	Wetenschappelijke Raad voor het Regeringsbeleid (Netherlands Scientific Council for Government Policy)

References

Cole S, Miles I (1978) Assumptions and methods: population, economic development, modelling and technical change. Pages 51–75 in Freeman C, Jahoda M (Eds.) World futures, the great debate. Martin Robertson & Co., London, UK.

Van Keulen H, Wolf J (Eds.) (1986) Modelling of agricultural production: weather, soils and crops. Simulation Monographs. Pudoc, Wageningen, The Netherlands.

Kolata G (1986) Asking impossible questions about the economy and getting impossible answers. Science, 234.

Van Lanen H A J, Hack ten Broeke M J D, Bouma J, De Groot W J M (1992a) A mixed qualitative/quantitative physical land evaluation methodology. Geoderma 55:37–54.

Van Lanen H A J, Hendriks C M A, Bulens J D (1992b) Crop production potential of rural areas within the European communities, part V: Qualitative suitability assessment for forestry and fruit crops. Working Documents W 69, Netherlands Scientific Council for Government Policy, The Hague, The Netherlands.

Van Latesteijn H C (1993) A methodological framework to explore long-term options for land use. Pages 445–455 in Penning de Vries F W T, Teng P S, Metselaar K (Eds.) Systems Approaches for Agricultural Development. Kluwer Academic Publishers, Dordrecht, The Netherlands.

Rabbinge R, Van Latesteijn H C (1992) Long term options for land use in the European Community. Agricultural Systems 40:195–210.

De Wit C T, Huisman H, Rabbinge R (1987) Agriculture and its environment: are there other ways. Agricultural Systems 23:211–236.

De Wit C T (1992) Resource use efficiency in agriculture. Agricultural Systems 40:125–151.

WRR (Netherlands Scientific Council for Government Policy) (1988) Government and future research. Reports to the Government No. 34, The Hague, Sdu Publishers, The Hague, The Netherlands.

WRR (Netherlands Scientific Council for Government Policy) (1992) Ground for choices. Four perspectives for the rural areas in the European Community. Reports to the Government No. 42, Sdu Publishers, The Hague, The Netherlands.

Natural resources and limits of food production in 2040

F.W.T. PENNING DE VRIES[1,2], H. VAN KEULEN[1,3] and
R. RABBINGE[2,4]

[1] *DLO-Research Institute for Agrobiology and Soil Fertility, P.O. Box 14, 6700 AA Wageningen, The Netherlands;* [2] *C.T. De Wit Graduate School for Production Ecology, Wageningen, The Netherlands;* [3] *Section Animal Productions Systems, Department of Animal Husbandry, Wageningen Agricultural University, Marijkeweg 40, 6709 PG Wageningen, The Netherlands;* [4] *Department of Theoretical Production Ecology, Wageningen Agricultural University, P.O. Box 430, 6700 AK Wageningen, The Netherlands*

Key words: agricultural productivity, arable cropping, diet, food consumption, food production, food security, hunger, grassland, intensive farming, irrigation, natural resources, soils

Abstract. Food demand is estimated for the 15 major regions of the world for the year 2040. It is compared with the potential food production in these regions, which is derived from the area with soils suitable for cropping and grazing, the amount of irrigation water available, and the farming system used. All farmers are assumed to employ the best known techniques for sustainable farming. Two alternative production systems are explored: optimum productivity per unit of land, with intensive use of chemical inputs and energy to produce top yields ('HEI'), and agriculture in which environmental damage per unit area is minimised ('LEI'). In the latter system, legumes provide all nitrogen, agriculture is more diverse, and hectare yields are lower. Farming could occur at a smaller scale than in HEI-farming, with strong integration of arable farming and animal husbandry, but these aspects play no role in this study.

Comparing 2040 scenarios of demand and supply of food shows that most regions can avoid to run into food security problems, but that in Asia situations could develop where a moderate or affluent diet is out of reach of its population, even when maximum use is made of all natural resources.

When HEI-agriculture is practised, all regions can produce food required for an affluent diet, except for East, South and West Asia. Also Southeast Asia and West and North Africa come close to the lower limit. A diet much less expensive provides the only option for escape. The three regions with the least leeway will carry almost half of the global population. Europe, the former USSR, the American regions and Central Africa are well off and need only a part of the suitable land to feed their populations.

Practising LEI-agriculture, only South Asia will have food shortage. In this heavily populated region, there is no way out via less expensive diets or lower population growth (both already at a minimum). Europe could grow all its food on less than half of its suitable soils if the LEI system goes with the low food demand scenario. Only the former USSR, North and South America, Central Africa and Oceania can consider to offer its population an affluent diet.

Introduction

The UN projected population growth into the next century and expects the global population to stabilise around 2040. The world will then carry 1.5 to

J. Bouma et al. (eds.), Eco-regional approaches for sustainable land use and food production, 65–87.

2.2 times more human beings than in 1990, many of whom will require 2–3 times more food. Can the earth provide enough food by socially acceptable ways of farming, without sacrificing its natural resources?

Two decades ago, the maximum global food production was estimated to be 50 billion ton (Buringh *et al.* 1975), enough to feed at least 30 billion persons. Better knowledge of soils, fresh water resources, and crop performance allows us to improve this estimate. We added also rangelands as a potentially major source of food. Global changes in soil and climate make people wonder whether food production is threatened. Furthermore, questions are posed whether top yields achieved at experiment stations can be achieved at large scale and maintained (World Resources Institute, 1994). Recent reports suggest a ceiling to global food supply that support populations much below 30 billion, some even as low as the current population of just 5 billion (Pinstrup-Andersen 1993; World Resources Institute 1994; Brown and Kane 1994).

Our objective is to investigate in a quantitative manner whether the natural resources allow food security for the future populations in 15 major regions. We will discuss the results of computations for alternative production and demand scenarios, compare them with other reports, and comment on implications. Both production systems considered are agro-ecologically sustainable. How to stimulate farming communities to adopt the production systems proposed is not discussed here.

We do not address effects of global changes in climate or soil because of lack of sufficiently reliable information and because farming is likely to be sufficiently flexible to adapt to possible changes.

A full technical report on methods and basic data is published by Luyten (1995); highlights were presented in the wider context of use of natural resources for industry, transportation, and recreation (Wetenschappelijke Raad Regeringsbeleid 1994). In other articles, we explored implications for soil science (Penning de Vries *et al.* 1995) and zoomed in on China (Luyten *et al.* 1995).

Outline of the approach

We compute the amount of plant biomass required to feed the future population in each of 15 regions, and compare that with the amount of food that could be produced from an agro-technical point of view in those regions in a sustainable manner, while using the natural resources efficiently. Our 15 regions (figure 1) are those distinguished in the UN population study (we have added the very small region of the Caribbean to Central America, and grouped the four European regions into one). These regions differ significantly in many respects (table 1).

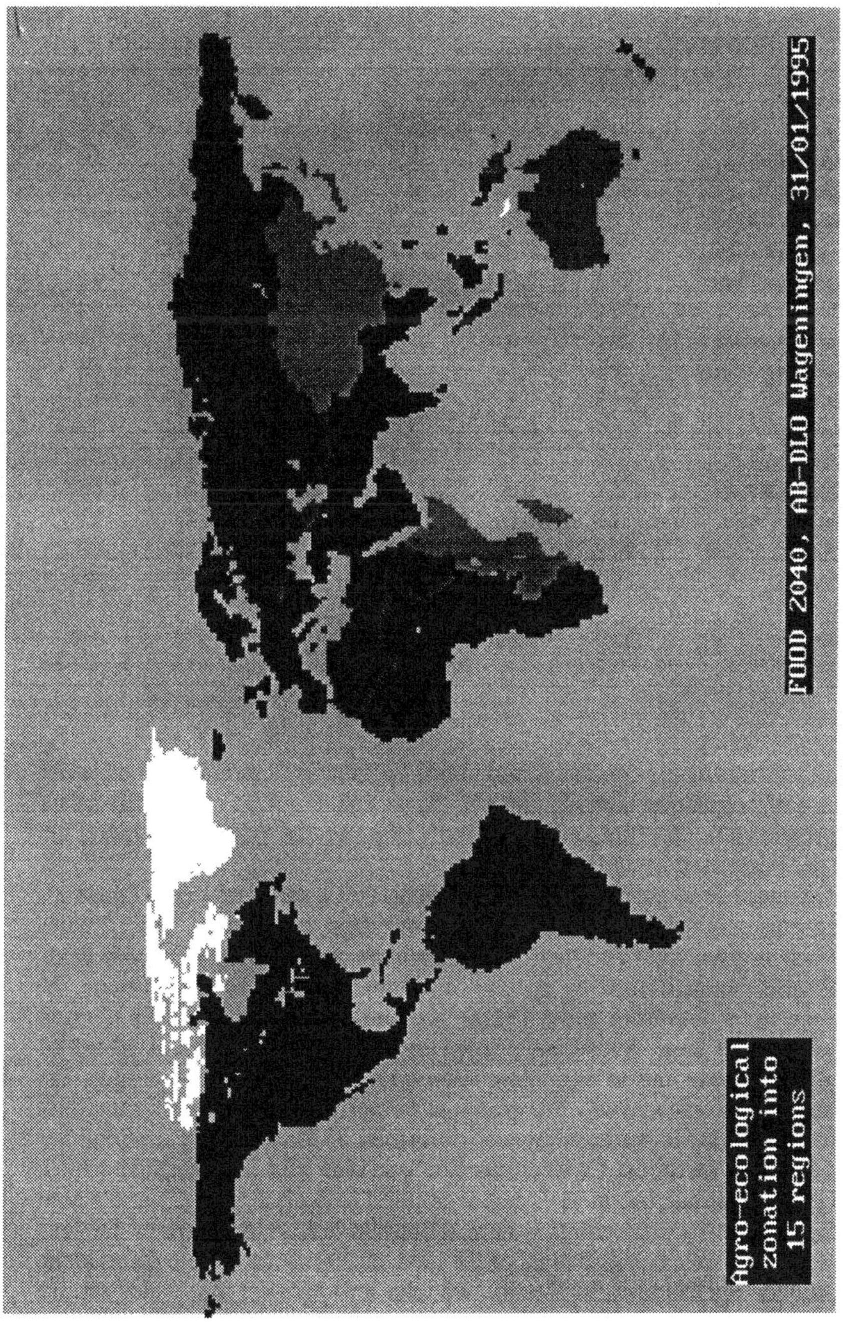

Fig. 1. World map of the 15 regions considered in the paper. Other data in table 1.

Table 1. Key characteristics of the 15 regions.

#	Region	Total land area [M km^2]	Avrg. land suitability [frac.]	Avrg. crops per year [#]	Available irrig. water [km^3 yr^{-1}]	Popul. in 1990 [million]	GNP per capita [k US$]
1	South America	16.8	0.82	2.3	3150	297	1.6
2	Central America	2.3	0.69	2.4	410	151	1.5
3	Northern America	15.9	0.56	1.3	730	276	18.3
4	Northern Africa	7.9	0.70	2.2	150	141	1.1
5	Western Africa	5.9	0.74	2.9	550	194	0.5
6	Central Africa	6.3	0.86	2.2	1380	70	0.4
7	Eastern Africa	5.9	0.80	1.9	1250	197	0.2
8	Southern Africa	2.6	0.74	1.5	270	41	1.5
9	Oceania	7.9	0.77	2.4	390	27	8.7
10	Southeast Asia	3.5	0.58	2.7	290	445	0.6
11	Easthern Asia	11.0	0.52	1.4	430	1336	2.6
12	Southern Asia	6.5	0.60	2.4	620	1201	0.3
13	Western Asia	4.1	0.66	2.4	170	132	2.4
14	(former) USSR	20.9	0.38	1.1	480	289	8.7
15	Europe	4.6	0.72	1.5	160	98	11.1
	World	122.0	0.64	2.0	10430	5293	3.6

Six features have been combined in this approach: Demand for food is the product of population size (1) and per capita consumption of carbohydrates and proteins (2). Medium and the extreme of projected values of demand for food are been used in our analyses.

Maximum food production is approximated in a series of steps (figure 2), taking into account four natural resources: crops, land, water and climate. For each region, total production results from aggregation of yields from small units characterised by specific combinations of soil and climate. In total, we used data from some 15500 land units, over 700 weather stations and about 100 large river basins, applying the concept that it is better to use all basic data and to aggregate subsequently, than to use averages (De Wit and Van Keulen 1987). We compute potential production for situations where all farming is practised by well-informed, skilled farmers applying the most efficient methods. Crop production per unit area is quantified with a crop growth simulation model, that is applied in both a high (3) and a low external input (4) production system. Potential arable land area is far less than total land area because most of the land surface is unsuitable for modern mechanised agriculture (5). Finally, we take into account that water supply for irrigation is limited (6).

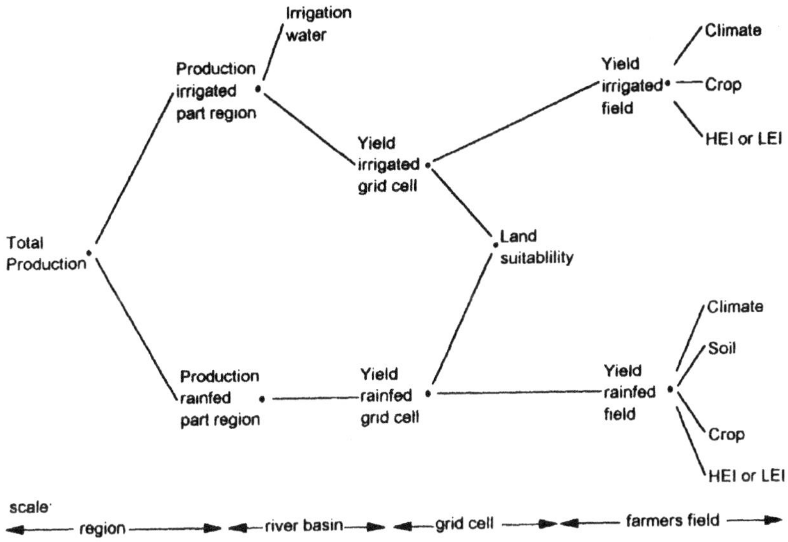

Fig. 2. Steps and aggregation level to compute food production by region.

We address two alternative types of farming, based on contrasting views with respect to the 'best' use of land (cf. WRR 1994). Different is these view is the approach to the environment. Similar is that all farmers are assumed to farm using 'best technical means', i.e. the best techniques currently available for each agro-climatic zone (cf. De Koning *et al.* 1992). Different levels of agriculture and use of natural resources result from the contrasting views. In the first view, one aims at optimum productivity of land and it is expected that environmental damage can be limited to acceptable levels. Agricultural practice is characterised by a high degree of mechanisation and heavy use of fossil fuel. The losses of inputs to the environment are minimal per unit of product, so that this high input farming is ecologically sustainable. We call this global and market oriented view of agricultural production systems the 'high external input' scenario (HEI).

In the second view, such intensive practices can never be sustainable. Therefore, the production system is designed to minimise loss per unit area and impact on the environment (we will return to this in the discussion). This is realised by replacing all N in chemical fertiliser by biological N fixation, elimination of biocides, restricted use of mechanisation, intensive recirculation of nutrients, and a 'local' consumption of the products. In this scenario, crop yields are lower, but product quality and prices higher, less fossil energy is used, and agriculture can be more integrated with nature development. We call this the 'low external input' scenario (LEI). P and K

Table 2. The population size in 1990, and the low, medium and high population estimates for 15 regions of the world for 2040 (United Nations 1992).

		Population	Population estimate 2040		
#	Region	1990	Low	Medium	High
1	South America	297	481	558	663
2	Central America	151	250	296	347
3	Northern America	276	274	328	398
4	Northern Africa	141	277	343	419
5	Western Africa	194	466	635	798
6	Central Africa	70	190	240	286
7	Eastern Africa	197	537	679	842
8	Southern Africa	41	89	100	123
9	Oceania	27	32	37	45
10	Southeast Asia	445	658	820	1005
11	Easthern Asia	1336	1503	1770	2098
12	Southern Asia	1201	1965	2408	2889
13	Western Asia	132	249	324	399
14	(former) USSR	289	323	369	419
15	Europe	498	437	498	563
	World	5293	7730	9404	11291

fertiliser cannot be replaced biologically, and are assumed to be supplied to allow high rates of N-fixation. Because of this use of inputs, LEI-farming is not identical to LEISA-farming (Reijntjes *et al.* 1993).

To compare expected food consumption and potential production, we express both in grain equivalents (GE). GE is a theoretical food unit. In the production process, it refers to the quantity (in kg) of dry grain that would be produced if only one type of crop were grown (a cereal), plus the amount of grain that needs not to be produced because of feed (grass) harvested from land unsuitable for arable farming; feed requires conversion via animals for human consumption. In the consumption process, GE refers to the amount of cereals (in kg) needed as raw material for the food consumed, plus the 'opportunity cost' to grow food that cannot be produced via 'grain' (e.g. fruit).

The approach

(1) For the projected size of the regional populations (table 2), we followed a recent UN-report (1992). We have chosen the year 2040 as target date since the low, medium and high projections then stabilise at global level.

Table 3. Average relative cost of food, expressed in grain equivalents (GE).

Food item	kg GE per kg dry mass	kg GE per kg fresh material
Plant products	2.0–3.5	0.8–1.2
Dairy products	6.3–7.2	2.4–3.3
Meat products	17.0–19.0	8.5–9.5

Table 4. Food composition in three diets. Note: our data on 'grain use per capita' are different from other authors (e.g. Crosson and Anderson 1992; Brown and Kane 1994). Our definition includes the opportunity cost (in GE) of all other food crops, such as tuber, oil seed and leguminous, fruits and vegetables, while other data refer to cereals *sensu stricto*.

Diet	GE [kg cap^{-1} yr^{-1}]	Energy [kJ d^{-1}]	Animal protein [g d^{-1}]	Plant protein [g d^{-1}]
Vegetarian	475	10.0	8.6	66.7
Moderate	875	10.0	31.2	50.0
Affluent	1530	11.5	63.2	28.9

(2) Human food is extremely variable in composition, and we cannot address even a fraction of that diversity. Food items vary strongly in energy and water content. Part of the food comes from animal sources, often formed in an inefficient conversion process from plant biomass (table 3). Therefore, the amount of basic food that needs to be grown depends strongly on the composition of our diet. To compare requirements for different populations with the production capacity of the land, we express all consumption in GE (for details, see Luyten 1995). Plant biomass (in GE) needed for a productive life consuming a largely vegetarian diet (grains, tuber crops, pulses, some milk) is only about one third of the quantity needed for a diet with a considerable proportion of animal products, including meat (table 4). We carry out our calculations for three diets: an ample and healthy vegetarian diet, a moderate diet (with some meat, similar to that in Japan or Italy), and an affluent diet (such as that of the USA in the 1970s). The affluent diet will mostly be found in rich societies with many pets, for which the feed is included.

The medium and extreme projections of demand for food per region are given in table 5.

(3) Food production. The natural resource base is characterised by soil, climate, plant genetic properties and surface water. We do this at as small a scale as soil data permit (i.e. a grid cell of 1° × 1°). We base our calculations on two crop types (cereal and grass) for reasons of simplicity, albeit that a temperate variety of the cereal ('wheat') and a tropical one ('rice') are

Table 5. The maximum, medium and minimum amount of food required in 2040, by region and total (in GE yr^{-1}).

#	Region	Minimum (veg. diet, low pop.)	Medium (mod. diet, med. pop.)	Maximum (affl. diet, high pop.)
1	South America	228	489	1016
2	Central America	118	259	532
3	Northern America	130	287	610
4	Northern Africa	131	300	642
5	Western Africa	221	556	1223
6	Central Africa	90	211	438
7	Eastern Africa	255	595	1290
8	Southern Africa	42	88	188
9	Oceania	15	33	68
10	Southeast Asia	312	718	1541
11	Easthern Asia	713	1551	3217
12	Southern Asia	932	2109	4428
13	Western Asia	118	283	611
14	(former) USSR	153	323	642
15	Europe	207	436	864
	World	3668	8238	17309

distinguished to account for climatic adaptation. Since all major agricultural crops produce biomass (dry matter) at a rate in the order of 200 kg ha^{-1} d^{-1} at full soil cover, total production is hardly affected by this simplification. A difference between the commodities is in the harvest index (HI, fraction of total biomass harvested), which is 0.4–0.45 for modern varieties of cereal crops, and 0.6–0.7 for permanent grassland. The simplification of using two crop types only is acceptable when conversions between food types produced and food types required for specific diets are possible.

Maximum production for wheat, rice and permanent grassland are computed considering local soil, climate and HEI-management practices. The cereal crops are grown in healthy rotations, with up to three crops per year when climate and soil permit. Varieties grown are similar to the best currently available (HI = 0.45); post harvest losses are set to 10%. Permanent grassland grows on soils unsuitable for arable crops. For grassland, we applied a rather low value of the HI (0.6) to reflect that some of these areas are difficult to reach for animals or have top soils that are easily damaged. Also here we assumed 10% post harvest losses.

For each grid cell, crop yield for the HEI-system is computed with a simple simulation model (SIMFOOD; figure 3). It is based on the well-tested concept

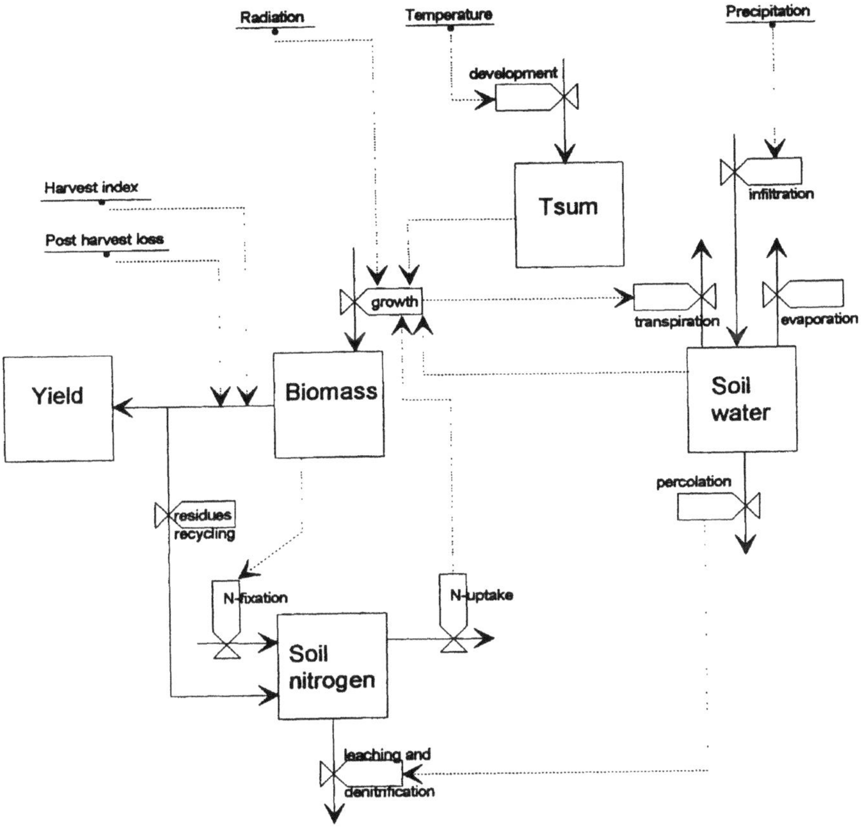

Fig. 3. Flowchart of the simulation model SIMFOOD. Rectangles stand for quantities, valve symbols for rates and underlines variables are constants or driving variables. Daily growth of biomass is computed from intercepted radiation for the duration of a growing season, which is monitored as a temperature sum. Final yield equals biomass times the harvest index, minus post harvest losses. The soil water balance model is used in rainfed conditions, and to determine the demand for irrigation water. When the relative water content is low, the growth rate is reduced. For LEI-simulations, the soil N-balance is invoked. Low soil N leads to reduced growth rates. High rates of percolation cause leaching and denitrification.

of conservative efficiencies of use of radiation (about 3 g dry biomass MJ^{-1} absorbed radiation), water (about 2 g biomass kg^{-1} water transpired, Monteith 1990) and nutrients (about 115 kg biomass kg^{-1} N absorbed, Sinclair 1990). For irrigated conditions, inputs are key crop characteristics (development rate, maximum photosynthesis), radiation and temperature; for rainfed conditions we used the same data plus soil data (water holding capacity) plus rainfall and evaporation. The model also computes the amount of supplementary water required to fully irrigate an arable crop (grassland is not irrigated), the amount of fertiliser N required to grow the cereal and grass crop without

Table 6. Typical yields of crops and permanent grassland (in GE, t ha^{-1} yr^{-1}) in temperate and tropical zones for the HEI and LEI production systems. In tropics, up to three crops per year are grown.

GRAIN:	HEI	LEI
zone		
tropics, irrigated	16–20	4–6
tropics, rainfed	4–12	2–4
temperate, irrigated	7–11	2–4
temperate, rainfed	4–9	2–3

GRASSLAND:	HEI	LEI
zone		
tropics, irrigated	21–26	7–11
tropics, rainfed	5–16	4–7
temperate, irrigated	9–14	4–7
temperate, rainfed	5–12	4–5

exhausting the soil, the associated NO_3-leaching, the required biocide input and the labour requirement. Typical yields for the HEI system are shown in table 6; the large difference between tropical and temperate zones is due to the number of crops per year, while the difference between rainfed and irrigated crops is due to the lower number of crops per year and the effects of temporary drought stress; grass yields exceed cereal yields because a larger fraction of grass is used.

In rainfed situations, the standard duration of the simulated crop does no justice to farmer's practices that fully exploit the growing season. But our model accounts for this discrepancy as it 'initiates' in the rainfed situation as many crops as in the irrigated case, utilising all available water, and adds the yields of all these crops in a year, so that their total yield corresponds with that of an adapted cropping system.

Arable cropping requires better soils and flatter land than grass production. We compute therefore first the production of arable crops on soils suitable for them, and then we compute grass production on remaining soils suitable for agriculture.

(4) In the LEI-system, crop production is ultimately limited by the supply of N in irrigated conditions, and by N or drought in rainfed situations. In SIMFOOD, production in each grid cell is related to the N absorbed. Uptake by the crop is one of the processes by which N is removed from the soil (leaching and denitrification being the other), and these are in balance with the processes by which N is supplied (fixation, recirculation, manure). Under appropriate farming and range management techniques, N-supply equals 270 kg N ha^{-1} $rotation^{-1}$, most of which originates from N-fixation by legumes; crop

residues and animal manure are recycled completely. We assumed arable cropping to be practised in a 2:1 cereal-legume rotation (the legume having the same growth characteristics as the cereal), and the rangeland to contain sufficient N-fixing legumes. N-supply per crop has been taken constant for all soils and climates, but when crop growth is limited by moisture supply, N-supply is reduced accordingly. Though agriculture in LEI-systems is also supposed to be carried out by skilful and knowledgeable farmers, the 'best technical means' are slightly different: HI is a little lower (0.4) and post harvest losses higher (0.2) because biocides are not used, the turn-around time between crops is longer (2 vs. 4 weeks), the N use efficiency is 120 kg kg^{-1} (vs. 112 in HEI). Permanent grassland comprises a mixture of grasses and legumes, and because of the higher quality of the LEI-pastures we assume that they are exploited better (HI = 0.7). Post harvest losses are also set to 20%. The fraction of N lost to denitrification and leaching is related to soil texture and to the precipitation surplus (De Koning *et al.* 1992).

Yield levels in the LEI-system are roughly one third of those in the HEI-system (table 6). The difference between irrigated and rainfed crops and grasslands is small because N-supply is usually the main limiting factor. The use of rangeland for food production is neglected in many studies (e.g. Buringh *et al.* 1975). However, vast amounts of land unsuitable for cropping are potentially valuable resources of animal products.

(5) Not all land is equally suitable for arable cropping. Slopes may be too steep, soils too stony and shallow, too saline, etc. Although LEI-farming may be less demanding than HEI-farming, we have not made a distinction. Arable cropping is more demanding than permanent grassland. Suitability is therefore judged per grid cell for cereal crops and grassland separately. It is expressed as the fraction of the area where crops can be grown without soil-related restriction; the remainder is not cropped, but can be used for other purposes (nature, recreational areas, infrastructure). Suitability for agriculture could not be taken directly from a source, so we used a global database designed for climate impact studies (Zobler 1986). It specifies the relative suitability of grid cells with respect to slope, soil phase and soil texture; by expert judgement we attributed values of 0 to 1 to each characteristic and defined 'suitability' as the product of the three factors. The world's average suitability for arable crops (weighed according to area) is only 0.31. An additional 0.33 is suitable for grassland. Suitability is distributed highly irregular at small and large scales (figure 1, Penning de Vries *et al.* 1995): that of South America stands out as a relatively high value, while that of South East Asia and the former USSR are low.

The soil water holding capacity depends on soil texture: high for clay and loamy soils, low for sandy soils; the texture of the dominant soil type is applied to the entire grid cell. In our simple water balance model, soils are not layered or cracked, well drained, and 0.6 m deep; there is no run off, and no watershed level storage of water.

(6) Irrigation. The irrigable area per river basin was calculated as river discharge and demand of arable crops for full irrigation (we choose not to irrigate grassland). Data were used on the current water discharge of 95 major river basins (those exceeding 3.10^4 km^2) that together occupy about 0.6 of the total land surface and carry currently 0.9 of the global population. River discharge was adjusted to account for unavoidable flood runoff (about half of the total), and corrected for projected household and industrial water use (assuming first priority for these uses) and for current extraction in irrigation schemes (Delft Hydraulics 1992). Of the irrigation water for cropping, only about half is actually used by plants (the 'best technical means' in irrigation schemes (Doorenbos and Pruitt 1977; the current water use efficiency is around 0.35), while 0.25 of the water flows back to the river and is not re-used to prevent salinization. This approach provides a conservative estimate of the quantity of water available for crop growth. Based on the reasoning that water is always used at maximum efficiency, we apply irrigation water first to crops with the lowest demands to attain maximum yields (i.e. in the most humid zones in each basin and on the best soils); only under ample water supply, drier zones are also irrigated.

As a world average in the HEI-scenario, 0.64 of the land suitable for arable cropping can be irrigated. In North Africa and Oceania is water scarce and less than 0.25 of the suitable area is irrigable, but also South and East Asia, West and South Africa have little surface water (< 0.5 irrigable). Demand for water for crop production is roughly proportional to yield. Per unit area, LEI-production systems need much less water, and large areas do not even require additional water, so that on a much larger area crops are grown without water shortage: 0.91 of the total area, i.e. all crop land in all regions except for ones mentioned with major shortage in the HEI-system.

The regional and global food situation

The HEI-scenario

The absolute maximum regional and global quantity of food that can be produced in a sustainable manner would be achieved if water were never in short supply (e.g. when desalinised sea water would be freely available). If all land suitable for arable cropping is irrigated (and no use is made of grassland), the HEI-production system could supply an affluent diet to over 50 billion persons, though even then, food would have to be imported in East and South Asia under the high population scenario.

It is unlikely that availability of water will increase dramatically. With maximum use of all fresh water supplies, 0.51 of the global area suitable for cropping can be irrigated. The distribution of irrigeable land is highly irregular: Asia has already much irrigable land, and major potentials for irrigation

Table 7. Maximum and irrigated food production (GE, Mt yr^{-1}) and the maximum area irrigated (Mha) in the HEI and the LEI situation. The total food production is the sum of contributions by irrigated and rainfed arable crops, and that of rangeland.

#	Region	HEI system			LEI system		
		Total prod. [Mt yr^{-1}]	Irrigated prod. [Mt yr^{-1}]	Irrigated area [Mha]	Total prod. [Mt yr^{-1}]	Irrigated prod. [Mt yr^{-1}]	Irrigated area [Mha]
1	South America	20373	11837	697	6877	3804	852
2	Central America	1853	949	55	811	284	55
3	Northern America	6418	2396	250	3252	1519	480
4	Northern Africa	1798	648	35	1066	290	58
5	Western Africa	3546	1449	73	1503	788	138
6	Central Africa	7505	4691	301	2672	1467	318
7	Eastern Africa	5594	3169	169	1892	1057	254
8	Southern Africa	1304	654	35	616	326	89
9	Oceania	4137	1020	55	2238	821	150
10	Southeast Asia	3670	1394	78	1185	368	78
11	Easthern Asia	4056	1949	236	2261	740	236
12	Southern Asia	3442	1594	98	1836	931	179
13	Western Asia	1245	772	45	658	305	59
14	(former) USSR	4524	1645	218	2459	1113	432
15	Europe	2792	1011	110	1348	375	120
	World	72256	35175	2457	30673	14188	3499

exist in South America, Central and Eastern Africa, North America and the former USSR. A substantial proportion of the unirrigated land receives sufficient rain to produce a decent rainfed crop. On a global basis, potential irrigated production exceeds rainfed production considerably (35 vs. 8 billion ton GE), whereas maximum grassland production could provide another 29 billion ton. Details are shown in table 7.

In Oceania there is still so much rangeland that the maximum rainfed production equals that of irrigated production. In absolute terms, South America has by far the highest food production potential of all regions: due to its large capacity for irrigation, and it also has the highest rainfed production potential. Obviously, realisation of this potential requires agricultural use of areas currently under rain forest. A second substantial area for irrigated cropping is Central Africa, while East Africa comes third. Clearly: there are enormous potentials in areas with few people (table 1).

Different regions (or indeed, all countries within regions) can simultaneously apply the HEI or the LEI production system (or anything in between). The global sums in the tables 8 and 9 have therefore only a reference value.

The LEI-scenario

Maximum hectare-yields in the LEI-scenario are lower than in the HEI-scenario, but this is largely compensated by a larger area that can be irrigated (0.9 of the cropped land). Indeed: N is the limiting production factor in this situation, not water.

If water shortage were completely eliminated, about 20 billion persons could be sustained on a vegetarian diet. With the current water supply, maximum global food production is 31 billion ton, or about one third of the maximum for HEI, using the same natural resources (table 7). Rainfed crop production is 20 × smaller than irrigated production, and only of importance in Oceania, and West and North Africa. Feed production from rangelands (16 billion ton) is still significant. Again, differences among regions are large. Barring food imports, Asia and to a lesser extent Europe, cannot produce the food to provide an affluent (or even a moderate) diet to its population in all but the lowest population scenario. North Africa's production largely originates from rainfed crops and grassland, which is also significant in South and West Asia.

Supply versus demand

What picture emerges when future production potentials and future demands are confronted? Excluding large-scale unilateral export of food from regions, the ratio of potential supply over expected demand indicates the potential relative food security (table 8). Ratios vary from values below 1.0 to over 100. Generalisations are risky, as each region has its own balance of resources and food demand.

Under HEI-farming, the projected medium demand can be satisfied in all regions, though in some regions (Asia), all suitable cropping and grazing land is needed. Results of a similar study for Europe also showed that not all land will be needed for agriculture in the future (WRR 1992).

Practising LEI-farming, Asia and to a lesser extent Europe, cannot produce the food to provide an affluent diet (or even a moderate one in Asia) to its population in all but the lowest population scenario, assuming massive food transfers do not take place. North Africa's production largely originates from rainfed crops and grassland, which is also significant in South and West Asia.

Before examining all ratios individually, their number can be reduced by focusing at some scenarios. To do so, we return to the contrasting views on agricultural development that are the basis of the current study. The view

Table 8a. Ratio of potential demand and supply of food, by region, and global total. Table 8a gives all ratios, table 8b shows the selection of ratios with values of 2.0 or less, and table 8c shows the most likely scenarios.

#	Region	HEI system			LEI system		
		veg.diet low.pop.	mod.diet med.pop.	affl.diet high.pop.	veg.diet low.pop.	mod.diet med.pop.	affl.diet high.pop.
1	South America	89.2	41.7	20.0	30.1	14.1	6.8
2	Central America	15.6	7.2	3.5	6.8	3.1	1.5
3	Northern America	49.3	22.3	10.5	25.0	11.3	5.3
4	Northern Africa	13.7	6.0	2.8	8.1	3.5	1.7
5	Western Africa	16.0	6.4	2.9	6.8	2.7	1.2
6	Central Africa	83.2	35.6	17.1	29.6	12.7	6.1
7	Eastern Africa	22.0	9.4	4.3	7.4	3.2	1.5
8	Southern Africa	31.0	14.8	6.9	14.6	7.0	3.3
9	Oceania	270.7	126.9	60.6	146.5	68.7	32.8
10	Southeast Asia	11.8	5.1	2.4	3.8	1.7	0.8
11	Easthern Asia	5.7	2.6	1.3	3.2	1.5	0.7
12	Southern Asia	3.7	1.6	0.8	2.0	0.9	0.4
13	Western Asia	10.5	4.4	2.0	5.6	2.3	1.1
14	(former) USSR	29.5	14.0	7.0	16.0	7.6	3.8
15	Europe	13.5	6.4	3.2	6.5	3.1	1.6
	World	19.7	8.8	4.2	8.4	3.7	1.8

that maximum land productivity should be achieved and some environmental damage is acceptable is likely to concur with the view that 'better diets are deserved', that economic development in less-endowed countries should originate completely from local production, which grows slowly, so that poverty and high population growth rates pertain. In other words, the HEI production system is more likely to be combined with the medium-high food demand scenario. By the same reasoning, a view leading to LEI-agriculture might concur with vegetarian-moderate diets, faster economic development of poor countries due to better terms of trade, and therefore lower population growth rates. LEI-agriculture could then be compared to the low-medium demand scenario.

It is interesting to observe that, by coincidence, the potential supply/demand ratios for the globe and by region are about equal for these HEI and LEI-scenarios. Three groups of regions can be distinguished: those in the danger zone (ratios close to 2 or less), those with a capacity to produce more than 10 x the potential demand, and those in between. For much of Asia:

Table 8b.

#	Region	HEI system			LEI system		
		veg.diet low.pop.	mod.diet med.pop.	affl.diet high.pop.	veg.diet low.pop.	mod.diet med.pop.	affl.diet high.pop.
1	South America						
2	Central America						1.5
3	Northern America						
4	Northern Africa						1.7
5	Western Africa						1.2
6	Central Africa						
7	Eastern Africa						1.5
8	Southern Africa						
9	Oceania						
10	Southeast Asia					1.7	0.8
11	Easthern Asia			1.3		1.5	0.7
12	Southern Asia		1.6	0.8	2.0	0.9	0.4
13	Western Asia			2.0			1.1
14	(former) USSR						
15	Europe						1.6
	World	19.7	8.8	4.2	8.4	3.7	1.8

the supply/demand ratio is always in danger zone. The situation is even worse than the ratio reflects since the 'option' of a fully vegetarian diet might not exist (rangelands contribute to food production as much or more than arable cropping). The Americas, Central Africa and Oceania are consistently in the second group, implying that there is ample scope for alternative land use (e.g. for rain forests). While for the regions in the middle group ample food can be produced, parts of the regions may be much more limited in their options of land use.

With respect to the current world food situation, it is recognised that if all food were equally distributed, no one would go hungry (Smits 1986). In fact, as many as one billion persons are hungry because of unequal income distribution that keeps food inaccessible to the poor.

A supply/demand ratio of 1.0 reflects situations where food security is met if food is distributed very efficiently. Particularly for the LEI scenario, where massive food transports are not in line with the environment friendly attitude, a somewhat higher ratio is required to achieve full food security for all households. Economic studies are to refine issue. Supply/demand ratios of 2.0 and more indicate that food could be produced on a smaller areas than all

Table 8c.

#	Region	HEI system Affluent diet High population	LEI system Moderate diet Med. population
1	South America	20.0	14.1
2	Central America	3.5	3.1
3	Northern America	10.5	11.3
4	Northern Africa	2.8	3.5
5	Western Africa	2.9	2.7
6	Central Africa	17.1	12.7
7	Eastern Africa	4.3	3.2
8	Southern Africa	6.9	7.0
9	Oceania	60.6	68.7
10	Southeast Asia	2.4	1.7
11	Easthern Asia	1.3	1.5
12	Southern Asia	0.8	0.9
13	Western Asia	2.0	2.3
14	(former) USSR	7.0	7.6
15	Europe	3.2	3.1
	World	4.2	3.7

suitable soils, with less intensive production techniques, with less productive crops or varieties, or that land is available to grow bio-energy crops or bulk export crops.

Discussion

Issues related to food demand

The range in increase in the demand for food till 2040 is about 6-fold. Although not in all regions the full range has to be considered (growth in some regions is slow and consumption is already high, other regions may choose vegetarian diets), probably 2.5–3.5 times more food will be needed at the global level. More than half of this increase could result from more 'expensive' diets. It is desirable, therefore, that food technologies are developed to produce 'expensive' food (in terms of GE) considerably more efficiently (cf. table 3).

Simulated yields per unit area in the under HEI system (table 6) correspond reasonably well to experimental yields in optimal conditions, as mentioned already by Buringh *et al.* (1975), and has been confirmed in related studies for many agro-ecological zones (Penning de Vries *et al.* 1989). A spot-check in West Africa (Kayes, Mali) showed SIMFOOD-yields of 22 t ha^{-1} (3 crops irrigated rice), and 7 t ha^{-1} for a rainfed crop; these values compare favourably with observations (Penning de Vries and Djitèye 1982).

In real production systems, many crop species are grown in addition to cereals: (sweet) potato, cassava, soybean, vegetable crops, fruits, etc. Yet, 75% of the worlds arable crop yield (by weight of dry product) originates from cereal crops, 7% from tuber and root crops, 8% from pulses, vegetables, and fruits, and 4% from oil crops (FAO 1993). Light, water and nutrient use efficiency of non-cereal crops are roughly similar to those of cereals, and differences in HI are compensated by differences in energy content. Hence, grain production is a fair approximation of potential food production. We did not evaluate the benefit of biotechnological breakthroughs. These may lead to better local adaptations, quality and crop protection, but we do not expect breakthroughs in higher efficiency in use of light or water as nature has been selective in these respects for aeons.

Top yields at experimental farms have not really increased for decades, and there is some concern that top yields are unsustainable (Yoshida *et al.* 1972; Pinstrup-Andersen 1994; IRRI 1990). It is crucial that top yields do not decline. Fortunately, recent indications are that inability to reproduce previous record yields was only a temporary setback, caused by the fact that soil and crop environment changed much more than was anticipated (Kropff *et al.* 1993). Global average hectare yields of cereal crops have not increased over the past years (Pinstrup-Andersen 1994; Brown and Kane 1994). The main reason for stagnating production is falling food prices, but since this is unrelated to yield potentials, it does not affect results of our study.

Simulated yields for the LEI-production system are more difficult to compare with observed values, as experimental data for this type of agriculture are scarce. Indirect evidence confirms the values that we computed: average wheat yields in the Netherlands around 1900, when little fertiliser was used and cereals were grown in rotation with legumes, were about 2 t ha^{-1} (Spiertz *et al.* 1992); in Thailand, with a low rate of fertiliser application (FAO 1993), average rice yields in rice-legume cropping systems are about 1.8 t ha^{-1}, which for 2–3 crops leads to production level as shown (table 6). Yet, our yield estimates are probably on the conservative side, since in the cases quoted the soil P-level was probably sub optimal for N-fixation. Predicted crop yields depend heavily on the assumed N-fixing capacity of leguminous crops, which is subject to discussion (Caporali and Onnis 1992). It has been suggested that biotechnology might increase this significantly (e.g. by giving cereals the capacity to fix N). An increase in N-use efficiency is not expected as selection pressure for this feature has been high for ages. Doubling N-input would hardly increases production on irrigated land (6%) since the irrigated area decreases (more productive crops need more water), but the rainfed area grows and productivity of plots would go up, so that in total production could be up by 59%.

Our computations are based on the concept that all farming systems are in an equilibrium situation in 2040, and that a maximum of nutrients harvested and removed from the field are recycled. But to achieve this stable situation, build-up of crop nutrients in the soil is required, among others of inorganic P. Saturating the soil with P has already occurred in some countries (e.g. The Netherlands, Wijnands 1992), but massive amounts are still needed in areas where little fertiliser has been applied or mining is even still ongoing (Smaling 1993). Leguminous crops in particular require high levels of soil P to achieve maximum rates of N fixation. In a rough approximation, we estimate that in the order of 1 t ha^{-1} of P is required to permit optimal production and growth. Such a value, multiplied with the area of potential agricultural land, leads to the potential need for 8 billion ton of P. The identified commercially exploitable P-reserves are estimated at 4–10 billion ton, most of which as rock phosphate in Africa (Tisdale *et al.* 1985). Apart from the question whether production of fertiliser P would release a harmful quantity of fluorine or bring about cadmium pollution (De Wit, personal communication), all known stocks would become exhausted. When the 'stable' situation is reached, agriculture continues to require small amounts of P to compensate for incomplete recycling. Hence, there is a distinct possibility that there is insufficient P in the world to feed the 2040 population when only LEI-production methods are practiced.

Errors of approximation are made because of the rather large area of our smallest soil units (10^4 km^2) and fresh water resources (river basins). The study would be more informative when performed for countries, and for regions with natural rather than administrative boundaries. A finer level of detail was not attempted by lack of basic data, the eventual size of the study, and because we aimed at a global overview. However, we do intend to zoom in on specific areas and issues (e.g. Luyten *et al.* 1995).

Suitability classification of soils has a major effect on the results of this study, while water holding capacity and soil depth are very important in rainfed conditions. Our knowledge of soils shows important weaknesses with respect to the extent of global cover, extrapolation of point observations to grid cells, definition of 'suitability', definition of soil characteristics compatible with crop models, and handling of preferential flow (Penning de Vries *et al.* 1995). We computed the relative suitability of land for modern agriculture (per grid cell) as the multiple of relative values for slope, soil 'phase' and texture. This is only correct when these features are distributed independently;

otherwise this study underestimates suitable land area and potential food production.

Our model does not take into account any heterogeneity in soils and feedbacks at the watershed level. These phenomena will probably affect yield negatively.

It may be argued that some soils that are not suitable for HEI-production may be used for LEI-systems, where mechanisation is not crucial. When we eliminated the reducing effect of slope on agriculture, about the irrigated area increased by 14% and rainfed production even more (34%), leading to an overall global increase of 16%. The largest relative benefit is in the Asian regions. This shows the importance of good hill-side agriculture in LEI-production systems.

Our procedure assumed maximum efficiency of external inputs, including fertiliser and water. This implies that humid areas are irrigated first (with only just enough water), and drier areas only when irrigation water is left. We have not yet compared the resulting distribution of irrigated land with the actual distribution.

Current total irrigated area in Asia amounts to 1.6 M km^2 (FAO 1993), while 0.7 M km^2 is potentially irrigable (Crosson and Anderson 1992); these are below our estimate for Asia (4.5 M km^2 for HEI, 5.5 M km^2 for LEI), mainly because we assume use of all available water and a higher water use efficiency. Our estimates of the maximum irrigable area in South America (7–8 M km^2) and Africa (6–8 M km^2), however, exceeds current irrigated areas (FAO 1993) 5–10 fold. This is presumably due to the fact that we neither address the economics of new realising irrigation systems nor the rate with which this can be implemented. Fifty years at a growth rate of 2–3% $year^{-1}$ (World Resources Institute 1990; Delft Hydraulics 1992) would only triple or quadruple irrigated land. This indicates that our estimates of potential food production cannot be realised without special and major efforts, but at the same time that there are probably still very important possibilities to expand irrigation and boost food production. Autonomous growth is definitely not sufficient.

We assume that all farmers use the currently known 'best technical means' of production and resource use. Although this may sound 'conservative' as better techniques will emerge, it already implies that an enormous amount of adaptive research is needed to adapt known technologies to local conditions.

In the HEI-scenario, transport of inputs (fertiliser, seed) and outputs (food, manure) comprises only a small fraction of the fossil energy input, most being required for production of N, P and K fertiliser, mechanical operations and processing. In the LEI-scenario, even less use of fossil energy is implied, as more localised production requires less transportation (but long distance transport of P-fertiliser is often unavoidable). Animal products and the feed required to produce them are generally transported over longer distances than vegetarian products. But as fossil energy is not part of this study, we did not explore further the implications.

Conclusions

This explorative study indicates that natural resources are available in the world to increase food production very significantly. The 15 regions are very different in their potential demands by 2040 and in their potential production capacities.

When HEI-agriculture is practised, all regions can produce food required even for an affluent diet, except for East, South and West Asia, and also Southeast Asia and West and North Africa come close to the lower limit. A diet much less expensive (in terms of basic food products) provides the only option for escape, though the contributions of pastures to food supply will be needed. The three regions with the least leeway will carry almost half of the global population. On the other hand, Europe, the former USSR, the American regions and Central Africa are well off. Depending on the level of consumption chosen, Europe can grow its food on 0.3–0.6 of the suitable area, North America on 0.2 of the land, and South America and Oceania on an even smaller fraction.

When LEI-agriculture is practised, only South Asia will have food shortage. In this crowded region, there is no way out via less expensive diets or lower population growth (both already at a minimum). Moreover, our model computes that grass is harvested and converted into food, which is not in agreement with the assumption of the vegetarian diet, making shortage worse. If the LEI system goes with the low food demand scenario, then Europe could again grow all its food at less than half of its suitable soils. Only the former USSR, North and South America, Central Africa and Oceania can consider to offer its population an affluent diet. A major challenge to Asian science is to develop management techniques that allow expansion of the area of soil suitable for efficient farming, e.g. on hill sides. It should be explored whether global P-reserves would be sufficient for large scale LEI-farming.

Though people in many countries nudge already towards the middle of the range of diets (including East Asia), this study provides the warning that it might be impossible to follow this course till the end, and that some countries cannot afford its people the choice between a vegetarian or moderate diet and the affluent one. To permit more individuals a full choice of diets, food technology should help by increasing drastically the efficiency of producing socially acceptable diets at low biomass cost. Currently, three times as much biomass is required to produce an affluent diet as for a vegetarian diet. This ratio is too high. Aqua culture of fish and shrimps provides an opportunity to produce animal protein that is insufficiently exploited (ICLARM 1992). These animals can be grown from plant biomass, crop residues and waste.

This technology provides therefore an option to valorise food materials, and hence to contribute to raise the efficiency of making animal products (cf. tables 3 and 4). Seafood is unlikely to become more important, in absolute quantities, as its catch is already close to its global ceiling (World Resources Institute 1994). Abiotic ways of producing food have not yet emerged.

It should be explored whether global P-reserves are sufficient for the high productive HEI and LEI-systems considered. If not, than production levels remain significantly below the values currently considered.

Acronyms

GE	Grain Equivalent
HEI	High External Input farming
HI	Harvest Index
LEI	Low External Input farming
LEISA	Low External Input Sustainable Agriculture

References

Brown L R, Kane H (1994) Full house. The World Watch Environmental Alert Series. Norton and Co., New York. USA.

Buringh P, Van Heemst H D J, Staring G (1975). Computation of the absolute maximum food production of the world. Report Dept. of Tropical Soil Science, Wageningen Agricultural University, Wageningen, The Netherlands.

Caporali F, Onnis A (1992) Validity of rotation as an effective agroecological principle for a sustainable agriculture. Agric., Ecosyst., Env. 41:101–113.

Crosson, Anderson J (1992). Global Food, resources and prospects for the major cereals. World Bank, Washington, USA.

Delft Hydraulics (1992) Global water availability for future irrigated agriculture. Internal Report, Delft, The Netherlands.

Doorenbos J, Pruitt W O (1977) Guidelines for predicting crop water requirements. FAO Drainage and Irrigation Paper 24, FAO, Rome, Italy.

FAO (1993) Yearbook Production 1992. Vol. 46. Food and Agriculture Organization, Rome, Italy.

ICLARM (1992) ICLARM's strategy on living aquatic resources management. International Center for Living Aquatic Resources Management, Manila, Philippines.

IRRI (1990) Strategy towards 2000. International Rice Research Institute, Los Banos, Philippines.

De Koning G H J, Janssen H, Van Keulen H (1992) Input and output coefficients of various cropping and livestock systems in the European Communities. Working Document W62, Wetenschappelijke Raad Regeringsbeleid, The Hague, The Netherlands.

Kropff M J, Cassman K G, Penning de Vries F W T, Van Laar H H (1993) Increasing the yield plateau in rice and the role of global climate change. Journal Agricultural Meteorology 48(5):795–798.

Luyten J C (1995) Sustainable world food production and environment. Internal Report AB-DLO, Wageningen, The Netherlands.

Luyten J C, Shi Qinghua, Penning de Vries F W T (1995) The limits of consumption and production of food in China in 2030. AB-DLO, Wageningen, The Netherlands (in prep).

Monteith J L (1990) Conservative behaviour in the response of crops to water and light. Pages 3–16 in Rabbinge R, Goudriaan J, Van Keulen H, Penning de Vries F W T, Van Laar H H (Eds.) Theoretical production ecology: reflections and prospects. Simulation Monographs 34, Pudoc, Wageningen, The Netherlands.

Penning de Vries F W T, Djitèye M A (1982) La productivité des pâturages sahéliens. Une étude des sols, des végétations et de l'exploitation de cette ressource naturelle. Agricultural Research Report 916, Pudoc, Wageningen, The Netherlands.

Penning de Vries F W T, Jansen D M, Ten Berge H F M, Bakema A (1989) Simulation of ecophysiological processes of growth in several annual crops. Simulation Monograph 29, PUDOC, Wageningen, The Netherlands, and IRRI, Los Banos, Philippines.

Penning de Vries F W T, Van Keulen H, Luyten J C (1995) The role of soil science in estimating food security 2040. In Wagenet R J, Bouma J, Hutson J L (Eds.) The role of soil science in interdisciplinary research. Special publication of the Soil Science Society of America. (in press).

Pinstrup-Andersen P (1994) World food trends and future food security. Internation Food Policy Research Institute, Washington, USA. 25 pp.

Reijntjes C, Haverkort B, Waters-Bayer A (1992) Farming for the Future, an introduction to Low External Input and Sustainable Agriculture, MacMillan, London, UK.

Sinclair T R (1990) Nitrogen influence on the physiology of crop yield. Pages 41–55 in Rabbinge R, Goudriaan J, Van Keulen H, Penning de Vries F W T, Van Laar H H (Eds.) Theoretical production ecology: reflections and prospects. Simulation Monographs 34, Pudoc, Wageningen, The Netherlands.

Smaling E (1993) An agro-ecological framework for integrated nutrient management. PhD thesis, Wageningen Agricultural University, Wageningen, The Netherlands.

Smits A P (Ed.) (1986) Food security in developing countries: Kuikduin seminar March 1985. Raad van Advies voor het Wetenschappelijk Onderzoek in het kader van Ontwikkelingssamenwerking (RAWOO), The Hague, The Netherlands.

Spiertz J H J, Van Heemst H D J, Van Keulen H (1992) Field crop systems in North western Europe. Pages 357–371 in Pearson C J (Ed.) Ecosystems of the world. Elsevier, Amsterdam, The Netherlands.

Tisdale S L, Nelson W L, Beaton J D (1985) Soil fertility and fertilizers. MacMillan Publishing Co., New York, USA.

United Nations (1992) Long range world population projections: two centuries of growth (1950–2150). Department of International Economic and Social Affairs, New York, USA.

Wijnands F G (1992) Evaluation and introduction of integrated arable farming in practice. Netherlands Journal Agric. Sciences 40:239–249.

De Wit C T, Van Keulen H (1987) Modelling production of field crops and its requirements. Geoderma 40:253–265.

World Resources Institute (1994) World Resources 1994–1995. World Resources Institute, Oxford University Press.

WRR (Wetenschappelijke Raad Regeringsbeleid) (1992) Ground for choices: four perspectives for rural areas in the European Community. Netherlands Scientific Council for Government Policy, Sdu Publishers, The Hague, The Netherlands.

WRR (Wetenschappelijke Raad Regeringsbeleid) (1994) Persistent risks: a permanent condition. Netherlands Scientific Council for Government Policy, Sdu Publishers, The Hague, The Netherlands. (in press).

Yoshida S, Parao F T, Beachell H M (1972) A maximum annual rice production trial in the tropics. Int. Rice Comm. Newslewtt. 21:27–32.

Zobler L (1986) A world soil file for global climate modelling. NASA Technical Memorandum 87802. NASA, Scientific and Technical Information Branch, USA.

Prospects for world food security and distribution

P. PINSTRUP-ANDERSEN and R. PANDYA-LORCH
International Food Policy Research Institute (IFPRI), 1200 17th Street, Washington, DC 20036, USA

Key words: agricultural development, food consumption, food production, food security, hunger, malnutrition, poverty

Abstract. Despite global availability of food, the world is not food secure: in the late 1980s, over 780 million people were undernourished, more than 180 million children underweight, and around 1.1 billion people living in poverty. As almost 100 million people are added to the world's population every year in the next quarter century, it is clear that whether enough food is available to feed all people and whether all people have access to the available food in needed quantities depends on actions the world community takes today. IFPRI projections suggest that the gap between production and consumption will widen in all developing countries by 2020. The better-off countries will be able to fill the gap through commercial imports, but the poorer countries will lack sufficient foreign exchange to avert widespread food insecurity. Agricultural development, by producing more food and ensuring better access to food via employment creation and income growth, is the key to improving world food security. The long lag time between investment in agricultural development and corresponding improvements in food security require that a commitment be made now to improve world food security, otherwise many more people will go hungry and malnourished in coming years.

Introduction

In the next quarter century, almost 100 million people are expected to be added to the world's population annually, increasing it by 2.3 billion to reach 8 billion in 2020 (United Nations 1993). Will there be enough food available to feed this unprecedented population increase? And even if enough food is available, will all people have access to sufficient food to lead healthy, productive lives? What are the prospects for future world food security? Will the world be able to assure food security for all of its population or will the current situation of food surpluses co-existing with widespread hunger and food insecurity persist?

Following a brief discussion of food security concepts, this paper assesses the current world food security situation and considers what the trends tell us about prospects for the future, examines the major factors likely to affect and determine food security in coming years, and presents scenarios and prospects for future food security.

J. Bouma et al. (eds.), Eco-regional approaches for sustainable land use and food production, 89–111.

Food security concepts

The world is food secure when each and every person is assured of access at all times to the food required for a healthy and productive life. Food security is jointly determined by availability of food and access to food. Availability of food does not guarantee access to food, but access to food is contingent on there being food available. National, regional, or local availability of food is a function of food production, stockholding, and trade. National access to food from international markets is determined by world food prices and foreign exchange availability. Household availability of food requires that food be available at local or regional markets, which is determined by market operations, infrastructure, and information flows. Access to food by households and individuals is usually conditioned by income: the poor commonly lack adequate means to secure their access to food.

Food security at any level does not guarantee food security at any other level. For example, household food security does not necessarily mean that all individuals in that household have access to the needed food some members of the household may be denied their full share of the needed food. Intra-household inequality in distribution of food, with women in particular eating less than their share of household food, is observed quite often. Similarly, regional or national food security does not necessarily lead to household or individual food security the available food may not be equally distributed and households or individuals may not have equitable access to it. And, of course, global food availability does not mean universal food security there may be marked national, regional, household, and individual differences in access to food.

Current world food security situation

Despite impressive food production growth in recent decades such that enough food is available to meet the basic requirements of every person in the world, the world is not food secure. If available food energy were evenly distributed within each country, there were still 33 countries that were unable to assure sufficient food energy (2,200 calories per person per day) for their populations during 1990–1992. Of these countries 25 are in Africa, 4 in Asia, 3 in Latin America and the Caribbean, and 1 in the Middle East.

Available food is neither evenly distributed nor fully consumed among or within countries. Over 780 million people in the developing world, 20 percent of the population, were chronically undernourished or underfed in the late 1980s (FAO 1994); more than 180 million preschool children were underweight in 1990; and many hundreds of millions of people suffered from diseases of hunger and malnutrition (United Nations ACC/SCN 1992). Availability of global food supplies does not ensure that everyone has access to

Table 1. Prevalence of chronic undernutrition in the developing world, 1969–1971 and 1988–1990. (Source: FAO 1993)

Region	1968–1971		1988–1990	
	No. of Persons (million)	Percent of Population	No. of Persons (million)	Percent of Population
Sub-Saharan Africa	94	35	175	37
Near East/North Africa	42	24	24	8
East Asia	497	44	252	16
South Asia	254	34	271	24
Latin America and the Caribbean	54	19	59	13
Developing countries	941	36	781	20

food. Considerable progress has been made in reducing hunger and food insecurity since 1969–1971 when 941 million people, 36 percent of the developing world's population, were undernourished. Most of the decline has taken place in East Asia where the number of undernourished people went from 497 million people in 1969–1971 to 252 million in 1988–1990 (table 1). Nevertheless, East Asia is home to one-third of the world's undernourished people. Another one-third, 271 million, is located in South Asia. With two-thirds of the developing world's undernourished, South and East Asia remain the main areas of concern. However, a new 'flash-point' for hunger and food insecurity is emerging: Sub-Saharan Africa. The number of undernourished people in that region almost doubled in two decades from 94 million in 1969–1971 to 175 million in 1988–1990, and the proportion of the population that is undernourished rose from 35 percent to 37 percent. In all other developing regions, by contrast, the proportion of the undernourished population declined.

Underweight preschool children are another indicator of food insecurity. Their numbers rose during the 1980s from 164 million in 1980 to 184 million in 1990, although due to population growth their share of the preschool children population declined slightly from 37.8 percent to 34.3 percent (table 2). One-third of all preschool children are still underweight in the developing world. More than half of the underweight children are in South Asia where they make up 59 percent of the region's preschool children population. The prevalence of underweight preschool children is also high in Sub-Saharan Africa, 29.9 percent, and South East Asia, 31.3 percent, but only 7.7 percent in South America.

In addition to energy deficiencies, micronutrient deficiencies are widespread in the developing world. For example, iron deficiency affects about 1 billion people, particularly women of child-bearing age and children (United Nations ACC/SCN 1992). About 370 million reproductive-aged women in

Table 2. Prevalence of underweight children in the developing world, 1980–1990. (Source: United Nations ACC/SCN 1992)

Region	1980 Number Underweight (million)	1980 Percent Underweight	1990 Number Underweight (million)	1990 Percent Underweight
Sub-Saharan Africa	19.9	28.9	28.2	29.9
Near East/North Africa	5.0	17.2	4.8	13.4
South Asia	89.9	63.7	101.2	58.5
Southeast Asia	22.8	39.1	19.9	31.3
China	20.5	23.8	23.6	21.8
Middle America/Caribbean	3.1	17.7	3.0	15.4
South America	3.1	9.3	2.8	7.7
Developing countries	164.0	37.8	184.0	34.3

Table 3. Poverty in the developing world, 1990. (Source: World Bank 1992)

Region	Number of Poor (million)	Percent of Population below Poverty Line	Percent Share of Total Poor Population
Sub-Saharan Africa	216	47.8	19.1
Middle East/North Africa	73	33.1	6.4
South Asia	562	49.0	49.6
East Asia	169	11.3	14.9
Latin America and the Caribbean	108	25.5	9.5
Developing countries	1133	29.7	100.0

Note: The poverty line is $370 annual income per capita in 1985 purchasing power parity dollars. In 1990 prices, the poverty line would be approximately $420.00 annual income per capita.

the developing world were anemic in the 1980s. About 14 million preschool children suffer eye damage and between 250,000 and 500,000 children go blind each year due to deficiency of Vitamin A. Iodine deficiency disorders affect more than 210 million people of all ages.

Hunger is a major consequence of poverty, which is significant and persistent in many developing countries. In 1990, an estimated 1.1 billion people lived in households that earned a dollar a day or less per person (table 3). Fifty percent of these absolutely poor people live in South Asia, 19 percent in Sub-Saharan Africa, 15 percent in East Asia, and 10 percent in Latin America and the Caribbean. Almost one-half of the population of South Asia and

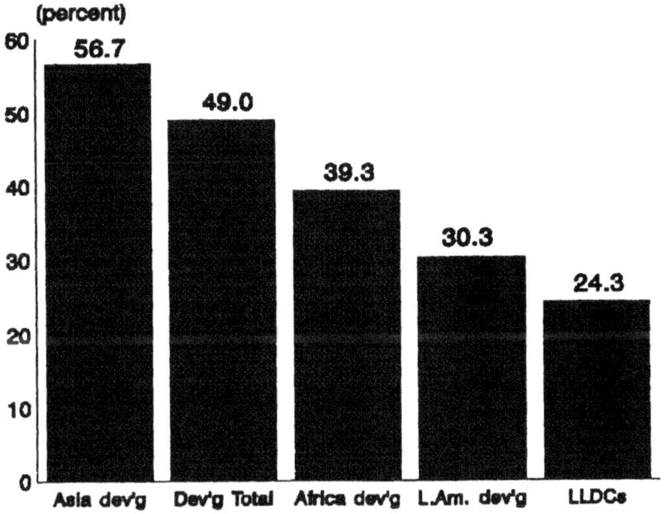

Fig. 1. Percentage change in total food production, 1979–1981 to 1991–1993. (Source: FAO 1994)

Sub-Saharan Africa, and one-third of the Middle East and North Africa, live in poverty.

The estimates presented here catch those who are already food insecure and show symptoms or consequences of food insecurity. There are many hundreds of millions of people worldwide who are at the borderline and live with the risk of food insecurity: their incomes are so low that any sudden shock such as loss of employment or price fluctuations could tip them into food insecurity. These vulnerable people must also be taken into account when considering the state of the world's food security.

Earlier it was noted that food security is jointly determined by availability of food and access to food. Thus far, we have considered access to food. Turning to availability of food, food production growth in recent decades has been impressive. During the period 1961–1993, cereal production worldwide more than doubled from 877 million tons to 1,894 million tons; in developing countries it almost tripled from 396 million tons to 1,089 million tons (FAO 1994). Between 1979–1981 and 1991–1993, food production worldwide increased by 29 percent (figure 1). In developing countries as a group, food production increased 49 percent, with particularly large increases of 689 percent in China. Even in African developing countries, where concerns regarding the future food situation are greatest, food production increased 39 percent.

However, food production growth barely kept up with population growth. Per capita food production for the world increased by less than five percent between 1979–1981 and 1991–1993 (figure 2), while in developing countries

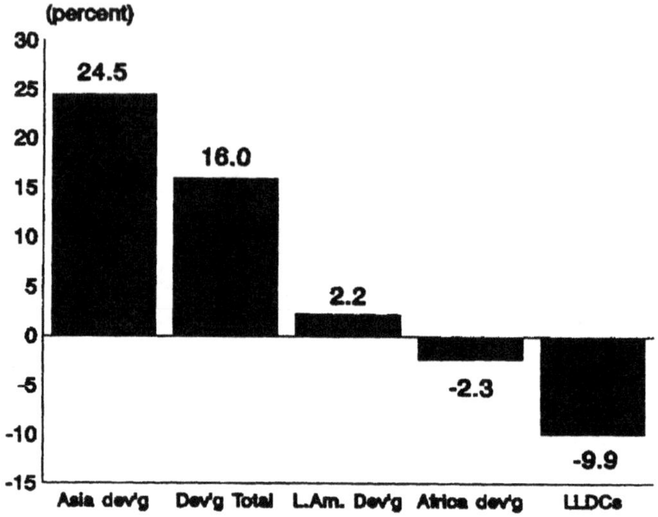

Fig. 2. Percentage change in per capita food production, 1979–1981 to 1991–1993. (Source: FAO 1994)

it increased by 16 percent. Food production performance varied widely among regions: while per capita food production increased by 42 percent in China and by 25 percent in Asian developing countries, less food was produced per person in developing African countries during 1991–1993 than in 1979–1981. The least-developed countries experienced a 9.9 percent reduction in food production per capita. Growth in food production did not keep pace with population growth in 56 developing countries. Thus, the encouraging global food production trends are, upon further examination, somewhat less encouraging in terms of their implications for food security.

Of note on the food production front is the role of yield increases, which have been the source of 92 percent of the increased cereal production in the developing world between 1961 and 1990; area expansion contributed only 8 percent (World Bank 1992). While cultivated area is still increasing in most developing countries, it is doing so at a low and declining rate. Yield trends in developing countries climbed steadily upward for the three major cereals of rice, maize, and wheat between the 1960s and late 1980s, particularly in Asia where during 1961–1991 maize yields almost tripled from 1.2 tons to 3.4 tons per hectare, wheat yields increased five-fold from 0.5 tons to 2.5 tons per hectare, and rice yields doubled from 1.7 tons to 3.6 tons per hectare (FAO 1992). In Africa, however, yield performances of major cereals have been poor and variable; Africa has a long way to go to catch up with Asian yields.

However, yield growth rates in some areas are stagnating and, in a few cases, falling. In Asia, for instance, the annual rate of increase in rice yields has declined from about 3 percent between the mid-1970s and early 1980s to less than 2 percent in the late 1980s (Rosegrant and Svendsen 1993); since 1989, rice yields have stagnated at around 3.6 tons per hectare (FAO 1992). Annual yield growth for wheat in Asia have also slowed from 6.2 percent in the early 1960s to 2.7 percent in the 1980s (Rosegrant and Svendsen 1993); since 1989, wheat yields have remained around 2.5 tons per hectare (FAO 1992). A slowdown in the rate of increase of yields of major cereals raises concern since increased yields will have to be source of increased food production in the future. Most cultivable land in Asia, North Africa, and Central America has already been brought under cultivation, and physical and technological constraints are likely to restrain large-scale conversion of potentially cultivable land in Sub-Saharan Africa and South America. The option of area expansion as a source of food production increases is rapidly disappearing, and even Africa will have to rely mostly on increased yields to expand food production.

Another cause of concern on the food production front is the levelling off during the 1980s and early 1990s of grain production per person for the world and for the developing countries as a group, after steady increases during the 1950s, 1960s, and 1970s. Since 1980, the trend in grain production per person for the world and for developing countries as a group can be best characterized as constant. If corrective actions are not taken soon, this trend could turn downward, with potentially adverse repercussions not just because the additional population needs adequate food, but because factors in addition to population growth are pushing up demand for grain. While future demand for grain for direct consumption is expected to grow at a rate only slightly above population growth, the expected growth rate in world feedgrain demand is more than twice the expected population growth rate (Paulino 1986). Once incomes increase beyond a certain level, demand for feedgrain increases rapidly; most developing countries have incomes still below the level where feedgrain use increases rapidly.

In sum, the world is not food secure today. Food insecurity, hunger, and malnutrition are widespread, particularly in South and East Asia and Sub-Saharan Africa. Poverty is the main force restraining access to food. Global availability of food has not yet translated into availability of and access to food by all people.

Factors influencing future food security

Food security depends on both the supply and the demand for food. Population growth, urbanization, income change, relative food prices, and changes in preferences associated with changing lifestyles are among some of the major driving forces that will influence food consumption. It is expected that in Sub-Saharan Africa, population growth and urbanization will be the overwhelming

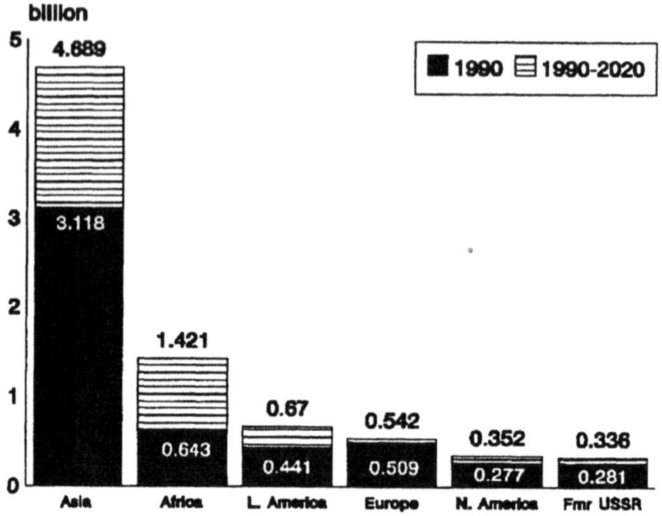

Fig. 3. Population, 1990 and 2020. (Source: United Nations 1993)

driving forces, a combination of income and population growth will account for most of the growth in food consumption in South Asia and Latin America, and income growth and changing lifestyles will play the biggest role in East Asia. Some of the major driving forces that will influence food production in come years will be investments in agriculture, considerations related to natural resource management and the environment, and government policy.

Population growth and urbanization

Almost 100 million people will be added to the world's population each year in the next three decades, the largest annual population increase in history. Most of this increase (93 percent) will take place in the developing world, whose share of global population is projected to increase from 78 percent in 1995 to 83 percent in 2020. Over this period, the absolute population increase will be highest in Asia, 1.6 billion, but the relative increase will be greatest in Africa, where the population will more than double from 0.6 billion to 1.4 billion (figure 3). Population growth of this magnitude will severely constrain efforts to increase income and improve welfare while at the same time it will greatly increase the need for food. Sub-Saharan Africa is particularly unlikely to be able to cope with the projected large population growth without significant increases in poverty and decreases in the overall standard of living in the absence of a massive influx of aid from the rest of the world. Failure to significantly reduce population growth rates in Sub-Saharan Africa within the next 20–30 years will render moot any development efforts.

Table 4. Annual growth in real per capita income, by region, 1980–2000. (Source: World Bank 1992)

Region	1980–1989 (percent)	1990–2000 (percent)
Sub-Saharan Africa	–0.9	0.3
East Asia	6.3	5.7
South Asia	3.1	3.1
Latin America	–0.5	2.2
Middle East and North Africa	–2.5	1.6
Developing countries	1.2	2.9

Urbanization is accelerating in the developing world. Already about 36 percent of the population of low- and middle-income countries lives in urban areas, up from 25 percent in 1970 (World Bank 1994). The urban population in these countries grew at an annual average rate of 3.7 percent during 1980–1992 while total population grew at 1.9 percent per year (World Bank 1994). Urban population growth is particularly rapid in East Asia and the Pacific and in Sub-Saharan Africa. Rapidly growing urban population will place severe pressures on food marketing systems, including transportation, storage, processing, grading, and market information. In many developing countries, particularly Sub-Saharan Africa, massive investments in physical infrastructure in both rural and urban areas will be needed to support the feeding of the urban population. Urbanization will accelerate the dietary transition. Increasing opportunity costs of women's time; changes in food preferences caused by promotions, advertising, and changing lifestyles; and changes in relative prices associated with rural-urban migration will change food demand from basic staples such as sorghum, millet, maize, and root crops to cereals such as wheat and rice (which require less preparation), fruits, foods of animal origin, sugar, and processed foods. Rapid increases in demand for livestock products will have severe repercussions on the demand for cereals.

Income growth

Economic growth in the 1980s was disappointing in most developing countries. While real per capita income grew by 1.2 percent per year for developing countries as a group during 1980–1989, growth was negative in several regions, notably Sub-Saharan Africa, Middle East and North Africa, and Latin America (table 4). Prospects for the 1990s appear better: High rates of growth in Asia are expected to continue while elsewhere growth is expected to substantially improve, even in Sub-Saharan Africa where the growth rate is projected to be positive.

Future economic growth depends on internal policies as well as on the international environment. The extent to which structural adjustment and economic reforms currently underway in many developing countries are successfully completed at an appropriate speed and sequence is of paramount importance for future economic growth in these countries.

Future economic growth will also depend on the international environment, including trade distortions by developed countries, and access to external aid. Import restrictions for agricultural and nonagricultural products in the developed countries, along with domestic agricultural subsidies and implicit and explicit export subsidies for agricultural products, are of particular concern. Trade liberalization is an integral part of structural adjustment and economic reforms in most developing countries. Failure to achieve sizable reductions in agricultural protection in developed countries would make it difficult for developing countries to benefit from their own trade liberalization.

Investment in agriculture

In low-income developing countries, agriculture is the most effective and frequently the only viable lead sector for overall economic growth. Agricultural growth stimulates economic growth in nonagricultural sectors, which, in turn, results in increased employment and reduced poverty. Unfortunately, international financial support to agricultural development has decreased sharply during the 1980s from about US$12 billion in 1980 to about $10 billion in 1990 (figure 4). Agriculture's share of total development assistance declined over this period from about 20 percent to 14 percent. Reversing the downward trend in external assistance to agriculture and, correspondingly, changing the recent mode of ignoring the agricultural sector are imperative to jump-start the economies of many low-income countries.

Agricultural growth and development must be vigorously pursued to alleviate poverty through employment creation and income generation in rural areas, to meet the growing food needs, and to stimulate overall economic growth. Various investments are needed to facilitate agricultural and rural growth. Of critical importance is yield-increasing production technology; area expansion is no longer feasible or profitable in many parts of the world, and future increases in food production must come mainly from higher yields. Besides higher yields, crop varieties that are more tolerant or resistant to pests and droughts and new varieties and hybrids better suited for various agroecological conditions reduce risks and uncertainty and enhance sustainability in production. Accelerated investment in agricultural research and technological improvements is the only viable option for meeting future food needs and demands at reasonable prices without irreversible degradation of the natural resource base.

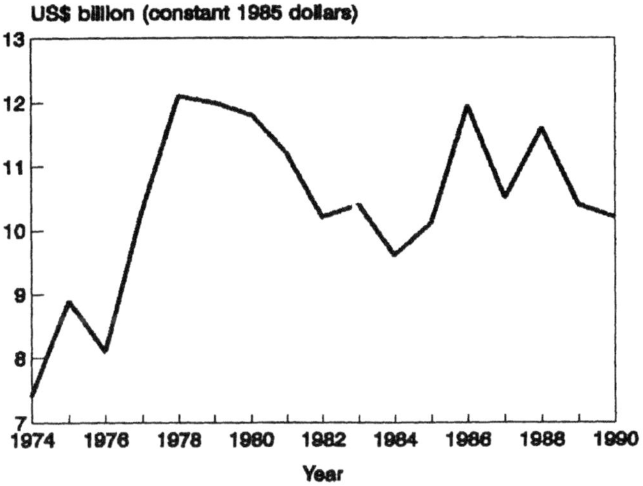

Fig. 4. Total assistence to agriculture, 1974–1990. (Source: World Bank, Agriculture and Rural Development Department; Organization for Economic Cooperation and Development, Statistics Department; Inter-American Development Bank Annual Report; African Development Bank Annual Report; Asian Development Bank Annual Report; European Union, Statistics Department (Lomé countries); EU Annual Report 1990; and Food and Agricultural Organization of the United Nations; Statistics Department)

In addition to land, water is another natural resource becoming increasingly scarce in many production environments. Efforts to meet future food needs and demands while promoting sustainable economic growth and poverty alleviation, including agricultural research and government policy, must pay increasing attention to the efficiency of water use, sustainability of water management practices, and investment in establishing and maintaining irrigation facilities.

Improved rural infrastructure is needed to facilitate access to markets for both outputs and inputs. Farmers need access to modern inputs such as improved livestock, improved crop varieties and hybrids, fertilizers and pest control measures, and credit and technical assistance to meet food production goals. These are essential components of a successful strategy to meet food production and development goals. Efforts to enhance farmers' access to modern inputs must recognize the role of women in farming and marketing and design the programs accordingly.

Natural resource management and environmental considerations

Almost 2 billion hectares of land worldwide (17 percent of vegetated soils) have been degraded in the past 45 years, of which 300 million hectares have suffered such strong degradation that their original biotic functions are damaged and reclamation may be costly if not impossible (Oldeman *et al.* 1990). About 17 million hectares of forests are cut down each year; about two-thirds of the land clearance is for conversion to agricultural use by farmers (Forest Resources Assessment 1990 Project 1991). About 70 percent of the world's degraded lands are found in Asia and Africa, but human-induced degradation as a proportion of total agricultural land is most severe in Central America and Mexico where one-quarter of the vegetated land is degraded. Overgrazing, deforestation, and overexploitation for fuelwood account for almost three-quarters of human-induced soil degradation worldwide, while faulty agricultural practices are responsible for 28 percent of soil degradation (Oldeman *et al.* 1990). Overgrazing, deforestation, and overexploitation for fuelwood result, to a large extent, from poverty and lack of opportunities for agricultural intensification. It would appear that, since 70 percent of soil degradation worldwide is caused by these activities, there is a close relationship between poverty, lack of agricultural intensification, and degradation.

As population, poverty, and food demands continue to grow, failure to develop and implement appropriate technology in production and marketing will lead to either more food insecurity and hunger, for which the current generation of poor people will pay, or to further degradation of the natural resource base, for which future generations will pay. Tradeoffs between meeting future food demands and maintaining production capacity can be avoided and sustainability in food production can be assured only if investment in appropriate research and technology is accelerated, if relevant externalities are internalized into production and consumption decisions or effectively dealt with by government policy, and if poverty is significantly reduced or eradicated.

Agricultural intensification production of more food on land already under cultivation is key to simultaneously alleviate poverty, meet current and future food needs, and avoid degradation of natural resources. Yield-enhancing agricultural technology is the most promising avenue to sustainable agricultural production. Higher yields on land with high production potential will reduce the pressure on fragile land and, together with better definition and distribution of land ownership and user rights, will reduce deforestation and desertification. Agricultural research and technology can help develop production methods that will help maintain land quality and productivity over time. Of course, inappropriate or mismanaged agricultural intensification, such as excessive use of water or pesticides or untimely application of fertilizers, can lead to environmental degradation, but agricultural intensification per se need not degrade the environment. The most serious environmental problem in developing countries is poverty. Rural poverty, pressured by increasing population densities and inadequate agricultural intensification, is responsible for much of the forced exploitation and consequent degradation of environmentally fragile lands. Such degradation will not be prevented by reducing agricultural intensification but rather by alleviating poverty, which is likely to require accelerated intensification.

Government policy

During the 1980s, government policy in most developing countries was dominated by the need to deal with economic crises. During the first half of the decade, many countries relied on demand-contracting stabilization policies, which caused severe hardship on both the poor and the middle-class. More recently, the emphasis has been on supply-enhancing structural adjustments and policy reforms. While policy reforms frequently aim at market liberalization and privatization and reductions of inefficiencies and rents captured by the public sector, much more attention needs to be given to the removal of distortions in input and output markets, asset ownership, and other institutional and market distortions that are adverse to the poor. Market and government failures in food pricing need to be addressed.

More attention needs to be paid to adjustments and reforms that improve the productivity of human resources (health, education, and nutrition) and enhance access by the poor to productive resources such as land and capital. Improved human resources and enhanced access by the poor to productive resources will contribute to reduced poverty and food insecurity as well as to higher economic growth. Policies that will expand investment in rural infrastructure, primary health care, education, agricultural research, and improved production technology, technical assistance, and credit are urgently needed to enhance earnings, food security, and nutritional status among the rural poor and to reduce unit costs of food production for the benefit of both rural and urban poor. Proper price policies play an important role in both production and consumption of food. For example, policies that increase food prices to stimulate expanded production may negatively affect rural poor who are net buyers of food.

Policies are needed to pursue the triple goal of poverty alleviation, increased productivity in food production, and sustainability. No attempts will be made to detail such policies here but such a list would include policies to improve water management, expand agricultural research, and deal effectively with externalities resulting in land degradation and deforestation. In addition, policies are needed to protect the rural and urban poor from both chronic and transitory food insecurity in the short run. Such policies would include targeted food and cash transfers of various types, labor-intensive public works program, and emergency relief. International food aid may have an important role to play.

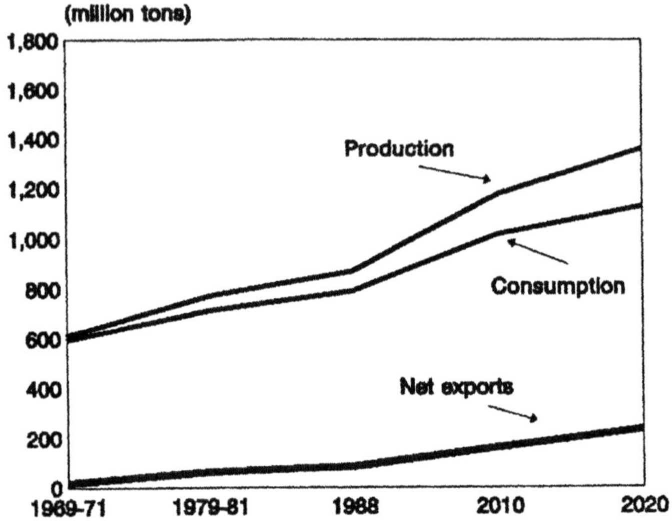

Fig. 5. Production, consumption, and trade of cereals of developed countries.

Prospects for future world food security

Strong disagreements prevail on what the future world food situation is lil to be. Some predict rapidly increasing food scarcity and real prices in the r few years. Others predict that regardless of what the world does today, wc resources will be incapable of supporting the needs of future generations. others argue that there are no serious problems ahead.

Preliminary IFPRI projections* till the year 2020 suggest that develop countries as a group will continue to produce more cereals than they consur (figure 5). Their net cereal exports are expected to almost triple in volum from 82.2 million tons in 1988 to 230 million tons in 2020. Developir countries are likely to increase their cereal production from around 812 millic tons in 1988 to 1,516 million tons in 2020, but will still consume more tha they produce (figure 6). Despite production increases of over 700 million tons the gap between cereal production and consumption is expected to widen fron 82 million tons to 231 million tons over this period and can only be fillec by net imports, which will become increasingly important. The share of total cereals consumed that originates from net imports is projected to rise from 10 percent in 1988 to 15 percent in 2020. Maize and other coarse grains will constitute 53 percent of net cereal imports in 2020, up from 31 percent in 1988, as the volume of net coarse grain imports increase five-fold during this period. Net imports of wheat are projected to almost double during 1988–2020 from 56 million tons to 106 million tons, but the share of wheat imports

* For information on the model, see Appendix 1.

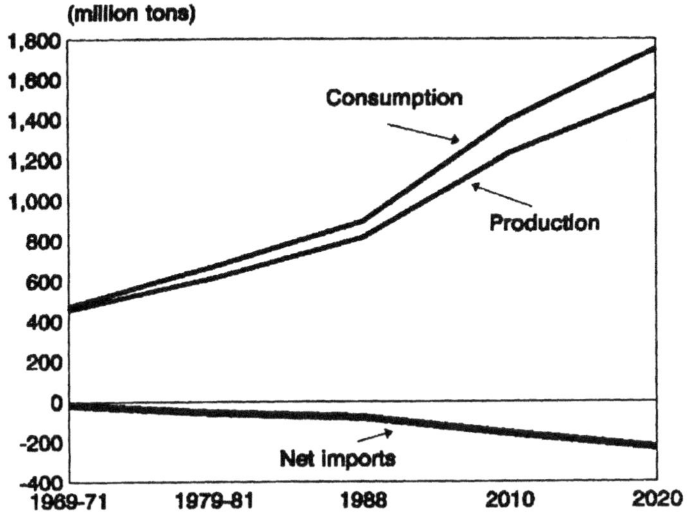

Fig. 6. Production, consumption, and trade of cereals of developing countries.

in net cereal imports is expected to decline from 68 percent to 46 percent. Net imports of rice are, by comparison, minor, rising from less than a million tons in 1988 to 3.3 million tons in 2020.

Until now, cereal production of developed countries has exceeded that of developing countries, but this pattern is not expected to continue for long. By 2020, it is projected that developing countries' cereal production will exceed 1,515 million tons, more than the 1,358 million tons produced by developed countries. Between 1988 and 2020, cereal production in developing countries will increase 87 percent while in developed countries it will increase 56 percent. Consumption, on the other hand, is projected to increase 95 percent in developing countries and 43 percent in developed countries. Developing countries will account for 59 percent of the increase in cereal production but 72 percent of the increase in cereal consumption during 1988–2020.

The gap between production and consumption is expected to widen in all developing regions (table 5). Asia is projected to have the largest net imports in 2020, followed by North Africa and the Middle East. Sub-Saharan Africa's net imports are projected to increase 330 percent during 1988–2010. Net imports are a reflection of the gap between production and demand. The gap between production and need will be even wider. Many of the poor are priced out of the market, even at low food prices, and are unable to exercise their demand for needed food.

The better-off developing countries, notably large parts of East Asia, will be able to fill the gap between cereal production and demand through commercial imports, but the poorer countries will lack sufficient foreign exchange to import food in needed quantities. It is the latter group of coun-

Table 5. Projected net cereal imports, 1988–2020. (Source: Preliminary results from projections generated from the International Food Policy and Trade Model (IFPTSIM) developed at IFPRI by Keij Oga and extended by Mark Rosegrant and Mercedita Agcaoili)

Region	1988	2010	2020
Latin America	10.7	15.6	25.9
Sub-Saharan Africa	10.7	32.4	46.3
Middle East/North Africa	39.9	65.3	81.1
Asia	21.2	49.2	87.0

tries, including most of Sub-Saharan Africa and parts of South Asia, that will remain a challenge for the world community, requiring special assistance to avert widespread hunger and malnutrition. Population growth in Sub-Saharan Africa will outstrip growth in food production for a long time to come, unless more is done to accelerate agricultural growth. Between now and 2000, population is likely to grow at 3 percent per year while food production is likely to grow at 2 percent or less per year. By 2000, the production shortfall could be as much as three times the current level of imports (Von Braun *et al.* 1990). The region is not likely to have the necessary foreign exchange to import such large amounts of food. Neither is it likely that enough food aid will be forthcoming to make up the difference.

The other regions, while not as badly off as Africa, show disquieting trends. Cereal yields are increasing at a much slower rate in Asia than they did in the past. This is happening for a number of reasons, including a dramatic reduction in investments in irrigation facilities in Southeast Asia and in agricultural research and development throughout Asia, and falling wheat and rice prices. At the same time, income growth and urbanization in Asia are expected to accelerate the current trend toward diet diversification. This will increase the demand for livestock products and, hence, for feedgrain.

Projections for future food supply and consumption are extremely sensitive to assumptions made about future yield increases. The projections reported here are based on the assumption that yields will continue to rise at the rate of the mid- to late-1980s. Slower yield growth rates could have a major impact on food availability and prices. Recent stagnation in rice and wheat yields in Asia should be of particular concern in this regard.

Food prices have been on a downward trend for many years (figure 7). During the 1980s, real food prices dropped by 6.5 percent per year, with further decreases of 4.9 and 1.0 percent in 1991 and 1992, respectively (World Bank 1992a). Falling real food prices reflect successful supply expansions and lack of purchasing power among a large share of the population. Poor people cannot express their food needs as market demand. More than one billion people earn less than a dollar a day. Clearly, they are not in a position to convert their food needs to effective market demand. Since price is a

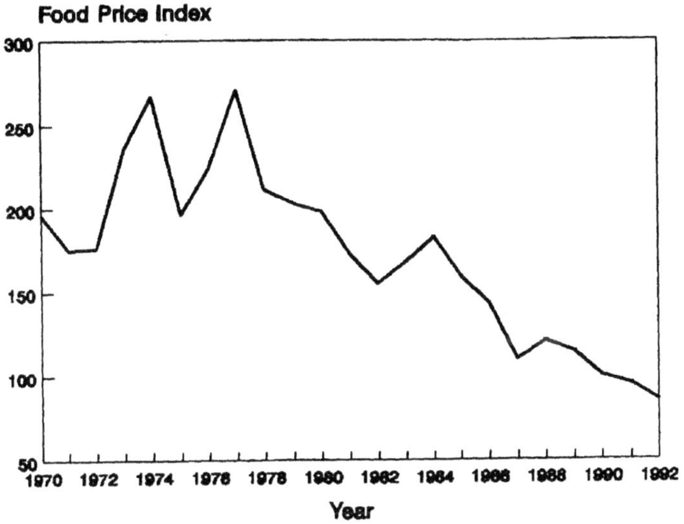

Fig. 7. World food price index, 1970–1992 (in constant US dollars, 1990 = 100).

product of both food supplies and economic demand, low prices indicate the persistence of poverty and a lack of sufficient purchasing power as well as increasing food production.

Do the projected shortfalls mentioned above suggest that the world is headed toward a global food shortage and rapidly increasing real food prices? While it is difficult to predict where food prices are headed, there is no reason to expect significant increases in international real food prices during the next 10 years, if yield growth rates of the 1970s and 1980s continue. The recently concluded GATT agreement will reduce agricultural subsidies in the European Union and the United States, but at a rate and speed far below what was visualized at the outset of the Uruguay Round more than eight years ago. Nevertheless, the agreement will place upward pressure on international food prices. However, the effect will be much less than the 15–20 percent price increase projected for total liberalization. It is likely that Eastern Europe and the former Soviet Union will increase agricultural production considerably faster than demand during the next 10 years (Tyers 1994). This will place downward pressures on world prices. Food shortages in Sub-Saharan Africa are unlikely to significantly influence international demands and prices due to lack of foreign exchange, unless increasing amounts of food aid are made available.

Global food insecurity, hunger, and poverty are not expected to diminish much in the near future. The number of undernourished people is projected to decline from 781 million in 1988–1990 to 637 million in 2010 (table 6), with the largest decline in East Asia followed by South Asia. However, the number of undernourished people is projected to increase 70 percent in Sub-Saharan

Table 6. Prevalence of chronic undernutrition in the developing world, 2010. (Source: FAO 1993)

Region	Number of Persons (million)	Percent of Population	Percent of Undernourished
Sub-Saharan Africa	296	32	46.5
Near East/North Africa	29	6	4.6
East Asia	70	4	11.0
South Asia	202	12	31.7
Latin America and the Caribbean	40	6	6.3
Developing countries	637	11	100.0

Table 7. Prevalence of underweight children in the developing world, 1975–2000. (Source: United Nations ACC/SCN 1992)

Region	1975	1990	2000	
			Pessimistic Scenario	Optimistic Scenario
	(million)	(million)	(million)	(million)
Sub-Saharan Africa	19	28	38	30
Near East/North Africa	5	5	5	3
South Asia	91	101	110	100
Southeast Asia	24	20	17	15
China	21	24	30	24
Latin America	8	6	6	3

Note: Underweight children are those children under five years of age with weight below −2 standard deviations from the meant weight-for-age.

Africa to 296 million people, 32 percent of the region's population. By 2010, almost half of the developing world's undernourished people will be located in Sub-Saharan Africa, up from 10 percent in 1969–1971.

Projections by the UN ACC/SCN suggest that regardless of whether optimistic or pessimistic scenarios are employed, the number of underweight preschool children in Sub-Saharan Africa is likely to increase during the 1990s (table 7). The situation in South Asia is also rather precarious, with almost half of the preschool children still projected to be underweight in 2000. However, the picture is not completely bleak. In East Asia, for instance, the number of poor is projected to decline from 169 million to 73 million during 1990–2000 (World Bank 1992b).

Table 8. Poverty in the developing world, 2000. (Source: World Bank 1992)

Region	Number of Poor (million)	Percent of Population below Poverty Line	Percent Share of Total Poor Population
Sub-Saharan Africa	304	49.7	27.7
Middle East/North Africa	89	30.6	8.0
South Asia	511	36.9	46.2
East Asia	73	4.2	6.6
Latin America and the Caribbean	126	24.9	11.4
Developing countries	1107	24.1	100.0

Poverty in the developing world is not expected to diminish much in the near future (table 8). The total number of poor people is projected to remain around 1.1 billion in 2000, although regional shifts in the distribution of total poverty are anticipated. Sub-Saharan Africa will increasingly become a new locus of poverty: the number of poor is expected to increase by 40 percent between 1990 and 2000, and the region's share of the developing world's poor is expected to increase from 19 percent to 27 percent over this period.

In sum, the outlook for food production in developed countries is favorable. They will continue to produce more cereals than they need, driving down world prices. However, even at lower prices, the outlook for the group of developing countries, particularly for some groupings such as Sub-Saharan Africa, is not favorable. It is quite likely that many people will continue to go hungry, even at lower prices.

Conclusions

What the world's food security situation will look like in coming years depends on what we do today. If the international community continues to act complacent, as it has done in recent years, there is no question it will get worse many more people will go hungry and malnourished, more children will be underweight, the prevalence of diseases of hunger and malnutrition will overshadow all other health problems, and the magnitude of emergency relief needed will accelerate. If the global community can exercise some foresight and mobilize the needed leadership to pursue a common path toward assuring world food security, then perhaps there is some hope for the world's hungry and food insecure.

However, the global community is splintered on the need and the means to do something to improve the world's food security situation. There is a lack of a common vision, unlike during the 1960s when a shared vision and appropriate action resulted in the green revolution and the consequent access by millions of people to food. The world food security situation today would have been considerably worse if the green revolution had not taken place. Today's complacency regarding world food security is misplaced, and the recent cutbacks by donors in investment to agriculture in developing countries are an inappropriate and potentially costly response to these trends.

Today, the challenge is both to grow more food and to assure that the food is accessible by all people at all times. The 1960s and 1970s were dominated by concerns to grow more food, while the 1980s were characterized by concerns of poor distribution. The time has come to stabilize the pendulum and to focus on both more food and better access. Even though today there is enough food to feed the world, that does not help the many people who cannot get access to it. We have to focus on both better distribution of available food and increased production of food to meet the needs of the ever-growing population and to generate the incomes needed by the poor to convert their food needs into effective demand, recognizing that available food will not be evenly distributed.

Agricultural development is key for both producing more food and ensuring better access to food via employment creation and income growth. Most of the world's food insecure and hungry are rural-based and even those who do not grow their own food are dependent on rural employment and income created by agricultural activities. Agricultural development must be encouraged through renewed support to agricultural research for productivity-enhancing and sustainable technological innovations for both high- and low-potential areas; through creation and maintenance of rural infrastructure such as roads; through appropriate policies; through provision of appropriate inputs such as fertilizer and improved seeds and to credit to facilitate access to these inputs'; and through public sector investment in education, health care, and nutrition to enhance human resources and improve overall well-being.

If a sustainable balance between world food production and food needs (as opposed to food demand) is to be achieved in the coming years, five conditions must be met: (1) economic growth must resume in those regions, especially Sub-Saharan Africa, where growth slowed in recent years in most of these countries, accelerated economic growth will occur only if agricultural growth is accelerated; (2) effective policies to reduce population growth and to slow rural-to-urban migration must be adopted; (3) resources must be committed to development of rural infrastructure, to expand international and national agricultural research, to provide credit and technical assistance, and to give farmers access to modern inputs; (4) macroeconomic and agricultural input and output market reforms must be successfully concluded and the appropriate role of the state as well as appropriate public policies must be identified and implemented; and (5) measures must be developed to manage natural resources and to prevent environmental degradation.

There is a long lag time between investment in agricultural development and corresponding improvements in food security. A commitment must be made now to improve world food security, otherwise we will pay many times over in the years to come for our complacency today.

Acronyms

GATT	General Agreement on Tariffs and Trade
IFPRI	International Food Policy Research Institute
UN	United Nations

Appendix 1

These projections are generated from the International Food Policy Simulation (IFPSIM) model developed at IFPRI by Keiji Oga and extended by Mark Rosegrant and Mercedita Agcaoili. The model is a market equilibrium model of foodgrains and grain-fed livestock products that solves for prices, demand, production, and trade by major countries/regions and for the world. The model uses a system of supply and demand elasticities, incorporated into a series of linear equations, to approximate the underlying production and demand functions.

In the crop models, domestic production is determined by the area harvested and yield per hectare. Harvested area is specified in terms of a partial adjustment function of the crop's own-price, cross-price effects with other crops, and the past year's harvested area. The yield function in the model is typically a function of past yield level and a trend factor which is used to reflect technology developments. For the major grains, this yield function is expanded to include not only the trend factor which reflects technology development but also the effects of output and input prices, including fertilizer and labor. Livestock production is modeled similarly as the food crops.

Total domestic demand for a commodity is the sum of demand for food, feed, and other industrial uses. Total food demand and demand for other industrial uses are per capita demand for a commodity multiplied by population. Per capita demand is a function of real per capita income, real price of the commodity, real prices of competing commodities, and a time trend representing changes in tastes and preferences. Feed demand, on the other hand, is a derived demand relating changes in the volume of output of livestock products to changes in the composition of feed rations via the feed demand elasticities. Feed demand elasticities determine the rate with which one feed can be substituted for another in response to a movement of relative prices.

Prices are endogenous in the system and are generally specified as the sum of world prices, in domestic currency terms, the assistance to the commodity measured by the producer subsidy equivalents and consumer subsidy equivalents (PSE/CSE), and a margin allowing for factors such as transport costs or quality differences. The PSE/CSEs are the standard policy structure found in IFPTSIM. However, for some countries and commodities, policy instruments which explicitly shift demand and supply relationships are also used. The US acreage reduction program, for example, is reflected in the model through a production control mechanism.

Trade is the difference between domestic supply and total demand and, as such, does not permit separate identification of imports and exports in cases where a country is an exporter and importer of the same commodity. Stocks are not explicitly modeled because markets are assumed to be in equilibrium in the intermediate run.

Finally, the market clearing condition is defined by equating the sum of net trade balances to zero at the world level. When an exogenous variable is shocked, world reference prices are assumed to adjust to a new equilibrium, where each adjustment is passed back to the effective producer and consumer prices via the price transmission equations. Changes in the domestic prices subsequently affect the supply and demand of commodities, necessitating their iterative adjustment until world supply and demand balance and world net trade is again equal to zero.

The base year data is a three-year average centered in 1990 for 17 commodities and 32 countries/regions. The price and income elasticities as well as data for the policy variables are taken from various sources, including OECD and FAO. For many Asian countries, however, the price and income elasticity parameters are updated using results from more recent IFPRI studies. Growth rates in yield, population, and per capita income generally reflect the trends estimated in 1982–1990. But for most countries, these rates are adjusted to incorporate the impact of structural adjustments and other policy reforms which alter effective relative factor prices and incentive structures. Income growth rates in Asian countries, for example, are slightly higher than present estimates. Yield growth rates in Eastern Europe, the former Soviet Union, and India are larger than in the past because of projected stronger developments in the agricultural production sector.

References

Von Braun, Paulino J, Hopkins R F, Puetz D, Pandya-Lorch R (1993) Aid to agriculture: reversing the decline. Food Policy Report. International Food Policy Research Institute. Washington D.C., USA.

Von Braun, Paulino J, Paulino L (1990) Food in Sub-Saharan Africa: Trends and policy challenges for the 1990s. Food Policy 15 (December): 505–517.

FAO (1992) FAO agrostat-PC (computer disk), population, production, and total balance sheets domains. Food and Agriculture Organization of the United Nations. Rome, Italy.

FAO (1993) Agriculture: Towards 2010. Food and Agriculture Organization of the United Nations. Rome, Italy.

FAO (1994) FAO agrostat-PC (computer disk), production domain, version 3.0. Food and Agriculture Organization of the United Nations. Rome, Italy.
Forest Resources Assessment 1990 Project, Food and Agriculture Organization of the United Nations. 1991. Second interim report on the state of tropical forests. Paper presented at the Tenth World Forestry Congress, Paris, September 1990.
Oldeman L R, Van Engelen V W P, Pulles J H M (1990) The extent of human induced soil degradation. In Oldeman L R, Hakkeling R T A, Sombroek W G (Eds.) World map of the status of human induced soil degradation: an explanatory note. International Soil Reference and Information Centre (ISRIC). Wageningen, The Netherlands.
Paulino L (1986) Food in the Third World: past trends and projections to 2000. Research Report 52. International Food Policy Research Institute. Washington D.C., USA.
Rosegrant, Mark W, Svendsen M (1993) Asian food production in the 1990s: irrigation investment and management policy. Food Policy (February): 13–32.
Tyers R (1994) Economic reform in Europe and the former Soviet Union: Implications for international food markets. Research Report 99. International Food Policy Research Institute. Washington D.C., USA.
United Nations ACC/SCN (United Nations Administrative Committee on Coordination Subcommittee on Nutrition) (1992). Second report on the world nutrition situation, Vol. 1, Global and regional results. Geneva, Switzerland.
United Nations (1993) World population prospects. New York, USA.
World Bank (1992a) Global economic prospects and the developing countries 1992. Oxford University Press. New York, USA.
World Bank (1992b). World development report 1992. Oxford University Press. New York, USA.
World Bank (1994) World development report 1994. Oxford University Press. New York, USA.

Response to sections A* and B

W. TIMS

Centre for World Food Studies (SOW-VU), De Boelelaan 1105, 1081 HV Amsterdam, The Netherlands

Introduction

This comment addresses two issues: the eco-regional approaches for food production and the estimation of resource availabilities for meeting global food needs in the coming decades. Papers have been presented on both these topics and I have been requested to take those as my starting point. With the exception of the paper by Pinstrup-Andersen and Pandya-Lorch, these are agro-technical and discuss mainly approaches and options regarding the use of physical resources to make food available to future populations. Future demand for food is dealt with rather cursorily, which pains the economist. Even more regrettable is the absence of prices in the analytic approaches and in the options presented for the future. As economists are convinced that prices play a central role in determining future paths of development, you will understand that my comments will elaborate on issues of demand and prices.

As I have not received all of the papers in advance, I will not be able to focus my comments on issues which I consider central in this workshop. The original idea behind the invitation as a respondent concerned an exchange on issues of methodology. As it is exactly those papers which are missing, I therefore have to refocus my intervention. Thus, I will address the issues raised by the papers I was able to read in advance; those concern future agricultural production and the options for meeting future food needs. Towards the end I would like to say a few words about research priorities.

Food security

Initially this workshop referred in its title to the concept of food security. In the programme I received I noted that it has been replaced by 'sustainable land use and food production'. I may therefore seem to bark up the wrong tree when raising the question: What do we mean by food security? But even the

* From section A, only the paper by Quiroz *et al.* was available in advance for review.

J. Bouma et al. (eds.), Eco-regional approaches for sustainable land use and food production, 113–121.

title change cannot justify leaving aside the question; it needs to be raised in explicit terms, as food security is the ultimate objective of food production, with sustainable land use as a major condition. In addition, divergent views are held about the meaning of the food security concept. I prefer the view expressed by Pinstrup-Andersen and Pandya-Lorch: the world is food secure when each and every person is assured of access at all times to the food required for a healthy and productive life.

Today, the world produces enough food to meet the needs of all; in case the supplies would be well distributed, there would be no world food problem, certainly not when there was sound reason to expect ample scope for expanding production further in the decades ahead. Penning de Vries *et al.* conclude that "... natural resources exist to increase food production very significantly". As their conclusion appears to apply to the next 40–50 years, one can only conclude that the essence of the world food problem for the two next generations is the inequitable distribution of food. In addressing the world food problem, that issue appears to be the central one. I like to expand a bit further on the reasons why it is the core issue and what implications this has, also for our research.

I will suggest that the papers before us, except for the one by Pinstrup-Andersen and Pandya-Lorch, assume the food security problem away and simply focus on something else. They design an analytical approach to tackle other problems and then try to find answers. I conclude that these studies are not useful for finding solutions to the world food problem. They may be interesting from numerous other points of view, but that is no consolation for an audience which expects to hear about solutions to the world food problem. Simply laying the problem aside by assumption leads to scenarios which are not only irrelevant but also misleading, particularly for the prospects of the poor and their food demand.

The studies of Penning de Vries *et al.* and of Van Latesteijn assume the poor away. They make estimates of food demand which will provide every world citizen in the year 2040 with adequate food. It may be a vegetarian or an affluent diet or something in-between, but everybody is assumed to get at least enough. Given past trends in reducing poverty and hunger, that is a rather bold assumption: absolute numbers of people below an imaginary poverty line have continued to increase over the decades; also, even the most optimistic estimates of the World Bank expect their absolute number to decline only marginally over the course of the current decade. Thereafter, maybe by more, and faster; but one has to be a wild-eyed optimist to assume a total eradication of hunger by 2040.

On any realistic assumption about the prevalence of hunger and malnutrition in 2040, the levels of world food demand will therefore be lower than the levels assumed in the scenarios in the paper of Penning de Vries *et al.* In that case, production will grow more slowly as well; if supplies would continue along past trends, prices of food relative to all other goods and services would continue their secular decline. The long-term trends of past decades suggest that this is not just a figment of the imagination.

Demand prospects

In the world of today, this process of declining prices has been given additional impetus by government policies. Some farmers – in the European Union, the United States, Japan and some other countries – have been able to expand their output irrespective of demand trends as they were shielded against lower market prices. They could continue to invest and to produce independently from declining world prices. Their increasing surpluses are still today being sold internationally at whatever prices and with whatever subsidy this may take, thus contributing significantly to the long-run price declines of agricultural products in international markets.

At the same time, governments of developing countries did engage in opposite policies: reducing domestic agricultural prices through public monopolies which squeezed large resources from their farming sectors, leaving little or no incentive to produce for the market and in fact promoting rural out-migration. As domestic prices were kept even below the depressed levels of world market prices, negative implications for domestic agricultural production were quite substantial.

Excessively low prices in developing countries and in world markets send wrong signals to farmers, governments and consumers, particularly in the developing countries: it suggests to them that food can be produced abundantly elsewhere at low prices: if they cannot match those, they should not be in the farming business. But in fact, a large part of those abundant and cheap supplies survive artificially, i.e. behind quite high protective walls.

The Uruguay Round of trade negotiations did lead to an agreement to reduce agricultural protection. But it will be a slow process and will take a long time to have visible effects on world market prices. Those will therefore in all probability remain low, if not declining, discouraging investments in agriculture in the developing world, discouraging technological progress and casting doubt on the wisdom of research spending on agricultural technology. In the developed countries production will grow at a snails pace when those countries dismantle current protection; at the same time, in the developing countries production growth may accelerate somewhat, if and when those countries make the transition to open markets and less price or trade intervention by their governments. But on the whole, liberalization is not expected to give much of a boost to world agricultural production. Its geographic distribution will change, and so will the pattern of international trade flows.

Supply prospects

Turning again to the paper by Penning de Vries *et al.*, one feels reassured to learn that, even when using a low external input (LEI) assumption, all regions in the world will possess the physical resources to produce more than enough food to feed their populations well; even more so in the variant with high external inputs (HEI). In fact, the situation in all regions will be even more comfortable as it is unlikely that overall development will be fast enough to generate the full employment and the income levels which are the conditions for providing access for all people to adequate food. A positive outcome, in terms of the papers presented, but a negative one in terms of food security which this seminar should be addressing. For the economist interested in food security as defined earlier, this positive outcome regarding the abundance of resources will be a reason no longer to take account of physical or ecological ceilings: those are too far away to be relevant. Those variables can be removed from the models without any loss of relevance.

Whether that would be wise is to be doubted as the projected resources are rather crudely estimated. This is the next topic. Best use of land is described by Penning de Vries *et al.* as "... maximum productivity of labour and land and minimum food prices". This is a rather cryptic statement which appears to lack internal consistency. Leaving aside the question whether maximum productivity of land and labour can be reached simultaneously – i.e. a situation without trade-offs – the role and meaning of minimum food prices is totally unclear. Minimum for whom? Probably for farmers, to cover some imagined level of costs, as if there would be an agent other than the market to set that price. In reality, it is the farmer's income that functions as the dependent variable between costs of production and market prices, at given technology. The authors thus are assuming an income level first – without its basis – leading to a cost estimate which in turn is translated into a price by some agent like a government. A world-wide Common Agricultural Policy? I thought we were in the process of moving towards market conformity.

At the same time, the paper gives no indication whatever about the investment costs of meeting all food needs of all people by 2040. There are assumptions about irrigation, crop and livestock technologies and other investments which are undoubtedly very large and even beyond what can be expected in terms of their prospective rates of return. These latter may be assumed to be adequate under the assumption of 'minimum food prices', but those are certainly not the prices that will prevail: actual prices will undoubtedly be lower and squeeze out a sizeable number of technically feasible projects.

At least, one would hope that the projections of available resources in 2040 to produce food are sound from an ecological perspective. But the way in which the process is described to estimate the suitability of soils – mainly on the basis of NASA-data – and the way climate is treated – leaving out a major part of the rainfall data – makes one wonder. Also the allocation of water, which is not priced, to crops is done in a manner which has little to do with economic rationality: i.e. by water consumption rates with first service crops with lowest need. If water were to become very expensive, this might be correct, but there is no basis to derive such an assumption from the paper. With those questions looming over the optimistic conclusions of the authors, economists obviously will do well to preserve their interest in physical and ecological variables as determinants of food production: limits may in fact be reached much earlier than projected and should therefore be incorporated in the analysis. At the same time, those in the business of constructing eco-regional models will do wise to start humanizing their approaches by taking account of the actual situation of poor farmers and consumers, the investment costs of what they propose and the relative prices of food which are associated with each of their scenarios.

The use of explorations

Turning to the paper by Van Latesteijn I would like to be brief. The paper discusses forecasting methods, scenario techniques and, what he labels, explorations without reference to the considerable literature on these important topics; it is difficult to take that major part of his text seriously when the sourcing is so thin. Further in the paper, there is a description of the method used in the European study, without indicating why a choice was made for multiple goal programming. It is obviously a useful technique, but there are others that have been applied successfully as well, also for studies of European agriculture. Particularly the modelling work undertaken by a Dutch team from the Agricultural Economics Research Institute, the Central Planning Bureau and the Centre for World Food Studies might have lent itself for comparison as it has its roots soundly in theoretical concepts and was used intensively in the policy discussions on reforming European agricultural policies. Van Latesteijn's conclusion that "The generation of information on what is possible is a valuable contribution of scientific research to the policy debate" is to be questioned, both on methodology grounds as more specifically with regard to the information generated by the research reported here.

The need for a focal unit

The paper by Quiroz *et al.* is interesting as it raises some of the most relevant questions for a regional approach to sustainable and more productive agriculture, particularly regarding the adoption behaviour of farmers, the sustainability of alternative technologies and the incentives to poor farmers to engage in conservation practices. But the approach presented in the paper is designed too narrowly to provide relevant answers: comparisons are between several crop rotations and cattle holding, but their combination in one household is not addressed: it is as if farm households specialize in just one of the alternatives. Nor is any attention given to other (non-agricultural) sources of income which may give rise to a different allocation of the household's limited resources.

In describing its approach, the paper states "Optimal alternatives ... were selected" and the definition of 'optimal' is stated as "... maximum profitability with minimum pasture and soil erosions". In the economists' vocabulary, profitability is the difference between the expected values of inputs and outputs, both measured at expected prices. Is this what the author has in mind? If so, why would in that case the condition of minimum pasture be introduced? At the prices generated by the model exercise, one may expect to obtain as an outcome whether pasture fits into a production pattern designed towards maximum profitability. Also with regard to soil erosion, one would expect this to be endogenous to the model and measured along with other costs at some (preconceived or projected) price per unit of soil loss. As the aim of the exercise is stated to be the calculation of an internal rate of return, one would expect the exercise to be done with the least number of *ad hoc* assumptions; certainly for the long term, neglect of relative price changes between inputs and outputs will severely limit the relevance of the exercise.

There is a rapidly growing literature on the adoption behaviour of farmers with respect to new technologies. The findings on determinants of technical progress are, however, not recognized by the authors, who operate on the basis of alternative logistic adoption curves, with or without the project. If the project does not promise to raise incomes, it would be rather unlikely that the project would have any effect on rates of adoption. If there are expected benefits, probably their size will determine the rate of adoption. In both cases the lack of careful scrutiny of farmers' behaviour and its replacement by an entirely arbitrary assumption is to be deplored. One would like to suggest, as an overriding conclusion regarding this paper, that more attempts need to be made to model household behaviour in a comprehensive manner, taking actual production patterns of households as the starting point and then tracing the changes which will be needed over time to achieve both higher incomes and more sustainability for typical households.

Markets and public goods

Let me now come to the paper by Pinstrup-Andersen and Pandya-Lorch. I agree with the way the food problem is formulated and dealt with by them. The question is, whether it is true as stated in the paper that "Agricultural development is key for producing more food and ensuring better access to food via employment creating and income growth". My doubts arise from the observation that even rapid agricultural growth will only absorb a tiny part of the additional labour supplies entering the markets. Raising the incomes of the poor will mainly need to come from non-agricultural employment in both rural and urban settings. Therefore I hesitate to put my eggs in the basket of agricultural growth as the engine of development.

The second conclusion of the authors concerns the need to promote lower rates of population growth and urbanization. Although endorsing this policy priority, one should not loose sight of the need for non-agricultural employment as the main objective, whether in towns or in rural areas. There is increasing awareness of the importance that attaches to non-agricultural income sources of farm households and the role that migration movements play in this respect.

The fifth condition the authors mention relates to the management of natural resources. It is not elaborated in the paper. Against the background of what was said above, I like to suggest that options are generated for dealing with ecological and conservation policies in a market-oriented fashion. The increasing scarcity of water needs to be reflected in a price to be paid, preferably to rightful stakeholders. In principle, all citizens of a country have an equal right to this and other natural resources, possibly even including clean air and a clean environment. Users are to pay for uses beyond their own personal entitlement to those who find their entitlement eroded. Some of those markets will neither come about on their own volition, nor operate properly unless a government develops systems to simulate these and organizes the payments to be made and received. It may well be that an important source of income for the poor may be created along this path.

Returning now to the employment prospects: relative price declines for agricultural products may be a good thing: scarce resources are re-allocated to non-agricultural sectors which need to grow faster in order to absorb the labour that cannot be accommodated in agriculture. Whether this sectoral migration can succeed in reducing poverty and hunger sufficiently is not in all cases assured; two problems may be mentioned which may hamper or retard the process and where the government may need to act. One is that large segments of the agricultural sector may not be adequately connected through markets with the rest of the economy: price signals will not reach farmers and rising demand in the rest of the economy will not be met by domestic supplies. In this case, governments need to attach a high priority to investments in both physical and institutional infrastructure to achieve better integration and to ensure a fuller response of farmers to market signals. The other is where the options for non-agricultural development are so limited by resource constraints that labour surpluses continue to accumulate in both rural and urban areas and a process of impoverishment becomes dominant.

The first problem, market integration, is the dominant one in sub-Saharan Africa, with the second one of rising labour surplus and impoverishment threatening that region already now and in the future. At present, the labour surplus situation is observed most clearly in Bangladesh, other countries of

South Asia and in parts of China, although the lack of infrastructure and well-functioning markets plays there as well. In the Asian case, adding external resources and instituting employment programmes to provide an income and access to food may even in the long term be necessary palliatives.

In fact, these approaches are consistent with the ones proposed by the authors in their first, second and fourth points of the concluding section. I have doubts regarding the third, proposing to commit (I presume public) resources to research, credit, technical assistance, and better access of farmers to modern inputs. At the least, the statement should be made with more nuance, asking the question what can be left to markets, what is to be done by governments and donors, and were the authors want to express reasons for being in doubt about the proper choice. Thus, credit should be left to markets, with access to credit freed from the government imposed constraints on a majority of their farmers. Nor should there be reasons for input subsidies, except indirectly, when for the sake of competitiveness set-up costs of new market parties are to be funded.

Technical assistance probably belongs largely in the public sector as extension services (non-rival markets) are part of institutional infrastructure, in particular when the aim is its rapid expansion and enhanced effectiveness. Research may belong only partly in the public domain, depending on an assessment of the character of its outputs: in a number of cases private markets within proper rules of conduct prescribed by the government may function efficiently. In any case, it is not appropriate to label research results as public goods by assumption and to get away with that.

The research agenda

There is ample scope for further research. Probably not by way of explorations: it may tell us how much snow there is on high mountains to assess the world's capacity for skiing, without addressing either the demand side or the investment costs. In the present case of the world's food prospects, the first need is for attention to farmers, consumers and their behaviour, to investment costs and opportunities and to the expected price developments in each of the scenarios. Without those, explorations can be misguiding in setting further research priorities, at a time when we can do without additional confusion in a field where the lives of many people and the future of many more are at stake.

As to the CGIAR agenda, an important issue will be what parts of its work programme need to be funded in the public sector and which parts lend themselves for full exposure to markets. This will require an assessment of the characteristics of the system's outputs in terms of non-rivalness, jointness of consumption, risks of free-ridership and excludability. In some cases when these characteristics are found to apply, measures can be proposed to set rules which permit markets to operate smoothly; in others, this may not be possible and the outputs need to be labelled public goods. The distinction between types of outputs by their market characteristics will undoubtedly help to design an appropriate institutional structure for future research and for its sources of funding.

In any case, the research should increasingly centre around studies of farm households, each with farming as one of its activities, but among a number of other options to use its scarce resources and with trade-offs which affect the choices in the field of agriculture. Also, account must be taken of the character of production decisions in a number of cases as derived decisions, with consumer preferences and avoidance of risk being main determinants of crop choices.

Households interact through markets and research cannot truly answer policy questions unless it traces those market relations, with price and income formation and the responses to those by households and by governments. I do not suggest that the CGIAR system undertake such studies entirely on its own, nor that each Institute should run a complete programme on these matters. Within the CGIAR complementarities can be identified and fostered, which in turn can (re)generate the interest in research funding by present and future CG-participants on the donor side.

Response to section B

P.S. TENG, M. HOSSAIN and M.J. KROPFF
International Rice Research Institute (IRRI), P.O. Box 933, Manila, Philippines

Response to paper by Van Latesteijn

In general, the author is to be commended for clearly explaining a complex but innovative approach to explore land use options. The results offer options for different desired scenarios, without specific recommendation of one. This is closer to the real world situation, where ultimately, many political decisions are made with consultation and weighing the pros or cons of each available option. The advantage of the approach over more conventional economic approaches appears to be its ability to not stipulate an optimal solution. The author puts this succinctly when he stated "the focus of research is no longer on developments that might be expected, but on developments that might be feasible". In short, the approach explores, and does not choose.

The paper devotes much space to explaining the procedures used for the analyses and not enough is devoted to discussion on findings. The usefulness of the paper could be enhanced by explaining why the different scenarios were selected and how to formulate other scenarios of interest.

The LP procedure, even though it is multiple goal, is nevertheless a 'static modelling' technique. Its limitations include pre-setting assumptions and constraints. For example, high yields are assumed to be those obtained by the best technical means and it would be very cumbersome to incorporate the influence of changing technology to accommodate several yield scenarios. Also, the price – production relationship remains the same while the influence of external demand or supply of commodities (such as may occur with GATT) is set and not allowed to change. In essence, the entire EC region is treated as one big farm divided into *n* regions. Farmer behavioral response, or politician behavioral response to individual goals as affected by the results of other goals, is not included in the analysis.

The procedure used requires prior analysis of factors that determine agricultural development at the regional level. In a sense, there is a selection of key factors to explore. Thus, results will have to recognize that other factors, not included, could play a role in influencing land use. However, a strength of the approach also appears to be that there is no assumption of exchangeability

J. Bouma et al. (eds.), Eco-regional approaches for sustainable land use and food production, 123–129.

between labour, capital and nutrients and all factors are treated equally. This allows a 'freer' exploration of possibilities.

One point that has to be repeatedly kept in mind is that the approach used by the author is for analysis purposes. The next step is specification of a time path for implementation using an economic model.

Response to paper by Penning de Vries et al.

The authors are to be commended for taking a systematic approach, using empirical data, to make an objective estimate of food demand in relation to the production capacity of different agro-ecological zones for supplying food (measured as grain equivalents). However, like most studies of a global nature, this study was also hampered by availability of data and the scale of aggregation. Given the state of current technology (such as geographical information systems, networking, etc.), global projections should account for as much detail as possible, in particular of local influences on the supply-demand equation. This is what John Naisbett calls the post 'New Age' paradigm of *Think Locally, Act Globally* (Naisbett 1994). The authors have tried to do so in their study, but it still falls short because of the small number of spatial units that were used in the aggregation procedure.

The ERA-meeting organizers must have put this and the paper by Pinstrup-Andersen and Pandya-Lorch together to contrast approaches. The first is very 'technocratic' while the second has a more socio-economic flavour.

Global projecting is a process full of potential pitfalls. One is the need to generalize assumptions or average values. Although the authors are to be commended for noting that "it is better to use all basic data and to aggregate subsequently than to use averages", they nevertheless could not apply this guideline to some of the variables used, such as crop yield and consumption demands. For example, we know that in Asia, which grows > 90% of the world's rice, even in rainfed environments, there is a big difference between actual yields of the favourable rainfed, the drought susceptible rainfed and the submergence susceptible rainfed. External input is only one source of influence on rice yields in these environments. Hence, a more logical division for calculating rice supply would have been to use water as a first determinant, then external input (HEI or LEI) as a second determinant.

The population scenarios used are very reasonable, i.e. low, medium and high. As population growth is a key factor in the food demand calculation, the authors are to be commended for taking this into account. However, it would have added to the discussion had there been some explanation of the three scenarios used in relation to other projections of population growth. For example, the World Bank (1992) also provided three scenarios for population growth based on a rapid decline in fertility (2040 – 8 billion), a base scenario at present rates of growth (2040 – 10 billion) and a scenario with slow fertility decline (2040 – 12 billion). These scenarios are highly germane to the question of quantity of food demand and to shifts in the type of food. For example, the authors showed that in eastern, southern, and Southeast Asia, there is likely to be a food deficit. Yet, this part of the world is generally recognized as the most dynamic in terms of economic growth and its accompanying purchasing power. Any statement on food sufficiency for these regions will have to deal with (i) the purposeful reduction in actual production capacity, as has happened in Malaysia where the government has set a goal of 65% self-sufficiency in rice; (ii) the substitution of cereal demand for meat and vegetables, a phenomenon that accompanies increased affluence, and (iii) the ability to import from areas such as North America and Australia, which have in recent years developed crop varieties closer to Asian tastes.

As an addendum, the role of GATT and APEC in influencing food production and prices has yet to be factored in and will likely affect food production.

The authors take a rather arduous route to come to the same conclusion that was arrived at by Smits (1986), "that if all food were equally distributed, no one would go hungry". If this was taken as a given, and it appears to be supported by at least one sector of protagonists in the global food supply-demand debate (see Avery 1994), then a more appropriate hypothesis to be explored in the paper should have been whether this can be done without degradation of the natural resource base. The authors define the natural resource base as soil, climate, plant genetic properties and surface water. A more encompassing definition could have been soil, water, biodata and climate. Degradation of the natural resource base will be a key factor in determining the production capacity of much of the global agricultural lands. To assume minimal to no degradation is to ignore an important influence on both supply of and demand for food. In tropical Asia, 2.7 million ha are deforested each year, mainly in the uplands (FAO 1991), resulting in extreme erosion. The uplands are an important ecosystem because of their role as watersheds for much of the lowland rice, even though rice is not a major component of the farming systems of the uplands. Degradation of the lowlands in intensively cropped systems may also be exacerbated in the coming years, as exemplified by the phenomenon of 'yield decline' and decline in total factor productivity (Cassman *et al.* 1994) in rice and wheat. In the humid tropical eco-region of Asia, loss of agricultural land to urbanization is another relatively recent phenomenon that is likely to have profound effects on food supply. By 2040 it is estimated that in Asia, more than half the population will live in Megacities, compared to the approximately one-third of urban population in 1990. On the surface, these factors appear at odds with each other and it will take detailed analyses to make even projections at the Asian level look realistic.

Much of the decision on food sufficiency (security) is based on calculating the relative ratio of supply/demand, but the explanation of why the ratio value of 2.0 is selected as the 'break-even' point is not clear. However, to recap the authors findings, under an HEI scenario with moderate diets and a medium rate of population growth, only southern Asia would have problems meeting demands, while under an LEI scenario with moderate diets and medium population increase, all of Asia would have problems. This perhaps illustrates a weakness in 'closed' system projections that exclude the role of imports. It is unrealistic to expect that in this dynamic region of economic growth there would not be importation. Conversely, that all regions of Africa showed no food security problems under the same scenarios would further suggest a need to re-examine the scenarios. What is more likely to happen in Asia is the occurrence of medium population growth with a shift to affluent diets, while Africa may be anticipated to have medium population growth with moderate diets. A recent article by Avery (1994) pokes fun at the food shortage pronouncements of L. Brown regarding Asia and he argues that economic growth, trade, and high-yield farming will make up for any potential food shortages in the region. Avery (1994) cites in particular the impact of freer trade in mobilizing the production capacity of 20 million ha of farmland set aside in the USA due to current grain surpluses, and the 30 million ha in Argentina diverted to pastures.

The study could also be made more realistic by taking into account availability of key resources to increase yields such as nitrogenous fertilizers. HEI agriculture may not be that sustainable if there is another increase in petroleum prices. For example, to tap the potential of rice plants to yield an average of 12 tons ha^{-1} will require almost doubling of applied nitrogen unless more efficient techniques of application and utilization are found. Thus, instead of a HEI versus LEI scenario, it would have been more useful to apply several scenarios of N availability at different price structures.

The authors refer to the role of new technologies in influencing future cereal grain supply. It would have added greatly to the paper, and to the study, if in fact scenarios had been presented of yield gains to be anticipated from various technologies, such as high end biotechnology and low end Integrated Pest Management (IPM). Using rice as an example, to meet the anticipated increase in demand in 2050 would require that rice yields increase in irrigated areas from 4.9 tons ha^{-1} in 1990 to 9.3 tons ha^{-1} in 2050 (90% increase), and corresponding yields in the rainfed areas increase from 1.9 tons ha^{-1} in 1990 to 4.5 tons ha^{-1} in 2050 (137% increase) (Zeigler *et al.* 1994). If there is to be no increase in rainfed rice yields, then irrigated rice yield would have to increase to 12 tons ha^{-1} in 2050 (145% increase relative to 1990)! These enormous increases in rice yields will be difficult to achieve with conventional breeding to raise the yield plateau, and at the level of the eco-region, decisions may have to be made to mount a concerted effort involving mechanisms such as consortia to accelerate efforts aimed at either reducing current gaps between attainable and actual (which vary from 2–6 t ha^{-1}) yields, or at increasing yield potential. Part of the international dialogue on eco-regional approaches must take into consideration not only analyses but suggestions for future action to be taken at expanded time and space horizons (borrowing terminology from the Club of Rome). Increasing the potential yield may be difficult to achieve even with biotechnology in the rainfed environments since many of the traits are multigenetic and their inheritances are as yet not well-understood. Conservative estimates of preharvest losses (gap between actual and attainable yield caused by biotic stresses) which have not been reduced through breeding range from 15–30% annually. At the eco-regional level, Asia accounts for over 80% of all pesticides (insecticides, fungicides, herbicides) sold world-wide. Therefore, the situation exists in which substantial losses are being incurred in spite of high inputs of pesticides and presence of host plant resistance in most genotypes. Crop improvement versus natural resource management strategies could assist the scientific community concerned about eco-regional issues such as IPM policy and implementation.

Response to paper by Pinstrup-Andersen and Pandya-Lorch

This is an excellent paper which succinctly reviews recent developments on the production, consumption (nutrition) and distribution of food in the developing world, the future prospects, and their implications for policy regarding research and development. The authors are to be commended for pulling together a lot of material and synthesizing them into a very logical and well-argued format. Many of the points regarding distribution (access) versus availability (production), however, have been articulated by previous writers. The sentence "global availability of food has not yet translated into availability of and access to food by all people", sums it all up. What the authors have missed is the opportunity to explore implications by eco-regions, taking into account projected changes in the parity purchasing power of different countries in the major eco-regions. This would have added greatly to the paper and pointed to some policy possibilities for specific eco-regions. The paper argues that except in Africa, the prospect for food grain production is encouraging, but the gap between production and consumption is expected to widen, because of the problem of distribution and lack of purchasing capacity of the poor.

The paper appears to be optimistic regarding the recent developments and the future prospect of sustaining the growth of food production in Asia. This is mainly because in reviewing the recent growth in grain-food production the authors fail to distinguish between the 'once-for-all' effect of the policy changes introduced in many Asian countries since late 1970s, and the developments of factors that sustain long-term growth. For example, China had a rapid growth in food production during the 1980–84 period, because of the policy changes introduced since 1978, and Vietnam had a rapid growth during the 1987–91 period due to the economic liberalization policies. In both countries the growth has slackened after the initial acceleration. The authors have used 1979–81 as the base for assessing the recent developments in Asia. China's performance during 1979–84 has influenced the indices since it is a big country. Since 1984, the rice production in China has remained almost unchanged, and that of Asia increased at 1.3% per year, much slower than the growth of population.

The authors have rightly mentioned that since 1989 rice yields have stagnated at around 3–6 tons ha^{-1}. This however gives a misleading impression about a large yield gap that is available for exploitation in the future, since Japan and South Korea achieved a yield of 6.5 tons ha^{-1} and have stagnated at that level. A disaggregation of the rice ecosystem into irrigated and rainfed, and yields achieved by farmers in those ecosystems would have cleared this misunderstanding. Farmers in China where almost all rice area is irrigated, have already achieved an average national yield of 6.0 tons ha^{-1}. In India, where the average yield is less than 3.0 tons ha^{-1}, farmers have already achieved 5.5 tons ha^{-1} in irrigated Punjab and Tamil Nadu. Thus, in the irrigated ecosystem farmers have almost reached the yield plateau. The future growth would be limited unless research makes another breakthrough in shifting the yield potential. In the rainfed ecosystem, the growth in rice yield has been slow, and is around 2.0 tons ha^{-1} because technological progress for this ecosystem has been limited. In countries with a large proportion of area under the rainfed ecosystem, such as in South Asia, the growth may slow down unless there is an acceleration of investment in irrigation, drainage, and flood control for the transformation of the rainfed into irrigated ecosystem, and/or the scientists succeed in developing high yielding varieties resistance to droughts, floods, and problem soils, which are difficult scientific challenges.

A comment on interpretation of macro or national level food production figures is needed. In the Asian region, some countries have purposely reduced area under food crops and total food production in lieu of export income generating activities. Malaysia has a stated goal of 65% self-sufficieny even though productive capacity could be much higher. Other countries are contemplating substitution as land resources have higher economic worth for non-agricultural activities, e.g. Thailand. An added value to the analyses done in this paper would have been to examine actual production rather than potential. What are the food security implications for the world, if divided into eco-regions of different production capacity, if government and multilateral policies could be put in place? Surely, one of the aims of the 'eco-regional paradigm' is to get concerted action, albeit harmonized, on common issues that will benefit the most people in an eco-region.

The authors are to be commended for making the forceful point that food availability is necessary but not a sufficient condition for food security, because of the lack of purchasing capacity for acquisition of the food. They rightly mention that in spite of the remarkable economic progress in Asia, hunger and malnutrition is widely prevalent in many countries which have declared self-sufficiency in grain-food production at the national level. In fact, poverty is acute in regions which had low yields in grain-food because of the unfavourable production environments. Since production of staple grain-food is the major source of employment and income at the low-levels of economic development, the purchasing capacity of the poor could not be increased without raising the productivity in the grain-food sector. Thus, to help alleviate the food security problem, research must focus on problems of the specific regions where the poor live and work. Hence, the need to move from global to eco-regional approach to research, the theme of their conference. The paper could have made this point more clear.

A key factor that was not addressed in the paper is what it means to expand irrigation to the favourable rainfed areas, and to improve existing irrigation infrastructure. For the humid, tropical Asian eco-region, many irrigation systems date from the post-Second World War era, and are suffering signs of decline and inefficiency in water utilization, with accompanying increases in biotic/abiotic stress effects on potential yield. About 25% of rice area is in the rainfed lowland, with great potential for spectacular yield increases equivalent to the first Green Revolution, if converted to irrigated area. This would require investments in irrigation, which the current price and demand for rice do not appear to justify. Given that only 2–3% of global rice production is traded, food security for the marginal environments of Asia may have to rely on improved infrastructure for water control brought about through new investments.

Acronyms

EC	European Community
GATT	General Agreement on Tariffs and Trade
HEI	High External Input farming
IPM	Integrated Pest Management
LEI	Low External Input farming

References

Avery D T (1994) Feast or Famine? Far Eastern Economic Review 157(48):40.

Cassman K G, De Datta S K, Olk D C, Alcantra J, Samson M, Descalsota J, Dixon M (1994) Yield decline and nitrogen balance in long-term experiments on continuous, irrigated rice systems in the tropics. Advances in Soil Science (in press).

Meadows D H, Meadows D L, Randers J, Behrens III W W (1972) The limits to growth. Pan Books, London, UK.

Naisbett J (1994) Global Paradox. William Morrow & Co., Inc., New York, USA.

Zeigler R S, Hossain M, Teng P S (1994) Sustainable agricultural development of Asian Tropics and Subtropics: emerging trends and potential. In proceedings of IFPRI Eco-regional Workshop, November 1994, 7–11, 1994, Airlie House, VA, USA (in press).

SECTION C

Linking different scale levels in ecological and economic studies

Agro-ecological knowledge at different scales

L.O. FRESCO*

Department of Agronomy, Wageningen Agricultural University, P.O. Box 341, 6700 AH Wageningen, The Netherlands

Key words: scales, aggregation, agro-ecological processes

Abstract. Scale is a real and quantifiable phenomenon, not just a conceptual construct. Characteristics of agro-ecological processes depend on scale. Observed ecological heterogeneity in the field may often result from inadequate data resolution. Scaling up or down depends on a proper identification of different processes that are specific to each scale, and differs from a cumulation of lower or higher scale data or variables. There is an optimal scale level at which each process can and must be studied. In order to improve scale-sensitivity in agro-ecological research key processes must be identified that provide an entry into linking scales together. Possible 'hidden effects' of agricultural technology at other scale levels than the one at which technology aims, present a special analytical challenge.

Why do we bother about scales?

Scales, however difficult to comprehend, are essential to take into account in describing and understanding agro-ecology. The primary reason is that the precise characteristics of agro-ecological processes depend on scale. Take, for example, the mutual relationships between climate, soil and vegetation. Vegetation belts across the globe, i.e. at global scale, reflect primarily climatic zones. At the scale of the plot, vegetation is a function of weather and soils, and at that same scale, vegetation in turn influences both micro-climate and soils through evapotranspiration and the breakdown of its biomass. At the scale of the landscape, vegetation patterns shape the hydrology and the meso-climate, but together these aggregate to factors such as surface roughness, albedo and evapotranspiration that ultimately affect global change (cf. also Holling 1992). In other words, the relationships between soils, climate and vegetation are scale-specific: depending on the scale, a factor can be seen either as an exogenous driver or an endogenous variable, and they are linked across different scales.

The second reason why scales must be dealt with is that observed ecological heterogeneity may often result from inadequate data resolution or grain (Kolasa and Pickett 1991). When characteristic values of a phenomenon are scattered or have a substantial spread, it is usually necessary to study the same phenomenon at a smaller (i.e. less detailed) or a larger (i.e. coarser)

* Comments by Dr A. Veldkamp are gratefully acknowledged.

J. Bouma et al. (eds.), Eco-regional approaches for sustainable land use and food production, 133–141.

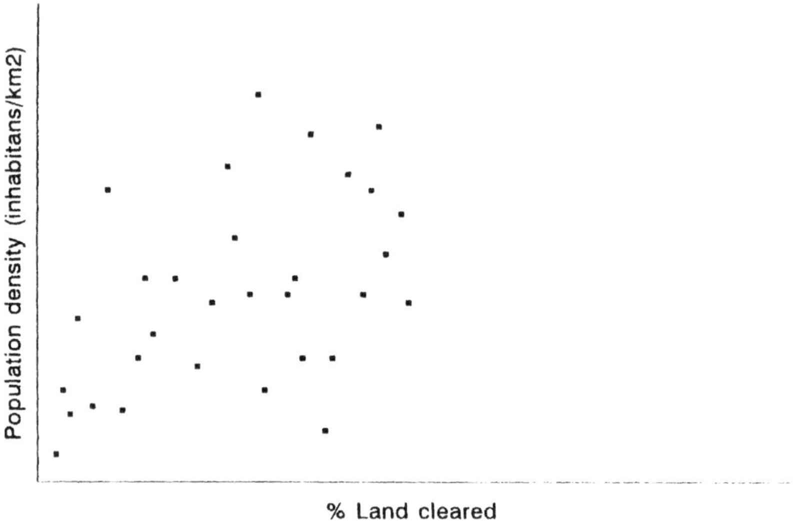

Fig. 1. The relationship between population density (people/km^2) and deforestation rate (km^2 year^{-1}) in the Brazilian Amazon.

scale to reduce heterogeneity in observation. An example is provided in figure 1 (based on Turner *et al.* 1993), which represents the relationship between deforestation rate and population density in the entire Brazilian Amazon. The scattering of the values indicates clearly that this is not the scale at which the phenomenon can be studied, although there may be good evidence from previous work that population density acts as a driving force for deforestation in certain districts. In order to falsify a null-hypothesis that there is no relationship, data must be collected and reassembled at a different scale, possibly both more and less detailed scales. In other words: scale is the spatial and temporal 'observation window' through which we look at the real world. As a corollary, the process we want to study and its data resolution determine the appropriate scales.

Obviously, there is always a degree of heterogeneity that cannot be 'explained away' by looking at the correct scale. Firstly, this may be the result of measurement errors and natural random variation. Secondly, this heterogeneity exists because every individual in a population displays some unique features and the only way to remove that type of individual variability would be to limit the study to the individual, i.e. the most detailed scale only. However, if we want to explain rather than solely describe a phenomenon, processes at both higher and lower scales must be studied also. This is further discussed below.

The third, related reason why scales are important, is that many statements and statistics are of little value unless they are made in reference to specific scales. Yield – the penultimate variable in agricultural science – is a good example: it may refer to a single plant, a crop (population of plants), a field, a farm, a region, a country, a continent or the world. A recent extensive review of data sets ranging from global through regional to individual plant level show that coefficients of variation in yields change with scales of aggregation and length of aggregation, but not entirely in a systematic manner (De Steenhuijsen Piters 1995). In this review, coefficients of variation change from 128% (rainfed sorghum, Jodhpur region, India, 1954–1970) to 3% (world cereal yields, 1971–1982). Also, scales are relevant because we are interested in prediction and therefore in representativity and in sampling strategies. All of these are scale specific, since variability is a function of scale.

Furthermore, risk is a phenomenon with clear spatial and temporal scale dimensions. This is partly due to statistical aggregation, which generally leads to decreasing risk at less-detailed scales, but there are also real spatial scale effects. A theoretical sigmoidal relationship could be formulated of the spatial dimension of risk of crop loss due to frost or drought as a function of latitude (with drought as the operational factor in the lower latitudes (as in the Sudan-Sahel) and frost in the higher latitudes (in North America)). The sigmoidal curve indicates that at increasing distances from the equator risk increases disproportionately. To put it simply in the case of the Sudano-Sahel: a shift of 300 km to the north in Mali leads to greater crop losses than an identical shift closer to the equator, say in Liberia. Risk prediction and avoidance are thus both geographically and scale specific.

We may conclude that scale is a real and quantifiable phenomenon, not just a conceptual construct. Understanding scale effects is essential in agricultural research, even if they have been generally grossly ignored. Perhaps the most convincing case for 'scale-mindedness' occurs in the classical example of ecological sustainability. In its simplest sense, sustainability may be interpreted as no net loss of nutrients, i.e. the nutrients exported in the crop product must be replenished from elsewhere. In many instances of historical or modern ecological agriculture, 'elsewhere' implies transfers of manure or organic matter from outside the field where the crop is grown. By studying only the nutrient flows at the field level, our picture is distorted: replenishment of one site necessarily means a loss from another site, thus a higher spatial scale must be taken into consideration. Similarly, in the short run, agriculture on soils derived from young alluvial or volcanic material may be practised without compensation for the loss of nutrients. However, in the long run, i.e. at other temporal scales, replenishment of nutrients is essential if the rate of depletion is higher than the rate at which fresh nutrients become available (Fresco and Kroonenberg 1992). Based on this approach, it may be calculated how long the soil nutrient reserve may support a given land use (Smaling and Fresco 1993).

On the nature of agro-ecological scales

Because every plant, every vegetation, every land unit displays some unique features (and this applies, *a fortiori*, to farms and human communities), 'scaling up', in the sense of cumulating and generalising data, results in loss of detail. The degree to which this may be acceptable depends on the purpose of the study. True 'scaling up', however, concerns the identification of different processes that are specific to each scale. It is not just a cumulation of lower scale data or variables. For example, when studying stomatal behaviour, we are not interested in the movement of individual electrons and water molecules; we take temperature as a proxy at the relevant scale level. At the higher scale, i.e. of the whole plant, stomatal behaviour in itself becomes of less interest and only the gross results of transpiration and respiration are taken into account in crop growth models. The art is in relating phenomena to processes at both higher level and lower level scales. In other words, there is an optimal scale level at which each process can and must be studied. At this level, variability is minimal (but not necessarily zero). If we take aggregation in this sense, it implies that a process studied at a given level cannot just be an accumulation of lower level data, but requires specific data relating to that process. Scaling up can therefore not be carried out in a meaningful way if it only means accumulation without identification of the optimal scale.

Because ecological phenomena are hierarchically structured (from cell to tissue to individual to population etc.), the point then becomes to identify the relevant scale hierarchies for agro-ecosystems. There have been many attempts to formulate hierarchies for the agricultural sciences (e.g. Conway 1987), mostly encompassing both socioeconomic and biophysical units, although it may be necessary to identify separate hierarchies and related processes before trying to combine them across disciplines (Stomph *et al.* 1993). Ideally, we like to think of scales as nested: higher level units comprising lower level ones. As a corollary, exogenous factors (constraints) at one level become endogenous (limitations) at a higher level. Or, to put it differently, the question is: at what scale does a driver become a constant? Yet, nesting scales are easiest to identify with respect to biophysical units (i.e. crop or vegetation and pedon at the lowest level, field combinations and facets at an intermediate level, up tot landscape elements and entire landscapes at the highest level). These levels correspond to clearly identifiable units – in fact the distinction between the land related (pedon, facet) and the vegetation related lower level units is only logical in analytical terms, and the two are necessarily combined at the higher levels. In contrast, as soon as human factors are introduced (farmer, farm household, administrative units), the biophysical nestedness is interrupted: the farm household is not physically made up of cropping systems as lower level units, but of household members, as shown in figure 2. Furthermore, the boundaries of the socio-economic units at higher levels, which are a function of administrative and political distinctions, do

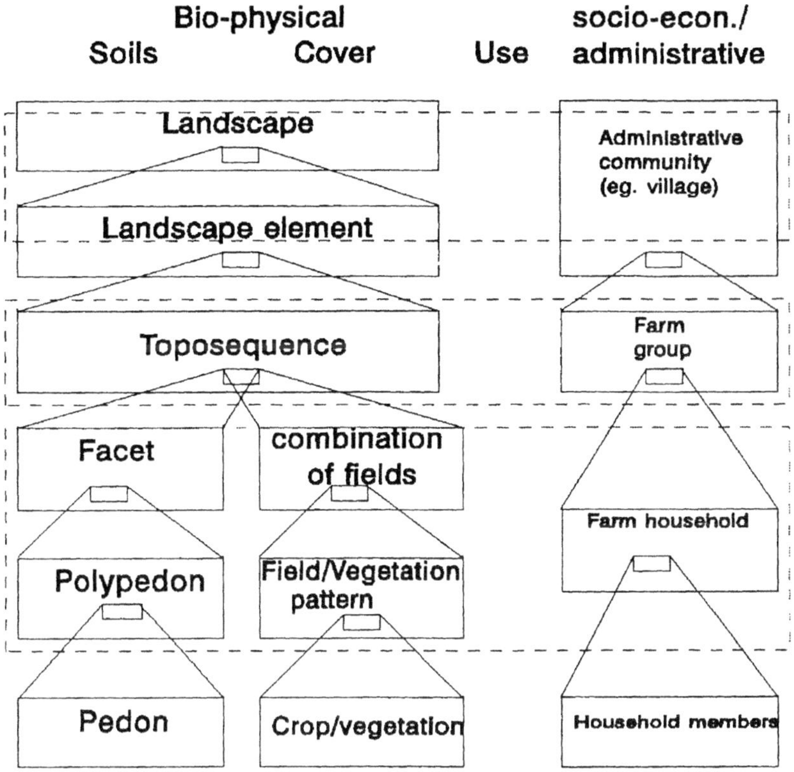

Fig. 2. Parallel hierarchies of biophysical (left side) and socio-economic (right side) systems relevant for land use. At each level, land use systems are identified as combinations within broken lines. (Source: Stomph *et al.* 1994)

not correspond with the boundaries of biophysical units at similar scales. The combined nestedness of human and biophysical scales does not correspond, because their basic units do not match. Needless to say, because we deal with agro-ecological systems, we cannot ignore these elements of human control and decision making that so interrupt our neat biophysical scales.

An additional issue is to what extent 'natural' scales exist in agro-ecology. While a single plant and a plant or animal population are identifiable with clear spatial and temporal boundaries and specific internal processes (respiration, for example), this becomes more difficult when dealing with the biological community and higher levels. For example, when dealing with crop pests, the boundary of the crop-cum-pest-cum-predator community is not just the field, but also the environment of the field. How far this environment extends, depends on the behaviour of the pests and predators, as well as on specific landscape features (natural boundaries like rivers or bare rocks) as well as on meso-climate. It is clear that the definition of the spatial and temporal domains of such a system is at least to some extent subjective and not entirely absolute or natural.

We may conclude that in order to operationalise the concept of eco-regionality we need to determine, for each type of agricultural system, its specific set of scales and therefore the relevant spatial and temporal units at which it must be studied. Furthermore, the scales must be linked which puts great demands on data collection. This also implies that the scales of agro-ecological processes are not necessarily identical, even within the same eco-region. If shifting cultivation as well as intensive lowland rice cropping occur in the same region, as is the case in Sri Lanka and other island of South East Asia, processes and spatio-temporal scales, for example, may differ considerably. For example, the residual effect of inorganic fertiliser on rice yields may be 1–2 years and a proportion of the nutrients may be transported through irrigation water to a few adjacent fields, whereas the restoration of soil fertility on swidden fields may require approximately 20 years and supposes the availability of an additional area at least 9 times the size of the field (if 2 years of cropping before 18 years of fallow is the rule). From the point of view of nutrient management, the spatial and temporal scales of shifting cultivation are larger than those of rice fields. However, if the rice fields form part of a mixed farm and are also fertilised with animal manure, the spatial scale of the process determining nutrient availability increases to include grazing areas for the animals. As a result, the input-output relations of nutrients vary according to the system, and nutrient use efficiency in the rice crop must be calculated with reference to the pertinent scale, as has been shown for nitrogen in different rice cropping systems (Fresco *et al.* 1994).

Concrete problems of scale in eco-regional studies

If international and national agricultural research aims to restructure itself on an eco-regional basis, solutions have to be found for a number of concrete problems where scale characteristics are particularly significant. A few of these practical applications are listed below.

Characterization studies in a given area immediately confronted us with two basic scale-related questions: what are relevant data at what scale and how do we select our samples to best characterise agro-ecosystems? The tendency, not least promoted by farming systems researchers, has been to strive towards exhaustive coverage of data for a very limited number of samples and an equally limited number of seasons. In other words, full descriptions of farm households and agricultural practices in two or three villages over two or three seasons. As a result, the data are difficult to compare and to interpret, especially after several years, and the entire exercise is generally very costly. So called 'quick and dirty' surveys while cheaper, only yield qualitative data. These issues may be resolved to a considerable degree if a hierarchical approach is taken to identify relevant variables and corresponding units. This could mean, for example, rapid characterization of land units at a coarse scale, followed by more in-depth work on soil profiles and geostatistical sampling at the cropping systems scale. In this approach, existing census data on agricultural production, reassembled by land units may be combined with more detailed soil and crop survey data, thus linking scales together.

A related problem lies in data extrapolation and in defining more or less uniform 'recommendation domains', i.e. in classifying environments with respect to agricultural technology (Harrington and Tripp 1984). This is not only a problem with respect to ecological processes, such as the well-known fact that results from Wischmeyer run-off plots cannot be extrapolated to determine sediment harvest and erosion at larger landscape units. The implication of the scale-dependency of processes influencing agricultural change is that a hierarchy of recommendation domains must be constructed. In other words, units may appear uniform only at a given scale, i.e. place and time, and for a given agricultural technology. For example, the recommendation domain for the introduction of virus resistant varieties of maize may encompass very large units, even subcontinents irrespective of relief, farm size or even climate, while the recommendation domain for mechanized weed control may be limited to large farms in flat alluvial areas with specific soil and economic characteristics. The construction of a scale hierarchy is also relevant to modelling: a general, non-scale-specific mass balance (input-output) model may be applied for initial identification of nutrient flows at the continental scale, while at national and regional level the model introduces several scales (Smaling 1993).

In order to improve scale-sensitivity in agro-ecological research we need to search for key processes that provide an entry into linking scales together. At the same time, we must identify possible 'hidden effects' of agricultural technology at other scale levels than the one at which technology aims, such as the effects on regional groundwater resources of pesticide application at plot level. This means that we must follow a multi-scale approach. An interesting area related to sustainability may be the investigation of agro-ecological trade-offs or substitutes and their effects. For example, zero tillage may have positive effects on erosion and run-off reduction, and be a better alternative than anti-erosion ridges, but at the same time the herbicides required to suppress weeds on zero-till plots affect groundwater and biological communities at various scales. Although the ecological processes of erosion and pollution seem to occur at similar landscape scales, the trade-off involve socio-economic factors at different scales opposing the individual farmer and regional land and water management authorities. To complicate matters even further, the substitution of anti-erosion ridges by zero-tillage entails also a temporal analysis, since substitutes may not be active on similar time scales. A similar analysis may also be made for the substitution of fertilizer by crop residue incorporation.

Finally, we need to improve the conceptual framework used in eco-regional research allowing to translate patterns at one scale to the next and use the understanding gained at one level to do so. An interesting area for future research would be, in my view, the functional similarities between mixed cropping at the plot scale and the patterns of diversity of land covers – forests and grass strips between fields – at the scale of a valley. Although truly scale-neutral models do not really exist yet, we might envisage the study of water and nutrient flows at both scales through similar models fed by scale-specific data. Such studies across scales may be helpful in identifying scale-specific features.

To conclude: I have argued that the concept of eco-regionality means scale-sensitivity as well as multi-scale modelling (in the broad sense of the word model). Its operationalisation requires that we define ecoregions at different scales. The inherent scale-sensitivity of eco-regionality implies that this concept can only be defined in a meaningful way with respect to the relevant temporal and spatial scales of the agro-ecosystems prevailing in the region. In other words, within the subcontinental ecoregions as defined by TAC/CGIAR, a hierarchy of ecoregions may be distinguished each with its specific processes influencing land use. Within each level of this ecoregional hierarchy, agro-ecosystems are not necessarily uniform, but display an acceptable degree of heterogeneity for the purpose of agricultural technology development and testing.

Acronyms

CGIAR	Consultative Group on International Agricultural Research
TAC	Technical Advisory Committee

References

Conway G R (1987) The properties of agroecosystems. Agricultural Systems 24:95–117.

Fresco L O, Kroonenberg S B (1992) Time and spatial scales in ecological sustainability. Pages 155–168 in Land Use Policy, July 1992.

Fresco L O, De Ridder N, Stomph T J (1994) Magnitudes of disturbance in the evolution of agro-ecosystems: N-flows in rice-based systems. Pages 137–149 in Struik, P C, Vredenberg W J, Renkema J A, Parlevliet J E (Eds.) Plant production on the threshold of a new century. Kluwer Academic Publishers, Dordrecht, The Netherlands.

Harrington L, Tripp R (1984) Recommendation domains: a framework for on-farm research. CIMMYT Economics program working paper 2/84. International Center for the Improvement of Maize and Wheat (CIMMYT), Mexico-City, Mexico.

Holling C S (1992) Cross-scale morphology, geometry, and dynamics of ecosystems. Ecological monographs vol 62, no 4. The Ecological Society of America.

Kolasa J, Pickett S T A (Eds.) (1991) Ecological heterogeneity. Ecological studies no. 86, Springer Verlag, New York, USA.

Smaling E (1994) An agro-ecological framework for integrated nutrient management, with special reference to Kenya. PhD thesis. Wageningen Agricultural University, Wageningen, The Netherlands.
Smaling E, Fresco L O (1993) A decision support model for monitoring nutrient balances under agricultural land use (NUTMON). Geoderma 60:235–256.
De Steenhuijsen-Piters B (1995) Diversity of fields and farmers: explaining yield variations in northern Cameroon. PhD thesis. Wageningen Agricultural University, Wageningen, The Netherlands.
Stomph T J, Fresco L O, Van Keulen H (1994) Land use system evaluation; concepts and methodology. Agricultural Systems 44:243–255.
Turner II B L, Moss R H, Skole D L (Eds.) (1993) Relating land use and global land cover change: a proposal for an IGBP-HDP Core Project. IGBP-HDP, Stockholm, Sweden.

Aggregating economic knowledge for use at national, regional, sector, and farm level

W.G. JANSSEN*

International Service for National Agricultural Research (ISNAR), P.O. Box 93375, 2509 AJ The Hague, The Netherlands

Key words: aggregation, decision hierarchy, disaggregation, economic analysis, methodology, micro-economic information, NRM policies

Abstract. Decisions concerning land use, agricultural production and environmental management are taken at many different levels, ranging from the national to the farm level. For proper decision making, policy makers at each level should dispose of relevant, reliable and consistent information. National or regional policy makers require information on what happens at farm or regional level. Can such information be supplied by aggregating farm level knowledge into sector, regional or national models? The author explores this question in three steps.

First the methodological issues in aggregation of knowledge are discussed. Aggregation problems for individual data, response functions and systems models are described and conditions for correct aggregation are formulated. Spatial and temporal aggregation is discussed and juxtaposed with the exploration and analysis of disaggregated data. Five frequently occurring aggregation biases are discussed.

Secondly the use of knowledge by policy makers is treated. Different levels of policy making and their specific requirements are distinguished. Problems of integrating farmers and policy makers perspectives are reviewed. Political and agro-ecological boundaries for aggregation purposes are compared and the friction between more detailed model specification and the user perspective is explored.

Finally six suggestions for improving the integration of policy making and (dis)aggregated analysis are made: early interactions between policy makers and analysts; more empirical studies on policy making; a shift from theory development to empirical application; the adoption of participatory action research elements in agro-ecological research; a focus on interventions instead of (eco)systems; and the integration of disciplines along the different hierarchical levels.

Introduction

Agricultural and natural resource management policies are defined at national, sector, regional and farm level. For proper decision making it is required that the information available at each level is reliable, relevant (Pinstrup-Andersen 1993) and consistent between the levels. Such information can be acquired

* The author has received useful suggestions for literature search and has benefitted from discussions with Sjoerd Duiker, Louise Fresco, Peter Goldsworthy, Larry Harrington, Peter Jones, Arie Kuyvenhoven, Jean François Merlet, Roger Norton and Steve Vosti. Their help is greatly appreciated. Some of the comments made by Roger Norton during the symposium have been incorporated in the final version.

J. Bouma et al. (eds.), Eco-regional approaches for sustainable land use and food production, 143–163.

from other decision levels, both higher and lower, or from the level itself at which a possible decision may be taken. The issue is to assemble the best information set for each different problem, taking into account the cost of assembling such information.

Information on natural resources and their management often comes available at the farm level. Crops, animal husbandry activities, rotations, input use, cultural practices etc. are defined at the farm level, and if policy decisions at national, regional or sector level are to be taken, it appears logical to base these decisions on actual knowledge obtained at the farm level. The characteristics and behavior of these individual farms now have to be aggregated in the variables that are relevant at those levels. For example, at the regional level it is more important to know that 2% of the forest area is yearly encroached by settlements, than that a certain farmer chops down one hectare of forest in his property per year to expand his cultivation. Yet, it is obvious that the two processes are related.

Aggregation thus appears obvious as a way for improving policy information and for integrating decision levels. To many outsiders this process appears quite straightforward. Surprisingly, aggregation and integration of analytical and decision levels is also considered rather lightheartedly in various research proposals. For example, Palm *et al.* (1993) link seven hierarchical levels in their proposal for a slash and burn research project; Kropff *et al.* (1994) link rice genome studies with continental rice production. Andriessen *et al.* (1994) define four characterization levels for agro-ecosystems in West Africa, but have not come around to implement the most detailed one. The conceptual application of aggregation and decision levels does not always seem to take account of the theoretical issues involved in aggregation concepts (Day 1963) on the one hand, or the realities of policy making on the other (Röling 1994).

In this paper I will try to review some methodological issues of aggregation and some issues concerning the use of aggregated data in policy making. I will conclude with suggestions on further and improved integration of policy making and (dis)aggregated data analysis. I will try to concentrate on socio-economic data, but in all honesty I am not very convinced of the relevance of the distinction of socio-economic and agro-ecological information. When reviewing studies such as the one done by Jones *et al.* (1991) for defining CIAT's strategic plan, the work of WRR (1992) on land use scenarios for Europe, of Gallopin (1992) on land use scenarios in Latin America, and of Veeneklaas *et al.* (1994) for Mali, I am delighted with the implicit integration of the disciplines and with the arrival of 'macro-agronomy' as an evolving field of science that competes with economics for the attention of policy makers.

Methodological issues in aggregating knowledge

Aggregating what?

Three types of knowledge, at increasing levels of analytical complexity, can be aggregated. These are:

1) Individual data, such as farm size or the area planted per crop.
2) Response functions, such as supply or demand equations.
3) Systems models, such as programming approximations of farm types.

The problems that arise in the first and second categories are also relevant in the second and the third. However in the higher categories additional problems arise. Here I will first discuss the methodological problems in aggregation. The question of relevance of the construct that results after aggregation will be dealt with later.

Individual data can be aggregated as long as they refer to strictly the same quality or function. For example, irrigation water for farmer A and irrigation water for farmer B can be safely added. Drinking water can only be added if the interest is only in the physical availability of water. If the interest is in the value of the water resources (as will often be the case in agro-ecological analysis), the data on irrigation water and drinking water have to be multiplied first with their price. Unless the physical quantities can be weighted by unchanging base prices, aggregates cannot be compared over regions or time (Nelson 1991). Many of the valuation and ideological conflicts in Natural Resource Management (NRM) policies stem from the different estimates about how such relative prices will change.

Aggregation of individual data depends on a consistent classification framework. The classification framework has to reflect the purpose of the analysis. Aggregation over classes depends on unambiguously defined and unchanging weights (often prices).*

Response functions can be added into meaningful aggregates if they have similar linear specifications (Theil 1954). Problems arise with non-linear functions, or when empirically estimated response functions would have different specifications. Lewbel (1994) shows that for distributed lag models (frequently used for supply analysis) the aggregate behavior does not mimic the behavior of the individual agent. He warns for the use of 'representative agent response functions'. Rather than working with 'representative functions', the response function should be re-estimated at the concerned level of analysis.

* Aggregation of individual data will not be given significant attention in this paper. The discipline of statistics has dealt with the subject in an extensive manner, for example in sampling theory. Within economics, index number theory is well developed (e.g. Hicks 1946).

Requirements for aggregation of response functions are rather stringent. In order to meet these requirements response functions may be defined in very simple terms (e.g. demand as a linear function of only price and income), omitting other variables that would not be significant in every individual demand equation (e.g. prices of substitutes). The value of aggregating individual data is that it allows to distinguish detail, but part of that detail gets lost anyway: the higher the aggregation level, the more simplistic the representation of the individual actors that can be accommodated in the analysis (Tripp *et al.* 1993). The choice of specification greatly defines the ability to aggregate, thereby biasing the analysis from the empirically best specification to the one that can be manipulated best.

Systems models are even more difficult to aggregate. Theil (1954) treats the problem of simultaneous systems of response equations (e.g. a market equilibrium with demand and supply equations). To aggregate such simultaneous systems, he proposes to first derive the reduced (micro) form from the structural (micro) form, then to aggregate the reduced (micro) forms to a reduced (macro) form and finally to extract the structural macro form. The problem is in the last step of extracting macro structural equations from the macro reduced equations, where identification problems occur (the structural equations cannot be derived any longer from the reduced form equations in a unique manner).

Linear programming models (or their more elaborate derivatives) have often been the subject of aggregation (Hazell and Norton 1986). Three basic conditions should be met to allow mathematically correct aggregation of programming models (Day 1963):

1) Technological homogeneity. Enterprises should have the same production possibilities, the same type of resources and constraints, the same levels of technology and the same managerial ability.
2) 'Pecunious' proportionality. The expectations about returns per activity should be proportional to the average expectations.
3) Constraints proportionality. The constraints vector for each individual enterprise should be proportional to the average constraint vector.

These conditions are very difficult to meet. Schipper *et al.* (1994) and Wossink (1993) apply clustering techniques to try to satisfy the requirements. However, rather than complying with the previous conditions, such clustering procedures minimize the deviation from the ideal conditions. It would be highly pretentious to assume that such stringent conditions can be met in empirical situations.

Aggregating system models can only be done at a certain cost in terms of representativeness and detail. The challenge for the analyst is not to eliminate all aggregation bias, but to assure that the bias present in the model does not affect the principal purpose of the analysis. Again this implies that aggregate (farm) models can only be built once the purpose of analysis has been defined.

Aggregation dimensions

Spatial aggregation. Agro-ecological approaches have been principally concerned with spatial aggregation issues. Normally this implies that data from the farm level are integrated in (sub)regional models. Ideally these (sub)regional models should afterwards be integrated in national models. Such a spatial model of integration allows for good insight in the resource management modifications that take place. Land use and water use changes can be neatly identified and often localized (e.g., by integrating a GIS system; Alfaro *et al.* 1994). However, this aggregation approach is often quite deterministic, i.e. the modeled behavior is a function of the environment, but does not influence it.

A traditional aggregation approach, currently receiving less attention is by commodity markets. Cropping systems can be integrated in supply models, these can be confronted with demand models, and the results of the simulated or real market confrontation can be translated for the different producer and consumer categories (Lynam and Janssen 1992). Such commodity models are better able to capture the incentive structure of farmers and the recursive nature of the economy than agro-ecological models. They may also correspond better with the available policy instruments (e.g. prices of outputs and inputs, credit policies, extension, public storage systems, commodity research, land tenure, infrastructural improvements).

The spatial integration approach is the proper answer to the concerns about sustainability and natural resource management, by allowing the modelling of resource modifications.* More efforts are urgently needed to integrate markets in the existing regional concepts. Some work has been done. Hazel and Norton (1986) have done an excellent job on integrating market mechanisms in aggregated sector models, but the approach remains laborious. Wood and Pardey (1994) are trying to adapt commodity market models to agro-ecological classifications. At CIAT work has been started on the interaction of market characteristics and sustainable farming (Castaño 1994). Prices and sales prospects remain key incentives for farmers and the integration of market systems and agro-ecological systems remains a principal challenge. If agro-ecological models can include production as well as consumption modules, this would greatly enhance their utility.

* Spatial approaches are well suited to identify externalities of resource use. These externalities have to be integrated in the agro-ecological analysis to obtain a correct estimate of costs and benefits of resource use. This can be considered as another dimension of the aggregation problem, also because the externalities may be felt and measured at other hierarchical levels. Externalities have been extensively discussed in the NRM-literature (Crosson and Anderson 1993). In this paper I have not given further attention to externalities.

Aggregation over time. The definition of the proper time frame of analysis and the aggregation of results over the years is a first issue of interest. Agro-ecological approaches are often concerned with slow and long lasting processes of change. For proper economic analysis choice of the time horizon is vital, and associated with the time horizon the question of the discount rate comes up (Arndt 1993). Time horizons will normally be longer than in traditional agricultural research and the choice of discount rate strongly influences the outcomes of the analysis. Pierce *et al.* (1990) argue that the choice of discount rate is not the real difficulty, but the identification of all possible future benefits and costs.

Two suggestions can be made for dealing with the longer time horizons and the subsequent uncertainty: The first approach is to organize an expert and stakeholder consultation on the expected changes. Initially an exhaustive list of the possible effects of the policy changes should be developed. For each effect the expected benefits and costs should be assessed, preferably by applying empirical knowledge and existing models from other situations. For the benefits and costs that cannot be treated in this manner a subjective, consultative process aimed at obtaining a consensus on expected benefits and costs should be started. Then it should be considered how the subjectively obtained benefits and costs can be integrated in the existing models and how improved measurement methods can be developed. The approach is somewhat opportunistic but also relatively fast, and will lead to iterative improvements.

The second approach is to refocus traditional benefit-cost analysis. Benefit-cost analysis is normally based on balancing stock values (such as building up assets) with flow values (recurrent benefits and costs). Agricultural economics has long tried to improve the estimates of recurrent benefits and costs, but should maybe focus more on estimating stock values (e.g., by contingent measurement approaches; Crosson and Anderson 1993). For example, if the price difference of degraded and rehabilitated land is known, it is not necessary to predict yield and net revenue differences over time. By focusing more on stock values, the time horizon of analysis can be shortened.

In certain conditions aggregation of benefits over time in Net Present Values or Internal Rate of Returns may disguise cash flow problems that occur in the initial years of a project. Project may have a very acceptable rate of return, but lack of income in the first years may limit the interest of participate. Implementation considerations would then suggest to maintain a disaggregated analysis over time, simultaneously to the calculation of project returns.

An additional issue which is often overlooked is the aggregation of sequential farmer decisions in technological packages (Rae 1977). Programming models normally assume that farmers take their decisions for the year on one moment, based on average expectations. However many decisions (e.g., the decision for fertilizing at flowering for legumes) depend on earlier events (proper rainfall). With sequential decision making farm plans change considerably from unconditional decision making, but the cost in terms of calculation time and methodological complexity are high (Antle 1983). In the LEFSA (Land Evaluation and Farming Systems Analysis for Land Use Planning) approach (Fresco *et al.* 1992; Stomph *et al.* 1994) sequenced operations are identified, but the link with sequenced decisions has not been operationalized.

Integration of time and space. After the previous discussion the methodological complexity of this type of integration should be obvious. In fact, I will not elaborate on it, except for stating that this integration is so complex that it may not even be possible to evaluate how good it is in an objective sense. Quality of aggregation can only be defined towards a certain objective, and in order to do so the analyst has to define specifically the geographic boundaries and the relevant time interval for the problem at hand (Fresco and Kroonenberg 1992). Aggregation over many hierarchical levels with different time and space horizons appears unfeasible. Levels of analysis should be selected with care based on the policy problem at hand. Where (dis)aggregation is pursued, awareness and recognition of the bias that may be produced should be explicit.

Exploring disaggregated information

Agro-ecological analysis has often been concerned with identifying homogenous zones, where possible policy interventions (research, pricing, legislation) would have a similar effect. Within these homogenous zones, the response of individual representatives is studied and the outcomes of these studies are then generalized and aggregated to the level of the homogeneous zones and beyond for prediction purposes. The aggregation process itself is difficult, as explained before. It can also be questioned whether our understanding of how farmers and other actors react to certain incentives is sufficient and whether we can aggregate at all. After all, assumptions on the behavior of farmers in aggregated programming models is normally not based on empirical observations of behavior, but on a normative technique, normally specified to maximize net revenues (or a complex approximation) under a more or less binding set of constraints. There is increasing evidence that farmer behavior is often not driven by such profitability considerations (Van der Ploeg 1990)

For prediction purposes and for supporting policy decisions, the aggregation question cannot and should not be avoided. At the same time we should use the available information for cross sectional analysis, in order to understand better how people react and why (Hazel 1993). Cross-sectional analysis in space may even improve our understanding of the dynamics of land use over time, as shown by Carter and Jones (1990) for cassava cultivation in

Africa. Analysis across areas with different policy regimes may also improve our insights in the actual effect of certain interventions (Carter *et al.* 1994). Though often neglected, cross-sectional analysis of disaggregated empirical data is the logical partner for aggregate predictive research.

Aggregation bias

Imperfection in aggregation cannot be avoided and should be accepted as a fact of life. Possible aggregation imperfections should be anticipated, specially when using programming models for policy analysis. To conclude the review of methodological issues in aggregation, I have listed what I consider five possible biases that appear in model aggregation. Some of these biases (specifically the first and the fifth) can be linked with the aggregation conditions of Day (1963). Some others are mainly linked with the recursive nature of economic behavior (the second and third), whereas one bias (the fourth) is principally concerned with researcher subjectivity.

1) Mathematical aggregation bias. Conclusions regarding resource utilization in aggregated programming models may be more favorable than at the disaggregated level, because at the aggregated level resource combinations can be realized that are not accessible to the individual farms. In theoretical terms, the third criterion of Day (1963), as discussed before, has not been met in such a situation. Results at the aggregate level will always be more favorable than at the individual level (Hazel and Norton 1986). Rabbinge and Van Ittersum (1994) present an example where the resource utilization at the aggregate level is less favorable than at the disaggregated level, but this is caused by a violation of the first criterion of Day (homogenous technologies).

2) Interaction bias. Certain parameters that are exogenous at the lower levels, become endogenous at higher hierarchical levels. Market prices, migration trends and labor availability are examples. If these interactions are not properly included, model results at lower levels tend to be too extreme. To decide at which level a certain parameter becomes endogenous, the size of the parameter in the model should be compared with the size of the parameter in the price defining market system. As a rule of thumb, below 5% participation, price endogeneity should not be an issue. Between 5 and 15%, endogeneity may become relevant, dependent on market functioning and market structure. Above 15% endogeneity should be considered. Caution with the application of such a rule of thumb is warranted however. The objectives of one type of analysis may require more precision than those of another.

3) Control bias. Programming models at the farm level normally have a normative perspective. They optimize resource utilization subject to an objective function and a set of constraints. Aggregating such models to the regional level allows to define the optimum resource utilization mix for the region, but, apart from running into interaction problems, it may not provide many clues about how to reach this mix. The control over farm resources that is presumed by the use of a programming model is absent. At the regional level the policy question is not only how to allocate regional resources, but also how to steer resource users towards such an optimum mix. Econometricly estimated supply-demand models may be more suitable for such a purpose, but this suggests again respecification of the model, rather than aggregation.

The opposite problem may also play. In farm models, certain government controlled prices may have been kept constant, since they lay outside the influence of the individual farmer. At the regional level however, such prices may be changed by the government. Decisions that are exogenous at one level are endogenous at the next, and the other way around.

4) Resolution bias. Model specification at the aggregated level will tend to concentrate on the information that comes available from lower level. When detailed land suitability work has been done, the aggregate model will often emphasize land suitability as a defining variable for land use, whereas credit availability may have been the more vital one. Models at the higher hierarchical levels may be derivatives of lower hierarchy models, rather than proper representations of the problem at hand. Also, in the process of aggregation of systems models (e.g., on land use), one may forget about non-modeled behavior. Regional land use may be evaluated by aggregating different farm models, but the most significant impact may well come from urbanization. Resolution bias is particularly dangerous for improved natural resource management, since it implies the reduction of resource use from various types of users to only one (often farmers). By omitting the implications to other utilizations of the resources, policy conclusions from such models may become one sided and politically hard to swallow.

5) Simplification bias. Aggregation requires comparable model specifications at the lower hierarchical levels, which normally involves a certain simplification. To satisfy the technological homogeneity constraint of Day, non-homogeneous activities may be excluded or treated in a highly simplified manner. These models will overestimate the effect of policy variables that are included in the models, at the cost of those excluded. For example, Janssen (1986) found an average price elasticity of supply for cassava of 2.0 using quadratic programming models for different farm sizes, but only 1.3 when using an econometric specification.

Aggregated knowledge and policy makers

Levels of policy making

From an economic perspective, four levels of policy making will be distinguished. These are the national or macro level; the regional level and the agricultural sector level (both meso); and the farm level (micro). Other levels could be distinguished: the municipality or subregional level; the commodity level; the international level. For reasons of conciseness, I have left these out, even though they are very relevant to understand policy implementation issues (municipality level) or legislative arrangements concerning trade and environmental protection (international level).

Farmers have been included as a (micro and private) policy agent, because in fact they are the ones that control natural resources to a large extent. However, where as farmers have a private rationale, the other policy makers follow a public rationale. Aggregation from farm level to the regional or sector level thus provides an initial problem in the shift of rationale. Aggregated programming models provide optimum solutions to private problems, but not necessarily to public problems. The public policy maker will certainly learn from knowing these private optimum conditions, but in order to define the best policy interventions he will need to evaluate how the private optimum changes in response to changing policy variables.

We thus face a simultaneous optimization problem at two levels. While at the farm level a private optimum is pursued, at the policy level a public optimum is pursued. The private optimum can be easily modeled either in implicit or explicit terms, but the public optimum is normally obtained by comparing alternative policy scenarios and choosing the best one. The role of the policy analyst is to help define the possible range of policy options and to show the implications of each set of options (Norton 1995).

Policy decisions at different levels

As must be clear now, policy makers at these four levels are concerned with different problems.* Also, these policy makers dispose of different policy instruments for realizing their objectives, they need to deal with different actors and they can use different analytical tools. Table 1 provides a very brief summary description (by no means complete) of some of the typical

* Policy consistency between different hierarchical levels is another concern for decision makers (Norton 1995), though somewhat outside the scope of this paper: If the extension service in a certain region is stimulating orange trees as a means to control erosion, charging an export levy on the exports of oranges may not be very helpful. My impression is that policy consistency for natural resource management is often poorly considered. NRM policies often rely heavily on regulation, while presently the overall trend in most economies is towards liberalization.

Table 1. A summary of policy making processes at different hierarchical levels.

Policy level	Policy goals	Policy instruments	Policy actors	Analytical tools
Country (macro)	economic growth employment equity balance of payment public spending ecological stability economic stability	interest rate money supply exchange rate minimum wage tax system budget deficit	cabinet min. planning trade unions business community finance community	macro-economic models input-output tables national accounts international benchmarking
Agricultural sector (meso)	farm income food prices food self-reliance competitiveness public spending	output prices input prices credit research extension market regulation market structuring	min. agriculture farm organizations agro-industries	partial equilibrium models general equilibrium models demand systems domestic resource costs
Region (meso)	resource quality living climate natural conservation accessibility employment	development plans (including infrastructure pollution norms nature reserves user rights (water) settlement incentives	regional government resource management authorities	spatial equilibrium GIS environmental impact regional programming
Farm (micro)	income leisure risk reduction income spread subsistence	production plan labor allocation sales strategies land allocation capital availability investment plans	farmers farmer's families	farm programming cash flow analysis budgets port-folio analysis decision support rules

objectives, instruments, actors and analytical tools at each level of policy making.

At the macro level and at the agricultural sector level, incentive management (interest rate, exchange rate, input and output prices) is a principal policy approach. At the regional level, there is some potential for incentive management (e.g., settlement subsidies), but the onus is with regulation and structural policy.* Policy instruments are complementary. If the analytical tools for the different policy levels are tailor made (as follows from earlier conclusions in this paper), the potential for integrating and aggregating from the regional policy level to the national policy level may not be very high.

Policy decisions made at higher levels in the hierarchy tend to influence conditions at lower levels of the hierarchy, but the opposite is much less true. For understanding farmer behavior, interest rates, taxation systems, foreign exchange regimes and money availability may be relevant, but for understanding (and defining) macro-economic behavior it will be less important to know agricultural employment, commodity prices or the effectiveness of the research system. These latter variables will have been reconciled with

* There is a growing debate on the need of incentive management at the micro level. A good example is the pricing of irrigation water. Successful incentive management requires control over the supply or demand of the resources or goods for which the prices are being managed. For an irrigation scheme, such control may be confined to a watershed, or for pumping to an isolated irrigation scheme. In most cases (specially for agricultural commodities) incentive management, even at the local level, will imply the management of prices at the national level. Incentive management will often favor the development of commodity market models, as opposed to regional land use models.

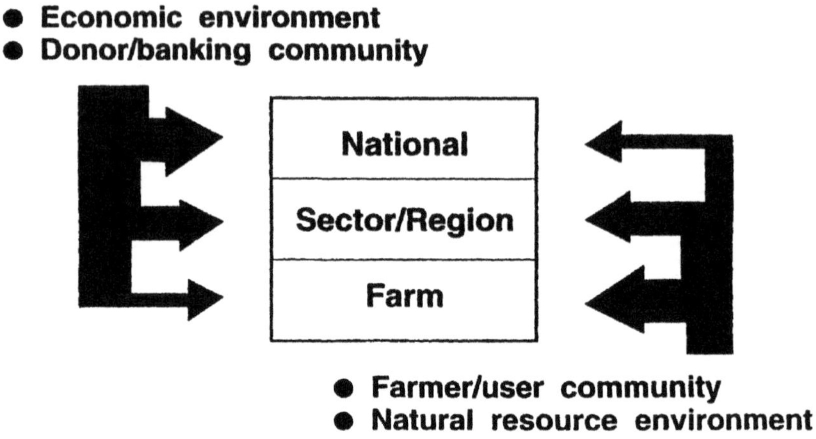

Fig. 1. Information use for (agricultural) policy management.

many others (such as industrial, services and public sector employment) at the macro-economic level. Figure 1 provides a graphical representation of this notion.

Policy decisions and the need for (dis)aggregated knowledge

For the policy maker a principal question is to define what detail he requires to make the right decisions. Of the many variables that may be affected by the decisions, typically only a few will be singled out for more detailed analysis. These variables may not be agro-ecological at all, for example if the problem concerns institutional reform. Choosing the right variables for in depth study, and the right amount of detail is key. This would thus suggest a redefinition of the problem on how to manage knowledge at different hierarchical levels. The question is not how to aggregate information, but where to disaggregate from the sector or macro level down to regions, commodities, farm size categories or other variables.

The analytical tools that are being used at the different levels may be related in a mathematical sense (programming models and simultaneous systems may be used at all levels), but are specified in completely different manners. In macro-economic models, the agricultural sector may be reduced to a few coefficients of an input-output matrix, and regions may not be distinguished at all. Knowledge of detailed information will be useful, but the methodological requirements at the macro level are so different from the sector or regional level that 'condensation' may be the approach for digesting detailed knowledge rather than aggregation. 'Condensation' starts with identifying the information requirements of the analytical approach, rather than with reviewing what information is available.*

Policy decisions at different levels do not resemble hierarchical aggregates. At the national level, the decisions are not only bigger but, most of all, different. Our capacity to deal with these different decisions in one integrated model is lacking and policy advice will continue to depend on an analytical patchwork approach, where models are defined for specific problems and linked at an ad hoc basis. My perception is that the aggregation of analytical policy support models (e.g., regional development) into macro policy support models is often unfeasible, but also irrelevant because of the different objectives that they have to serve. Rather the knowledge requirements are defined by the policy question at hand. These knowledge requirements will then define the model specification at each possible level of aggregation. The principal challenge is model (re)specification, instead of knowledge aggregation.

The distinction of hierarchical policy levels is more relevant for disaggregating the policy context and for defining at which level certain decisions are made than for providing aggregation pathways. A good understanding of the hierarchical context of policy making clarifies the decision autonomy, available at each level and helps to focus policy support models on the variables that can be defined there.

Stepwise versus integrated decision making

Though for a slightly different field of policy making (research priority setting), Lynam (1994a) has proposed the development of an integrated decision making system for the CGIAR. Such a system would be built on a uniform database and a uniform methodology that would allow to evaluate all the different activities that the CG could be involved with. Though decision making in the CG-system is far less complex than in any country, developing such a system may well be a Sisyphus effort.** The aggregation problems that appear in the development of such a system have already been discussed, but are aggravated by problems of managing it and of the decision makers comprehension.

Managing such an integrated decision support system requires analytical skills, which cannot be found in a single person. Even if a functioning group for managing such a system can be developed, the utility of global uniform knowledge collection will be questionable. Decisions are normally not driven by the desires of the decision makers alone, but are urged upon them by changing conditions elsewhere (the donor community for the CG; international markets or employment trends for countries). Decisions may need to be taken under time pressure and may require only a minor part of the base

* In this respect it helps to address the resolution bias that was discussed earlier.

** Figure from Greek mythology who filled his days rolling a large stone up a hill. When he would almost be on top, he would loose control over the stone, and would have to start again.

information. Using an integrated system may be more expensive and slower than required.

It is also questionable if the decision maker at the higher level, even if he is completely sincere and good willing, wants to be concerned with the actual developments at the lower levels. He may not be familiar with the issues at hand at the lower levels. Moreover, he may explicitly wish to leave these decisions to other people, both because he is aware of his own limitations and because by delegation he improves motivation and thereby quality of the decisions.

Policy making is normally a stepwise process (Janssen 1993). There is a challenge to integrate these steps, but this may concern a procedural integration rather than a knowledge or analytical integration (KARI 1994). Using consistent data across levels (and consistently aggregated data) will improve such integration, but using uniform analytical models may not.

Policy making and aggregation boundaries

The definition of homogeneous agro-ecological zones has been a topic of substantial interest (e.g. Jones *et al.* 1992) and has been the basis for many agro-ecological studies. The logic would be to use these homogeneous zones also for aggregation purposes and for developing policy support models (e.g. Schipper *et al.* 1994). The problem with this approach is that the agro-ecological homogeneity does not always coincide with political divisions. Agro-ecological zones may extend themselves across national or provincial borders, or may be confined to part of a province or department. Agro-ecological zones are a construct of agricultural and ecological scientists, but not necessarily of policy makers.

In case agro-ecological zones extend beyond borders the question of socio-economic and policy homogeneity comes up. Pricing regimes may be different, credit availability may change, extension organized in a different manner. For policy interactions, subdivisions of the agro-ecological zone will be required. Where the agro-ecological zone is confined to part of a province, policy makers will want to assess the effects of policy changes for the whole of the province and not only for the agro-ecological zone.

Agro-ecological zones do not necessarily coincide with provinces or other administrative units, thereby reducing the relevance of the concept to regional policy makers. For national policy makers aggregation by agro-ecological zone may even be less relevant. As observed by Norton (1995), national policy makers tend to distinguish farms by land tenure categories or by the type of production and marketing system. Farm size, production systems and market access should be included in the set of criteria that lead to the definition of agro-ecological zones.

Overlaying socio-economic boundaries on the agro-ecological boundaries seems the obvious solution for including additional variables, but this may lead to excessive fragmentation. Carter (1988) designed a classification for cassava production in the Atlantic Coast of Colombia that was homogeneous for agro-ecological and socio-economic conditions, and found more than 20 (suitably called) micro regions for a production area of less than 100,000 ha.

The suggestion would be to define homogeneity in the expected response to policy interventions, rather than in current agro-ecological or socio-economic variables. The challenge then becomes how to define such homogeneous policy responses. Norton (1995) suggests that relative factor endowments (e.g. ratios of land to labor, ratio of flat to hilly land, irrigation water to land) form important response variables to policy change. I would not have other a priori suggestions, except that defining these response variables will require intensive interaction between policy makers and scientists at the start of the project.

The silence of the user

Whereas data and lower level models may still be aggregated into meaningful regional models, how are the opinions and interests of resource users integrated? Fresco (1994) concludes that the integration of users in land use planning is notably poor. Of course, user concerns and objectives may have been expressed in the analytical models, but whether this has happened to an agreeable extent is not always clear. Space has often several users, with different requirements (e.g., common property land; or wheat production plots with stubble grazing by nomadic herds; or regions with a recreation, wildlife and agricultural production function). If policy decisions at the regional level are taken, how are these different users considered?

Scenario building aims to incorporate the interests of different users in a range of possible future outlooks. Scenarios are normally expressed in terms that are fairly abstract for the average users, and it is questionable whether they can recognize their interests. Also, not the outcomes of one or another set of objectives are interesting but the trade-offs between them. Few policy makers will fare on models only in this highly political arena. Models may provide ideas about the trade-off, but not about the strength with which different user groups can pursue their objectives (Monke and Pearson 1989). From a policy advice perspective the question is to invest in further detail of the analytical structure of policy advice models, or in consultations with users. Substantial benefits may be obtained by making the entire policy formulation process more participatory and by moving beyond a mere dialogue between policy analysts and policy makers. Benefits of participation can be categorized in three groups: improved policies by considering more relevant aspects; improved chance of approval of the policy; and improved chances of implementation (Norton 1995).

Decision support models need to be timely available and need to be accessible to the policy maker. Aggregating farm level models into regional or sector models may improve the quality of information, but also extends the time required for model development and reduces the accessibility of the model. When certain decisions come up routinely, the integration of disaggregated knowledge in model development may be justified by the frequent use of the model afterwards. However, if a decision model is developed in response to a specific policy question, timeliness may constrain the amount of detail in the model.

Policy decisions are often needed to react to a changing environment and the quality of the decision depends also on the timing. For example, with the successful conclusion of the Uruguay round of the GATT, new land use options may become available in certain countries. However, if a certain country takes too long to evaluate these options, it may find its possible niche in the world market being absorbed by other countries. Agro-ecological research has been characterized by a holistic systems perspective. While recommendable from an analytical point of view, we should recognize that the approach is slow and that the time required to arrive at conclusions about possible interventions is long. A compromise between holistic quality and speed, which is acceptable to both the analyst and the policy maker, should be found.

Improving the integration of policy making and (dis)aggregated analysis

As discussed in the methodological section, aggregation of knowledge is certainly no simple mechanic process. Aggregation is only feasible under rather strict conditions that often will not be met. At the same time, decision makers at higher hierarchical levels are not concerned with the same problems as at the lower levels. They do not need more aggregated knowledge, but other knowledge. And their decision making perspective (e.g. for a province or department) may not concur with the methodological requirements for proper aggregation (homogeneity, as defined for example in agro-ecological zones). What is the potential under these circumstances to contribute to policy making with information on micro-economic behavior?

The potential is enormous, certainly with the current developments in information technology. Nevertheless, the principal progress to be made is not in the improved manipulation of micro-economic data. Realization of the potential hinges most of all on a better understanding of the policy context, regarding the type of decisions that are relevant at macro, meso and micro level.

1) Early interaction of policy makers and researchers. Sometimes there is too little concern about whether the data collected are those that are actually needed for policy analysis. Data collection may respond to the desire of the researcher to obtain a 'best representation'. Logic would suggest however, that the research process begins with the need for information, then moves to the development of an analytical framework, and once that is achieved, collect and analyze data (Pinstrup-Andersen, 1993). Policy makers should thus have a principal say in the analytical framework and the question thus becomes how to integrate their perspective. Contract research is one way of obtaining the perspective; personnel exchange between ministries and universities another; to include user and policy consultation as an obligatory step in the development of research proposal may be a third suggestion. For further suggestions on how to improve the problem focus of research, see Janssen and Goldsworthy (1994).

Once the analytical framework is defined, it can be decided at what level of (dis)aggregation information can be collected. The purpose of the analysis drives data collection and decisions on aggregation, rather than that the availability of information drives the desire to aggregate towards other levels of decision making. The type of data to be collected may be agroecological in nature, but may also concern detailed micro level data on savings, investment behavior, marketing waste, farm tenure etc., depending on the type of problem.

2) Empirical studies on policy making in Natural Resource Management. For many reasons interactions with policy makers at the initiation of the study may not be as intensive as desired. In order to anticipate the issues that may come up, the actual policy making in other cases may be analyzed. A good understanding of the policy making process in NRM may identify where the knowledge base is weak and where information from lower hierarchical levels may improve decision making. Studying comparable cases may also provide conclusions that are directly relevant, without requiring much further analysis for the case in question. Where information from other cases is absent, the question may be asked if an analytical approach through modeling is sufficient, or if a pilot project could be developed (Harrington and Ashby, forthcoming). Pilot projects can be particularly useful to evaluate the non-analytical issues of policy making such as the bureaucratic and political implications (Grindle and Thomas 1991). For such pilot projects, replicability should be a principal guideline in their design.

3) More empirical work, less agro-ecological theory. Progress in the economics of agricultural resource management will depend as much on solid empirical work as on theoretical development, and maybe even more so (Lynam 1994b). Conceptual thinking about the hierarchical systems structure of agriculture has taught us about the complexities and the interdependencies but has paralyzed our ability to conclude. Rather than considering the many hierarchical levels and the many interactions, it may be useful to have a modest start: Identify concrete problems, and review where the data are available to study them. Limit the analysis to the hierarchical level of the problem, and the next level above and below (Rabbinge and van Ittersum 1994).

4) Integrate participatory action research elements in agro-ecological analysis. With the recognized complexities of agro-ecosystems it is very difficult to identify objective criteria, such as statistical significance, for the quality of research. Also, the quality of research depends very much on whether the models are actually being used for policy advice. The search for refinement and for further use of micro information in model development will not be controlled by objective validation standards (Dent and Blackie 1979). Rather validation standards are user defined, in this case by the policy maker that the models are targeted to (Whyte 1991). A set of criteria that may be used to evaluate agro-ecological models would include: transparency; participation; simplicity; theoretical logic; robustness of outcomes; potential to discriminate effects of policy interventions; cost of development and application.

5) Study interventions, not just systems. The interest of policy makers is not so much with the best state of an agro-ecological system, but with the effect of the type of interventions they have in mind. By building the analysis around the proposed interventions, applicability of the results will be enhanced and the needs for data collection can be more clearly defined. The number of hierarchical links that should be included can be diminished and the attention can be focused on those that are directly affected by the proposed interventions. Such emphasis on interventions should be embedded in a systems framework, but a large part of this systems framework may be left in qualitative, descriptive terms and may not require extensive data collection and quantitative analysis. Such a 'reduced systems perspective' will obviously produce less comprehensive results than a full systems perspective, but the chance that results will be produced and used in a timely manner is much bigger.

6) Pass on Sisyphus' stone. The analytical tools that are being used at the macro, meso, and micro level are quite different and the questions that have to be answered with them also. Earlier in this paper I observed that linking between these levels may be better called "condensation" than aggregation. Defining rules for such condensation is difficult, partly because data requirements are so problem and situation specific. Not many economists or researchers of other disciplines will be able to grasp the specific issues of more than two levels. The proper integration of different hierarchical levels thus requires people (or even teams and institutes) with different backgrounds. Division of labor along the hierarchical levels is essential. Otherwise the stone will never reach the top of the hill and Sisyphus will be left all alone, probably at the foot of the hill, may be in the middle, but certainly not on top.

Acronyms

CIAT	Centro Internacional de Agricultura Tropical (International Center for Tropical Agriculture)
GATT	General Agreement on Trade and Tariffs
LEFSA	Land Evaluation and Farming Systems Analysis
NRM	Natural Resource Management

References

Alfaro R, Bouma J, Fresco L O, Jansen D M, Kroonenberg, S B, Van Leeuwen A C J, Schipper R A, Sevenhuysen R J, Stoorvogel, J J, Watsen, V (1994) Sustainable land use planning in Costa Rica: a methodological case study on farm and regional level. Pages 183–202 in Fresco L O, Stroosnijder L, Bouma J, Van Keulen H (Eds.) The future of the land: mobilizing and integrating knowledge for land use options. John Wiley & Sons, New York, USA.

Andriessen W, Fresco L O, Van Duivenbooden N, Windmeijer P N (1994) Multi-scale characterization of inland valley agro-ecosystems in West Africa. Netherlands Journal of Agricultural Science 42:159–179.

Antle J M (1983) Sequential decision making in production models. American Journal of Agricultural Economics 65:282–290.

Arndt H W (1993) Sustainable development and the discount rate. Economic Development and Cultural Change 1993:651–661.

Carter S E (1988) The design of crop specific micro regions and their contribution to agricultural research and rural development initiatives in South America: the case of cassava. PhD dissertation. University of Newcastle upon Tyne, UK.

Carter S E, Jones P (1990) A model of cassava distribution in Africa. In Janssen W. (Ed.) Trends in CIAT commodities 1990. CIAT working document no. 74. International Centre for Tropical Agriculture, Cali, Colombia.

Carter S E, Bradley, P N, Franzel, S, Lynam J K (1994) Dealing with spatial variation in research for natural resource management. mimeo, TSBF-Programme, Nairobi, Kenya.

Castaño J (1994) Market system effects on the sustainability of agricultural systems in Andean hillsides. PhD research proposal, Dept. of Marketing and Market Research, Wageningen Agricultural University, Wageningen, The Netherlands.

Crosson P, Anderson J (1993) Concerns for sustainability. ISNAR Research report no. 4. International Service for National Agricultural Research, The Hague, The Netherlands.

Day R H (1963) On aggregating linear programming models of production. Journal of Farm Economics 45:797–813.

Dent J B, Blackie M J (1979) Systems simulation in agriculture. Applied Science Publishers Ltd., London, UK.

Fresco L O (1994) Planning for the people and the land of the future. Pages 395–398 in Fresco L O, Stroosnijder L, Bouma J, Van Keulen H (Eds.) The future of the land: mobilizing and integrating knowledge for land use options. John Wiley & Sons, New York, USA.

Fresco L O, Huizing H, Van Keulen H, Luning H A, Schipper R A (1992) Land evaluation and farming systems analysis for land use planning. FAO working document. Food and Agriculture Organization of the United Nations, Rome, Italy.

Fresco L O, Kroonenberg S B (1992) Time and spatial scales in ecological sustainability. Land Use Policy, July 1992:155–168.

Gallopin G (1992) Science, technology and the ecological future of Latin America. World Development 20:1391–1400.

Grindle M, Thomas J (1991) Public Choices and Policy Change. John Hopkins, Baltimore ML, USA.

Harrington L, Ashby J (forthcoming) Research strategies for Natural Resource Management. CIMMYT, Mexico, D.F. and CIAT, Cali, Colombia.

Hazell P B R (1993) New challenges for resource and environment data for policy research. In Von Braun J, Puetz D (Eds.). Data needs for food policy in developing countries: new directions for household surveys. Occasional Paper. International Food Policy Research Institute, Washington, DC, USA.

Hazell P B R, Norton R D (1986) Mathematical programming for economic analysis in agriculture. MacMillan Publishing Company, New York, USA.

Hicks J R (1946) Value and Capital. Oxford University Press, Oxford, UK.

Izac A-M N, Palm C, Vosti S (1993) Procedural guidelines for characterization and diagnosis: alternatives to slash and burn project. Mimeo, International Centre for Research in Agroforesty (ICRAF), Nairobi, Kenya.

Janssen W G (1986) Market impact on cassava's development potential in the Atlantic Coast region of Colombia. International Centre for Tropical Agriculture (CIAT), Cali, Colombia.

Janssen W G (1993) An analysis of the decision process in agricultural research. Paper presented at the ISNAR-EDI seminar on structural adjustment and agricultural research. Nairobi, Kenya, March 21–26, 1993.

Janssen W G, Goldsworthy P (1994) Multidisciplinary research programs: What, why and how? mimeograph. International Service for National Agricultural Research, The Hague, The Netherlands.

Jones P G, Robison D M, Carter S E (1991) A GIS approach to identifying research problems and opportunities in natural resource management. In: CIAT in the 1990s and beyond: a strategic plan, supplement. CIAT publication no. 198. International Centre for Tropical Agriculture, Cali, Colombia.

KARI (1994) Priority setting into the 21st century: a position paper by the priority setting working group. Discussion paper. KARI, Nairobi.

Kropff M J, Penning de Vries F W T, Teng P S (1994) Capacity building and human resource development for applying systems analysis in rice research. Pages 323–339 in Goldsworthy P, Penning de Vries F W T (Eds.) Opportunities, use and transfer of systems research methods in agriculture to developing countries. Kluwer Academic Publishers, Dordrecht, The Netherlands.

Lewbel A (1994) Aggregation and simple dynamics. American Economic Review, 1994:905–918.

Lynam J K, Janssen W (1992) Commodity research programs from the demand side. Agricultural Systems 39:231–252.

Lynam J K (1994a) Integrating research planning, priority setting and input evaluation within the CGIAR. In Collinson M P, Wright Platais K (Eds.) Social science in the CGIAR. CGIAR Study paper no. 28, World Bank, Washington, DC, USA.

Lynam J K (1994b) Sustainable growth in agricultural production: the links between production, resources, and research. Pages 3–27 in Goldsworthy P, Penning de Vries F W T (Eds.) Opportunities, use and transfer of systems research methods in agriculture to developing countries. Kluwer Academic Publishers, Dordrecht, The Netherlands.

Monke E A, Pearson S R (1989) The policy analysis matrix for agricultural development. Cornell University Press, London, UK.

Nelson J A (1991) Quality variation and quantity aggregation in consumer demand for food. American Journal of Agricultural Economics 73:1204–1212.

Norton R D (1995) Comments on: "Aggregating economic knowledge for use at national, regional, sector and farm level" (pages 333–351 in this volume).

Pierce D W, Barbier E, Markandya A (1990) Sustainable development: economics and environment in the third world. Edgar Algar, London, UK.
Pinstrup-Andersen P (1993) Policy issues and problems of data collection. In Von Braun J,Puetz D (Eds.) Data needs for food policy in developing countries: new directions for household surveys. Occasional Paper. International Food Policy Research Institute, Washington, DC, USA.
Van der Ploeg J D (1990) Labor, markets and agricultural production. Westview Press, Boulder CO, USA.
Rabbinge R, Van Ittersum M K (1994) Tension between aggregation levels. Pages 31–40 in Fresco L O, Stroosnijder L, Bouma J, Van Keulen H (Eds.) The future of the land: mobilizing and integrating knowledge for land use options. John Wiley & Sons, New York, USA.
Rae A N (1977) Crop management economics. Granada Publishing, London, UK.
Röling N (1994) Platforms for decision making about ecosystems. Pages 385–393 in Fresco L O, Stroosnijder L, Bouma J, Van Keulen H (Eds.) The future of the land: mobilizing and integrating knowledge for land use options. John Wiley & Sons, New York, USA.
Schipper R A, Jansen H G P, Stoorvogel J J, Jansen D M (1995). Evaluating policies for sustainable land use: a sub-regional model with farm types in Costa Rica (pages 377–395 in this volume).
Stomph T J, Fresco L O, Van Keulen H (1994). Land use system evaluation: concepts and methodology. Agricultural Systems 44:243–255.
Theil H (1954) Linear aggregation of economic relations. North-Holland Publishing Company, Amsterdam, The Netherlands.
Tripp R, Buckles D, Van Nieuwkoop M, Harrington L (1993) Land classification, land economics and technical change: awkward issues in farmer adoption of land-conserving technologies. Paper presented at: "Seminario/taller internacional para la definicion de una metodologia de evaluacion de tierras para una agricultura sostenible en Mexico. El Batan, Mexico, August 10–13, 1993.
Veeneklaas F R, Van Keulen H, Cissé S, Gosseye P, Van Duivenbooden N (1994) Competing for limited resources: options for land use in the fifth region of Mali. Pages 227–247 in Fresco L O, Stroosnijder L, Bouma J, Van Keulen H (Eds.) The future of the land: mobilizing and integrating knowledge for land use options. John Wiley & Sons, New York, USA.
Whyte W F (Ed.) (1991) Participatory Action Research. Sage Publications, London, UK.
Wood S, Pardey P G (1994) Supporting agricultural research policy and priority decisions: an economic-ecological systems approach. Pages 45–66 in Goldsworthy P, Penning de Vries F W T (Eds.) Opportunities, use and transfer of systems research methods in agriculture to developing countries. Kluwer Academic Publishers, Dordrecht, The Netherlands.
Wossink A (1993) Analysis of future agricultural change: a farm economics approach applied to Dutch arable farming. PhD-dissertation, Wageningen Agricultural University, Wageningen, The Netherlands.
WRR (Wetenschappelijke Raad Regeringsbeleid) (1992) Ground for choices: four perspectives for rural areas in the European Community. Netherlands Scientific Council for Government Policy, Sdu Publishers, The Hague, The Netherlands.

SECTION D

Case studies of eco-regional approaches at rational and regional scale

A systems approach to analyze production options for wheat in India

P.K. AGGARWAL, N. KALRA, S.K. BANDYOPADHYAY and S. SELVARAJAN

Indian Agricultural Research Institute, New Delhi 110 012, India

Key words: agro-ecological zonation, eco-regional, crop growth model, production potential, systems approach, wheat, yield gap

Abstract. Systems tools such as simulation models, geographical information systems, databases and optimization techniques can be used to understand dynamic interactions among components of a production system. They can quantitatively analyze the agro-ecological properties of different land evaluation units in terms of biophysical and socio-economic factors, and their interactions. In this paper, we illustrate this for determining potentials, constraints and opportunities for further increase in productivity of wheat, a major food crop, in different regions of India. The whole country has been considered as a mega eco-region and its districts as sub eco-regions because most planning is done following these administrative boundaries.

In a large number of districts spread over the states of Punjab, Uttar Pradesh (U.P.), Bihar, Assam, Rajasthan, and Madhya Pradesh (M.P.), potential yields were 7 t ha^{-1} or more. Most districts in U.P. have a yield potential between 6 to 6.5 t ha^{-1}. The potential yield was between 5 and 6 t ha^{-1} in middle latitudes and states of West Bengal and M.P. Economically optimal levels of N fertilizer application in irrigated environments were estimated for all locations based on current price ratios of N fertilizer and grain, native soil fertility, simulated crop response to N fertilizer and other costs related to transport, harvesting and market forces. A comparison of optimal and actual N applications showed that in Ludhiana district of Punjab, N application is more than the simulated optimal whereas in other districts it is at par or lower. The estimated yields corresponding to the profit maximizing amount of N apllication (henceforth refered as optimal economic yields) were generally 200 to 500 kg ha^{-1} lower than the potential yield irrespective of the location. The small difference between potential and optimal economic yield is due to distorted but favorable price ratios at present. In rainfed environments, optimal economic yields would be still lower.

At most locations, there was a large yield gap. At higher latitudes, the main wheat belt of India, yield gap of 2 t ha^{-1} was common even in well-irrigated regions. Almost 35–50% of the gap could be ascribed to delayed sowing, common in a number of districts. Factors such as limited and timely availability of irrigation and fertilizers, cropping pattern and access to credit and other services are some of the other principal causes of yield gaps.

It is concluded that crop growth simulation models together with databases of physical, biological and socio-economic attributes, geographical referencing and optimization techniques can help in setting up information systems to estimate crop production potentials, yield gaps, resource requirements for different agricultural strategies, assess potential environmental impacts, generate thematic maps and tables, and thus help in productivity related agro-ecological characterization.

J. Bouma et al. (eds.), Eco-regional approaches for sustainable land use and food production, 167–186.

Introduction

Increasing population, urbanization and income growth are resulting in rapidly increasing demand for food in India as well as many other developing countries (Pinstrup-Andersen and Pandya-Lorch 1995). It is estimated that the total food grain requirement of India will increase from the present level of approximately 180 Mt to 211 Mt by 2000 and 269 Mt by 2010 (Kumar *et al.* 1995). There is, therefore, a need to know the agricultural productivity potential in different agro-climatic zones of the country to determine whether India can produce this much, where, and how.

Although there is now great pressure to increase production, recent trends indicate a significant slow down of the growth rate in area, production as well as yield (figure 1). In future, expansion of cropped area is very unlikely, further production increase has to come, therefore, from increase in yield per unit area. It is also of concern to note that off-late that the seed based technology of sixties which helped in meeting food demand in the past is now leading to problems of environmental degradation. There are scattered reports that some areas where high yielding technology was adopted in the sixties are now showing signs of greater fluctuation in water table, leaching of nutrients and loss of soil fertility (Singh *et al.* 1987).

In order to identify the constraints limiting productivity at present and opportunities for sustainable increase in future, it is important to analyze the various factors constituting a production environment. The latter is made up of natural resources such as soil fertility, germplasm, level of inputs usage, opportunities allowed by climate, interactions with climatic variability, services providing assistance to farmers such as credit facilities, and market forces. Interactions among these factors often make decision making a difficult process in many production systems of today. Such situations need an eco-regional approach to apply inter-disciplinary knowledge to help focus on region specific problems and their optimal solutions (Rabbinge 1995).

In view of the considerable diversity in India's physical, biological and socio-economic production environment, it is also necessary to demarcate homologous agro-ecological zones. Such zones are appropriate as a framework to develop data inventories of environmental resources which may show spatial and temporal variations. They help in linking physical, biological and socio-economic attributes to the performance of a crop. Extrapolation of research results from selected sites to throughout the homogenous zone can be made more effectively through such tools.

Several attempts have been made to classify India's diverse climates for making effective land use plans. Planning Commission of Government of India carved 15 agro-climatic zones based on physiography and climate (Government of India 1987). Subsequently, these were sub-divided into 127 agro-ecological zones by the National Agricultural Research Program (NARP) based on physiography, rainfall, soils, cropping pattern and administrative

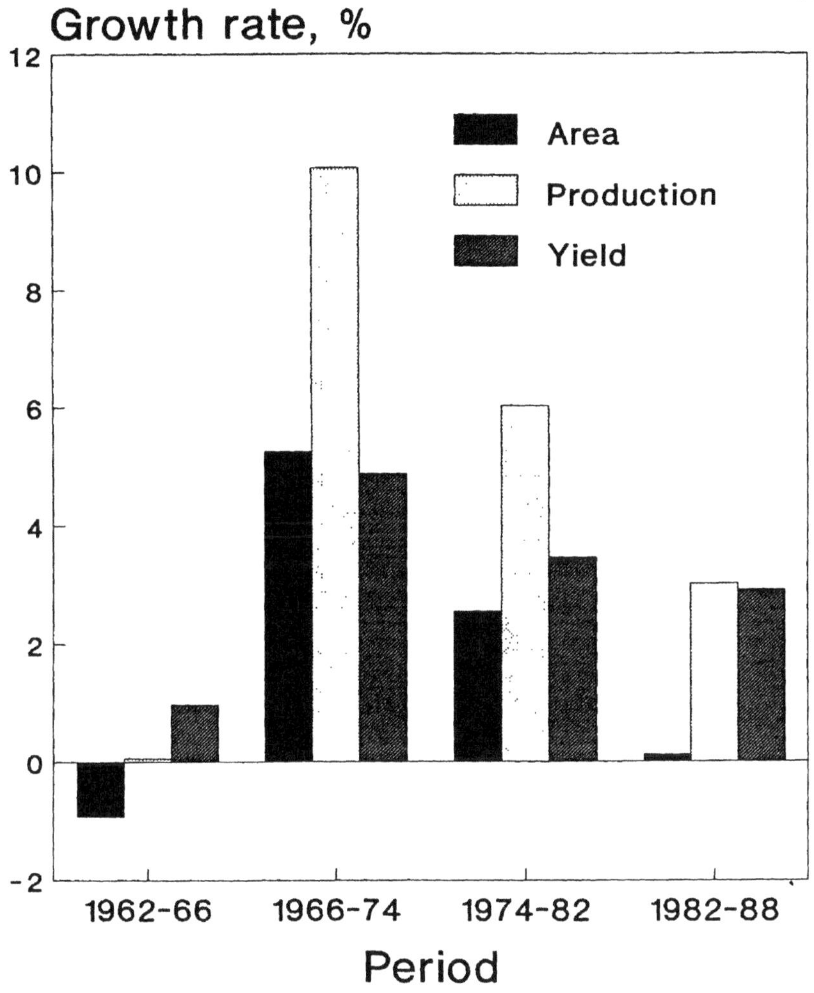

Fig. 1. Change with time in growth rates of area, production and yield of wheat in India.

boundaries (Ghosh 1990). The National Bureau of Soil Survey and Land Use Planning superimposed soil texture map, physiography, climate (aridity-humidity) and computed length of growing period to demarcate 21 zones in whole of India (Sehgal *et al.* 1990). These agro-ecological zoning efforts have been principally based on considerations of rainfall and to some extent on soil texture. Major quantitative output of these analyses is growing period. From a purely ecological view point, this may be adequate but when the primary interest is in the knowledge of agricultural potential of a region and its development, one needs to know much more, for example, the yield potential of crops, optimal crop duration and major constraints limiting yield. Aggarwal (1993) argued that the growing period based agro-ecological zoning is appropriate only when agriculture is monoculture and rainfall is the principal variable determining crop yield such as in large tracts of semi-arid tropics. With the increased availability of irrigation resources (> 70% of Indian wheat in now irrigated) and increased cropping intensity during last few decades, there is now a need to reconsider the main variables effecting crop growth and yield and hence agro-ecological zoning.

The major biotic and abiotic factors affecting crop growth and development in any production system are radiation, temperature (yield determining), water, nutrition (yield limiting), and pests and diseases (yield reducing) (Rabbinge 1993). In addition, productivity is also determined by many other factors such as variety, its physiology and crop management which interact with weather and soils to influence yield level. Systems tools such as simulation models, GIS, databases and optimization technologies now offer exciting opportunities to understand dynamic interactions among various components of production systems. Work is now in progress at our institute to use these approaches for determining potentials, constraints and opportunities for major crops in various production environments of India.

We consider India as a mega eco-region. Since most planning is done following administrative boundaries, districts are chosen as the primary land evaluation units. This is the smallest unit for which dependable primary data on cropped area, productivity, etc. is readily available. In this paper, we illustrate our approach for documenting and analyzing current production environment of wheat, a major food crop, in different regions of India and to determine options for further increase in productivity. This is largely an explanatory type of study aimed at understanding limitations and opportunities due to biophysical factors. More specifically, the objectives are:

1. To characterize current production environments in terms of physical, biological and socio-economic attributes that effect growth and yield of wheat.
2. To determine the potential and optimal economic yields of wheat in different land evaluation units and to demarcate regions with similar yields.
3. To quantify the magnitude of the yield gap and its principal causes in different regions.
4. To determine if future climate change will cause shifts in wheat producing areas.

Table 1. Basic statistics on wheat for major wheat producing states of India.

State	Total[1] area Mha	Product-ivity[1] t ha^{-1}	Irrigated[2] area % of total	Fertilizer[1] use kg ha^{-1}	Growth rate[3]		
					Area	Production	Yield
Uttar Pradesh	8.66	2.05	88	86	0.82	3.14	2.30
Punjab	3.25	3.59	95	156	0.71	2.93	2.20
Haryana	1.86	3.18	98	114	1.53	5.05	3.47
Bihar	1.97	1.64	81	55	0.40	2.95	2.54
Madhya Pradesh	3.20	1.21	39	30	–0.23	2.79	3.03
Rajasthan	1.65	2.06	90	21	2.32	5.72	3.32

[1] 1989–90 data
[2] 1987–88 data
[3] 1985/86–1990/91

Current yields

Wheat is grown in diverse agro-climatic conditions from 11 to 35°N, from 72 to 92°E and from almost sea level to very high elevations. The major wheat producing areas are between 20 and 32°N comprising the states of Punjab, Haryana, Uttar Pradesh (U.P.), Madhya Pradesh (M.P.), Rajasthan, Bihar, West Bengal and Maharashtra. The wheat production has increased tremendously from 12.3 Mt in 1965 to 56 Mt in 1992 (figure 2). This has been possible due to increase in area under wheat from 13.4 to 23 M ha and yield from 0.92 to 2.44 t ha^{-1}. Most of this increase in yield was largely due to adoption of modern, semi-dwarf, high yielding cultivars, increased water and nutrient applications and support of suitable government policies and economic incentives.

Table 1 lists wheat area, percent wheat area under irrigation, fertilizer consumption, and average yield for some key wheat producing states of India. In Punjab and Haryana, almost the entire area is irrigated and relatively average fertilizer application is high. This results in higher yields compared to other states. There is nevertheless considerable variation at the district level. In Punjab, yield varies from 2 to 4 t ha^{-1} depending upon the region. U.P. state alone accounts for 1/3 of the total wheat area of the country. Most of this area is irrigated, average fertilizer use is modest and crop yields are relatively low (2 t ha^{-1}). In western parts of the state, yields are more than the eastern parts. In Rajasthan, although 90% wheat is irrigated, fertilizer consumption is very low (18.9 kg ha^{-1}) and yet the productivity is comparable to U.P. M.P. has a large proportion of rainfed wheat, fertilizer consumption is low and so are the yields.

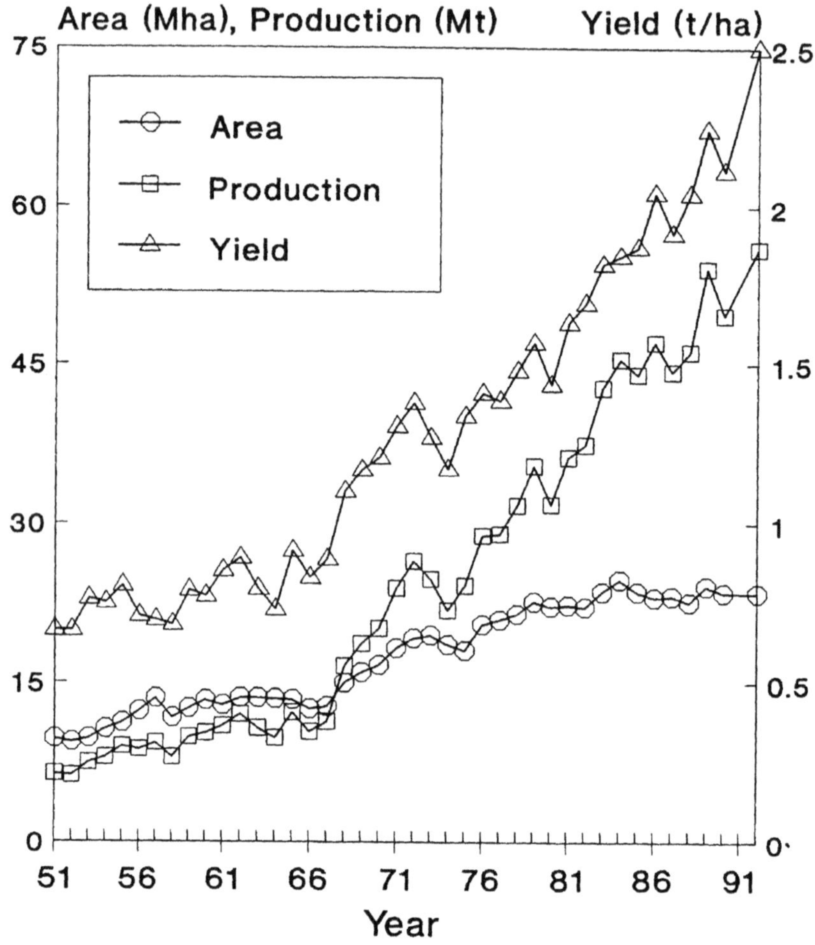

Fig. 2. Change with time in area, production and yield of wheat in India.

There has been a rapid diffusion of modern varieties all over India particularly in irrigated areas. Locally adapted varieties are continuously developed and released by wheat breeders to maintain disease resistance.

Area growth at present is almost negligible in all states except Rajasthan (table 1). Growth rate in yield is low (2.2–2.3%) in Punjab and U.P. and modest (2.5–3.5%) in other states (table 1).

A significant portion of the yield variation in different parts of India could be related to climate. The temperatures during the crop season vary a lot depending upon the latitude. From south to north, mean January temperature varies by 15°C. Temperatures in Punjab and Haryana states are relatively lower compared to U.P., Bihar, M.P. and Rajasthan. Photothermal quotient (radiation/temperature), related to growth and yield of wheat crops (Fischer 1985), is almost similar in most states of northern India but is relatively lower in eastern and central regions (Aggarwal 1994). Seasonal rainfall is about 100 mm in many parts of Punjab and northern U.P. and is less than 50 mm in most parts of central and eastern India. Because rainfall is limited, crops depend upon stored soil moisture and irrigation for meeting their water requirements.

Potential yields

In view of the large differences in actual yield in different regions, it is important to know the opportunities allowed by the climate of that region. Potential yield is the integrated expression of the influence of radiation and temperature on crop growth and development of a particular crop/variety. The production system is characterized by adequate water and nutrient supply and absence of all yield reducing factors such as pests and diseases. Potential yield can be interpreted as the upper limit that can be achieved by the current varieties in a no constraint environment.

The methodology used for estimation of potential wheat yields is described by Aggarwal (1993). A crop model WTGROWS (Aggarwal *et al.* 1994) was used to estimate the yield potential of wheat. The performance of the model has been satisfactorily evaluated in terms of dry matter growth, water and nitrogen uptake and yield in diverse agro-environments (Aggarwal *et al.* 1994a). The model has earlier been used for determining potential yields and optimal management practices (Aggarwal and Kalra 1994). Long-term average monthly weather data of 138 locations below 600 m elevation from all wheat producing regions of India were used. The daily weather data was estimated by linear interpolation. Because wheat is grown in clear sunny weather in almost all parts of India, simulated potential yields obtained from the interpolated daily values (from monthly means) were only marginally different from those obtained by using several years of measured daily values (results not shown).

Deterministic crop growth models such as WTGROWS require several inputs relating to crop/variety, soil physical properties, weather and crop management. The input values used could be significantly uncertain due to random and systematic measurement errors and spatial and temporal variation observed in many of these inputs. The effect of uncertainties in input values on simulated grain yield was determined by Aggarwal (1995) using Monte Carlo simulation techniques. The results showed that there was a 80% probability that the bias in the grain yield was always less than 10% in potential and irrigated production systems. Thus, the potential, irrigated and optimal economic yields simulated using WTGROWS are likely to have only a small bias due to uncertainties in model inputs.

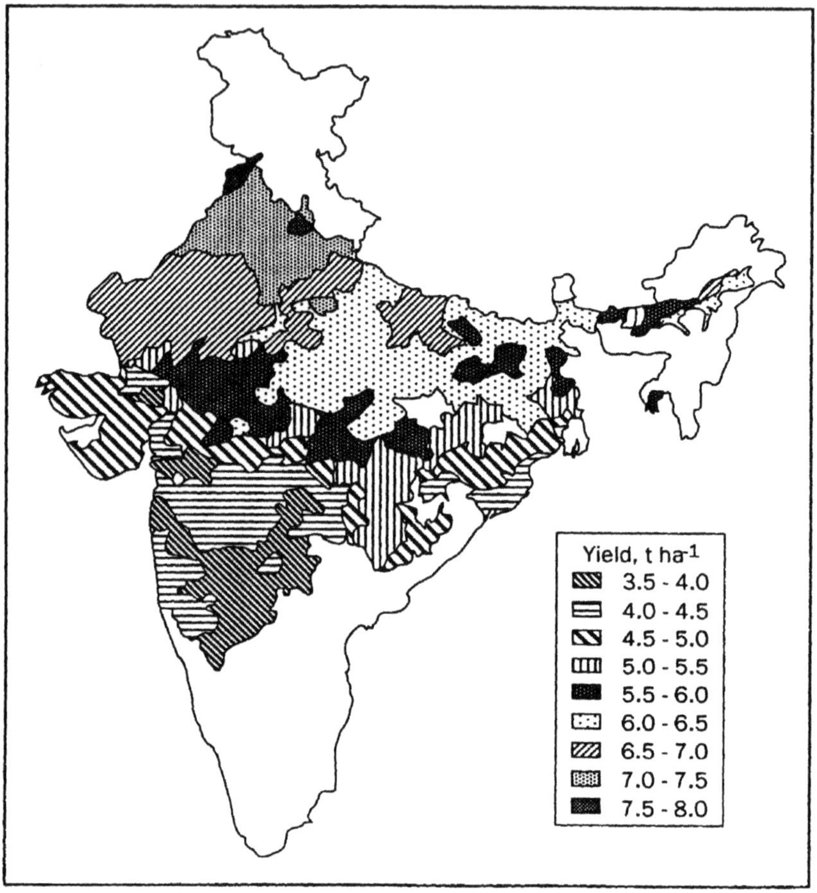

Fig. 3. Potential yields (t ha^{-1}) of wheat in India. Only those districts are labelled where wheat cultivation is significant.

The simulated potential grain yield varied between 2.5 to 8.0 t ha^{-1} depending upon the location. The yield potential was between 7.0 and 7.5 t ha^{-1} in north western regions lying above 28°N (figure 3). Between 24 and 28°N, potential yield varied from 5 to 7 t ha^{-1}. The potential yield varied from 4.0 to 6.5 t ha^{-1} between 20 and 24°N. In this region, inland positions generally had higher yield than the coastal areas. For latitudes below 20°N, potential yield was between 3.5 and 5.0 t ha^{-1} and generally increased with longitude.

Potential yield for each district was interpolated from individual location yield. Digitized maps comprising district boundaries, soils, weather stations and agro-ecological regions were stored in GIS packages ARC/INFO and IDRISI. In a large number of districts spread over the states of Punjab,

U.P., Bihar, Assam, Rajasthan, and M.P., potential yields were 7.0 t ha^{-1} or more. Most districts in U.P. have a yield potential between 6.0 to 6.5 t ha^{-1}. The yield was between 5.0 and 6.0 t ha^{-1} in middle latitudes and states of West Bengal and M.P. In most districts of Gujrat, Maharashtra and Orissa, potential yield was only 4.0 to 5.0 t ha^{-1}. Agroclimate of southern states of Andhra Pradesh, Karnataka, southern and coastal Maharashtra did not allow yield potential to exceed 4.0 t ha^{-1}. These geographical trends in yield potential were apparently related to changes in mean temperature changes with latitude/location (Aggarwal and Kalra 1994a).

The above results are based on mean weather data and for locations below 600 m.a.s.l. In practice, one may find a few locations where yield potential is different as suggested by the above analysis. Such examples may be true because the emphasis in the present study was to establish trends of climatically potential yields across locations and not absolute yield potential of an individual location. The yields are also expected to increase with elevation and to vary due to yearly variation in weather. The effect of year-to-year variation in weather on potential yield was simulated by Aggarwal and Kalra (1994) for few contrasting locations. At all locations, climatic variation caused fluctuations in yield which exceeded by more than 2 t ha^{-1}. For most locations, the difference in grain yields between 75 and 25% probability level was between 0.5 and 1.0 t ha^{-1}. The relatively small yield variation is apparently due to clear weather in the post-rainy wheat growing season (the rainfall during wheat season is small in most crop seasons).

Optimal N application and economic yields

Potential yields as estimated above may not be economically optimal. The later depends on cost of cultivation and net benefit: cost ratio. Normally, optimal levels are defined considering the whole production system and component crops, and opportunity costs outside agriculture but this analysis refers to cultivation of wheat alone. We recently made studies to determine economically optimal levels of N fertilizer application in irrigated environments based on current price ratios of N fertilizer and grain, native soil fertility, N response and other costs related to transport, harvesting and market forces (Aggarwal *et al.* unpublished). The basic level of soil fertility is low in most Indian soils (Ghosh *et al.* 1978). Initial soil N profile in the model was calibrated to these low levels. Of course, this level is dynamic and would be very much dependent upon cropping history, but for simplicity, uniform level at the time of wheat sowing was assumed for all locations. The geographical distribution of soils was obtained from the agro-ecological zones map of India (Sehgal *et al.* 1990). The dominant soil textures of wheat growing regions are loam sand, sandy loam, black and red loam. For each of these textures, a representative water retention profile was used (Kalra *et al.* 1995). Simula-

Table 2. Actual and simulated optimal N application in few selected districts of Punjab and U.P. states (Aggarwal *et al.* unpublished). Actual application was estimated from rabi (wheat season) N offtake by the estimated share applied to wheat (0.9 for Punjab and 0.75 for U.P.).

District	N application, kg ha^{-1}	
	Actual	Optimal
Punjab		
Ludhiana	191	171
Amritsar	158	159
Patiala	110	169
UP		
Varanasi	116	175
Kanpur	119	179
Agra	143	162
Bareilly	150	170
Saharanpur	128	168

tions indicated that irrigated fields with no fertilizer application can produce between 1.5 to 2.5 t ha^{-1} depending upon the region. Indeed, Goswami *et al.* (1979) based on approximately 2000 experiments conducted over several seasons in many locations showed that this fertility level is sufficient to produce 1.5 to 2.0 t ha^{-1} yield depending upon the place without any fertilizer application. The crop model was then used to simulate fertilizer response for all locations used for determining potential yields. It was assumed that fertilizer is applied in two splits at sowing (2/3) and at crown root initiation (spike initiation) stage (1/3), and other nutrients and biotic factors are not a constraint. Marginal physical productivity was estimated from the quadratic crop N response curves. The current price ratio for N vs grain was estimated to be 4.66 considering the costs of urea, transport costs of fertilizer, cost of fertilizer application, gate price of grain, cost of additional harvesting and opportunity costs of investment (Aggarwal *et al.* unpublished). Based on this price ratio, the optimal N application rate was calculated.

The simulated optimal N rate varied between 75 to 185 kg ha^{-1} depending upon the location. In most parts of U.P., optimal N rate was between 150–170 kg ha^{-1}. In most districts of Bihar, Rajasthan, Punjab, Haryana and some parts of U.P., it was between 130–150 kg ha^{-1}. These variations were a consequence of interactions among native soil fertility, soil physical properties, and weather factors. A comparison of optimal and actual N applications showed that in Ludhiana district of Punjab, N application is more than the simulated optimal whereas in other districts it is at par or lower. In other states, actual N application was lower than the simulated optimal (table 2).

Optimal economic yields are defined as the level where profit of N use is maximum. These yields were estimated by inputting optimal N rate in the quadratic response curve of yield vs. N. Since prices are dynamic, the optimal N rate would not be the same every year. Nevertheless, this would not significantly alter estimates of grain yield in many irrigated areas where yield response to N fertilizer is near plateau. Many other factors such as phosphatic fertilizers application, pesticide and labor use would also determine optimal economic level of grain yield. Optimal levels would be different if the whole production system and opportunty costs of a farm were considered. A more appropriate approach would perhaps be to do a total factor productivity analysis considering biophysical and socio-economic factors together. Recent trends nevertheless indicate that fertilizers constitute a major share of the operational costs, particularly in wheat production in Punjab, Haryana and U.P. (Anonymous 1991a) and more than 70% of the total fertilizer applied to wheat is nitrogen.

The results showed that optimal economic yields were generally 200 to 500 kg ha^{-1} lower than the potential yield irrespective of the location. In few high latitude locations, the difference was greater than 500 kg ha^{-1} perhaps due to interaction effects with irrigation. The small difference between potential and optimal economic yield is due to distorted but favorable price ratio at present. It can be noted that the price of urea has decreased significantly in nineties compared to seventies and eighties (figure 4). Simultaneously there has been an increase in net price of wheat. Nevertheless, price ratio of N and grain has come down allowing farmers to apply more N. This greater application of N at the margin, however, results in only small gains in productivity and reduces overall technical efficiency of N use. The optimal economic yield was between 6 and 6.5 t ha^{-1} in some parts of U.P. and most of Punjab. In most parts of U.P., Bihar and Rajasthan the optimal economic yields were between 6 and 6.5 t ha^{-1}. In the state of M.P. these yields varied between 5 and 6 t ha^{-1}. It must be remembered that these estimates are for well irrigated fields with no biotic and abiotic constraints. Such situations are not very common. In rainfed environments, such as those in large areas in M.P. and Bihar, optimal economic yields would be lower than those estimated by us. Also, optimal economic yields are dynamic because of temporal and spatial variations in prices, soil physical properties and biophysical factors.

Yield gap

Actual grain yields obtained by farmers on a regional basis are reported every year by the Government of India. The smallest reporting unit is a district. Data for all districts was collected for the years 1988–89 and 1989–1990 (Anonymous 1990, 1991). The maximum of the two years yields was

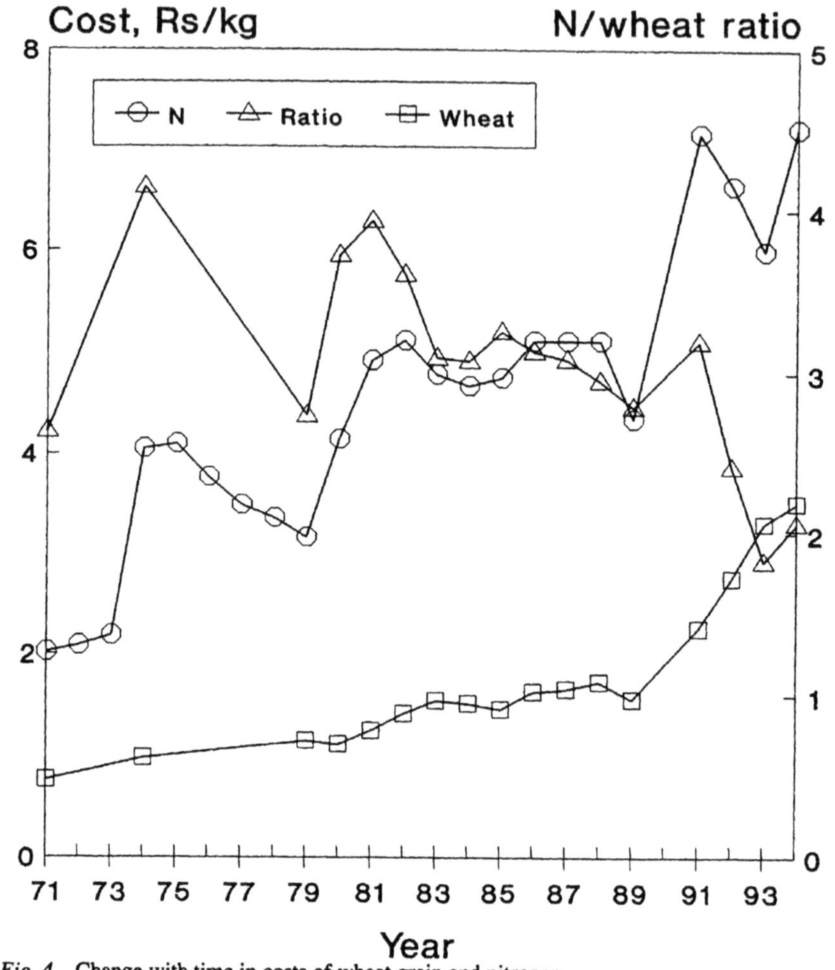

Fig. 4. Change with time in costs of wheat grain and nitrogen.

assumed to be actual maximum (attained) yield of that location. There are indications that at present actual yields are 10–15% more than these estimates but recent detailed data was not available. The difference between optimal economic yield and actual yield was considered as the yield gap.

In order to determine yield gaps at different locations, potential, optimal economic and actual yields at these locations were plotted against latitude (figure 5). It is apparent that yields increased with latitudes although the rate of increase varied. For latitudes below 18°N optimal economic yield was 3.5 t ha^{-1} and the actual grain yield was 1.2 t ha^{-1} indicating a yield gap of 2.0 t ha^{-1} or more. From 18 to 25°N, optimal economic and potential

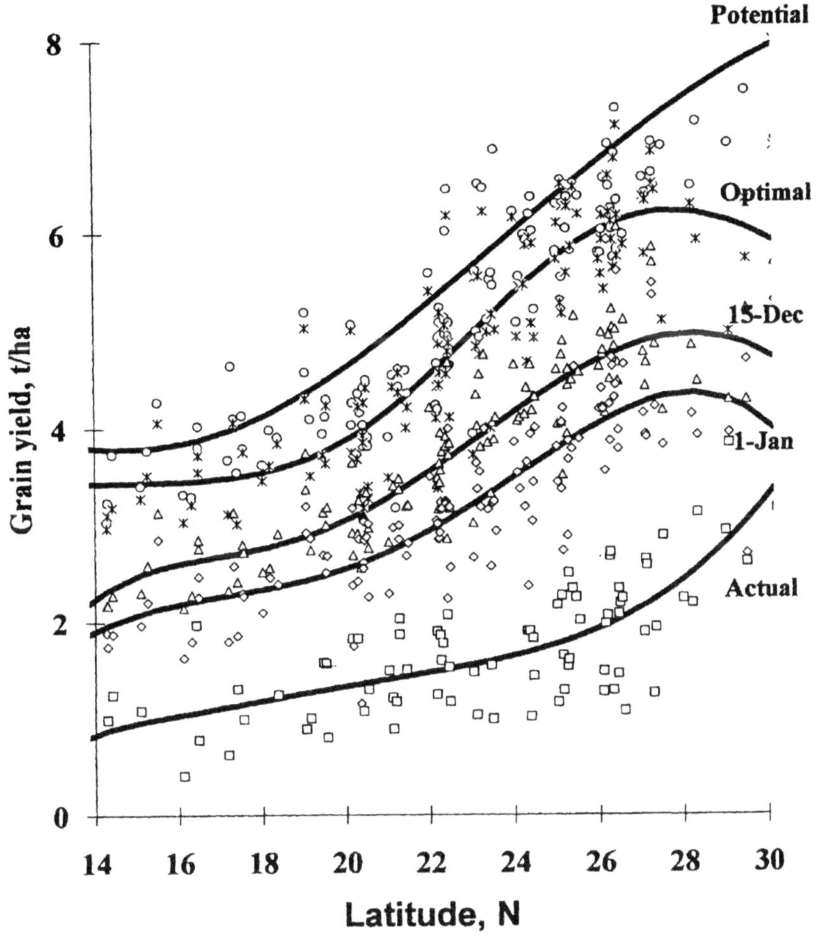

Fig. 5. Potential, optimal economic and actual grain yields of wheat as a function of latitude. Also shown are the simulated yields on 15 December and 1 January sowings to illustrate the contribution of late sowing to yield gap.

yields showed a steep linear increase with latitude. Actual yields however showed a small response to latitude – ranging from 1.0 t ha^{-1} at 18°N to 1.5 t ha^{-1} at 25°N. This indicates that there was a large yield gap which increased with latitude. Most of this area is in central India and is rainfed or receives limited irrigation which apparently might be the major yield reducing factor. At higher latitudes, which is the main wheat belt of India, potential yield was always greater than 6 t ha^{-1} and optimal economic yield was about 6.0 t ha^{-1}. Actual yields here showed a distinct linear increasing trend with

Table 3. Magnitude of yield gap in irrigated districts. Only those districts where 90% or more wheat area is irrigated are considered. Optimal economic yields and yield gaps would generally be lower than those shown here if the limitations due to frequency and timing of irrigation availability and biotic factors are also considered. See text for more details.

State	No. of districts	Area	Current		Optimal economic		Gap	
			Yield	Production	Yield	Production	Yield	Production
		Mha	$t\ ha^{-1}$	Mt	$t\ ha^{-1}$	Mt	$t\ ha^{-1}$	Mt
U.P.	18	3.04	2.47	7.53	5.32	16.20	2.84	8.64
Punjab	9	2.72	3.76	10.23	5.02	13.66	1.26	3.43
Haryana	12	1.82	3.41	6.20	4.70	8.55	1.29	2.35
Rajasthan	11	0.52	2.42	1.26	5.48	2.85	3.06	1.59
Gujrat	4	0.05	2.40	0.12	4.60	0.23	2.20	0.11
Others	21	0.11	1.45	0.16	2.20	0.24	0.75	0.08
India	75	8.26	3.09	25.50	5.05	41.70	1.96	16.20

latitude; yields increased from 1.5 t ha^{-1} to 4.0 t ha^{-1} as the latitude increased from 26 to 31°N. Nevertheless, at all latitudes there was a considerable yield gap. In general, the results showed a yield gap of at least one t ha^{-1} at all places.

Magnitude of yield gap was estimated for fully irrigated districts (> 90% irrigated area) of major states. In such districts of U.P. optimal economic yield was 5.32 t ha^{-1} and actual yield was only 2.47 t ha^{-1} indicating a yield gap of 2.84 t ha^{-1}. This yield gap if bridged can result in additional production of 8.64 Mt (table 3). Yield gap is rather low (1.2–1.3 t ha^{-1}) in most districts of Punjab and Haryana but nevertheless closing this gap can result in additional production of over 5 Mt. There is a large yield gap of over 3.0 t ha^{-1} in Rajasthan. Bridging the yield gap in other districts where irrigation potential is relatively lower provides sufficient opportunities to meet future demands. As pointed out earlier as well, these yield gaps were obtained from optimal economic yields in a no constraint environment. Consideration of factors such as limited and timely availability of irrigation, cropping pattern, access to credit and other services, availability of other inputs would reduce optimal economic yield levels and thus yield gap. There is a need to understand principal factors that cause these yield gaps.

Principal causes of yield gap

Date of sowing

Experimental studies as well as simulation results indicate that the first fortnight of November is the optimal time for wheat sowing (Aggarwal and Kalra 1994). But a large proportion of farmers sow wheat much after the optimal time. In U.P. and Bihar, a large proportion of farmers sow as late as December and January (WPD 1989). A major reason of this delay is that wheat is now being grown after rice, which matures in October–November. Simultaneously, there is often a long turn-around period. Aggarwal and Kalra (1994) determined the reduction in grain yield per day if sowing is delayed beyond the optimal time (November 15). The per day yield decrease was between 0.25% and 0.75% of potential yield when the latter was lower than 4 t ha^{-1}, between 0.5 and 1.0% for yield potential between 4 and 6 t ha^{-1} and between 0.75 and 1.0% for yield potential greater than 6 t ha^{-1}. Since rice, the crop preceeding wheat in many districts is generally more renumerative, farmers try to maximize its yield. Thus, even if wheat sowing gets delayed they make more profit from the cropping pattern.

The magnitude of yield gap that can be explained by date of sowing alone can also be gauged from figure 5. It is evident that almost 35–50% of the gap could be ascribed to sowing date effects. As much as 1.5 t ha^{-1} gap could be explained by one month's delay in sowing. In particular at higher latitudes a large proportion of the yield gap was explained simply by sowing date effects. In middle latitudes considerable area is unirrigated and therefore, a greater number of factors cause yield gap.

Irrigation

The dramatic increase in yield over the last two decades, particularly in northwestern regions of the country, was associated with enormous change in input structure. There was a intensified use of inputs particularly of irrigation, fertilizers and pesticides. During the last 25 years, percentage of irrigated wheat area has increased from 35 to 75%. But not all areas classified as irrigated receive optimal irrigation amount at recommended stages. In western U.P., for example, although general recommendation is 5–6 irrigations only a limited area receives more than three irrigations (Sinha *et al.* 1985). On the other hand in some other parts where both canal and groundwater are freely available, such as in Punjab, there is a tendency to over-irrigate because irrigation constitutes only a small portion of the total farm expenditure. Recent compilation of data shows that Punjab farmers spend only 5.8% of their total operational costs on irrigation whereas Haryana and U.P. farmers spend 10–12% (Anonymous 1991a). Such trends in availability and cost of irrigation cause significant effect on use efficiency of other inputs and result in decreased yields and large yield gaps.

Irrigation availability greatly affects N response. Aggarwal and Kalra (1994) found considerable interactions between climatic variability, irrigation availability and fertilizer response. These interactions were particularly pronounced at low levels of input (irrigation and N) usage such as in eastern U.P., Bihar and M.P. This has important implication. At low levels of water availability it is difficult to decide optimal levels of N fertilizer for maximizing yield returns in view of uncertainty of response until late in the season. This may also explain farmers reluctance to apply N to optimal levels indicated by simulation due to uncertainty of weather, availability of irrigation and hence returns to fertilizer use.

Nutrients

Consumption of inorganic fertilizers have increased over time in all wheat producing regions. Simultaneously there is a distinct decline in the amount of organic matter applied. There is large inter-state and intra-state variation in fertilizer use (figure 6). The NPK application to wheat has crossed 200 kg ha^{-1} level in Punjab whereas it is only about 100 kg ha^{-1} or less in states of U.P. and Bihar. In Rajasthan most farmers apply 10–15 kg ha^{-1} fertilizers to wheat. The application of nutrients varies a lot within most states. In general, in most states only 50% or less farmers apply recommended dosages (WPD 1989). Such suboptimal application of nutrients, often even in well irrigated areas, despite a favorable price ratio causes large yield gaps. On the other hand, there are few districts in Punjab where farmers apply more than optimal or recommended dosage (generally 120–150 kg N ha^{-1}).

The response of crops to N also depends on timely availability of irrigation, other nutrients and access to cash or credit to purchase inputs. At many places such as in eastern India, the farmers capacity to purchase inputs is limited. Credit facilities at such places are also limited. Subsidies on phosphatic and potassic fertilizers were recently reduced by the Government. This resulted in sharp increase in their prices and consequent reduction in their application. This imbalanced application will affect crop response to other nutrients including N and thus reduce optimal economic yields. Current price ratios of grain and N fertilizer are encouraging farmers in areas such as Punjab where input use is very high to apply more N to perhaps cover up the inefficiencies in the use of other inputs. This is very likely to increase problems of N leaching due to reduced N uptake efficiency at high N application rates and thus gradual deterioration of environment. Policy interventions are needed.

Yields in changed climate

In exploring options for future, it is also necessary to consider impact of projected climate change on yields. Recent estimates indicate that over much

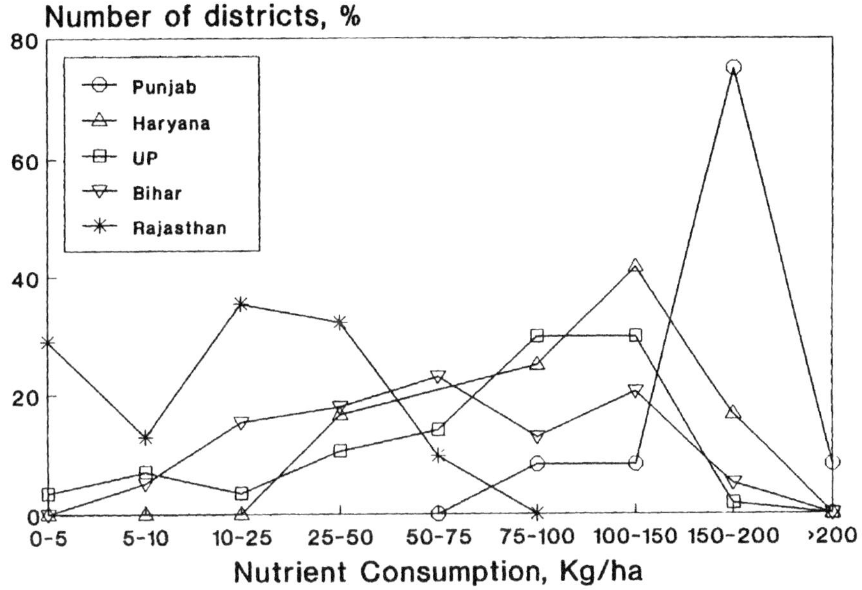

Fig. 6. Variation in the level of NPK use in different states. (Source: FAI 1990)

of the wheat growing regions of India, carbon dioxide (CO_2) level will rise to 425 ppm and temperature will be higher by 1.8°C (Houghton 1991). Aggarwal and Sinha (1993) studied the effect of increased CO_2 and temperature on yield of wheat in India. The effect of climate change was dependent upon the magnitude of temperature increase and production system. At 425 ppm CO_2 concentration and no increase in temperature, grain yield at all production levels increases significantly. In northern India, a 1°C rise in mean temperature had no significant effect on potential yields but irrigated and rainfed yields showed a small increase. An increase of 2°C reduced potential grain yield at most places. The effect on irrigated yield varied with location (figure 7). There was almost no effect in northern India but yields were reduced in central India by 10–15%. This shift in productivity unless accompanied with suitable research and policy interventions may reduce wheat production options in central India.

Conclusions

Crop growth simulation models together with existing databases of physical, biological and socio-economic attributes, geographical referencing and optimization techniques can help in setting up information systems that can be used for a number of applications. Such systems allow us to estimate

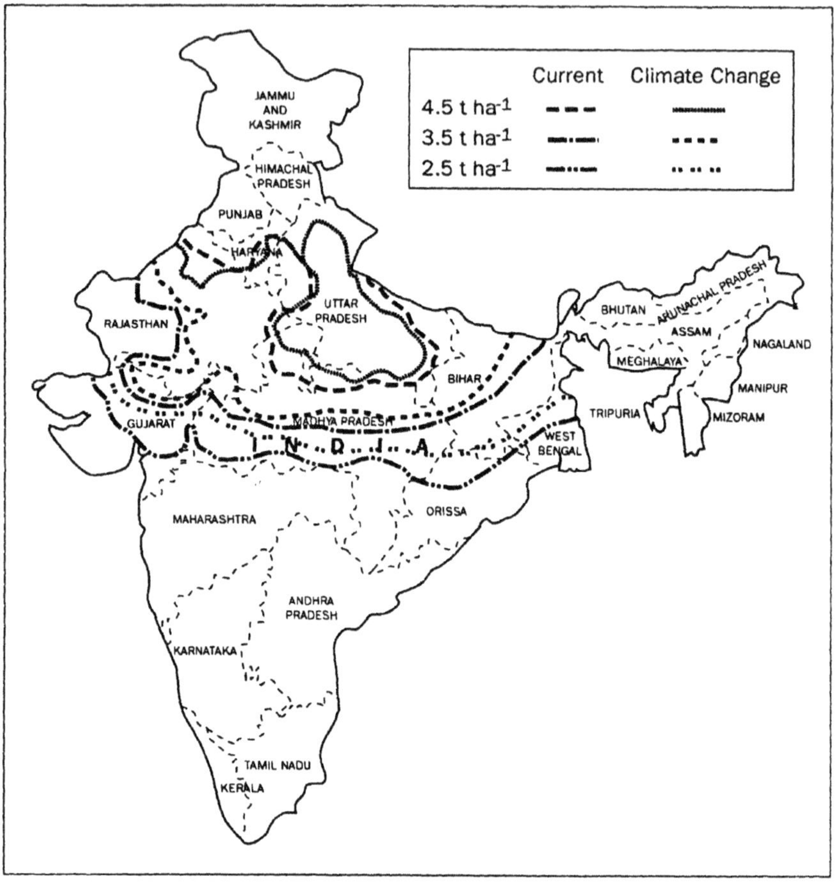

Fig. 7. Impact of climate change on shift in irrigated wheat productivity zones. Climate change scenario was 425 ppm CO_2 and a 2°C rise in mean temperature.

crop production potentials, yield gaps, resource requirements for different agricultural strategies, assess potential environmental impacts, generate thematic maps and tables, and thus help in productivity related agro-ecological characterization. The methods presented in this paper quantitatively analyze the agro-ecological properties of different land evaluation units in terms of biophysical and socio-economic factors, and their interactions and provide us an instrument to understand potentials, constraints and opportunities for agricultural development. Regions with greater potential for yield increase with a given amount of input can be identified. For example, results indicate that optimal economic yields in Punjab and U.P. state are similar and larger yield gaps in most districts of U.P. state are due to suboptimal input use and late plantings. Relatively small yield gap in Punjab indicates that research there must now focus on increasing yield potential and input use efficiency. Results based on single commodity analysis however may have limitations. In Indo-gangetic plains, most wheat is grown after rice. The latter fetches higher price and thus even if wheat sowings get delayed farmers make more profits per year. There is a need to analyze crop production and profits over whole farm which farmers try to maximize.

Once databases are stored in a GIS system, this methodology can rapidly generate new agro-productivity zones, or modify old ones as new concepts or technology such as new varieties become available. The databases and GIS can also be used to store spatial and temporal inventory of regional data on yields, fertilizer application practices and other input usage for ready and easy reference. The analysis done for this paper had limitations in terms of availability of precise databases, for example, on costs of grain and N, and soil physical characteristics in different regions. There is an urgent need to develop high quality databases relating to different aspects of agro-environment. The Government of India has recently set up database centers in each district to collect primary data on soils, weather, land use and crop productivity besides other activities such as urban development, education, water resources and health. These centers are connected via a satellite network, NICNET, based in New Delhi. Such databases will be extremely useful for strategic and tactical decisions relating to eco-regional development.

Acronyms

GIS	Geographical Information System
NARP	National Agricultural Research Program

References

Aggarwal P K (1993) Agro-ecological zoning using crop growth simulation models: characterization of wheat environments of India. Pages 97–109 in Penning de Vries F W T, Teng P S, Metselaar K (Eds.) Systems Approaches for agricultural development. Kluwer Academic Publishers, Dordrecht, The Netherlands.

Aggarwal P K (1994) Constraints in wheat productivity in India. Pages 1–11 in Aggarwal P K and Kalra N (Eds.) Simulating the effect of climatic factors, genotype and management on productivity of wheat in India. Indian Agricultural Research Institute, New Delhi, India.

Aggarwal P K (1995) Uncertainties in plant, soil and weather inputs used in growth models: Implications for simulated outputs and their applications. Agricultural Systems (in press).

Aggarwal P K and Sinha S K (1993) Effect of probable increase in carbon dioxide and temperature on productivity of wheat in Indian Journal Agricultural Meteorology 48(5): 811–814.

Aggarwal P K, Kalra N, Singh A K, Sinha S K (1994) Analyzing the limitations set by climatic factors, water and nitrogen availability on productivity of wheat. I. The model documentation, parameterization and validation. Field Crops Research (in press).

Aggarwal, P K, Kalra N (1994) Analyzing the limitations set by climatic factors, water and nitrogen availability on productivity of wheat. II. Climatically potential yields and optimal management strategies. Field Crops Research (in press).

Aggarwal P K, Kalra N (1994a) Simulating the effect of climatic factors, genotype and management on productivity of wheat in India. Indian Agricultural Research Institute, New Delhi, India.

Anonymous (1990) District-wise estimates of area and productivity of wheat 1988–89 (final). Agric Situation in India, Ministry of Agriculture, Government of India, India.

Anonymous (1991) District-wise estimates of area and productivity of wheat 1988–89 (final). Agricultural Situation in India, Ministry of Agriculture, Government of India, India.

Anonymous (1991a) Cost of cultivation of principal crops in India. Ministry of Agriculture, Government of India, India.

FAI (1990) Fertilizer statistics 1989–90. The fertilizer Association of India, New Delhi, India.

Fischer R A (1985) Number of kernels in wheat crops and the influence of solar radiation and temperature. Journal of Agricultural Science 105:447–461.

Ghosh S P (1990) Agro-climatic zone specific research. Indian perspective under NARP. Indian Council for Agricultural Research, New Delhi, India.

Ghosh A B, Hasan R (1980) Nitrogen fertility status of soils in India. Fert. News: 19–22.

Government of India (1987) Agro-climatic regions planning: An overview. Planning Commission, New Delhi, India.

Goswami N N, Shinde J E, Sarkar, M C (1987) Efficient use of nitrogen in relation to soil, water and crop management. Bull. Indian Society of Soil Science 13:51–67.

Houghton (1991) Scientific assessment of climate change: summary of the IPCC working group I report. Pages 23–44 in Jager J, Ferguson H L (Eds.) Climate change: science, impacts and policy. Cambridge University, UK.

Kalra N, Aggarwal P K, Bandyopadhyay S K, Malik A K, Kumar S (1995) Prediction of moisture retention and transmission characteristics from soil texture of Indian soils. Pages 26–35 in Lansigan F P, Bouman B A M, Van Laar H H (Eds.) Agro-ecological zonation, characterization and optimization of rice-based cropping systems. SARP Research Proceedings. AB-DLO and WAU-TPE, Wageningen, The Netherlands. (in press).

Kumar P, Rosegrant M W, Bouis H E (1995) Demand for foodgrains and other food in India. IFPRI Research Report, Washington, USA. (in press).

Pinstrup-Andersen, P, Pandya-Lorch R (1995) Scenarios for world food security and distribution (pages 89–111 in this volume).

Singh I P, Singh B, Bal H S (1987) Indiscriminate fertilizer use vis-à-vis groundwater pollution in central Punjab. Indian Journal of Agricultural Econmics 42:404–409.

Sinha S K, Aggarwal P K, Chopra R K (1985) Irrigation in India: Phenological and Physiological basis of water management in grain crops. Adv. Irrig. 3:129–212. Academic Press.

Sehgal J L, Mandal D K, Mandal C, Vadivelu S (1990) Agro-ecological regions of India. Tech. Bull. NBSS Publ 24.

WPD (1989) Project Directors's Report. Pages 1–156 in Annual report, All India Coordinated Wheat Improvement Project, IARI, New Delhi, India.

Options for sustainable agricultural systems and policy instruments to reach them

A. KUYVENHOVEN, R. RUBEN and G. KRUSEMAN
Department of Development Economics, Wageningen Agricultural University, P.O. Box 8130, 6700 EW Wageningen, The Netherlands

Key words: farm household modelling, farm stratefication, market development, multiple goal planning, policy instruments, regional development scenarios, supply response analysis, sustainable land use

Abstract. Options for sustainable agricultural systems can be evaluated at regional level making use of scenario analysis. Two different approaches for the design and appraisal of regional development scenarios are presented: (a) comprehensive resource based planning, using multiple goal linear programming techniques, and (b) supply response analysis using econometric techniques and continuous homogeneous production function. Both modelling approaches, appropriately modified, are combined into an integrated framework for interactive evaluation of trade-offs between agro-ecological goals at regional level and the attainable socio-economic options at farm household level.

To gain insight in the effectiveness of different instruments of agrarian policy, micro-economic analysis of farm level response follows a household modelling approach based on production and consumption criteria. Farm types are distinguished according to their objective functions, and land use adjustments are calculated as induced by price changes. Response multipliers appeared to be highly dependent on the level of development of factor and product markets.

Selection of feasible policy instruments to influence resource allocation decisions by farmers should also take into account transaction costs related to the adjustment process of farming systems, and procedures for regional aggregation.

Introduction

Depletion of natural resources and environmental pollution generate increasing attention from both a bio-physical and socio-economic viewpoint. Growing public awareness and political consensus on the necessity to attain sustainable resource use call for deliberate government intervention and structural reform as supporting measures to induce corresponding private behaviour. The design and selection of suitable policy instruments proves, however, to be difficult, for the following reasons: (i) conflicts between policy objectives become evident, (ii) transmission of policy instruments into farmers decisions tends to be highly imperfect, and (iii) adjustment of farming systems is a costly and time-consuming process.

Policy instruments to influence farmers decisions on resource use and factor allocation need to be based on a thorough knowledge of structural

J. Bouma et al. (eds.), Eco-regional approaches for sustainable land use and food production, 187–212.

characteristics of farm households – their resource endowments and multiple objectives – and the market and institutional environment. Moreover, property rights and distributional aspects of agrarian contract choice also influence the scope for farmers response to policy incentives.

Sustainable land use can be fostered only when it coincides with farm level objectives in terms of income, food security and/or risk aversion. The decision making process on land use can be influenced by 'right' incentives, generating positive net private economic welfare effects when properly accounting for externalities. The selection of feasible instruments also depends on the extent of market and government failure.

The Wageningen cooperative research project 'Sustainable land use and food security' (DLV) aims at the development of an operational methodology for the formulation, exploration and evaluation of policy options for sustainable land use and food security at (sub)regional and farm level, based on the integration of agro-ecological and socio-economic criteria. Better understanding of the linkages between policy objectives at regional level and farm household responses at micro level enables selection of feasible instruments. The relevance of research results can be considered as an *ex ante* evaluation of the potential impact of various policy instruments, and the *a priori* selection of possible 'attractive' technical options for future research and extension.

The present contribution is structured along three different levels of analysis. First, procedures for the appraisal of prospects for sustainable development at regional level will be presented. Next, some results of farm household modelling approaches will be presented as a procedure to identify farm level reactions to agrarian policy instruments. Finally, the interaction of both types of analyses for the selection and evaluation of suitable policy instruments will be discussed.

Regional appraisal

The design and appraisal of rural development scenarios requires detailed knowledge on (i) the availability and quality of natural resources, (ii) access to factor markets for land, labour and capital, (iii) access to and spatial distribution of infrastructure and services, and (iv) the resource endowments structure of farm enterprises. These four elements should be combined into an analytical framework, that permits *ex ante* evaluation of the impact of agrarian policy instruments on regional resource use and allocation.

Regional scenarios can be defined as the different outcome for certain goal variables resulting from changing sets of policy variables (Thorbecke and Hall 1982). This analysis should be based on knowledge about the structural relationships between input factors (controllable variables and exogenous factors) and output factors (goal variables and unintended side effects).

Although the spatial definition of 'region' still leaves room for discussion (functional region, planning region, administrative region), this level of analysis permits an integral treatment of available policy options to exploit regional comparative advantages based on immobile factors, considering selective regional closure mechanisms for market clearance (Weaver 1981). Moreover, relations with higher level systems, e.g. foreign trade, government budget incidence, agro-processing or finance, should be included.

Approaches to regional planning

Two different approaches to regional planning may be distinguished: (a) comprehensive resource based planning, and (b) supply response analysis. Both conceptual frameworks share their focus on identification of policy instruments that permit land use adjustment, but their analytical procedures are quite different. The first approach (Fresco *et al.* 1992; Romero and Rehman 1989) focuses on the identification of potential land use options, based on the assessment of available resources. Main attention is given to boundary conditions as determined by bio-physical factors (soil types, climate) and market constraints. Selection of the preferred resource use can be realized subsequently in discussions with different stakeholders, based on the relative preference in objectives and their ability to realize them. Rational behaviour is considered a prerequisite for decision making.

The supply response approach (Askari and Cummings 1976) gives emphasis to farm level reactions to instruments of agrarian policy. Decisions on land use depend on resource endowments, (multiple) objectives and expected prices faced by individual actors. Therefore, actual land use may be different from potential land use due to market fragmentation, risk and uncertainty, and differential access to factor markets. Furthermore, aggregation of individual decisions at the regional level depends on market demand conditions.

Basic questions raised in the supply response approach are addressed in farm household modelling using a more accurate objective function in terms of utility, and by taking into account the impact of prices and public investment on production, consumption, factor allocation and marketing decisions (Singh *et al.* 1986). The effects of changes in input and output prices and transaction costs on the choice among alternative crops or production techniques can be analyzed in terms of supply response multipliers. These values reflect the prospects for adjustment of the production system and the related welfare effects for agricultural households.

Analytical procedures

The analytical instruments used in both approaches are clearly different in a number of respects. Resource based planning starts from multiple goal linear programming techniques (MGLP). This offers clear advantages for the exploration of future development options, but presents limitations because of the discrete nature of the technical alternatives derived from crop growth models or expert knowledge systems.

Multiple goal programming proved to be useful for identification of the technical outer boundaries of planning regions given different development goals (Veeneklaas *et al.* 1991; Erenstein and Schipper 1993). It can also be used for *ex ante* selection of feasible technical options for future agronomic or livestock research, thus improving efficiency in priority setting for agricultural research (Bakker 1993).

Major limitations of the multiple goal planning approach refer to the comparative static character of the analysis, and the absence of an economic analysis of motives and costs associated with land use adjustment at farm level. Moreover, methodological foundations of linear programming techniques in production economics – lack of economies of scale, absence of multiplier effects, and price exogeneity – pose serious questions for its application in economic analysis.* While multiple goal planning exercises intend to explore long-term strategic development options at regional level (Rabbinge and Van Ittersum 1994) or at farm level (Van Rheenen *et al.* 1994), the underlying assumptions almost prevent an appraisal of these options in economic terms.

The economic reasoning behind linear programming is based on the inseparability of resource allocation and valuation as two aspects of the same problem (Dorfman *et al.* 1958). The concept of Pareto optimality is used to select feasible solutions for which an increase in the value of one criterion can only be achieved by decreasing the value of at least one other criterion. This static optimization rule became easily linked to the theory of induced innovation (Hayami and Ruttan 1985) as a procedure to explain technological change from the relative scarcity relationships of production factors and relative factor prices. This reasoning received major criticism because prices do not always reflect competitive market conditions (market failure) and profit maximizing behaviour is only one element of the farmers' objective function (Beckford 1984; cited in Van der Ploeg 1991). Moreover, new institutional economics approaches dedicate more attention to property rights and transaction costs as major determinants of technological change in agriculture (Bardhan 1989).

Supply response models make use of econometric techniques and rely on continuous homogeneous production functions. Optimization takes place with regard to household utility derived from the expenditure system, expressed

* Several modifications have been introduced into linear programming to account for a non-linear objective function (quadratic programming), separable inputs (integer programming), risk analysis (stochastic programming) and multi-period analysis (recursive programming), but their applicability is limited because of difficulties to find efficient algorithms for solving the models (Romero and Rehman 1989). It is possible to address the problem of non-linearity through simulation with linear segments approaching a continuous function.

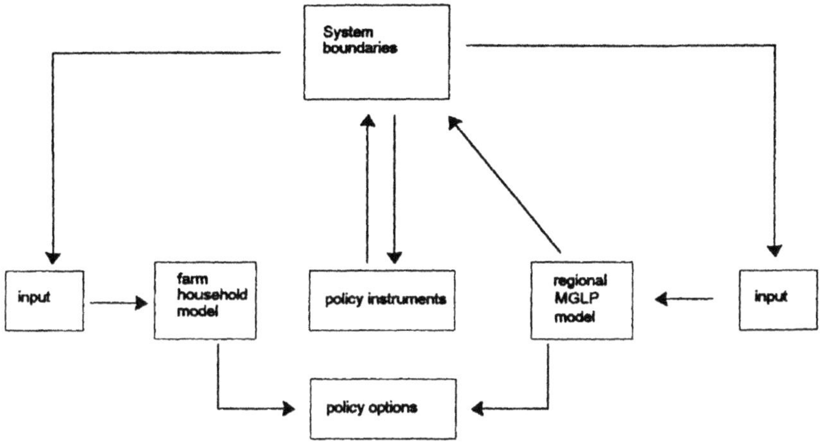

Fig. 1. Relations between regional multiple goal linear programming models and farm household models.

through a negative exponential functional form equation (Anderson *et al.* 1977). The estimation procedure permits analysis of adjustments in the farming system as a result of changes in the economic environment.* It may be used to analyze the implicit structure of objectives underlying current land use (Romero 1993), as well as for the simulation of the possible impact of specific policy instruments on the adjustments in land use and technology systems (Rabbinge and Van Ittersum 1994).

Based on the quantity and quality of the resource endowments at farm level and the household utility associated with consumption and leisure, optimal conditions for resource allocation, subject to certain constraints, can be identified. For analytical purposes, production and consumption are estimated separately through a recursive procedure: production decisions determine farm-level 'full profit' and household consumer demand and utility are subsequently derived (Strauss 1986). Decisions on technology choice in this approach depend on the (non)tradable character of the commodities involved and the properties of the relevant markets. The elasticity of peasant response to price change depends on substitution possibilities and household specific market transaction costs (De Janvry *et al.* 1991).

Integrated framework

Within the framework of the DLV research project, both modelling approaches, each addressing specific questions, have been combined into an integrat-

* Supply response models start their analysis with the current production system and indicate the pace and direction of adjustment as result of policy change. This procedure clearly contrasts with IMGLP analysis, where attention is focussed on potential land use independently of actual production systems.

ed framework (see figure 1). Multiple goal linear programming models are applied to explore the technical outer boundaries and regional development scenarios under different assumptions, and generate as dual solution the relevant shadow prices for constrained factors. Farm household models are used to analyze the feasibility of land use adjustment at farm level as induced by changes in the policy environment (prices or institutions).

The modelling approach used involves three basic characteristics: (i) separability between farm household production and consumption, (ii) econometric estimation of household demand combined with a linear programming approach of farm-household production, and (iii) incorporation of macro data into micro-economic household analysis.

The separation property of farm household models has been introduced to permit estimation of production decisions independent of consumption and labour-supply preferences (Singh *et al.* 1986). Separability has special implications for the treatment of labour, as both production and consumption (leisure) aspects are involved. To account for substitution among family labour, leisure and off-farm/hired labour, free mobility of labour and a uniform wage rate is required. Benjamin (1992) presents a test for separability in case of imperfect factor markets for labour, concluding that even under differential efficiencies of various categories of labour, the separability assumption cannot be rejected. Moreover, Fafchamps (1993) shows that labour allocation decisions of small farmers in the absence of a labour market are highly flexible in order to maintain control on exogenous risk parameters.

The assumption of separability also facilitates the use of different analytical procedures for the specification of the production and consumption side of farm household behaviour. In accordance with Delforce (1994), linear programming procedures are applied to analyze decisions on crop choice and technology selection, subject to resource constraints. The input-output coefficients used in the linear programming model include both current (non-sustainable) activities and potential (sustainable) activities, defined in terms of nutrient balance, biocide index and/or organic matter balance.* For the consumption model a negative exponential function can be estimated, based on cross-sectional household expenditure surveys.**

Integration of macro and micro data can be reached, linking outcomes of the regional model with parameters used in the farm household model. These interactions can function in two directions: (a) regional shadow prices for constrained factors that are used at the farm household level, and (b) farm level response multipliers that are used in the regional model to account

* The actual model still relies on assumptions of 'strong' sustainability, e.g. the absolute level of the sustainability indicators should not exceed a certain threshold level. Further refinement is required in order to permit for 'weak' sustainability, e.g. short-term substitution between capital/savings and sustainability indicators.

** Other alternatives include the use of Log-Linear (Lau *et al.* 1978) or Linear Expenditure Systems (Barnum and Squir 1978) or the derivation of an Almost Ideal Demand System (Deaton and Meullbauer 1980).

for dynamic adjustments. In both situations problems of aggregation and disaggregation occur (Hazell and Norton 1986). Micro-economic analysis focuses on decision-making behaviour of individual farm households within a partial equilibrium framework, thus ignoring interactions among households and feedbacks from the regional economic system. Regional analysis focuses on aggregate variables and assumes more or less homogeneous behaviour of micro-economic units, without considering structural change (Ruggles and Ruggles 1986).

Various procedures have been suggested to address the aggregation problem (Van Daal and Merkies 1984; Erenstein and Schipper 1993; Hazell and Norton 1986). First, relevant variables (prices, factor constraints) need to be specified as endogenous at the regional level and exogenous at the farm household level. Secondly, farm-level response multipliers are corrected for less than proportional reactions due to sectoral differentiation. Finally, the models should take into account transaction costs that reflect imperfect information between the regional and the farm household level.

For operational purposes, the aggregation problem could be solved, making use of an iterative procedure, where the regional model is treated as a single large farm, and response multipliers are calculated for representative farm households with known proportional weights in the regional agrarian structure. Classification of households is based on homogeneity in the objective structure, the availability of resources and the market constraints. For optimization specific rules for the sequence and weighing of objectives are required (Romero 1993).

Policy perspectives

The analytical framework described before can be used for different purposes. Within the framework of regional development scenarios, the so-called 'solution space' for agrarian policy can be determined, both from an agro-ecological viewpoint (*outer technical boundaries*), and from a socio-economic viewpoint (*feasible policy options*). It should be noted, however, that both solution spaces also interact, in such a way that relaxation of certain regional constraints can influence also the *attainable options* at farm level.

Trade-offs between development objectives* can be made explicit, as well as the consequences of choices made at one system level for the solution space at other system levels (Kruseman *et al.* 1993). Scenario analyses also permit exploration of boundary conditions at regional level (e.g. scarcity of land, labour or capital) that restrict the options for sustainable land use at farm level, or the alternatives that improve resource use efficiency.

* Discussions on criteria for 'sustainability' are mostly focused on the identification of threshold values for specific indicators (e.g. productivity, efficiency, stability or resilience). The analysis of interactions between these components permits weighing of certain results in terms of sacrifices of other variables. Therefore, trade-offs between objectives can be determined and substitution between different resources becomes a feasible option.

Decisions on land use and resource allocation in agriculture are (mainly) made by farm households as direct producers. These decisions can be influenced (indirectly) by the government through agrarian policies that modify the economic environment and thus the outcome of the production process. Also different forms of agrarian contract choice, e.g. the access to resources and/or markets, will influence resource allocation.

The most important criterion in the decision making process at farm household level is the economic viability both in the short and in the long run. The short-term perspective refers to optimization of the prospects for consumption of commodities and of leisure as components of household utility (Singh *et al.* 1986). Consumption is made possible by production of commodities and generation of income, making use of available resources (land, labour, capital and knowledge). The long-term perspective refers to decisions with respect to the maintenance of the resource base, consisting of natural resources, capital resources (savings) and human resources (nutrition). Long-term or strategic decision making involves weighing the relative importance of current income against future income streams, dependent on a time discount rate, as well as an assessment of the possibilities of substitution between natural and man-made resources.*

Agro-ecological sustainability goals are normally not included as separate variables in the utility function at farm household level. The major farm level goals are related to the objectives of food security, income generation and leisure. Risk criteria may be taken into account, based on the degree of fluctuation in these goals as a result of the instability of exogenous variables (prices, weather) and the occurrence of background risk (e.g. illness).

The prospects for sustainable land use at farm level depend on the relations between production and consumption, e.g. the proportion of the output reserved for maintenance or reproduction of the resource base. Reproduction of labour resources and knowledge is directly related to consumption (nutrition, education). Reproduction of the natural resource base depends on the relation between the use of soil and water resources, and capital and labour resources that can be mobilized to offset processes of natural resource degradation. Trade-offs appear when household consumption can only be maintained at the desired level at the expense of the natural resource base. Agrarian policy instruments should be identified that permit reconciliation of both objectives.

Operationalisation of this conceptual framework for the identification of feasible policy options requires simultaneous evaluation of trade-offs derived from modelling exercises at (sub)regional level (Veeneklaas *et al.* 1991; WRR 1992) and at farm level (Van Rheenen *et al.* 1994; Alfaro *et al.* 1994). Additionally, an appraisal of the effects of different policy instruments on

* Physical and human resources can be used to restore the natural resource base in order to maintain the long-term productive capacity, but then sacrifices have to be accepted at the consumption side (reduced short-term food security).

land use, household utility (in terms of consumption and leisure) and selected sustainability indicators is required (Kruseman *et al.* 1995). Integration of these approaches permits evaluation of resource use efficiency criteria at farm and regional level, based on the correspondence between agro-ecological and socio-economic objectives.

Farm household modelling approach

To assess the impact of policy instruments on farm household decisions regarding land use, a basic understanding of decision making processes at micro level is required. In farm household models relevant decision making processes are simulated in a quantitative way, to evaluate farm household response to changing external socio-economic circumstances.

The basic assumption underlying the theoretical framework of farm household modelling is that decisions on land use are taken by individual households based on their goals and aspirations, making use of the available resource endowments to undertake specific activities, subject to bio-physical and socio-economic constraints.

One of the aspects that complicates farm household model analysis is related to the existence of many different households with differentiated objectives. Although the interest is in the overall response of farm households to policy change, and not in the reactions of a specific household, the occurrence of a variety of farm types makes it necessary to distinguish among relevant households which show broadly similar reaction patterns. This implies that a relevant classification of farm household types has to be developed.

A second aspect related to modelling farm households for the evaluation of policies directed at promoting sustainable land use, is the need to incorporate sustainability criteria. Three types of processes negatively influencing agro-ecological sustainability can be identified (Kruseman *et al.* 1993): (1) processes related to specific land use systems having on-site effects (e.g. nutrient depletion); (2) processes related to specific land use systems with mainly off-site effects (e.g. biocide leaching); and (3) processes which cannot be attributed to specific land use systems (e.g. biodiversity loss). The latter have to be analyzed at regional level, using aggregate results from lower level analyses. The first two types of processes can be analyzed, at least partly, with farm household models. This implies that the model must account for sustainability criteria.

Econometrically-based continuous production functions, commonly used in farm household models, do not adequately account for synergistic properties of technological alternatives. Hence, the farm household model makes use of linear programming techniques to allow for technical input-output coefficients defined in terms of discrete production technologies generated by the bio-physical sciences (crop growth simulation models and/or expert systems).

These production technologies include bio-physical information, in terms of inputs to obtain desired outputs. In addition to the standard input-output coefficients, indicators on the extent of soil erosion, nutrient depletion and biocide emissions can also be included.* Sustainability criteria are incorporated in the analysis in two ways. In the first place, farm household objectives with respect to the reproduction of their resource base can often be equated to selected agro-ecological sustainability criteria. In the second place, and more important for linking model results to aggregate regional analyses, the side effects of production technologies can be included in the output information.

Policy instruments are intended to influence the socio-economic environment faced by farm households, encompassing markets, services and infrastructure. The approach should therefore take into account the degree to which farm households are incorporated in the market, the degree to which they have access to specific services, and the available infrastructure, e.g. roads or communication channels.

Farm stratification

To evaluate the impact of various instruments of agrarian policy on farm household behaviour, the development of a typology of agrarian households is important. A farm household type is an abstraction from the actual farm households. The function of the farm household type is to enable meaningful extractions from the diverse and diffuse point data on actual land use and farming systems. Different methods can be used to develop a typology of farm households. Two are mentioned here: (1) classifications based on resource endowments and (2) classifications based on differences in objective functions.

The most common way to start is by dividing households into categories based on relative resource availability and access to factor markets, i.e. land/labour ratios are commonly used. The resource-based farm classification is related to the aggregate analysis of Hayami and Ruttan (1985) with respect to technology choice in relation to land/man and capital/man ratios (Ruttan *et al.* 1978). This approach presents serious limitations for micro-economic stratification, as it does not account for agro-ecological differences within and between regions. Moreover, when using relative factor endowments, information must be available on relative factor scarcity, not only in an objective sense at regional level, but also as determined subjectively by household-specific variables, as market information and access to factor markets (labour, capital).

* In economic terms this refers to 'joint production'.

Alternatives proposed refer to production capacity (Higgins *et al.* 1982), i.e. land quality, in stead of farm size (Pingali *et al.* 1987; Binswanger and Pingali 1988), maintaining the notion that differences in resource endowments explain differences in land use. The relevant resources are land (which includes quantity, quality, legal status and spatial arrangement) and capital, in combination with labour and knowledge. When analyzing a larger region, it may be necessary to distinguish between differences in the socio-economic and administrative environment, or access to services (extension, credit, veterinary services, etc.), both related to relative proximity to regional centres. Other criteria that may be used refer to the relationship between the farm household and the socio-economic and administrative environment, viz. market access, access to services, availability of infrastructure, distance, using the concept of transaction costs (Goetz 1992). A final set of criteria may be derived from the social, political and legal structure of society. Often however, these very different criteria are interrelated: factor endowments, political influence and access to goods and services show a strong correlation.

The second classification principle proposed is based on differences in objective functions. The common practice among economists has been to use only the profit maximization objective. This function, however, is not sufficient to account for the technological choices made by farm households in developing countries. Differentiating between multiple objectives seems to harmonize with recent work by sociologists on farming styles (Van der Ploeg 1991). However, deriving the objective functions of farm households directly is impossible, let alone that weights can be attached to them. Alternatively, an indirect method for measuring and weighing multiple objectives is required.

To account for the existence of multiple goals that allows stratification of farms on the basis of divergent objectives, a procedure is needed to weigh different, often conflicting goals. Romero (1993) developed a procedure for weighing farm-level objectives. The methodology hinges on the determination of a number of tentative goals, whose implications are then compared to empirical data. The basic assumption underlying this technique is the possibility to identify tentative objective functions, for which the weights are calculated separately. This requires a specification of goal indicators in a completely independent way. The goals that are considered *ex ante* at the farm household level can be summarized in three vectors: (i) consumption utility maximization, (ii) risk management, and (iii) reproduction of the resource base. These can be translated into objective functions related to (i) consumption capacity, (ii) consumption capacity under uncertainty, and (iii) savings or consumption capacity in the future.

It appears possible to make a stratification based on differences in objective function, using less elaborate techniques, by taking into account differences among farms in terms of availability and access to productive resources (land area, soil quality, credit, family labour) and in terms of production strategies (risk taking, cropping mix, factor intensity) (Ruben *et al.* 1994).

Model structure

Farm household modelling approaches in our study follow the basic model presented by Singh *et al.* (1986), derived from earlier work, especially by Barnum and Squire (1978, 1979a, 1979b). The basic structure of the model, extended and adapted to suit a variety of purposes, is presented here to indicate its logic and to explain the modifications introduced by the DLV team.

For any production cycle, the household is assumed to maximize a utility function:

$$U = U(C^f, C^m, L^L), \tag{1}$$

where the consumption commodities are an agricultural staple (C^f), a market-purchased good (C^m), and leisure (L^L). Utility is maximized subject to a cash income constraint:

$$p^m * X^m = p^f * (Q - X^f) - w * (L^R - L^F), \tag{2}$$

where p^m and p^f are the prices of the market-purchased commodity and the staple, respectively, Q is the household's production of the staple and X^f is subsistence consumption (so that $Q - X^f$ is its marketed surplus), w is the market wage, L^R is total labour input (labour requirement to attain production Q), and L^F is total family labour input (so that if $L^R - L^F$ is positive, external (wage)labour is hired and, if negative, the family may be engaged in off-farm labour).*

The household is also confronted with a time constraint: it cannot allocate more time to leisure, on-farm production and off-farm employment than the total time available:

$$L^L + L^F \leq L^T, \tag{3}$$

where L^T is the total availability of household time.

The household faces a production constraint, in the basic model (one commodity, one technology) defined as a production function:

$$Q = Q(L^R, A), \tag{4}$$

where A is the household's fixed quantity of land.

The three constraints can be combined into a single constraint by substituting the production constraint into the cash income constraint for Q and substituting the time constraint into the cash income constraint for L^W:

$$p^m * X^m + P^f * X^f + w * L^L = w * L^L + \pi, \tag{5}$$

* This conditions holds assuming undifferentiated wage rates.

where

$$\pi = p^f * Q(L^R, A) - w * L^R \tag{6}$$

with π as a measure of farm profits. The left-hand side of equation (5) represents the expenditure system and the right hand side the production component of the farm household, based on the full income concept developed by Becker (1965). The model is usually solved by separating the production and consumption decisions, but for this econometric estimation a large and detailed data set is required. After calculating the partial derivatives, response elasticities can be calculated. This simple model can be extended to incorporate amongst others: various factor and non-factor inputs, multiple commodities, different goals, and differentiated factor prices.*

The basic model presented in this section has been adapted to suit the purpose of the modelling exercise. In the present case the main interest is in decisions concerning land use, i.e. the choice of land use systems and corresponding technology. In the first place this implies an extension of the simple production function to allow explicit analysis of technology choice. This is necessary for evaluation of the agro-ecological sustainability consequences of changing land use.

Secondly, the use of objective-based farm household stratification implies specification of multiple (conflicting) household level goals, pertaining to income (consumption), risk management, and sustainability (reproduction of the land resources, i.e. safeguarding consumption in the future).

Thirdly, the presence of missing market segments requires adaptation of the farm household model in such a way that the strict separability condition is relaxed, since this condition requires well functioning markets. This can only be done if there is a direct link between consumption goals and production activities. There are two important reasons for relaxing the separability criterion: (1) the need to evaluate the impact of conventional policy instruments in situations with missing markets; and (2) the need to evaluate the impact of market improvement as an alternative policy measure.

Finally, the model framework should be able to accommodate data-scarce situations common in most developing countries, since available human and financial resources often do not permit extensive data collection over a number of years necessary for the full econometric estimation of the farm household model.

* The most fundamental assumption is the separability of production and consumption. Other important conditions of the basic model – which could be relaxed in more extended versions – are: (1) risk is excluded; (2) production is confined to one crop year; (3) competitive markets are assumed; (4) factor endowments are fixed in the short run; (4) desired levels of savings are fixed; (5) off-farm income is endogenous; (6) perfect substitutability of family and hired labour under the assumption of perfect labour markets; (7) the households are price takers; (8) the objective is profit maximization; and (9) land is owned or rented at fixed rates, with no contractual arrangements leading to non-standard profit maximizing conditions.

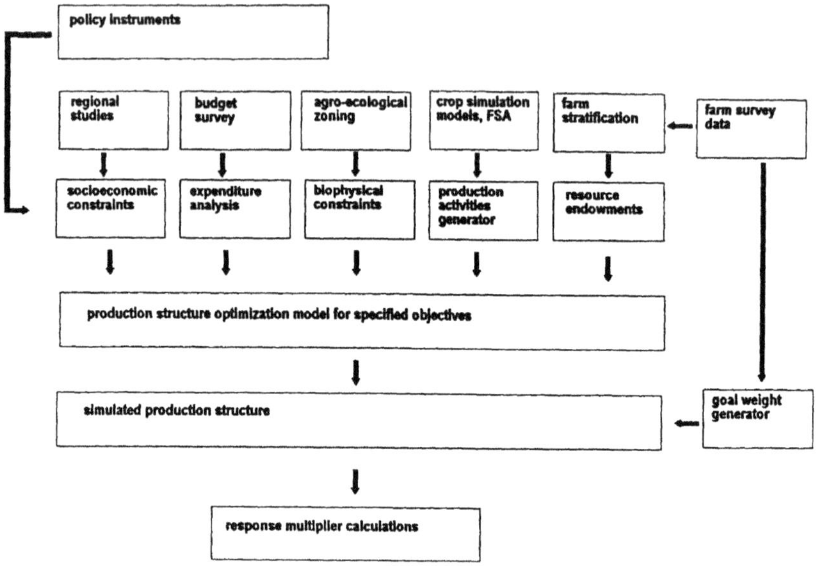

Fig. 2. Structure of the farm household model.

The model developed within the DLV project has a modular structure, for two reasons. First, the framework should be suitable for use in diverse locations, also when data availability is limited. Secondly, the incorporation of information from various disciplines leads to complex interactions. Including all relevant processes in a single mega-model would lead to several difficulties: the model would be extremely difficult to understand, and the implications of model results become increasingly difficult to comprehend. Also the data requirements for a mega-model will undoubtedly be more restrictive than in a modular approach. An additional advantage of the modular approach is that separate disciplinary teams can work on the various components.

The model framework enables the use of data from a variety of sources, making efficient use of available secondary information. These data often have to be transformed to fit into the syntax of the model core. The degree of detail and accuracy will vary; for specific situations, different approaches may be necessary to calculate the relevant parameters.

The model framework presented in this paper is summarized in figure 2. The core of the modelling approach is a linear programming model that optimizes specified goals, using a defined set of activity options, i.e. possible production technologies and consequently different values for relevant sustainability indicators. Optimization is subject to a number of constraints which are grouped as: constraints related to resource availability, including access to external resources; constraints related to the socio-economic environment, i.e. markets (access, prices), services (access) and infrastructure (transaction costs); and constraints related to the bio-physical environment (climate and soils).

The model parameters and the constraints are calculated in separate modules. The farm household model consist of the following components: (1) multiple objectives (consumption expenditures, risk analysis, reproduction of the resource base); (2) resource endowments (land, labour, capital, knowledge); (3) activity options (production technologies for agricultural activities); (4) socio-economic constraints (prices, market access constraints, service availability and access, transaction costs)* and (5) biophysical constraints (climate).

Traditional linear programming techniques give optimal solutions only for the specified objective function, implying the need for a routine to account for (weighted) multiple (conflicting) objective functions. After weighing the optimum outcomes for specified objectives, the simulated production structure of the specified farm household is obtained.**

Responses of farm households to changes in policy are expressed in terms of response multipliers,*** defined as:

$$\varepsilon_{\Lambda_m}^{\rho_p} = \frac{\mathrm{d}\Lambda_m}{\mathrm{d}\rho_p} * \frac{\rho_p}{\Lambda_m} \tag{7}$$

$$\frac{\mathrm{d}\Lambda_m}{\mathrm{d}\rho_p} = \frac{\delta\Lambda_m}{\delta\rho_p} + \frac{\delta\Lambda_m}{\delta U} * \frac{\delta U}{\delta\rho_p}, \tag{8}$$

where: Λ_m = relevant indicator: cultivated area, income, utility, fertilizer and biocide use, nutrient depletion; U = farm household goal: utility of consumption, risk-corrected utility of income, net present value of present and future income; ρ_p = relevant policy impact: output prices, fertilizer price, biocide price, transaction costs, wage rate and price for industrial goods, credit availability; $\varepsilon_{\Lambda_m}^{\rho_p}$ = response multiplier of impact on indicator m for a specified range of ρ.

The specification of the response multiplier in household models also takes into account the positive impact of the objective function effect which may dampen or even outweigh the effect of either standard consumer demand theory or standard agricultural response theory.

* These constraints can account for missing markets, which makes the problem of separability irrelevant.

** In some cases, when only one objective can be identified, or when a discrepancy between the model outcome and the actual production structure remains, it may be required to include a production structure adjustment module, based on calibration of the calculated production structure with statistical data (Kruseman *et al.* 1995).

*** These response multipliers resemble elasticities, but differ in the sense that since they are calculated with linear programming techniques for a specified range of policy change. Hence they do not correspond to the first-order Kuhn–Tucker conditions calculated on basis of a derived Lagrangean equation.

Results

The farm household modelling approach presented here was used in case studies in two different regions: (i) the Atlantic Zone of Costa Rica, where a model was used for diversified medium-size farms with cropping and livestock activities and access to markets for non-traditional export crops, and (ii) the Cercle de Koutiala in southern Mali, where the model was used for a single farm type: relatively well-endowed farm households, with more limited access to services.*

The models yield results with respect to the response of farm households to specific changes in the socio-economic environment. In both cases the final impact of policy instruments turned out to depend heavily on the way local factor and product markets function. In both models the same type of objective function was used. While labour is a constraining factor in the two settings, the labour market in the Atlantic Zone is well functioning, while it is missing in Mali. The market for non-factor inputs (fertilizers, biocides) is competitive in Costa Rica, while in Mali access depends on cash crop production. Access to rural financial markets is a limiting factor,** especially in Mali. As a result, the influence of price instruments on the production structure is very limited in the Mali case compared to Costa Rica. In the policy scenarios various price instruments are evaluated: output prices for selected cash crops and cereals, wage rates and input prices for biocides and fertilizers. Infrastructure development is evaluated in terms of changes in the margin between farm-gate and market-producer prices. Market development in terms of access to rural finance is defined in terms of the possibility to make in-depth investments (capital goods, perennial tree crops, etc.).

Table 1 highlights changes in production structure in Costa Rica and table 2 in Mali. The production structure is defined in terms of area allocated to different types of crops, e.g. cash crops, pulses and cereals. In Mali the production structure appeared to be relatively stable, and changes are only marginal; the area under cotton remains constant, not only because it is a profitable activity constrained only by rotational requirements, but also because it is the only activity where fertilizers are used. In the case of Costa Rica, the production structure is much more prone to change.*** In terms of the use of alternative technology (defined as agro-ecologically sustainable production systems), the reactions in Mali and Costa Rica are diverse. In Costa Rica,

* The work in Mali is ongoing and other farm types are being modeled at present.

** Limited functioning of rural financial markets refers to the absence of savings and insurance mechanisms outside the agricultural sector and/or a missing market segment for rural finance. Relaxation of financial constraints does not necessarily imply credit provision as policy device, but refers to the improvement of capital market conditions that may stimulate self finance and/or commercial lending operations.

*** Although it is not evident from table 1, there are shifts in the area under cash crops, since farmers have the choice between alternative cash crops.

Table 1. Changes in production structure in the case studies for Costa Rica (base scenario: 1990 circumstances, areas in hectares, new technology in % of the cropped area; policy scenarios: areas in hectares change and % new technology in point change compared to base scenario).

Scenario		area palmheart	area cassava	area plantain	area pulses	area fallow	% new technology
Base scenario		1.9	3.4	0.0	0.4	8.8	34
Price instruments							
cash crops							
palmheart	−10%	−1.7	−1.0	3.5	0.2	−1.2	−19
	−50%	−1.9	−1.0	3.5	0.6	−1.5	−21
plantain	−10%	0.0	0.0	0.0	0.0	0.0	0
	+10%	−1.6	−0.9	3.5	0.0	−1.1	−18
food crops							
maize	−10%	0.2	−0.3	0.0	0.0	0.1	0
	+10%	−0.1	0.1	0.0	0.0	0.0	1
inputs							
wage	−10%	0.1	−0.1	0.0	0.0	0.0	1
	+10%	0.0	0.0	0.0	0.0	0.1	0
fertilizer	−10%	0.1	0.0	0.0	0.0	−0.1	1
	+10%	0.0	−0.1	0.0	0.0	0.2	−1
biocides	−10%	0.1	0.0	0.0	0.0	−0.1	1
	+10%	0.0	−0.1	0.0	0.0	0.1	0
Infrastructure							
transport costs	−25%	−0.3	0.3	0.0	0.0	−0.1	1
	−50%	−0.2	0.2	0.0	0.0	0.0	1

Note: Model for diversified medium-size farms in the Neguev settlement, Atlantic Zone of Costa Rica.

technology choice depends on input and output prices and related transaction costs, while in Mali it depends mostly on access to factor markets.

Modification of cash crop prices has a profound effect on land use decisions, especially in terms of substitution between alternative cash crops. Palm heart is the most attractive crop, but is constrained by high operating costs. Plantain is the second best option and replaces palm heart as soon as the difference in net margin between the two cash crops changes sign. Since plantain uses more biocides (fungicides), overall sustainability is likely to be negatively affected in those scenarios where plantain substitutes for palmheart. Pulses, a food crop, and cassava, which is both a cash and a food crop, adapt along with the main cash crops as a consequence of more efficient resource allocation. The area under cereals is unaffected by the policy instruments

Table 2. Changes in production structure in the case studies for Mali (base scenario: 1990 circumstances, areas in hectares, new technology in % of the cropped area; policy scenarios: areas in hectares change and % new technology in point change compared to base scenario).

Scenario		area pulses	area cereals	area fallow	% new technology
Base scenario		1.9	6.1	0.5	30
Price instruments					
cash crops					
cotton	−10%	0.2	−0.2	0.0	−1
	+10%	−0.1	0.1	0.1	1
inputs					
fertilizer	−10%	0.0	0.0	0.0	−1
	+10%	0.0	0.0	0.0	0
biocides	−10%	0.0	−0.1	0.0	0
	+10%	0.1	−0.1	0.0	−1
Infrastructure					
transport costs	−25%	0.1	−0.1	0.0	−1
	−50%	0.2	−0.2	0.0	2
Market development					
access to	+25%	0.0	0.1	0.0	5
rural finance	+50%	0.0	0.0	0.0	7
	+100%	0.1	0.1	−0.1	15

Note: Model for farm households in the Kaniko area, Cercle de Koutiala, southern Mali.

selected. Improvement of rural finance does not effect production structure. A decrease in transportation costs improves the relative attractiveness of cassava. There is a slight influence of wages on the production structure, because farm households have a net demand for hired labour. The effect of food crop and input price changes on production structure is relatively small compared to the influence of cash crop prices. This can be attributed to the relatively strong market orientation of the households and to fairly high margins for their cash crops which are not drastically affected by changes in input prices.

By contrast, in Mali the area under cotton, the main cash crop, is unaffected by the policy measures. Cotton areas are constrained by rotation requirements, which are enforced by the CMDT (Compagnie Maliènne de Developpement des Textiles), the regional cotton board, which has a monopoly over input supply. There is some trade-off between cereals and pulses for various policy instruments. The strongest effect, however, is obtained through the improvement of the access to rural finance. Investment in animal traction and related

Table 3. Changes in goal variables in the case studies for Costa Rica and Mali (percent change compared to base scenario).

Scenario		Net income		Consumption utility		Nutrient balance/use		Biocide use	
		CR	Mali	CR	Mali	CR	Mali	CR	Mali
Price instruments									
cash crops									
palmheart	−10%	−4	–	−1	–	1	–	13	–
	−50%	−5	–	−1	–	1	–	15	–
cotton	−10%	–	−4	–	−2	–	2	–	−4
	+10%	–	4	–	1	–	−1	–	2
plantain	−10%	0	–	0	–	0	–	0	–
	+10%	3	–	1	–	2	–	13	–
food crops									
cereals	−10%	−9	−1	−2	0	−1	0	−2	0
	+10%	8	1	2	0	0	0	1	0
wage rate	−10%	1	0	1	0	1	0	−1	0
	+10%	−2	0	0	0	−1	0	−1	0
fertilizer	−10%	3	0	1	0	2	0	1	−1
	+10%	−4	0	−1	0	−2	0	−2	0
biocides	−10%	2	0	1	0	1	0	0	1
	+10%	−3	0	−1	0	−2	1	−1	−2
Infrastructure									
transport-	−25%	15	3	4	1	1	2	3	−3
costs	−50%	27	6	9	3	0	3	2	3
Market development									
	+25%	0	3	0	1	0	3	0	16
rural	+50%	0	5	0	2	0	7	0	16
finance	+100%	0	7	0	2	0	19	0	15

Note: CR = model for diversified medium-size farms in the Neguev settlement, Atlantic Zone of Costa Rica.
Mali = model for farm households in the Kaniko area, Cercle de Koutiala, southern Mali.

capital goods is responsible for this effect. The poor market integration for subsistence food crops causes lack of response to changes in cereal prices.

In table 3 changes in values of goal variables, both in terms of household, as well as regional objectives, are summarized. Income-related variables (net income and consumption utility) are more prone to change as a result of conventional price instruments in Costa Rica than in Mali. This can be explained by the differences in market orientation of farm households. Sustainability indicators, defined in terms of nutrient balances and biocide use, are more responsive to price instruments in Costa Rica than in Mali.

It can be concluded that different instruments have a differential impact on resource allocation and household welfare, depending on the development of local factor and product markets, and the market conditions faced by farm households. Cash crop prices influence technology choice and income-related parameters, but the intensity of response strongly depends on the degree of market integration of farm households. Prices of food crops only affect the production structure when household production is market-oriented, as in the Costa Rican case. Income effects are more pronounced than consumption effects, as part of the price increase is taken as leisure. The influence of wage rate modifications is directly linked to the functioning of the labour market; only in Costa Rica wages influence resource allocation and household consumption. The sign of these reactions indicates that substitution effects outweigh the income effect (Singh *et al.* 1986; De Janvry *et al.* 1991). In Mali, external wage labour opportunities are only to a limited extent available, and consequently the wage rate has no marked influence on farm household production or consumption.

The impact of input price modifications on factor use is limited to cash crop production in Mali,* while they lead to stronger reactions in Costa Rica due to shifts among different cropping systems. Given the poor market integration in Mali, it is not surprising that market development, such as the improvement of the functioning of rural financial markets, proves to be a suitable policy instrument to induce changes in production systems, in technology choice, as well as in income-related variables.

Considering nutrient (organic matter) balances and biocide use as agro-ecological sustainability indicators, simultaneous improvement in these indicators with farm household welfare requires a different set of policy instruments in each situation. While input price policies (e.g. fertilizer subsidies, biocide taxing) are effective for this purpose in Costa Rica, improved access to factor markets is a more suitable policy instrument in Mali. Finally, infrastructure improvement to reduce transaction costs is an effective instrument in both situations.

Policy scenarios

The concept of 'scenario' as used in production-ecological (explorative) studies refers to the possibilities and limitation pursuing different goals, considering also the perspectives of different groups of stakeholders. Based on the

* There exists a close correlation between input price changes and cash crop output prices, as fertilizer and seed distribution can be considered as an 'interlinked transaction' with cotton marketing, within the institutional framework of the Compagnie Malienne de Textiles (CMDT).

technical (input/output) coefficients for relevant land use activities, assumptions about off-farm employment options and information about prices, optimization can take place under various physical and market constraints. Different scenarios can be explored, depending on the goals that are pursued, e.g. maximum revenue, maximum employment generation, or minimum environmental damage (van Rheenen *et al.* 1994). Also considerations of uncertainty can be incorporated, taking into account different rainfall regimes (Veeneklaas *et al.* 1991). Finally, sensitivity analysis for the effects of changing relative prices offers information about the robustness of the results.

This type of scenario studies may be considered as an exploration of technical options, but are not intended for forecasting, nor can it be used to evaluate different paths to realize desired technical options (WRR 1992). Differences among various scenario outcomes may be used to illustrate the bio-physical constraints to land use modifications, and thereby provide a framework for assessing long-term strategic policy options. However, the scenarios exclude behavioral relations and institutional factors.

Policy scenarios can be defined, in contrast to exploratory studies, as the aggregate result of modifications in goal variables, caused by certain socio-economic interventions. Based on the seminal work of Tinbergen (1956), policy analysis focuses on the relations between exogenous input variables (instruments) and endogenous output variables (results), through a chain of connecting structural relations (markets and institutions). In addition to detailed knowledge on technical (input/output) relations, it also requires a set of assumptions about the possible reactions of individual farmers to policy instruments (micro-modelling), and a clear understanding of the transfer mechanisms to channel relevant information to the farmers (transaction costs).

Policy instruments can be classified according to their level of intervention into (i) farm-level policies (e.g. input subsidies), (ii) structural policies (investment in infrastructure, taxation) and (iii) trade policies (tariffs, quotas) (Colman and Young 1989). They may be categorized also according to their working sphere, as (i) legal or administrative regulation (maximum levels), (ii) price and fiscal regulation (taxes and subsidies), (iii) private arrangements (bilateral rules) and (iv) social regulation (information and transparency of exchange). For assessment of the impact of agrarian policy measures, distinction can be made among (i) price effects, (ii) production effects, (iii) consumption effects, (iv) trade effects, (v) public expenditure effects and (vi) distributional effects (Corden 1971; cited in Colman and Young 1989).

The appraisal of feasible policy options for regional development involves therefore the following components: (i) analysis of the rationality of present resource use and allocation, (ii) exploration of attainable future resource allocation, and (iii) analysis of attainable modifications in resource use under different scenarios of policy intervention. Analysis of the present situation offers information on the structure of the farm household objectives (relative weights of income, accumulation and risk objectives) and allows identification of the constraints at farm level. Future options resulting from technical studies will usually not be fully attainable, due to market and institutional imperfections and high transaction costs (transport, technical change, etc.). Policy scenarios may indicate the outcome under unchanged economic conditions, and the potential impact of changing economic conditions on farm level factor allocation and household utility. Furthermore, distributional effects should also be taken into account.

With respect to the evaluation of the impact of agrarian policies on sustainable resource use, welfare economics offers relevant criteria. The measurement of producer and consumer surplus may be considered a partial equilibrium measure for the evaluation of welfare consequences of resource allocation (Howitt and Taylor 1993). Dynamic welfare analysis also incorporates risk aversion criteria (risk premium) and resource use over time (implying discounting) into Pareto optimality. In case of market imperfections, second-best solutions can be derived. Equity aspects also play a role, as a potential Pareto improvement in natural resource use could be accomplished also through direct redistribution of assets, especially if no real obligation for compensation can be enforced (Bromley 1992).

The potential impact of agrarian policy instruments on farm-level resource allocation and sustainability indicators is also influenced by prevailing market and institutional arrangements. Binswanger and Rosenzweig (1986) demonstrate that crop choice and technology selection not only depend on relative prices, as market access is also influenced by factors like farm size, collateral availability, maintenance and supervision costs, and spatial dispersion. Interventions should therefore take full account of partial responses, e.g. in case of absence of contract choice alternatives (De Janvry *et al.* 1991; Hayami and Otsuka 1993).

The impact of policy instruments depends on the improvement of the market and institutional environment and overall resource availability. In land-scarce regions like Southern Mali, where factor markets for land and capital are not very well developed, instruments of price policy appeared to have limited influence on resource allocation and market deepening appears to be the device. By contrast, in a highly commercialized region like the Atlantic Zone of Costa Rica, modification of input prices and, to a lesser extent, transaction costs, can be considered as suitable instruments to improve sustainable land use while maintaining household consumption prospects.

Decision support systems for agricultural resource policies, as developed within the public choice framework, can be helpful to identify suitable instruments for specific circumstances. As technological change is highly dependent on demand conditions, the effects of price changes need to be analyzed for both production and consumption aspects. Other methodological problems that have to be faced when dealing with the selection of suitable policy instruments, refer to different transaction cost categories that should be taken into account for the evaluation of land use adjustment. In addition to information costs associated with access to markets, also adjustment costs related to modifications in the technical or operational requirements of new activities (training, marketing, etc.) need to be taken into account. Finally, aggregate evaluation procedures require refinement of the calculated micro-economic response multiplier for sectoral use.

Conclusions

Agrarian policies for regional development are commonly aimed at satisfying broad and conflicting objectives, defined in terms of economic growth, income distribution and conservation of the natural resource base. In conformity with these objectives, economic incentives are used to influence farmers' decisions on resource allocation.

Three levels of decision making are incorporated into the analysis. At policy level, institutional and administrative constraints and possibilities are studied to evaluate policy instruments in terms of their feasibility. The regional level comprises markets, services and infrastructure. This is the level at which policy instruments have their direct impact in terms of price adjustments, access to resources and services, improvement of infrastructure and functioning of markets. At the third, farm household, level, actual decisions on land use are made, in accordance with prevailing socio-economic conditions as determined by processes at the regional level.

The proposed methodological framework will be useful for land use planning exercises at the regional level and to support public debate on policy alternatives. Additionally, the approach allows an improved interaction between policy analysis and agricultural research. Using technological innovations as starting point, the method can be used to explore the possibilities to induce adoption of new technology through well directed policy measures. Similarly, analysis of policy impact may guide agricultural research in the development of appropriate technology, that has a better chance of adoption taking into account both production and welfare implications.

Further refinement of the methodology and analytical framework is required to enable better specification of adjustment costs and to develop aggregation procedures between farm and regional level. Testing the integrated modelling approach in different resource environments and validation in situations with different data availability levels, are the next steps in the DLV research programme.

Analysis of the institutional feasibility of different policy instruments and appraisal of the direct impact of policies at regional level in terms of price formation and resource availability, has a lower priority, since other groups and organizations are better equipped for this task. This final remark implies the need for close cooperation among different disciplines and organizations to increase the understanding of the complex processes related to sustainable agricultural systems, so that it will become possible to identify the possibilities to slow down the degradation of natural resources while maintaining food security.

Acknowledgments

Thanks are due to Herman van Keulen, Huib Hengsdijk and Henk Moll for their comments on an earlier version of this paper. Field work was realized in close cooperation with the WAU/CATIE/MAG *Atlantic Zone Programme* in Costa Rica and AB-DLO/IER *Production Soudano-Sahélienne* in Mali. For Mali use was made of farm survey data collected by IER/DRSPR/KIT Sikasso.

The DLV research programme is supported by the Netherlands Ministry of Agriculture, Nature Management and Fishery (LNV) and the Directorate General for International Cooperation (DGIS) of the Ministry of Foreign Affairs, through the Research Institute for Agrobiology and Soil Fertility (AB-BLO).

Acronyms

DLV	Duurzaam Landgebruik en Voedselvoorziening (Sustainable Land Use and Food Security)
MGLP	Multiple Goal Linear Programming

References

Alfaro R, Bouma J, Fresco L O, Jansen D M, Kroonenberg S B, Van Leeuwen A C J, Schipper R A, Sevenhuysen R J, Stoorvogel J J, Watson V (1994) Sustainable land use planning in Costa Rica: a methodological case study on farm and regional level. Pages 183–202 in Fresco L O, Stroosnijder, Bouma J, Van keulen H (Eds.) The Future of the land: mobilizing and integrating knowledge for land use options. John Wiley and Sons, New York, USA.

Anderson J R, Dillon J L, Hardaker J B (1977) Agricultural decision analysis. Iowa State University Press, Ames, USA.

Askari H, Cummings J T (1976) Agricultural supply response: a survey of the econometric evidence. Praeger Publishers, New York, USA.

Bakker E J (1993) Multiple goal planning as a tool for the analysis of sustainable agricultural production in Mali. Pages 189–198 in Goldsworthy P, Penning de Vries F W T (Eds.) Opportunities, use, and transfer of systems research methods in agriculture to developing countries. Kluwer Academic Publishers, Dordrecht, The Netherlands.

Bardhan P (Ed.) (1989) The economic theory of agrarian institutions. Clarendon Press, Oxford, UK.

Barnum H N and Squire L (1978) Technology and relative economic efficiency. Oxford Economic Papers 30:181–198.
Barnum H N and Squire L (1979a) A model of an agricultural household: theory and evidence. Johns Hopkins University Press, Baltimore, USA.
Barnum H N and Squire L (1979b) An econometric application of the theory of the farm household. Journal of Development Economics 6:79–102.
Becker G S (1965) A theory of the allocation of time. The Economic Journal 75:493–517.
Beckford G L (1984) Induced innovation model of agricultural development: comment. In Eicher C K and Staatz J M (Eds.) Agricultural Development in the third world. The Johns Hopkins University Press, Baltimore, USA, and London, UK.
Benjamin D (1992) Household composition, labour markets and labour demand: testing for separation in agricultural household models. Econometrica 60:287–322.
Binswanger H and Pingali P (1988) Technological priorities for farming in sub-saharan Africa. Research Observer 3:81–98.
Binswanger H P and Rosenzweig M R (1986) Behavioral and material determinants of production relations in agriculture. The Journal of Development Studies 22:503–538.
Bromley D W (1992) Land and water problems: an institutional perspective. American Journal of Agricultural Economics 64:834–44.
Corden W M (1971) The theory of protection. Oxford University Press, Oxford, UK.
Colman D and Young T (1989) Principles of agricultural economics: markets and prices in less developed countries. Cambridge University Press, Cambridge, UK.
Van Daal J and Merkies H Q M (1984) Aggregation in economic research: from individual to macro relations. Reidel Publishing Company, Dordrecht, Netherlands, and Boston, USA.
Deaton A and Meuhllbauer J (1980) An Almost Ideal Demand System. American Economic Review 70:312–326.
Delforce J C (1994) Separability in farm-household economics: an experiment with linear programming. Agricultural Economics 10:165–177.
Dorfman R, Samuelson P A, Solow R M (1958) Linear programming and economic analysis. McGraw-Hill Book Company, New York, USA.
Erenstein O C A, Schipper R (1993) Linear programming and land use planning. Wageningen Economic Studies 28, Wageningen, The Netherlands.
Fafchamps M (1993) Sequential labor decisions under uncertainty: an estimable household model of West-African farmers. Econometrica 61:1173–1197.
Fresco L O, Huizing H G J, Van Keulen H, Luning H A, Schipper R A (1992) Land evaluation and farming systems analysis for land use planning. FAO working document. FAO, Rome, ICT, Enschede, WAU, Wageningen, The Netherlands.
Goetz S J, (1992) A selectivity model of household food marketing behavior in Sub-Saharan Africa, American Journal of Agricultural Economics 64:444–452.
Hazell P B R, Norton R D (1986) Mathematical programming for economic analysis in agriculture. MacMillan Publishing Company, New York, USA.
Hayami Y, Ruttan V W (1985) Agricultural Development: an international perspective. The Johns Hopkins University Press, Baltimore, USA.
Hayami Y, Otsuka K (1993) The economic theory of contract choice: an agrarian perspective. Clarendon Press, Oxford, UK.
Higgins B (1982) Potential population supporting capacities of lands in the developing world. FAO, Rome.
Howitt R, Taylor C R (1993) Some micro-economics of agricultural resource use. In Carlson G A, Zilberman D, Miranowski J A (Eds.) Agricultural and Environmental Resource Economics. Oxford University Press, Oxford, UK.
De Janvry A, Fafchamps M, Sadoulet E (1991) Peasant household behaviour with missing markets: some paradoxes explained. The Economic Journal 101:1400–1417.
Kruseman G, Hengsdijk H, Ruben R (1993) Disentangling the concept of sustainability: conceptual definitions, analytical framework and operational techniques in sustainable land use. DLV report no. 2. AB-DLO/WAU, Wageningen, The Netherlands.

Kruseman G, Ruben R, Hengsdijk H, Van Ittersum M K (1995) Farm household modelling for estimating the effectiveness of price instruments in land use policy. Netherlands Journal of Agricultural Science (submitted).

Lau L J, Lin W L, Yotopoulos P L (1978) The linear logarithmic expenditure system: an application to consumption-leisure choice. Econometrica 46:843–868.

Pingali P, Bigot Y, Binswanger H P (1987) Agricultural mechanization and the evolution of farming systems in Sub-Saharan Africa. Johns Hopkins University Press for the World Bank, Baltimore, USA.

Van der Ploeg J D (1991) Landbouw als mensenwerk: arbeid en technologie in de agrarische ontwikkeling. Coutinho, Muiderberg, The Netherlands.

Rabbinge R, Van Ittersum M K (1994) Tension between aggregation levels. Pages 31–40 in Fresco L O, Stroosnijder, Bouma J, Van keulen H (Eds.) The Future of the land: mobilizing and integrating knowledge for land use options. John Wiley and Sons, New York, USA.

Van Rheenen T, Van Loon E, Rabbinge R (1994) Trade-offs between different scenarios for small farm enterprises in South East Java (draft), Department Development Economics, Wageningen Agricultural University, Wageningen, The Netherlands.

Ruggles R, Ruggles N D (1986) The integration of macro and micro data for the household sector. Review of Income and Wealth 32:245–276.

Ruttan V W (1978) Factor productivity and growth: a historical interpretation. In Binswanger H P (Ed.) Induced innovation. John Hopkins University Press, Baltimore, USA.

Romero A (1993) A research on Andalusian farmers' objectives: methodological aspects and policy implications. Paper presented to the VIIth EAAE congress, Stresa, Italy.

Romero C, Rehman T (1989) Multiple criteria analysis for agricultural decisions. Elsevier, Amsterdam, The Netherlands.

Ruben R, Kruseman G, Hengsdijk H (1994) Estimating the effectiveness of policy instruments for sustainable land use. DLV report no. 4. AB-DLO/WAU, Wageningen, The Netherlands.

Singh I, Squire L, Strauss J (Eds.) (1986) Agricultural Household Models: extensions, applications and policy. Johns Hopkins University Press, Baltimore, USA.

Strauss J (1986) Estimating the determinants of food consumption and caloric availability in rural Sierra Leone. In Singh I, Squire L, Strauss J (Eds.) Agricultural Household Models: extensions, applications and policy. Johns Hopkins University Press, Baltimore, USA.

Thorbecke E, Hall L (1982) Nature and scope of agricultural sector analysis: an overview. In Thorbecke E (Ed.) Agricultural sector analysis and models in developing countries. FAO economic and social development paper no. 5. FAO, Rome, Italy.

Tinbergen J (1956) Economic policy: principles and design. North Holland Publishing Company, Amsterdam, The Netherlands.

Weaver C (1981) Development theory and the regional question: a critique of spatial planning and its detractors. In Stohr W B, Taylor D R F (Eds.) Development from above or below? The dialectics of regional planning in developing countries. John Wiley and Sons. Chichester, UK.

Veeneklaas F R, Cissé S, Gosseye P A, Van Duivenboden N, Van Keulen H (1991) Competing for limited resources: the case of the fifth region of Mali. Development scenarios. CABO-DLO, Wageningen, The Netherlands, ESPR, Mopti, Mali.

WRR (1992) Ground for choices: four perspectives for the rural areas in the European Community. Netherlands Scientific Council for Government Policy. Sdu publishers, The Hague, The Netherlands.

Sustainable agriculture in the Sahel? (Integrated farming, perennials and fertilizers)

H. BREMAN

Research Institute for Agrobiology and Soil Fertility (AB-DLO), P.O. Box 14, 6700 AA Wageningen, The Netherlands

Key words: fertilizer use, integrating crops and livestock, overpopulation, research approach, rural development, Sahel, sustainable agriculture, systems analysis

Abstract. A case is made for the application of the eco-regional approach in vast regions with low quality of natural resources and limited availability of resources per inhabitant. Farmers will have little scope for choice in terms of options for increased sustainability of land use. Efforts to create more favourable socio-economic conditions will often be more useful than those directed towards the direct adoption of technical options by farmers.

Results of twenty years of Malian-Dutch research collaboration are presented and reviewed to show the usefulness of the eco-regional approach for the Sahel. Based on process studies, systems analysis and multiple goal planning, small margins for rural development are identified. Thus, trial and error in commodity oriented and farming systems research can be reduced, making land-use options more clear.

External inputs are essential for increased sustainability of production. Integration of annual crops and perennial plants, and of crops and livestock, are needed to allow inputs like fertilizer at farm level.

Introduction

The case to be presented is characterized by at least two particular conditions in favour of the eco-regional approach (ERA). One is the low quality of the natural resources, the other the limited availability of resources per inhabitant. As a consequence, the farmers have little scope for choice as far as options for increased sustainability of land use concerns. Continuous depletion and degradation of natural resources is more likely. Efforts to create more favourable socio-economic conditions will often be more useful than those directed to the direct adoption of technical options by farmers in a context which they hardly control in a positive sense. Decision makers, donors, development projects and research planners may be at least as important as target groups than farmers and extension services.

The disadvantage of an extreme case like the Sahel might be the judgement that the presented example is nowhere else applicable. The author hopes, however, that the relative simplicity of this extreme case inspires the readers enough to evaluate its value for comparable, but less extreme and more complicated situations. It will be shown that the ERA developed

J. Bouma et al. (eds.), Eco-regional approaches for sustainable land use and food production, 213–235.

lent itself admirably to the purpose. It concerns a Malian-Dutch research collaboration, which started twenty years ago. The partners are the Institut d'Economie Rural in Bamako, Mali, and the Research Institute for Agrobiology and Soil Fertility (AB-DLO) and the Wageningen Agricultural University (WAU), both in Wageningen, The Netherlands. Process studies of primary production on range lands, on rain fed and irrigated fields, are combined with eco-physiological studies in pots and climate rooms. Studies of free grazing livestock are combined with feeding experiments in stables. Syntheses of knowledge available, using systems analysis as a main tool, permit extrapolation of results, obtained on a limited number of sites, to the Sahel as a whole. Multiple goal planning helps to integrate agro-ecological and socio-economic knowledge. Strategies for rural development are elaborated, indicating which technical options might be propagated per agro-ecological zone to reach certain goals under different economic conditions, and which socio-economic measures might be taken to make those technical options accessible for farmers, which can push back resource depletion and degradation.

After a description of both special conditions for ERA, in itself a result of the research to be presented, the ERA developed will be outlined. Next, some results will be discussed, paying particular attention to the integration of animal husbandry and arable farming, and the need for perennial plants and chemical fertilizer to pursue sustainable production. In view of the goal of the conference and its participants, only the consequences for programmes and priorities of commodity research, farming systems research and socio-economic studies are discussed. Hopefully the presentation is convincing enough to stimulate other target groups to discover the messages for themselves. The results of the ERA in question are well documented and accessible (Anonymous 1991; Breman and Uithol 1984, 1987; Veeneklaas 1990).

Poor resource quality and overpopulation: reasons for ERA

Characteristics and quality of agricultural resource base

General information is presented by Penning de Vries and Djitèye (1991) and Breman and De Ridder (1991).

Climate

Its name defines the Sahel geographically: desert border. More generally isohyets are used; e.g. the region South of the Sahara with 100 to 600 mm yr^{-1} on average. The real particularity is however the extreme aridity of the dry season, a characteristic that applies also to the Sudanian savannah, the main agricultural zone of Sahelian countries (600 to 1200 mm yr^{-1}).

In both zones the temperatures are very high the year around, and the strongly concentrated rainfall concerns summer rain almost exclusively. As a consequence of these conditions, the potential evapotranspiration is at least 1000 mm yr^{-1} higher than in the North of tropical Australia, the region most similar to the Sahelian countries as far as climate is concerned.

Soil

The single and short rainy season starts and stops bluntly. Besides the rainfall distribution, the characteristics of dominant soils play a role. In the northern part (the driest part), deep sandy soils prevail. They are only superficially moistened, so the water is lost rapidly by evaporation. Going south, on a transect with increasing rainfall, loamy soils increasingly dominate. Limited water storage in this case is related to soil surface compacting, causing run-off, and frequent shallowness. Limited stocks of water and extreme low air humidity and high temperatures are at the base of the aridity of the dry season. The redistribution of rain water through run-off and drainage creates 'islands' of higher water availability in relatively low spots and depressions.

An even more important soil characteristic is its low fertility. The rather general tropical gradient from dry/nutrient-rich to humid/nutrient-poor is not pronounced in the region. The dominance of the poor sandy soils in the northern half is the explanation. Therefore, the gradient of water availability is linked to an overall and homogeneous low availability of nutrients. Nitrogen (N) and phosphorus (P) are the limiting elements. Only the soils of humid tropical savannas in Latin America and Australia have an even lower P availability, because of a high content of P-fixing aluminium and iron oxides.

Vegetation

Climate, reinforced by the soil distribution, causes a dominance of annual plant species. The Sahel is a region without a perennial climax vegetation. From the central Sahel southwards, perennial grasses and woody species are of increasing importance, but they are an unstable component of the vegetation. To explain the predominance of annuals, the competition with perennials in obtaining water, nutrients and light has to be understood. One factor, not mentioned yet, are the isohyets running almost parallel to the degrees of latitudes, which determine daylength. The annual species are photosensitive, having growing cycles linked to the average water availability at a certain latitude. With the exception of the run-on humidity islands, annuals and perennials compete for the same water and the same nutrients. In general, about 20% of the water is left in the soil at the end of the growing cycle of the annuals. One reason is the determination of the length of the cycle by daylength in stead of by water, the other the fact that production, so also the use of water, is low because of lack of nutrients. Perennial grasses, shrubs and trees benefit from the residual water, staying active during a part of the long dry season, reducing the length of the critical period. Internal transfer and storage of nutrients before senescence creates a stock that improves gradually their force of competition.

Droughts regularly nullify this advantage. The water availability becomes such that annuals use all during the rainy season; their cycle is too low for such years. Without any water during the dry season the perennials start dying. True, when the drought persists, annuals with a relatively long growing cycle are replaced by species with a shorter cycle. But these concern pioneer species during the first years, which continue growth as long as water is available.

Periods of good rainfall have to be long to restore the former situation: re-establishment of populations of perennials within annual vegetations, dominated by species with growing cycles that are relatively long and determined by daylength. Wind and migrating animals transport seeds southward from the period of seed ripening, while restoration of the situation requires a northward distribution.

Consequences for agriculture

The low contribution of perennials to the annual production of rangelands and crops, together with the high turnover rate of organic matter by the extreme climate, leads to very low levels of soil organic matter. This reinforces the water and nutrient shortage: important are the losses through run-off and through percolation, the latter mainly for the south and for run-on spots. The natural levels of production of rangelands and fields are very low, while the average efficiency of external inputs is also low (see section 'Prospective Malian-Dutch research: integrating crops and livestock'). Based on natural resources only, food self-sufficiency is not easy to reach, in view of the yield per man-year. This makes the step to cash crops risky and the realization of savings rare. Thus, the use of external inputs, which could theoretically modify the situation, remains limited. Farmers adopt more easily practices leading to soil depletion. In the favourable zone of Mali, the southern part, 40% of the income is already based on this source 'free of charge' (Van der Pol 1992).

Over-exploitation through over-population

The degree to which environmental factors limit an increase of agricultural production and rural development can be determined by an analysis comparing the production potentials of the natural ecosystem side by side with actual agricultural production and by analysing the possible difference between production potential and actual production. Agro-ecological research enables determination of the production level that agriculture (in the widest sense) can achieve without impairing the carrying capacity of the environment. Farming systems research (FSR) provides insight at the level of actual production and into the factors which determine this level. A comparison of both levels leads to an initial characterization of the situation in four categories (Breman

1990a):

- under-exploitation of natural resources;
- actual production equals the maximum production with respect to carrying capacity;
- over-exploitation of natural resources;
- production exceeds the potential of natural resources due to the introduction of external inputs, upgrading the carrying capacity of the environment.

The worst case, in view of the scope for rural development, concerns over-exploitation. Over-exploitation can be a matter of ignorance, obstinacy (doomsday mentality), and over-population. Though the actual population density of the northern and southern Sahel zone and the Sudanian savanna is only 1 (0–7), 13 (7–27) and 33 (7–66) persons per km^2, respectively, a comparison with the carrying capacity suggests a saturated northern Sahel zone, a seriously over-populated southern Sahel and an almost saturated, locally heavily overpopulated, Sudanian savanna (Breman 1992). Based on animal husbandry alone, carrying capacity is reached with respectively 1, 7 and 7 persons per km^2, based on arable farming alone with 0, 11 and 36, and based on integrated land use 1, 11 and 36 persons per km^2. The situation is still aggravated by the fact that the two other causes of over-exploitation, ignorance and obstinacy, also play their separate roles (Breman *et al.* 1990).

The Sahelian countries are not exceptional as far as over-exploitation of natural resources is concerned. The extreme poverty of the natural resources implicates, however, that over-population is reached at a very low absolute demographic pressure. This creates particular difficulties for the 'green revolution', the approach which has been successful in e.g over-populated areas of South-east Asia (Van Keulen and Breman 1990): the transition to intensive agriculture has to be made in Sahelian countries when land has yet almost no price, while the costs of infrastructure investments per man are very high.

A downward spiral of the environment and the economy is probably at equal or increasing pressure on the land. The indigenous production systems collapse by lack of space (decreasing herd mobility and disappearing fallow practice) and lack of labour. Once the ecological, capital extensive systems do not produce enough for survival, especially the young and strong family members try their luck elsewhere; labour is an important limiting factor for agriculture in the region (Veeneklaas *et al.* 1990). Ecological practices have to be neglected, the low levels of soil organic matter content decrease even more. As a consequence, the chances for intensification are decreasing further: degradation of the resources is expected, with decreasing efficiency of internal and external inputs, and with loss of knowledge about the agro-ecological environment (Breman 1990b).

ERA a fruitful approach

Intuition starting point

The analysis presented is far from being a commonly accepted one. Systems analysis, as presented in the former paragraph, is seldom at the base of research and development strategies. However, the analysis of programmes and actions allows the identification of other opinions concerning the relation between the potential of the natural resources and the actual production, and its interpretation.

A trial in a Sahelian country confronted the national policy and the World Bank policy with results presented in the section 'Over-exploitation through over population' (Breman 1989). Lack of knowledge, of farmers and by the population at large, had to be regarded as the main bottle-neck identified by the government. The natural resources were in general regarded as under exploited, with the exception of soil water. Analysis of the World Bank action plan led to the conclusion that water shortage and lack of knowledge were identified as the main bottle-necks. Over-exploitation of natural resources might have been regarded important in case of animal husbandry, ignorance being the identified cause. Systems analysis, however, indicated soil fertility to be the main limiting factor, leading to overexploitation by over-population, with unequal access to production factors as a second reason for inefficient resource use.

Commodity research

It will be clear that research programs initiated from the national and the World Bank viewpoints concentrate on improved resource use by farmers. Though research is not without any success, rural development could not be triggered yet in the region. The fact that the most successful commody research (CR) is related to the cash crop cotton, is more a confirmation of the results of the systems analysis than of the national and World Bank starting points: fertilizer use is an important element of the extension packages developed. Nevertheless, soil depletion is aggravated (Breman 1990b; Van der Pol 1992).

The rapid introduction of draught animals (e.g. Tyc 1976), as a result of effective farming systems research, is one of the reasons. The use of cotton seed and cotton cake for supplementary feeding of livestock outside the cotton region another.

The role of manure and fertilizers partially overlap (Pieri 1989). In FSR as well as CR a lot of attention is paid to the use of manure. The neglect of fertilizers in this case is one of the reasons behind the conclusion above: supposed lack of farmers knowledge drives research and rural development programmes. One does not realise that the request for manure in case of over-exploited natural resources in common use will increase rapidly the inequity in a society where inequity is already a bottle-neck for rural development (Breman *et al.* 1990). Possession of livestock will be the trigger.

Eco-regional approach

CR as well as FSR, in the described context, are looking for technical options to be transferred to farmers through training and extension. Economically feasible options to rapidly increase farmers productivity, without further undermining their resource base, are scarce in the situation of overexploitation of poor resources by sheer necessity. The efficiency of external inputs like fertilizer, which technically are able to increase the carrying capacity of the Sahel several times (Penning de Vries and Djitèye 1991), decrease with the decreasing quality of the land. Their use is not easily remunerative, while even more is needed: application of external inputs requires competitive agricultural production. In the situation described, food security is unattainable if the cultivation of cash crops is ignored (Breman 1990a).

The economic feasibility of the use of external inputs is only partially related to the quality of the natural resources. The economic and agricultural policy of governments easily dominates the role of the latter. The eco-regional approach, as described in the section 'Introduction', is an effective tool to weight both roles, and to identify the socio-economic measures which are indispensable to increase the access to technical options based on external inputs. Taking planners and decision makers as target groups may be the best help to farmers. The limited differentiation in rural development in a huge area like the Sahel is an extra argument.

The prospective character of ERA is of particular importance in the situation described. Depletion of natural resources, threatening the agro-ecosystems, are relatively slow processes. The same is true for the opposite, improvement of soil organic matter content and quality, and regeneration of vegetations. Besides, heterogeneity in space, and year to year fluctuations of production conditions mask negative and positive tendencies. Demonstrations of practices leading to more sustainable production, to convince farmers to stop certain practices or to adopt technical options, may require years. Isolated steps to more sustainable resource use and intensive agriculture are often not remunerative; whole packages (integrated development) of socio-economic measures, ecological interventions and agricultural techniques are needed (e.g. GEAU 1984). And even those packages are not immediately effective in connection with the speed of the ecological processes involved. In other words, prospective research is needed, before starting long term demonstrations and before convincing decision makers to change policies and farmers to invest, to take risks.

Analytical Malian-Dutch research: primary and secondary production

Primary Production Sahel (PPS)

General information is provided by Penning de Vries and Djitèye (1991).

Pin-pointing the bottle-necks of the Sahel problem has been the goal of the scientific research collaboration 'Primary Production Sahel' (PPS) (1976–1982), between the Institut d'Economie Rurale (IER, Bamako), the Department of Theoretical Production Ecology of the Wageningen Agricultural University, and the Research Institute for Agrobiology and Soil Fertility (Wageningen). The research concentrated on the understanding of yield and quality determining processes of the rangelands and on the identification of potential improvements. Climate, soil, vegetation and exploitation, as well as their mutual relations were studied, using modelling and simulation integrated with experimental research. Experiments and observations concerned rangelands and their exploitation systems between the 1100 and 200 mm isohyets (two transects of about 1000 km each), rangeland on different soils at 550 mm yr^{-1}, and pot experiments with Sahelian rangeland species in climate rooms. Interventions concerned fertilisation, exploitation, rain simulation, irrigation and fire. The main processes studied covered the water, the nitrogen, and the phosphorus balance, the vegetation dynamics, and the linkages between rangeland and livestock production.

Taking advantage of systems analysis

The research results triggered a series of other studies, taking advantage of the systems analysis approach. It concerned mainly desk studies: syntheses and interpretations of available knowledge, using elaborate models and insights obtained about the combined action of soils and climate, and the interference of primary and secondary production. The systems analysis allowed 'borrowing' and 'translation' of data from elsewhere, when the search for data from the region was not successful.

The analysis of constraints and potentials for fodder production, based on annual rangeland species, has been extrapolated to perennial grasses (Breman and De Ridder 1991) and to shrubs and trees (Breman and Kessler 1995), paying particular attention to their surplus-value through the internal recirculating of nutrients and niche differentiation. The extrapolation became a must during the elaboration of a practical manual for the evaluation of Sahelian rangelands and animal husbandry systems, to identify interventions for improvements (Breman and De Ridder 1991). Perennial grasses and woody species contribute little to the annual production of Sahelian rangelands, and are of very limited direct interest in connection to the production of cattle, studied by the PPS project (Penning de Vries and Djitèye 1991). But (i) their interest increases on flood plains and in the Sudanian savanna, where pastorales try to safeguard during the long dry season the weight gains of their animals obtained on rainy season pastures of annuals; (ii) small ruminants (goats in particular) use browse also for the rainy season growth; (iii) even in the real Sahel, the contribution of the perennial rangeland elements is not negligible in connection to environmental resistance, in view of their contribution to soil organic matter and soil cover.

The step from annual rangeland species to annual crops has been a relative simple one. A crucial difference is the fact that only generative plant parts of most crops are consumed by man, while vegetative growth of rangelands may form high quality fodder. In relation to crops, attention had to be paid to the length of the growing cycle and of the vegetative and generative development phases, making the linkage with water availability more complicated. Nevertheless, it became very clear that in arable farming also soil poverty limits production much more than rainfall, in the savanna zone as well as in the southern Sahel (SOW 1985; Van Duivenbooden and Cissé 1989).

An indispensable separate track has been the analysis of animal production. Primary production analysis has been a source of inspiration (Ketelaars 1990). Simple models have been elaborated for intake of digestible energy in dependence of fodder quality, for production of individual animals in relation to the intake, and for herd production in relation to the production of individual animals (Ketelaars 1991). The chosen approach permitted not only the derivation of secondary from primary production, it enabled us also to quantify the feed-back, the consequences of rangeland exploitation on rangeland production and quality (Penning de Vries and Djitèye 1991). The carrying capacity of Sahelian rangeland could be quantified in connection with production goals (Ketelaars and Breman 1991).

The developed models and the knowledge and experiences obtained became useful tools for FSR, and for elaboration of development strategies based on resource evaluation at regional or national level. The scientists concerned have been encouraged to use the tools by regular affirmation of the correctness of predictions and analysis. The following list is an illustration, serving at the same time to deepen the description of the region in section 'Poor resources quality and over-population: reasons for ERA', as an introduction to section 'Prospective Malian-Dutch research: integrating crops and livestock'.

- Based on a plant production model developed for the Negev desert and using soil, climate and plant parameters from the Sahel, Penning de Vries and Van Heemst (1975) predicted that rangeland production in the Sahel should be limited by nutrients, not by rainfall. Expected water limited production at 500 mm of annual rainfall should be 8000 kg ha^{-1}. Field experiments at the 500 mm isohyet showed production levels between 5000 and 12000 kg ha^{-1} of dry matter, depending on soil type, rainfall quantity and distribution, and vegetation composition (Penning de Vries and Djitèye 1991).

- The rangeland production of Mali has been estimated for the extreme dry year 1984: an almost linear increase from 0 kg ha^{-1} at the 250 mm isohyet to 2800 kg ha^{-1} at the 1200 mm isohyet has been obtained. The Malian-American PIRT (Projet d'Inventaire des Ressources Terrestres) registered for not or weakly degraded rangeland an increase from 0 kg ha^{-1} at the 250 mm isohyet to 2600 kg ha^{-1} at 1200 mm (Breman and Traoré 1987).
- The PPS project estimated the carrying capacity of Sahelian rangelands for the semi-nomadic transhumance system to be 3 and 7 ha TLU^{-1} (TLU = Tropical Livestock Unit) for a normal and a dry year. Before the drought of the seventies the stocking rate leaded to 3.7 ha TLU^{-1}, decreasing by drought starvation during the period 1972 to 1974 to 6 ha TLU^{-1} (Penning de Vries and Djitèye 1991).
- The PSS project predicted and showed a decrease of rangeland quality (nutrient content and digestibility) with increasing production going from the desert border to the savanna, and underlined the logic of the transhumance. Observations in Mauritania showed the highest stocking rates after the growing season on rangelands at the desert border with only 200 kg ha^{-1} dry matter. Ketelaars (1991) calculated the potential production of weaners in their first year at sedentary use of rangeland in case of limitation by fodder quality: maximum annual net live weight gain should be 98, 5 and 15 kg per animal for the northern Sahel, the southern Sahel and the Sudanian savanna, respectively. The last two figures could only be found experimentally; systems showing such low production levels are not viable. A live weight gain of 25 kg per animal per year should be the minimum. Low stocking rates, allowing for selective grazing, should be the strategy to obtain such gains on the low quality rangelands of the savanna. Weight gains up to 90 kg have been found on a ranch in northern Niger for young male zebus; the International Livestock Centre for Africa (ILCA) registered a negative net annual weight gain of 5 kg per animal at experimental sedentary grazing in the southern Sahel; farmers in the savanna zone of Mali obtain weight gains of 25 to 30 kg year.
- 90% of the arable soils in Burkina Faso do not contain more than 500 mg N kg^{-1} and 0.5 to 1 mg P kg^{-1}, which permits an average production of 830, 720 and 530 kg ha^{-1} of maize, sorghum and millet grain, respectively, following computer simulations (SOW 1985). The actual levels of production of the three cereals in that country were 850, 650 and 450 kg ha^{-1}, respectively. Simulations indicate that fertilization will be able to increase the millet production of a year with average rainfall to 1400 and 2100 kg ha^{-1} for the Sahel and savanna zone of Burkina; for sorghum those figures are 2200 and 3300 kg ha^{-1}. From Senegal

production levels of 3500 kg ha^{-1} of sorghum are known (Buldgen pers. comm.). Another proof of cereal production being limited by lack of nutrients like rangeland production is the millet production of ICRISAT-Niamey in the extreme dry year 1984. The fertilized fields produced 450 kg ha^{-1} at 260 mm of rainfall, equal to the average farm production of Niger, where the rainfall of the agricultural zone is 400 to 900 mm yr^{-1}.

– The theoretical maximum increase of cereal production through the influence of *Faidherbia albida* equals the field observations of 400 kg ha^{-1} yr^{-1} (Breman and Kessler 1995).

Requested by the 'Club du Sahel/CILSS', the situation of animal husbandry in Niger, Burkina Faso and Mali has been analysed, and strategies and programmes for future development have been proposed (Breman *et al.* 1991). Three main causes of the collapse of the transhumance system, a system more productive than livestock razing in arid USA and Australia (Breman and De Wit 1983), have been identified: overgrazing of the southern Sahel and the northern Sudanian zone, drought, and lack of equity as far as access to resources concerns. Increased cultivation of dry season rangelands is a crucial element of overgrazing. Sedentarisation of pastoralists increases the pressure on arable land, fallow periods are becoming very short and farmers need more and more livestock to concentrate the fertility of rangeland and waste land on their fields, the use of chemical fertilizers being not more that a few kg $ha^{--1}year^{-1}$ of N and P. In other words, the main point of animal husbandry is moving south, to regions where production conditions are worse; health problems are 'pinching', rangeland quality is low, and low use of fertilizer in crop production leads to a limited amount of mainly low quality agricultural by-products.

The price ratios of fertilizer, animal products and crops were such that more intensive use of fertilizer and other external inputs to intensify crop production seemed to be the best way to trigger development of animal husbandry (Wooning 1992). Only in this way the fast growing role to support arable farming through manure and draught, going together with decreased production of milk and meat, could be reversed. It might be that the devaluation of the F CFA* in January 1994 has created the possibility of direct intensification of animal husbandry, through fertilized fodder crops. Improved land use legislation and tenure is regarded as a condition *sine qua non* for both intensification of arable farming and animal husbandry in a sustainable manner.

The FSR in South Mali of the Malian IER adopted the systems analysis to clarify the potential role of animal husbandry in rural development in southeast Mali (e.g. Leloup and Traoré 1989). The situation described above

* Franc CFA, a currency unit in some African countries.

has been confirmed and specified. The fodder situation and land occupied by arable farming appeared to be crucial elements to understand the possession of livestock and their productivity. The higher the grade of occupation of the land by fields, the higher the animal density and the lower their production. One region was identified, where only a small fraction of the land was suitable for crop production. Here animal pressure was low enough to permit high selective grazing of the low quality savanna, leading to an animal production high enough for an income out of animal husbandry alone. The region with the highest occupation by crop land was indeed the region with the highest stocking rate. One village was found where, nevertheless, the animal productivity was not at the minimum level: an annual weight gain of weaners of 35 kg living weight (instead of the above indicated biological minimum of 25 to 30 kg) was observed. This appeared, however, to be a village where one third of the crop land was occupied by fertilized cotton production, while an important fraction of the cotton cake of their own crop was repurchased from the factory and used for supplementary feeding.

Development planning

The developed tools and the knowledge obtained have been applied to assist the Malian government in the elaboration of a land use and development plan for the Mopti region, located in the southern Sahel. Multiple goal planning through linear programming has been used in the search for optimal solutions in relation to the often conflicting goals (Veeneklaas *et al.* 1990). Examples for the region are the use of the flood plains for irrigated crop production or as dry season rangeland, the production of cash crops or food production for the nation, increased food security (even in dry years) or increased revenues? The methodology is treated in detail by Kuyvenhoven *et al.* (1995) and Schipper *et al.* (1995).

Two packages of goals have been optimised, producing two strategies, showing decision makers the consequences of choices to be made. In each case is specified, for all soil-climate combinations per agro-ecological zone, which production system should be propagated to reach the goals defined. The 'high revenue-high risk scenario' suggests to use as much resources as possible for fishery, animal husbandry and horticulture. The use of external inputs is not particularly advocated, the risk of resource depletion, desertification and hunger during droughts will be high. The 'high self-sufficiency and security scenario' suggests to intensify in particular cereal production. Intensive use of external inputs, N and P fertilizer in particular, is advocated to guarantee food production even in dry years, without undermining the natural resource base of agriculture.

Prospective Malian-Dutch research: integrating crops and livestock

General information is provided by Breman and Niangado (1994).

No ready-made solution

The indispensability of N and P fertilizer, while their use is often hardly or not remunerative, creates the situation in which CR and FSR have easily a trial and error character. This is dramatic in view of the poverty related to such a situation. The prospective character of ERA may offer relief. Priorities can be better defined, in connection with both the choice between socio-economic measures and technical options, and out of a variety of technical options to be developed.

Three regions have to be distinguished in this particular case; their borders being in the first place determined by soil, climate and socio-economic conditions:

a. Regions where external inputs, at least for certain crop- or animal productions, are economically feasible;
b. regions where the choice of production systems, the adoption of certain practices, and/or socio-economic measures create access to external inputs;
c. marginal land, where, at least at short and medium term, the use of external inputs will not be directly paid by agricultural production.

The scientific and technical programme 'Production Soudano-Sahélienne' (PSS)* is determining the borders between these regions for the sahelian-sudanian zone of West Africa, and tries to identify for region type 'b' the production systems, the practices to be adopted, and/or the socio-economic measures to be taken, which makes the use of N and P fertilizer economically feasibly. As far as the agro-technical elements are concerned, PSS limits itself to the optimal utilization of nutrients in animal husbandry. No direct attention is paid to the development of region type 'a'. But also here ERA and systems analysis would be very useful to elaborate a policy assuring that the external inputs will be used to make the production systems more sustainable, instead of contributing to the depletion and degradation of the environment, as is actually the case (Breman 1990b; Van der Pol 1992).

* PSS is a cooperation between the Institute of Rural Economy (IER) in Mali, the Research Institute for Agrobiology and Soil Fertility (AB-DLO) and the Department of Nature Conservation of the Wageningen Agricultural University Wageningen (DNC-WAU), both in the Netherlands. PSS is financially supported by the Directorate General for International Cooperation (DGIS), of the Netherlands Ministry of Foreign Affairs.

PSS, structure and tools

PSS is composed of four teams, of which three work in Mali and one in The Netherlands. Most work is done at three research stations of IER in the southern Sahel and the northern Sudanian zone. Processes are central, not the particular situation on three 'arbitrary' spots in the extensive Sahel in four arbitrary years. Modelling and simulation is used to generalise experimental results.

The '*support team*' in The Netherlands has the lead in modelling, gives or organizes back-stopping in general, and introduces the systems analysis in Mali.

The '*fodder production team*' is identifying crops and practices leading to the highest recovery of N and P nutrients, producing high quality fodder, in the hope that agro-ecological optimization increases economic feasibility. Elements out of ecological agriculture are studied and tested for their role to optimize fertilizer use (Breman 1990b). Recovery is measured and analysed for different crops and treatments, in three climatic zones on different soil types. Crop growth models are used to identify crucial production factors and to extrapolate the results to other soil and climate combinations and to other crops.

The '*fodder exploitation team*' in turn optimises the use of high quality fodder. It has been reasoned that the best use implies supplementation of livestock during the dry season, to improve the exploitation of low quality roughage. Most experiments concern stall feeding of young male cattle. Extrapolation to other animal classes, or to herds with their system-dependant composition, is done using zootechnical parameters synthesized by Ketelaars (1991) or measured by the project. Extrapolation to free grazing animals, the dominant practice in the region, is supported by studying animal behaviour with and without supplementation. Cattle as well as goats are studied, to include the role of browse.

For the '*systems modelling team*', calculation of cost/benefit ratios of optimal fodder production and use of supplements is only a minor activity. Crucial is the optimization of resource use in general, in connection with agro-ecological and socio-economic conditions, taking into account different goals, the conflicting ones included. Development strategies are formulated, using multiple goal linear programming. Different levels of integration are studied: the Malian-Sudano Sahelian zone, a case study at the level of a district, and farm level. In the last two cases collaboration is organised with the FSR team of IER from the district, and with the DLV project (Kuyvenhoven 1995). The outcome of the latter collaboration, increased understanding of farmers behaviour, will be an input in the strategy formulation at national level. The analysis at district and farm level will serve the FSR in their work with farmers. The judgement and comments of the scientists concerned help the PSS team to improve their work on the national level, planners and decision makers being their first target group. The Malian work will be generalized by studying the influence of varying economic conditions and of different soil and climate combinations on the outcome of multiple goal planning.

Recovery of nitrogen and phosphorus

Sustainability of agriculture is not assured by the compensation of export and losses of nutrients through chemical fertilizers in case of overexploitation of poor resources.

The required competitive production asks for soil improvement. The average recovery of N and P obtained for example in millet and sorghum production in the region is 37% and 16% (Van Duivenbooden 1992), while values of respectively 80% and 50% are potentially possible (Penning de Vries and Djitèye 1991). A crucial cause of losses by run-off (leaching, etc.) is the extreme low organic matter content of the soil. Manure production as such is no general solution any more. To increase organic C in the top 20 cm of a soil by 1 g kg^{-1}, an amount of 15, 30 or 55 t ha^{-1} yr^{-1} of straw, manure or compost, respectively, is required (De Ridder and Van Keulen 1990). 40 ha of rangeland are needed in the southern Sahel to maintain the fertility of 1 ha of crops through manure alone, without final degradation of the rangeland; 15 ha of rangeland is the figure for the north Sudanian zone (Breman & Traoré 1987). At least 10% of the Southern Sahel is in general already cultivated, however, and 20% or more of the northern Sudanian savanna.

A solution is worked out by the team based on three elements:

- The biological nitrogen fixation by legumes, to improve the P availability of the soil in a relatively cheap manner;
- the high biomass turnover of perennial species, woody species included, to improve the soil organic matter content;
- fertilizer use, to increase not only specific productions, but biomass also.

In all three cases, three *agro-ecological zones* are distinguished, each with two sub-zones:

i. production limited by insufficient vegetation, by natural origin (desert) or due to land degradation;
ii. production limited by water availability, due to relatively low rainfall or to water losses by run-off;
iii. production limited by nutrient availability, due to low soil fertility or to run-on (Penning de Vries and Djitèye 1991; Breman and Kessler 1995).

Natural run-off and run-on processes are reinforced by (mis)use of the land, a further sub-division of the agro-ecological zones 'i' and 'iii'. Infiltration of 250 mm yr^{-1} marks the transition between the zones 'ii' and 'iii'. On coarse sand, with good infiltration, this coincides almost with the 250 mm isohyet, the transition from northern to southern Sahel; with increasing run-off, in particular on loamy soils with surface crusts, the transition moves southwards. Water limited production may even occur in the savanna! Loss of soil organic matter by human land use, followed by loss of soil structure and increased run-off, is often a faster process than nutrient depletion, especially in animal husbandry. This turns nutrient limited production into water limited production, which is even more difficult to cure in an economic feasible way. The crucial role of soils implicates that the borders between the regions distinguished in the section 'No ready made soluation' do neither coincide with geographical borders nor with isohyets.

The use of legumes, with their nitrogen free of charge, does not appear to be the ideal solution on first sight. P fertilizer needs to be added to make the biological N fixation effective. The price of P (Wooning 1992), the competition of the rest of the vegetation, the annual character of the latter, and the decreased efficiency of N fixation at decreasing rainfall make rangeland improvement not feasible in the Sahel (Penning de Vries and Djitèye 1991). Even in the Sudanian savanna and using *Stylosanthes hamata*, a perennial species that behaves as an annual under dry conditions, a sustainable legume introduction appears questionable. Competition by other species is difficult to suppress, even where tenure is no bottle-neck for rangeland management (Coulibaly *et al.* 1994). Legumes will have better chances to enrich the production systems with P fertilizer when used as fodderbank, or as a fodder or cash crop in rotations. Perennity shows to be an important characteristic: if *Stylosanthes* is a perennial in good years, the production may be twice as high as that of cowpea (11000 against 6000 kg ha^{-1} at 700 mm rainfall). A maximum of 177 kg increase of legume production per kg P fertilizer has been observed (Koné and Groot, in prep.).

Field experiments suggest, however, that *Stylosanthes* will in most cases behave like an annual in the Sudanian zone; the aridity of the dry season appears to be too extreme.

Perennial grasses and woody species are studied in connection with their potential to make the use of fertilizers in general feasible. The role of woody species in agro-pastoral systems is elaborated for the main soil and climate combinations of the region (Breman and Kessler 1995). The minimum soil organic matter content required for sustainable agriculture in the region (3, 6 and 21 g kg^{-1} of C; according to Pieri 1989) can not be reached with the potential woody cover in the Sahel, without external inputs. From the Sudanian savanna southwards it is possible, but the tree density required will easily become harmful for crops by competition for light, nutrients and water. Precise definition of production goals and careful management are requested to take profit from woody species in agricultural systems. In agro-forestry for example, three situations were identified where the integration of woody plants with cropping has a synergistic effect in the region:

- On sandy soils, in the driest parts of agro-ecological zone 'i' (see above), or at run-on sites in zones 'i' where nutrients have become the limiting factor, woody plants oriented as windbreaks improve germination and establishment of crops. One condition is that the woody plants can effectively use subsoil water reserves.
- In the more humid parts of zone 'i', where run-off and leaching are significant processes, evenly distributed woody plants can lead to improved nutrient availability for crops at a cover of about 20%. Exploitation of the woody plant greatly reduces the beneficial effect.
- If, in particular in the last case, fertilizers are used, a three-fold benefit may be expected.

Optimization of supplementation

Innumerable combinations of roughages and supplements can be and are tested all over the region. To decrease the necessity of tests, methods are developed to be able to judge the value of all sort of combinations and to indicate the optimum amount and ratio of supplement and roughage in connection with their availability and price.

Detailed knowledge about the quality of fodders from rangelands (Breman and De Ridder 1991) is amplified by analysis of the digestibility and nitrogen content of crop-products and industrial agricultural by-products. Categories of feed are distinguished, defined by their quality and the inherent feed intake, expressed in relation to maintenance intake.

The influence of quality (digestibility and nitrogen content), structure (potential animal selection) and offered amounts of feed on intake of digestible energy of both supplement and roughage is studied. The experimental design results in series of amounts and ratios of supplement and roughage to obtain maintenance of livestock or well defined higher production levels (Kaasschieter *et al.* 1994). Millet straw and cotton cake, for example, can be used to maintain the live weight of young cattle in combinations 22 & 40, 30 & 25, 40 & 15, 60 & 5, etc. g kg^{-1} metabolic live weight per day.

If, for example, cotton cake is 4 times more expensive than millet straw, the cheapest combination is 68 g of millet straw and 2 g of cotton cake (millet straw alone has not enough quality to assure maintenance). With cotton cake twice as expensive as straw 50 & 8 g becomes the optimum ratio. If the target of supplementation is production instead of maintenance, e.g fattening at a level of feeding 1.4 times maintenance, 80 & 18 g straw and cotton cake per kg of metabolic live weight per day is the optimum ratio at a price off cake 4 times the price of straw; at a two times higher price only, 62 & 25 g becomes the 'economic' optimum.

'Economic' has been put between quotation marks, because it concerns a preliminary result. To translate it into a practical recommendation at least three questions has to be answered in an integrated way:

- Which level of production is the optimum in view of the farmers goals? A choice has to be made between maintenance feeding, higher production levels, or even lower ones. The goals may be, for example, milk production, fatting stock, or maintenance of draught oxen.
- Which animal category(ies) have to be supplemented in view of these goals, young or old, male or female?
- What is the availability of both (and other) fodder resources, in view of the own farm production, the market and the capital to invest?

The method developed by Ketelaars (1991), to link the fodder situation to the growth of weaners, and the growth of weaners to other animal categories or to a complete herd is indispensable to answer first two questions. It has to be realised, in relation to the first and the last question, that straw and capital may have other goals than animal husbandry alone in agro-pastoral systems. The use of straw to maintain the soil organic matter content as much as possible is only one example.

Optimal goal planning in rural development

The examples at the end of the last section represent only a few of the choices to be made by a sahelian farmer. Besides, planners and decision makers have the choices related to their own responsibilities and concerns. Answering has to be done very carefully in view of the restricted possibilities to make use of external inputs feasible in region 'b' (see section 'No ready-made solution'), while maintaining the internal ones. Still two other examples are presented, at farmers level and at the level of planning.

- Real pastorales show that even in the Sudanian savanna, with its low quality rangelands and the actual high stocking rates, remunerative animal production levels are possible. Where manure is the only production of farm animals, they are able to still produce milk at the end of the dry season. Two elements out of their production strategy are grazing as far as possible from villages and wells, to create the possibility of selective grazing, and day and night grazing. The consequences for farmers who like to reach the same productivity: a real labour investment in herding and almost no manure.
- More and more flood plains are turned into rice fields with (semi)controlled irrigation. Not only the benefits of the land will go to others than before, but the production value of the land changes. Often this change is not positive as supposed, based on a direct comparison of

the value of the area used for animal production or for arable farming. What is neglected is the loss of up to 10 ha of rainfed rainy season rangeland for the very effective transhumance for each ha of flood plain grazing into rice.

Before the devaluation of the F CFA, the value of the annual production on one ha of the best parts of the delta used for rice was 21000 F CFA, used for the transhumance it was 30000 to 40000 F CFA (Breman and Traoré 1987).

Using multiple goal planning as one of the tools, PSS hopes to contribute to the elaboration of policies for rural development. Where, at which soil/climate combinations which options for the use of external inputs are worthwhile to be tested and/or propagated? How do different social-economic conditions and development goals influence the choices? What packages of practices, like erosion control, introduction and management of perennial grasses and trees, reafforestation, etc. are also requested? Which socio-economic measures are required to give the use of external inputs a real chance? The first results are published in EMS (1995).

Marginal land

One of the results of the optimal goal planning will be the identification of marginal land, where, at least at short and medium term, the use of external inputs will not be directly paid for by the agricultural production (region 'c', session 'No ready-made solution'). The efficiency of the use of such inputs will also be known. This enables decision makers to analyse the costs of permanent support of the use of external inputs (e.g. of rock phosphate for leguminous crops and reafforestation): to create the conditions for a decent and productive life of farmers families in those areas, and protecting in this way the more productive adjacent region (region 'b') against desertification. The costs can be compared with the costs of urbanisation and the creation of employment out of agriculture, decreasing the intensity of exploitation of the marginal land below its carrying capacity.

After all, real scope for improvements in region 'c' will be created only by a fair distribution of means and possibilities among people at the national level, like real scope for improvements in most of the sahelian countries requires a fair distribution of means and possibilities at world level. The area of these countries occupied by marginal land of region 'c' is simply too vast. Politicians ultimately decide whether available effective options will be accessible.

Concluding remarks

Two particular conditions appeal for integration of ERA in agricultural research. One is the low quality of the natural resources, the other the limited availability of resources per inhabitant. If a situation of overexploitation is reached at low absolute population density, the feasibility of intensification of agriculture on the base of external inputs is marginal, though the use of these inputs is a must. The presented case shows the ability of ERA to identify priorities for research and policy in relation to situations and goals. Priority setting based on the analysis of constraints and potentials of both natural resources and the socio-economic conditions helps to detect the small margins for sustainable production. Improved tenure, integration of animal husbandry and arable farming, and the introduction or management of leguminous and perennial species appear a condition *sine qua non* for fertilizer use.

The reality of development and research programmes in the region is often at odds with the results of the ERA presented. Neglecting resource limitations and of overpopulation creates a fictive margin for increased production, for socio-economic development, based on the existing resources exclusively. The over-estimation of the potential of natural resources implies an under-estimation of the farmers: more can be done with the natural resources only. The top-down approach of the eventual results is therefore ingrained in those programmes. They lead, however, hardly ever to results that succeed to excite farmers, and if they do, the results often contribute to increased resource depletion. Examples are irrigation, genetic improvement of both animals and crops, animal traction, land use planning, animal health care, and pastoral hydrology.

ERA cannot replace commodity research, farming systems research, etc. It is only a tool that pays for itself through increased efficiency of research institutes. Nevertheless, the technology needed and the need for training and experience to become effective is not cheap at all. The problem might be addressed by pooling resources among countries with common interest, for research as well as for integrated land use and development planning. SPAAR could be the organising structure in the first case, CILSS in the other. Such an approach increases the quality and level of information and technology, as well as provides an effective mechanism to share solutions to common problems.

Acronyms

AB-DLO	Research Institute for Agrobiology and Soil Fertility
CILSS	Comité Inter-états de Lutte contre la Sécheresse dans le Sahel
ERA	Eco-Regional Approach
CR	Commody Research
FSR	Farming Systems Research

IER	Institut d'Economie Rurale (Rural Economy Institute)
ILCA	International Livestock Centre for Africa
PSS	Production Soudano-Sahélienne
PPS	Primary Production Sahel
SPAAR	Support Program for African Agricultural Research
TLU	Tropical Livestock Unit
WAU	Wageningen Agricultural University

References

Anonymous (1991) Recherches d'appui pour le développement rural au Sahel. Rapports et publications du Centre de Recherches Agrobiologiques. CABO-DLO, Wageningen, The Netherlands.

Breman H (1989) Milieu-aandacht in ontwikkelingssamenwerking. Pages 123–139 in Spaargaren G, Liefferink J D, Mol A P J, Brussaard W (Eds.) Internationaal milieubeleid. Sdu publishers, The Hague, The Netherlands.

Breman H (1990a) No sustainability without external inputs. Pages 124–133 in Beyond adjustment Africa Seminar, Maastricht, 30 juni 1990. CABO Pubication 778, CABO-DLO, Wageningen, The Netherlands.

Breman H (1990b) Integrating crops and livestock in southern Mali: rural development or environmental degradation? Pages 277–294 in Rabbinge R *et al.* (Eds.) Theoretical Production Ecology: reflections and prospects. Wageningen, The Netherlands.

Breman H (1992) Agro-ecological zones in the Sahel: potentials and constraints. In Blokland A, Van der Staay F (Eds.) Poverty and development. Analysis & Policy. Vol. 4. Sustainable development in semi-arid sub-saharan Africa. Ministry of Foreign Affairs, The Hague, The Netherlands.

Breman H, Kessler J-J (1995) Woody plants in agro-ecosystems of semi-arid regions, with an emphasis on the sahelian countries. Advanced Series in Agricultural Sciences, Vol. 23, Springer Verlag, Berlin, Germany.

Breman H, Niangado O (1994) Maintien de la production agricole sahélienne. Rapport mi-chemin du projet PSS. Rapports PSS no. 6. IER, Bamako, Mali, AB-DLO, Wageningen & Haren, DNC-WAU, Wageningen, The Netherlands.

Breman H, De Ridder N (Eds.) (1991) Manuel sur les pâturages des pays sahéliens. Karthala, ACCT, CABO-DLO et CTA, Paris, France.

Breman H, Traoré N (Eds.) (1987) Analyse des conditions de l'élevage et propositions de politiques et de programmes. Mali. SAHEL D(87)302. OCDE/CILSS/Club du Sahel. Paris, France.

Breman H, Uithol P W J (Eds.) (1984) The Primary Production in the Sahel (PPS) project. A bird's-eye view. CABO-DLO, Wageningen, The Netherlands.

Breman H, Uithol P W J (1987) Communication of the results of the 'Primary Production in the Sahel (PPS)' research project. Short Communication. Soil and Tillage Research 9:387–393.

Breman H. De Wit C T (1983) Rangeland productivity and exploitation in the Sahel. Science 221:1341–1347.

Breman H, Ketelaars J J M H, N'Golo Traoré (1990) Un remède contre le manque de terre? Bilan des éléments nutritifs, production primaire et élevage au Sahel. Sécheresse 2:109–117.

Coulibaly Y, Kaasschieter G, Coulibaly D (in prep.) Essais d'amélioration des parcours soudano-sahéliens par l'introduction de la légumineuse *Stylosanthes hamata* cv Verano. Expériences 1991–1992. Rapports PSS. IER, Bamako, Mali, AB-DLO, Wageningen & Haren, DNC-WAU, Wageningen, The Netherlands.

Van Duivenbooden N, Cissé L (1989) L'amélioration de l'alimentation hydrique et minérale par des techniques culturales liées à l'interaction eau/fertilisation azotée. Rapport final. No TS2–0010-NL(GDF). CABO Rapport No. 117. CABO-DLO, Wageningen, The Netherlands.

EMS (1995) Modélisation et politique de développement: perspectives d'un développement agricole durable. Cas du Cercle de Koutiala. Atelier de l'Equipe Modélisation des Systèmes (EMS). Projet PSS, Niono du 18 au 20 Septembre 1994. Rapport PSS no. 9. IER, Bamako, Mali, AB-DLO, Wageningen & Haren, DNC-WAU, Wageningen, The Netherlands.

GEAU (1984) GEAU, Gestion de l'eau. Tome I: Rapport principal sur la gestion de l'eau et l'expérimentation agricole dans le périmètre irrigué de l'office du Niger, Mali. Dept. of Irrigation and Civil Engineering, Wageningen Agricultural University, Wageningen, The Netherlands.

Kaasschieter G A, Coulibaly Y, Kané, M (1994) Supplémentation de la paille de mil (*Pennisetum thyphoides*) avec le tourteau de coton: effets sur l'ingestion, la digestibilité et la sélection. Rapports PSS no. 4. IER, Bamako, Mali, AB-DLO, Wageningen & Haren, DNC-WAU, Wageningen, The Netherlands.

Ketelaars J J M H (1990) Beslissen over nutriëntenstromen: een probleem voor mens en dier. Pages 65–70 in Spoelstra S F. 100 jaar IVO. Toekomst van de diervoeding en de kwaliteit van de dierlijke produktie. IVVO-DLO, Lelystad, The Netherlands.

Ketelaars J J M H (1991) La production animale. Pages 357–388 in Breman, H, De Ridder N (Eds.) Manuel sur les pâturages des pays sahéliens, Karthala, Paris, France.

Ketelaars J J M H, Breman H (1991) Evaluation des pâturages et production animale. Pages 255–288 in Breman H, De Ridder N (Eds.) Manuel sur les pâturages des pays sahéliens, Karthala, Paris, FRance.

Van Keulen H, Breman H (1990) Agricultural development in the West African Sahelian region: a cure against land hunger? Agriculture, Ecosystems and Environment, 32:177–197.

Koné D, Groot J J R (in prep.) Utilisation du phosphore et de l'azote par le *Stylosanthes hamata* et le *Vigna unguiculata* en zone soudano-sahélienne du Mali. Rapports PSS. IER, Bamako, Mali, AB-DLO, Wageningen & Haren, DNC-WAU, Wageningen, The Netherlands.

Kuyvenhoven A, Ruben R, Kruseman G (1995) Options for sustainable agricultural systems and policy instruments to reach them (pages 187–212 in this volume).

Leloup S, Traoré M (1989) La situation fourragère dans le Sud-Est du Mali. Une étude agro-écologique. Institut de l'Economie Rurale, Sikasso, Royal Tropical Institute (KIT), Amsterdam, The Netherlands.

Penning de Vries F W T, Djitèye M A (Eds.) (1991) (reprint from 1982 edition) La productivité des pâturages sahéliens. Une étude des sols, des végétations et de l'exploitation de cette ressource naturelle. Agric. Res. Rep. 918. Pudoc, Wageningen, The Netherlands.

Penning de Vries F W T, Van Heemst H D J (1975) Production primaire potentielle des terres non irriguées au Sahel: une première approximation. Pages 323–327 in Inventaire et cartographie des pâturages tropicaux Africains. Actes du colloque. Bamako, Mali, 3–8 mars 1975. ILCA, Addis Ababa, Ethiopia.

Pieri C (1989) Fertilité des terres de savanes. Bilan de trente ans de recherche et de développement agricoles au sud du Sahara. Ministère de la Coopération et du Développement & CIRAD-IRAT, Montpellier, France.

Van der Pol F (1992) Soil mining. An unseen contributor to farm income in southern Mali. Bulletin 325, Agricultural Development Section, Royal Tropical Institute (KIT), Amsterdam, The Netherlands.

De Ridder N, Van Keulen H (1990) Some aspects of the role of organic matter in sustainable intensified arable farming systems in the West-African semi-arid-tropics (SAT). Fertilizer Research 26:299–310.

Schipper R A, Jansen, H G P, Stoorvogel J J, Jansen, D M (1995) Evaluating policies for sustainable land use: a sub-regional model with farm types in Costa Rica (pages 377–395 in this volume).

SOW (1985) Potential food production increases from fertilizer aid. A case study of Burkina Faso, Ghana and Kenya. Volume I. A case Study prepared for the FAO. Centre for World Food Studies (SOW), Wageningen, The Netherlands.

SOW (1985) Potential food production increases from fertilizer aid. A case study of Burkina Faso, Ghana and Kenya. Volume II. Maps and Figures. A case study prepared for the FAO. Centre for World Food studies (SOW), Wageningen, The Netherlands.

Tyc J (1976) L'élevage en Afrique occidentale après la sécheresse. Le Courrier 37:23–30.

Veeneklaas F R, Cissé S, Gosseye P A, Van Duivenbooden N, Van Keulen H (1990) Compétition pour des ressources limitées: le cas de la cinquième région du Mali. Rapport 4. Scénarios de développement. Mopti, CABO-DLO, Wageningen, The Netherlands.

Wooning A (1993) Les prix du bétail, de la viande, des produits laitiers et des engrais dans les pays sahéliens. Rapports PSS No.1. IER, Bamako, Mali, CABO-DLO, Wageningen, IB-DLO, Haren, DNC-WAU, Wageningen, The Netherlands.

Response to section C*

R.D. NORTON
Proyecto PROMESA, El Salvador/Panama. Miami Express #N0004, P.O. Box 52–7948, Miami, FL 33152–7948, USA

1. Introduction

The paper by Janssen is wide-ranging and thoughtful and raises a number of penetrating issues. While it deals with methodological concerns, it is characterized by a pragmatic, problem-solving orientation. I find myself largely in agreement with his conclusions, such as the need to initiate the research process with a definition of the problem to be solved instead of with the data available, the usefulness of pilot policy projects, and the need to study possible policy interventions and not just agricultural systems.

Rather than attempt a systematic critique or review of the paper, I will take this opportunity to expand upon a number of his arguments and observations, in the interest of both clarifying them and developing some of their additional implications.

To make these comments more concrete, it may be helpful to think of the agricultural policy maker as a Minister of Agriculture, who often has authority over the management of renewable natural resources as well, and the macro economic policy maker can be visualized as an Economic Cabinet, usually represented by the Ministers of Finance and Economy, or their equivalents, and the President of the Central Bank.

When is aggregation a problem?

Aggregation is more of a problem for the analyst than for the policy maker. The latter is likely to think in terms of selected representative farms and make his or her own extrapolations on that basis. And those representative farms may not be the same as the ones that analysts typically use. Janssen comments: "Agro-ecological zones are a construct of agricultural and ecological scientists, but not necessarily of policy makers". The policy maker's selected farms typically are based on land tenure categories (farm size classes and the distinction between agrarian reform sectors and private sectors) and on

* From section C, only the paper by Janssen was available in advance for review.

J. Bouma et al. (eds.), Eco-regional approaches for sustainable land use and food production, 237–247.

the main alternative production systems: traditional exports, non-traditional exports, crops for the domestic market, livestock raising, and so forth. This does not mean we should abandon agro-ecological criteria in defining the representative farms, but it may mean giving greater weight to other criteria.

Of course in some countries the regional problem has a special importance, such as the distinction between jungle, mountains and coast in Peru, or the distinction among major island groups in Indonesia. Nevertheless, a spatially-based criterion often is complemented by, or is secondary to, these other criteria.

As is hinted in a couple of places in the paper, basing the analysis on the policy maker's perspective often means that the relevant question is the appropriate kind of disaggregation rather than aggregation. How many representative farms are needed, and of what kind? Will sector-wide supply functions suffice, or do we also need information on input use, employment and the like?

The policy maker also is likely to be concerned mainly about institutional issues and price levels, without worrying too much about projecting detailed consequences of reforms. In other words, a Minister of Agriculture may give priority to issues like making the agricultural extension service more effective, or restructuring the agricultural banking system, or convincing the Economic Cabinet to adopt policies that will result in higher producer prices of grains, without being much concerned with the need to make forecasts of the supply implications of better extension or higher farmgate prices.

There is no aggregation problem for prices. The 'national' producer price of corn is much the same for all classes of producers, subject to spatial variations in marketing costs. And in general there is no aggregation problem as long as we stay with static, descriptive analysis. Adding up numbers in a base period cross section does not pose methodological problems. To the extent that the underlying survey data are accurate, there is no error or bias in statements such as "60 percent of all coffee is grown by farms of less than 5 hectares in size", "farms with less than 2 hectares devote 90 percent of their land to grain crops", and "fertilizer use per hectare is three times as great on farms of 5–10 hectares as it is on farms of 0–5 hectares".

The question of aggregation error arises when we wish to forecast or make conditional projections of the implications of changes in policies or exogenous variables. For the sector as a whole, the initial conditions may be described by a column vector of outputs, inputs, prices and factor returns. The problem is that the incremental vector attributable to policy change usually is not proportional to the initial average vector. Hence Janssen's suggestion to seek aggregation categories on the basis of responses to policy-induced change: "to define homogeneity in the expected response to policy interventions". He leaves open the question of how to carry out this prescription, but experience suggests that this kind of homogeneity is likely to be related to similarity in the relative factor endowments, i.e. ratios of land to labour, or irrigation water to land, provided that the land tenure conditions are the same (renter or non-renter, member of a collective farmer or private farmer). It should be recalled that proportionality in the vectors of factor endowments is perhaps the most important of Day's criteria for unbiased aggregation. Of course, in applying this rule, a distinction should be made among the basic categories of land, especially between hillside land and flatter land, and lands that receive relatively high rainfall and those that do not, so here we start to introduce some important agro-ecological criteria.

When the analyst and policy maker are concerned about making projections of farmer behaviour, market structure gives rise to a fundamental aggregation issue, or multi-level issue. At the farm level, behaviour can be projected on the basis of the assumption that input and output prices are exogenous, but at the sector level prices are endogenous and failure to specify them that way can lead to significant errors in the projections of all other variables. That is the case because farmers' planting decisions depend above all on prices. This statement is also true of regional analysis, although it is often ignored. Since the demand elasticities for agricultural products typically vary from about 0.3 to 1.0, analysis of a region that supplies at least fifteen percent of national production in some commodities should incorporate the endogeneity of prices, in order to avoid significant bias because of the price factor.

This is another way of saying that sectoral and regional models are more useful to the extent that they include, explicitly or implicitly, both production and consumption modules. In the case of market-clearing linear programming models, the latter are represented implicitly, in aggregate form, by downward-sloping demand functions. The point gains force at the farm household level when it is desired to model impacts on the environment, for rural households' demand for energy is a major environmental concern in many countries.

In addition to aggregation (or disaggregation), there are two other fundamental multi-level concerns relevant for policy analysis and implementation: policy consistency and replicability of local-level experiences. Examples of policy inconsistency most often occur between macro-economic stabilization policies and agricultural trade and growth policies. In Latin America in recent years, many countries have used an overvalued exchange rate as part of their anti-inflationary strategy, and yet such a policy makes it politically very hard to implement a trade liberalization policy for agriculture, because it artificially cheapens (in local currency) the imports that compete with domestic products. Thus consumer resistance to the proposed liberalization becomes greater than otherwise. An overvalued exchange rate also leads to lower real agricultural prices within the country and therefore reduces real producer incomes, including those of small farmers, and so it may exacerbate rural poverty and increase rural-urban migration. The analyst can assist in promoting a resolution of this policy conflict by spelling out clearly the trade-offs involved. Establishing policy consistency sometimes requires only conceptual reasoning (non-empirical analysis), but it is important to do so.

Replicability is a concern because most of the work on sustainable agriculture has been done at a project (local) level, and not through sectoral policies. Thus, for example, many of those projects develop their own, *ad hoc* credit facilities for participating farmers, but once the project ends, thereby closing off access to that credit, farmers' rates of adoption of the sustainable practices usually drop sharply. In such a case, a fundamental reform in sector-wide credit mechanisms is needed to help bring about spontaneous and sustained adoption of the recommended practices on a wider scale. The analyst needs to be concerned about integrating local or regional practices with national macro-economic or sectoral policies.

The replicability issue can be addressed through the kind of pilot project that the author suggests, provided that those projects are used strictly to test out possible sectoral and macro policies, and to identify institutional barriers that may inhibit their implementation. This approach contrasts markedly with the usual focus of local or regional projects, which is development of that particular region. For a policy project of this nature, its raison d'etre is providing feedback into the national-level policy formulation process, and local development results may be viewed as useful by-products. More on this question later.

The nature of the policy problem and the role of the analyst

Some of the discussion in Janssen's second section ('Methodological issues in aggregating knowledge') can be summarized by saying that the aggregation problem has a theoretical dimension (as in non-linear response functions) and an empirical dimension (Day's conditions). It can be added that it also has a policy dimension. To illustrate, let us consider the complete policy problem in a simplified but basic form, in which policy goals are maximized subject to the reactions of decentralized decision makers, who maximize their own objectives:

$$\text{Max}_{z1,z2}\ f(y)$$

$$\text{Max}_x\ g(x, z_1)$$

$$Ax \leq b + z_2$$

$$Cx \leq y,$$

where: $f(y)$ is the policy objective function; $g(x, z_1)$ represents the decentralized decision functions (i.e. those of farmers and consumers); x is the set of decentralized decision variables (i.e. hectareage per crop, quantities

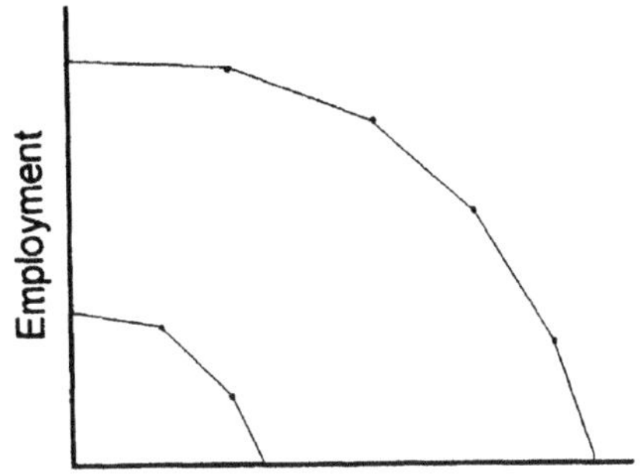

Fig. 1. The relation between employment and foreign exchange.

consumed); y is the set of aggregate variables of interest to the policy makers (total employment, total farm income, index of food prices, etc.); b is the set of initial resource endowments; z_1 is a set of policy instruments that influence resource endowments (i.e., public investments in irrigation, land redistribution); z_2 is a set of policy instruments that influence the objectives that influence the decision variables of decentralized agents, such as prices which influence farmer incentives).

This is a two-level optimization problem which in general does not have a solution because the existence of a global optimum cannot be guaranteed (Candler and Norton 1977), owing to the inability to be sure that the global solution space is convex. However, locally optimal solutions can be obtained and they can be useful in illustrating the difference between two kinds of attainable frontiers: the physically obtainable, and the economically obtainable. Explorations with numerical models of this sort reveal that the latter lies much inside the former, and the difference may be of the order of magnitude illustrated in figure 1.

Another implication of this way of stating the comprehensive policy problem is that the lower-level model's objective function g is *descriptive* and not *normative* from the policy maker's viewpoint. It does not necessarily represent something that he or she wants to maximize; rather it simply is a description of what motivates the reactions to policy change on the part of the decentralized agents in the economy. At the sector level, the descriptive objective function in agricultural sector models with endogenous prices is the one that leads to prices and quantities that clear the market, and in fact does not necessarily correspond to the goal of any decision maker, at any level. This is one of the few areas where it appears that the Janssen paper could use some clarification, stating as it does that "assumptions on the behaviour of farmers are often based on a normative modelling technique". The clues that he seeks to reaching the "optimum resource allocation mix" are provided by the values of the policy instruments in the vectors z_1 and z_2 in the above problem. In a stand-alone sector model, those values are varied parametrically over different solutions in order to explore the policy-attainable space illustrated above.

It should be pointed out also that the inner frontier's position is determined by three factors: (i) the nature of the decentralized decision problem, especially the decentralized objective functions g; (ii) the range of available policy instruments; and (iii) the limitations on the use of policy instruments, including the government's budget constraint. Clearly, it is the inner frontier that is relevant in a practical sense. But a one-level sector-wide or regional agro-ecological model that maximizes farm revenue (with fixed prices) or employment, or any other normative variable will reproduce the outer frontier.

This vision of the policy-cum-descriptive problem suggests that while the policy analyst should not try to guess the weights in the decision maker's objective function f, he or she has two basic roles to play: widen the set of policy options perceived by the policy maker, and show a fuller set of implications of each option. A Minister of Agriculture may wish to use the available policy instruments to promote wider planting of corn, and in those circumstances it could be important to demonstrate, for example, that such a policy may lead to displacement of crops that are more intensive in use of labour, or that increased corn plantings could lead to greater soil erosion on hilly lands.

Establishing a clear vision of the role of the policy analyst helps guide specification of the models and determine what is the appropriate degree of aggregation.

Levels of policy making

The author proposes four levels of policy making: national, sectoral, regional, and farm level. It may be suggested that the last one is not really a policy level; there the analytical challenge is to adequately describe farmers' reactions to policies determined at higher levels. And two other levels often recommend themselves for incorporation, especially when managing the rural environment is of concern: the international level and the community level. At the international level, there is an accelerating trend toward reaching binding agreements on trade policies and environmental policies. The community level is especially important for considerations of implementing environmental management policies, particularly for managing protected natural areas, coastal resources and watersheds.

Aggregation over time

The author suggests that it may be worthwhile to aggregate over time and to deal with the time issue by estimating stock values. However, there are cases when a full disaggregation over time is necessary. For example, the problem of insufficient interest on the part of peasant farmers in reforestation and agroforestry is attributable not only to externalities, but also to the distribution of benefits over time and the existence of cash flow constraints for each time period. In many cases, it appears that the NPV of reforestation for timber is quite high, but small farmers simply cannot wait for the benefits to accrue. In these circumstances, one proposal for a pilot project that has been made is to bring together investors and groups of small farmers, for the purpose of entering into an agreement under which the former would compensate the latter each quarter for the growth of standing timber volume, which is fairly readily measurable, thus providing an incentive for adequate management of the trees as well as allowing the small farmers to receive a different distribution over time of the benefits of reforestation. At the time of maturity of the trees, the investors would then receive the sales revenue from the timber, already having advanced a share of the profits in the former of the compensations to small farmers. In order to test and refine schemes like this, and related ones, careful numerical modelling over time is needed.

On the silence of the users

In this section the author raises the question of where those who are expected to use the models' results should be more fully involved in developing them. Except for the case of the unusual policy maker who has sufficient training in modelling, it should be recognized that models are inherently non-participatory exercises, limited to technical personnel. However, the basic point is that the entire policy formulation process should to be made more participatory than just a dialogue between modelers and a policy maker.

Resource users (farmers, fishermen and others) should be involved in the dialogue before the final versions of the policies are adopted. Wider participation serves three purposes: (i) it helps to produce better policies, because the resource users inevitably are aware of aspects of their reality that the model builders are not; (ii) it improves the prospects of eventual approval of the proposed policies, especially if they require the passage of new laws; and (iii) it improves the chances of the policies' implementation, because those who will be most involved in carrying out the policies will have become co-authors or co-sponsors of them.

A wider dialogue on policies can be facilitated by setting out in clear, non-technical terms the policy conclusions drawn from modelling exercises, and the basic reasoning behind them, and inviting representatives of resource users to critique and refine them. It should not be expected that the eventual results of this kind of dialogue can be run through a model again, because the dialogue inevitably will incorporate non-quantifiable judgments, but it should lead to better policy if properly carried out. A requirement of success in such a dialogue is that the discussion be carried out on realistic bases. For example, if subsidized credit is no longer possible, because the government budget will not permit it, or because it has been found that it generally is destined to low-productivity uses, the analysts need to be explicit about that from the outset and, in such a case, turn the dialogue toward ways to increase mobilization of financial resources for use in the agricultural sector.

Such a dialogue was carried out in Honduras in 1991 with representatives of all the national peasant farmer organizations and also of associations representing larger farmers, and the result was approval by the Honduran Congress in 1992 of a package of eight agricultural policy laws, the most comprehensive such package in the Western Hemisphere. Thus the silence of the users can be broken, but not at the modelling stage.

Policy-oriented specifications

To follow Janssen's prescription to analyze possible policy interventions, and not just systems for their own sake, it is important to define the realms of policy, and their limitations. Normally, policies are not defined at the crop level, because for the sake of economic efficiency corn policy should not differ from beans policy. In all countries there are programs at the crop level, designed particularly to promote adoption of improved technologies, but policies are usually more encompassing.

The typical realms of agricultural policy can be derived from a simple paradigm of the producer. In order to survive and prosper as a producer, he or she needs the following:

- Adequate economic incentives to produce.
- A resource base: land and water, preferably including full title to the land (adequate tenure conditions), and proper resource management, so that, for example, the soil does not erode away.
- Access to inputs, technology and markets.

It can be seen that these elements correspond, respectively, to the following components of an agricultural linear programming model: objective function, right-hand side, and the technology matrix plus the commodity balance constraints for inputs and outputs.

Accordingly, the main areas of agricultural policy are the following:

- Policies that affect real producer prices.
- Land tenure policies.
- Investment policies on irrigation and land improvement and settlement.
- Rural natural resource management policies.
- Agricultural financial policies (to facilitate access).

– Research and extension policies.
– Storage and marketing policies.

This is a slightly amplified list *vis-à-vis* the policy areas mentioned by Janssen. Another important caveat is that direct price controls are fewer and fewer these days, and that the true determinants of real agricultural prices are exchange rate policy, tariffs and trade policy. Thus parametrically varying prices in a model does not correspond to a possible policy scenario and is justified only if it is clarified that it is, for example, a representation of the sector repercussions of possible exchange rate movements. But in that case an exchange rate movement would affect the prices of all tradeables and not just one or two, so the parametric variation would have to be conducted simultaneously over many crops.

This list brings to the fore issues of resource ownership, such as for land, forest resources and irrigation systems, which are central issues for agricultural development in many countries.

Policy makers do not try to control cropping patterns directly. Thus a model's 'optimal cropping pattern' is meaningful only if it represents a simulation (with either a programming model or econometric techniques) of decentralized decision makers' responses to possible policy options. Looked at in another way, the main concerns of agricultural policy makers may be summarized as:

- Improving the efficiency of national institutions that serve the sector (credit institutions, research and extension institutions, grain storage facilities, irrigation systems, etc.).
- Eliminating subsidies that are regressive or untargeted and replacing them with subsidies that are targeted on the lower-income strata. Empirical work has shown that in practice a surprising number of interventions are regressive, including guaranteed prices, agricultural extension, and subsidized credit programs.
- Eliminating or improving policies that result in market distortions, such as erratic interventions in grains markets, or laws that prohibit supposed 'hoarding' of grains, thereby discouraging investment in needed grain storage capacity.
- Economic defense of the sector. In a world increasingly dedicated to trade liberalization, this policy objective is becoming less tenable, but even in developed countries the need to subsidize farming for a possibly long transition period is accepted as a legitimate national goal. In this case, the role of the policy analyst is to help develop non-distortive, non-regressive means of supplying the desired protection, with the highest volume of results per unit of fiscal expenditure.

Fig. 2. The relation between crop research, agro-ecological research and agro-policy research.

Criteria for model evaluation

Janssen mentions a set of criteria for evaluating agro-economic models. To that list perhaps the criterion of robustness of model results can be added. If the results vary significantly with minor variations in model specification or data, then the usefulness of the model is weakened. The same dictum applies to simpler analytic techniques. For example, it has been found that effective protection rates can vary greatly from year to year. Hence a proper use of them should require calculating them for period of at least three years, if not five, to average out short-term fluctuations.

The relevance of agro-ecological research to agro-economic policy

This question is a *leitmotif* that runs throughout Janssen's paper, so perhaps a few comments on it are in order. While a closer integration of the disciplines brings undoubted benefits, it is clear that many agricultural policy issues do not require an agro-ecological data base, and to pretend otherwise is to invite wasteful use of research funds. Reforming agricultural banks, strategies for privatizing state-owned grain silos and agro-processing facilities (to make farmers shareholders), and ways to restructure the legal and operational basis of production cooperatives are examples of policy issues that can be fruitfully dealt with in the absence of agro-ecological data.

The relation between crop research, agro-ecological research, and agro-policy research may be summarized in the figure 2.

The interest of this group lies in exploring the cross-hatched region. In a preliminary way, it may be suggested that the region – which is the set of areas in which agro-ecological knowledge is relevant to the formation of agro-economic policy – mainly comprises investigations of the following issues:

- *Determining comparative advantage.* This topic may seem a little arid, or even out of date, but in a world increasingly dedicated to free trade it is crucial. Lack of comparative advantage in particular crops will depress incomes of those who produce them, leading eventually to shifts in cropping patterns and/or increased rural-urban migration.

- *Making conditional predictions of producer responses to policy changes.* Here policy change refers to the three main areas of instruments that affect incentives, resource endowments, and access issues. Such simulations or projections can assist materially in the design of policies, provided they are carried out on a timely basis.
- *Helping design policies for sustainability.* Here one of the key contributions modelers can make is simply to document effects for which typically we do not yet have much information, such as the rates of soil loss or biocide accumulations that are associated with different cropping regimes, as is being attempted by Schipper *et al.* (section F, this volume).

Dealing adequately with these policy areas is of fundamental importance and by itself is a large undertaking for interdisciplinary work.

References

Candler W, Norton R D (1977) Multi-level programming and development policy. World Bank Staff Working Papers, no. 258, Washington, DC, USA.

Response to section D

K. ATTA-KRAH
International Centre for Research in Agroforestry (ICRAF), United Nations Avenue, P.O. Box 30677, Nairobi, Kenya

Introduction

The concept of eco-regional approaches (ERA) to research is a recent idea that has received considerable attention and is gaining considerable grounds in research circles, especially within the Consultative Group on International Agricultural Research (CGIAR). In the last 3–5 years, the CG-centers have been advocating for more ERA basis in research and also for systems-wide initiatives, which bring a number of CG-centers and relevant NARS to work together on common problems of cross-cutting importance. Of late also, there is an increasing number of research projects being described as eco-regional in nature, and therefore claiming to be adopting ERA in research. Without attempting to formulate a definition of ERA here, some (perhaps over-simplistic) views are offered at what ERA may be all about:

- ERA is basically an approach to research conceptualization, organisation and implementation (it is neither an analytical tool, nor a type of research).
- ERA requires that research should be generated on the basis of the constraints and potentials at the scale of agro-ecological zones (AEZ), and not just based on micro-site analysis and issues. Proper definition of the AEZ is therefore essential. Some degree of homogeneity is required for each AEZ identified for ERA-based research.
- Research mechanisms and tools such as characterization and geographical information systems (GIS) are useful for subdivision of AEZ's into homologous agro-ecological units, on which research is then based.
- ERA also invariably implies institutional collaboration and joint effort on common problems. It incorporates elements of resource-use efficiency, comparative advantage and research coordination across institutions operating within an identified AEZ.
- ERA advocates the creation of a 'common focus' to regional problems, and for concerted and coordinated approaches to problem solving through research.

J. Bouma et al. (eds.), Eco-regional approaches for sustainable land use and food production, 249–255.

- ERA requires a priority setting mechanism which enables all the key partners and institutions to determine what the regional priorities are, and how and by whom different aspects of the priority problems can be tackled.
- Sites chosen for ERA-based research should be representative for the eco-region under consideration, and results from such research should be capable of extrapolation to other areas within the eco-region.
- ERA-based research in a particular eco-region could involve several different types of research activities and methods: on-station research, on-farm research, commodity research, farmer participatory research, systems-oriented research, etc. ERA is simply the umbrella under which the research activities take place. What is important is that all the research is inter-related, and the results are made to come together through the use of tools such as synthesis, modelling, GIS, etc.

Three case studies have been presented which touch on ERA from different perspectives. The common element in all three cases is that the research is focused at addressing problems identified or conceptualized on an eco-regional scale and not just focuses on site-specific issues. All the papers are also interested in tools for wider application of research results across the eco-region.

Response to paper by Breman (section D, this volume)

This paper focuses on the Sahel region of Africa, and attempts to understand yield and quality determining processes that could help address the issue of sustainable agriculture in the Sahel, and also offers clues for identification of potential solutions. The author draws mainly on experiences gained during the 'Primary Production Sahel' (PPS) project (1976–1982) and from the follow-up project 'Production Soudano-Saheliénne' (PSS). The research activities presented are strongly process-oriented, and focus on:

- the water balance
- the nitrogen balance
- the phosphorus balance
- rangeland/livestock linkages

ERA is justified by the author on the basis of the relative homogeneity of the eco-region, even in its variation, and the scarcity of resources both for farmers and research groups. The study identified three sub-regions within the Sahel on the basis of returns to use of external inputs, which needed to be targeted quite separately in terms of sustainable agricultural research:

a. regions where use of external inputs makes clear economic sense;
b. regions where use of external inputs could make economic sense, provided certain production systems, and adoption of certain practices or socio-economic measures were in place;
c. marginal areas where external inputs were not likely to show economic gains.

It was decided to prioritize research efforts on region 'b'. Comment: Such sub-division of the region is important, and the need for ERA-based research to have a clearly defined targeted region and specific research themes is endorsed (by this respondent).

Research sites

Much of the research in both the PPS and the PSS projects was done at research stations of the Institut d'Economie Rurale (IER) in Mali. The questions that arise are:

- on what basis were these sites selected? (expediency, economy, suitability?)
- does the use of existing stations sometimes compromise on the requirements for benchmark sites?
- how do we strike a balance between expediency, economy and technical suitability in site selection?

A strategy for benchmark site selection in ERA research activity is required.

Institutional involvement

Three institutions are involved in the current PSS project:

- Institut d'Economie Rurale (IER, Mali)
- Research Institute for Agrobiology and Soil Fertility (AB-DLO, The Netherlands)
- Department of Nature Conservation, Wageningen Agricultural University (DNC-WAU, The Netherlands)

The question that arises is what is the link between the PSS project and other research programmes in the zone with similar goals, and supported by CG-centers? First, it is known that the ICRISAT Sahelian Centre (ISC) exists in the zone and undertakes farming systems research (presumably for sustainable agriculture purposes). The apparent weak links between the PSS project and ISC may be due to the commodity bias and the Farming System Research (FSR) focus of ISC. This however, does not justify the absence of collaboration. Secondly, ICRAF's Semi-Arid Lowlands of West Africa (SALWA) agroforestry-based research programme is concerned with issues of sustainable agriculture and involves collaboration with the same national institution, IER. From the perspective of ICRAF, a strong involvement of the PSS team in the SALWA research programme would be embraced. The expertise of the PSS modelling team is not easily reproduced elsewhere. SALWA would also gain considerably from an active collaboration with Netherlands modelling groups. In general, the SALWA project could draw on experiences of both PPS and ISC. A mechanism for better linking the three projects needs to be found.

An ideal ERA-approach to research for the Sahel should involve a consolidated, concerted and open agenda of research involving all the key players and partners. Such an approach is being conceived for the so-called 'Desert Margins Initiative' project which will have a strong base in the Sahel.

Conclusions

The research reported in this paper has considerable relevance for the Sahelian region. ERA is being used as a way of dealing with the complex and interactive problems of the Sahel. The strength of the PPS/PSS projects was/is the collaboration between local research institutions and Dutch research institutions, which has strengthened technical capabilities for modelling. Another strength is the critical mass of scientists that have been involved in these projects, and the long-term operation perspective of the research (10–15 years). The ERA perspective however, needs to be broadened by bringing in other research institutions in the region with similar goals and concerns. The links with FSR activities at ISC and with the SALWA project need to be strengthened. PSS involvement in SALWA research activities, now consolidated in Mali, is highly desirable.

Response to paper by Aggarwal *et al.* (section D, this volume)

This paper describes a commodity research perspective to ERA. The commodity in question is wheat, which is an important staple food crops in India. The research aims to utilise systems approaches to estimate productivity potentials of the crop in different agro-ecological zones of India, and to determine whether the required increases in productivity to meet projected future demands can be achieved. Yield gaps are analyzed between potential wheat production estimated by a deterministic model and actual production in the major wheat growing areas of India. Using management options available in crop growth simulation model – namely date of sowing, nitrogen application and irrigation frequency – the authors identified possible improvement which would reduce the calculated yield gap.

Interesting issues to be addressed in relation to ERA are the following:

- Does this commodity-based research fits the ERA approach? (see introduction of this response paper). Which is the eco-region of interest here?
- Two principal 'computer tools' are used that are interesting to ERA: GIS to delineate wheat production areas in India, and crop growth simulation models to determine wheat production potentials and to explore management options for yield increases.
- The paper mostly deals with a macro analysis of the production potential of wheat under irrigated conditions. The study covers only an area that presently produces 35 Mt (districts having greater than 10,000 ha area and more than 80% irrigation). It would be interesting to know how much potential exists to increase wheat yield outside this production zone, which probably is in the rainfed conditions. From this respondents point of view, it is debatable where rapid future increases are most likely to occur: in the less intensively managed areas, or in the already intensively managed areas?
- It was argued in the paper, justifiably, that crop growth simulation for a number of years, taking climatic variability into account, is necessary to arrive at an optimum fertiliser rate for wheat. Since irrigated wheat in India is mostly grown in multiple cropping systems, it is equally important to consider the amount of fertiliser applied to previous crops and the residual fertility for determining the optimum rate.

Response to paper by Kuyvenhoven *et al.* (section D, this volume)

The presented paper provides insights into some modelling methods and strategies for assessing the impact of alternative policy options, within an ERA framework. Also, the paper provides an economist's perspective to ERA: it addresses the issue of linkages between policy objectives at regional level and farm household responses at micro level. It is argued that such an approach will improve interaction between policy analysis on the one hand, and agricultural research and adoption of technology on the other, taking into account both production and welfare implications.

As the authors state, it is important that farm households are placed within a regional, national and international context and that policy formulation take this into account. Also, it is important to build models which can help guide policy in promoting sustainable land use management. The paper presents a good discussion of the various considerations in such model building. Some form of integration of existing approaches to modelling is needed to overcome shortcomings inherent in single approaches. The results of such an integration as presented in this paper seem to be sensible in a qualitative sense. However, some more explanation should be given on certain aspects to be convincing about the quantitative results, namely:

- There is a large leap from the topic of 'how to construct a model' to the section on 'results of simulations', and there is no discussion of the input data whatsoever. This leads to difficulty on the interpretation of the results on the reader's part.
- The authors did not mention how some concepts or factors were measured. For instance, among their variables are infrastructure and rural finance. Also, it is not clear what is meant by a percentage change in these variables. In the case of these broad level variables, I am also not sure how the authors found any variation in the sample of households from which to estimate reaction functions. More explanation is needed on these points.
- Cross-sectional data sets, especially if only a sub-set of households is used, are notoriously inadequate for deriving price responses due to lack of variation in prices (time series are usually required). It was unclear how regional level constraints were estimated using the data sets described.
- No indication is given of a minimum data set which would be required for an undertaking as reported in the paper. The authors referred to data limitations in their empirical work but did not mention how serious they were and what implications this had for the results. It would be important to know how sensitive the results are to missing data.

The authors could also comment on the cost effectiveness of the presented modelling approach for local-level planning. An analysis of the cost of building models versus their results needs to be made.

Acronyms

AEZ	Agro-ecological Zone
ERA	Eco-Regional Approach
FSR	Farming Systems Research
GIS	Geographical Information System
ICRAF	International Centre for Research in Agroforestry
IER	Institut d'Economie Rurale (Rural Economy Institute)
ISC	ICRISAT Sahelian Centre
PPS	Primary Production Sahel
PSS	Production Soudano-Saheliénne
SALWA	Semi-Arid Lowlands of West Africa

Acknowledgements

The assistance received from the following people at ICRAF in the review of the case studies is acknowledged: Peter Cooper, Steve Franzel, Frank Place, Katherine Snyder, Meka Rao and Chin Ong.

Response to section D

P.G. JONES*
Centro Internacional de Agricultura Tropical (CIAT), Apartado Aereo 6713, Cali, Colombia

The first question to be posed is: what is an eco-region? The Sahel may be one but, as Breman points out, it is an extreme case and even so a complicated area. Aggarwal *et al.* include the whole of wheat growing India, but tries to delimit regions on production statistics. Kuyvenhoven *et al.* neatly escape the question completely by suggesting that the 'spatial definition of region still leaves room for discussion'! This respondent seriously doubts that there are definable 'eco-regions' in a contiguous geographic sense. There exist various edaphic zones in the world that have been classified as to climate, soils and topography. What people do with them is another dimension, possibly the most important. It is also the most difficult to define or classify because of its variability.

CIAT embarked on Natural Research Management (NRM) research 5 years ago. It was evident that CIAT could not research all systems in all areas. Concessions were obtained to exclude mining, the deserts and the ice fields, and also restrict the study to Latin America. Then the agro-ecological studies unit helped to define three agro-ecosystems in which CIAT should work on NRM (Jones *et al.* 1991). CIAT chose to define an agro-ecosystem precisely as a combination of environment and what people are doing with it. This study resulted in three main agro-ecosystems:

1. *The Acid Hillsides.* Lands between roughly 1000 and 2000 m altitude with acid soils and a growing season of at least four months. This includes many of the coffee areas and regions growing annual crops, predominantly beans, maize and cassava. Some areas include intensive horticulture for local markets. Over half of the area is in pastures.
2. *The Savannas of Latin America.* Lands with acid soils in Venezuela, Colombia, Bolivia and Brasil with predominant grassy or at least grass understory vegetation. Extensive grazing of natural pastures is still the main land use. Improved pastures are gradually accounting for a larger proportion of the region. Cultivation of annual crops has expanded in some areas, particularly in Brasil. Heavy machinery can produce serious soil problems.

* This response expresses the personal view of Mr. Jones, not an offical CIAT view. The three corresponding papers have not been reviewed in detail.

J. Bouma et al. (eds.), Eco-regional approaches for sustainable land use and food production, 257–260.

3. *The Forest Margins*. The lands which are or have been tropical humid forest, which are legally accessible and are undergoing forest clearing. Forest is usually cleared by small scale colonists who put the land through a short cycle of annual cropping (1–2 years). The land is then turned over to pastures which proceed to degrade. The land is then often consolidated into more extensive grazing and the colonists move on to cut further forest.

The group of agro-ecosystems came out with some useful consistencies for CIAT. We could work on crops for acid soils that were in the most part free of serious drought.

In Latin America, there is a complex of agro-ecosystems beyond the range of CIAT's expertise. But, even within the CIAT agro-ecosystems there are interactions with serious consequences. In many countries, a mosaic of ecosystems allows trade offs between regions. Thus Brasil can choose to invade the Amazonian forest, or to intensify production in the savanna region of the Cerrados, or try to cope with foregoing both of these options. Colombia has fewer options. Most of Colombian Amazonas and Orinoquia have been ceded back to the indigenous tribes. The development of the savannas of Latin America is an option which will absorb few migratory people but has a major potential for national development (Pachico *et al.* 1994).

The interactions between agro-ecozones in Latin America are not just important, they are crucial. Migration between hillside areas and forest margins is a serious problem in some parts. Only the countries that hold savannas – Brasil, Colombia, Venezuela, Bolivia and Guyana – can develop them for intensified production to increase feeding capacity for an urbanized population. In some cases in Bolivia and Guyana, the savanna areas are so remote that little is expected from them in the near future. All other Latin American countries have no savanna option. In the case of countries without savanna option there is a different problem. Practically no country has the problem of overpopulation. Distribution of population and the legality of land holding are the major impediments to advance. Political instability is also an important factor.

As it comes about, GATT (General Agreement on Tariffs and Trade) will have a profound influence – we have already seen the flight of some Colombian flower producers to Mexico to benefit from the recent trade agreement between Mexico, Canada and the USA.

Changes can happen quite fast and not just in quantity but in quality. The example of wheat in India is a very good one, but planning for land use has to take into account much more rapid change, and do so more rapidly than previously thought. This change is not necessarily more of the same thing, the type of agriculture carried out may be changed. The Colombian savannas are a good example of this. For more than 20 years, the CIAT Pastures program has been searching for, and developing, advanced improved pastures for savannas. This has already had an effect – the size of holdings is coming down. But in a way this is still a case of more of the same. However, costs of renovating, or producing a grass legume pasture, are often high in relation to return and so implementation of improved pastures is a risky business.

A completely new and unexpected development was the production of acid tolerant rice varieties in CIAT. This produced unforeseen effects in that now a 3 ton upland rice crop will pay for moderate liming and fertilizer inputs, and allows management of grass legume pastures in a ley system. This is not more of the same. This is a change and has solved the problems of pasture management in a completely unforeseen way. Modelling systems is only useful if the full range of potential solutions is tested. We did not solve the problem of legume persistence in the pasture by producing a better legume. We solved it by growing rice.

The problem of aggregation is evident in the papers and addressed specifically by Janssen (section C, this volume). I would suggest that a novel way to access the subject would be the problem of 'disaggregation'. Let us look at the data available at each level of the system and work out what is left out. For example, if we take a core from an iceberg we may have a small sample of ice. When this melts, we have a small sample of water. If we probe the structure we will find that it is hydrogen and oxygen – going even further we would find protons, neutrons and electrons. These we might suppose to be elementary particles. The study of agricultural systems is somewhat similar. The agro-ecology of the world encompasses all. The agro-ecology of a region encompasses all except its external effects – usually political or economic. At the extreme we find that agro-ecology may be a garden tended by a gardener who is an external effect. The diversity of the gardens and the gardeners makes aggregation a problem. As Jansen suggests, one way is to categorize the gardeners and gardens.

Since the system looks different at each scale with different aspects included as internal effects, and other aspects as external, it is necessary to have different ways of categorizing the system at each level. Another important factor when considering systems at different scales is that solutions to problems at one scale are often found at a higher scale in the system. An example of this could be the sustainability, in terms of nutrient balance, of for example, a bean/maize relay system. Taken as a whole, the fields under beans/maize relay suffer an overall loss of plant nutrients. This system is therefore inherently unsustainable at the field level. But if we include the rest of the farm, a higher scaled view of the system, we might note that half of the enterprise was cattle. These produce farmyard manure, which can then be used to stabilize the inherently unsustainable beans/maize. We see then an interaction between scales, with a problem at one level solved from a higher system level. Agricultural subsidies at a regional or national level are another instance of an intervention from one system level to a lower one.

Although there are certain exceptions, in general it is difficult to assign a uniform geographical area to be an eco-region. Agro-ecosystems at various scale levels exist as a mosaic and undergo significant interaction. We have to be careful how we match our modelling and characterization to the scales involved and to be on the look-out for solutions which may not be intuitively obvious.

Acronyms

CIAT	Centro Internacional de Agriculture Tropical (International Center for tropical Agriculture)
NRM	Natural Resource Management

References

Jones P G, Robison D M, Carter S E (1991) A GIS approach to identifying research problems and opportunities in natural resource management. In CIAT in the 1990s and beyond: a strategic plan, supplement. CIAT publication no. 198, Cali, Colombia.

Pachico D, Ashby J, Sanint L R (1994) Natural Resource and Agricultural Prospects for the Hillsides of Latin America. Paper prepared for discussion at IFPRI 2020 Vision Workshop, Washington, DC, USA, November 7–10, 1994.

SECTION E

Institutional aspects

Founding a systems research network for rice

H.F.M. TEN BERGE[1] and M.J. KROPFF[1,2]

[1] *Research Institute for Agrobiology and Soil Fertility (AB-DLO), P.O. Box 14, 6700 AA Wageningen, The Netherlands;* [2] *International Rice Research Institute (IRRI), P.O. Box 933, Manila, The Philippines*

Key words: capacity building, NARC, research network, rice, simulation, SARP, systems analysis, training

Abstract. A systems research network for rice was established in eight Asian countries with 15 national agricultural research centres (NARCs) via the SARP project. NARCs are participating with interdisciplinary teams. A 'team establishment phase' covered six years, and included three international training programs. It was followed by an 'application phase', currently addressing six 'application pograms'. Mechanisms used in capacity building after the formal training phase are case studies, team visits by project staff, joint research planning and output analysis, short (3 months) and long (1 year) fellowships for exchange visits of team members, annual thematic workshops, PhD fellowships, support to local (national) training courses, a proceedings series for project publications, and the development and distribution of user friendly shells to facilitate the use of simulation models. Joint research programs were headed by 'theme coordinators'. The paper presents three examples of eco-regional analyses: (I) on the optimization of nitrogen use in rice; (II) on the effects of climate change on rice production; and (III) on agroecological zonation. It is concluded that systems research networks can provide a solid basis for a range of eco-regional studies on various topics, and for local in-depth studies.

Eco-regional analysis and systems research networks

The delineation and characterization of eco-regions is useful for targeting application domains of production technologies and development policies. It is also seen, now, as one of the first steps required in the process of research prioritization for agricultural development, because it is the basis of *ex-ante* assessment of the possible impacts (and their geographic extent) of changes in technology or in the socio-economic environment. Such assessments may be of explorative or predictive nature and are, of course, only possible after properly expressing the relevant production systems in the form of accepted models and their corresponding quantified parameters. Systems analysis – both in its model development and application phases – is therefore an indispensible element in the eco-regional approach to development research.

Constructing models for eco-regional analysis is team work. Good models enable the extrapolation of localized research findings across the eco-region, they may give guidance to local in-depth research, and should not require excessive data inputs. Workable models are a practical blend of empirical

J. Bouma et al. (eds.), Eco-regional approaches for sustainable land use and food production, 263–282.

and explanatory formulations, and their shape varies with their purpose. To arrive at models that are acceptable to a sufficiently broad users group for the purpose of eco-regional analysis, it is not only necessary to piece together different 'disciplinary insights' in a balanced manner, but also to cover – in the developing and testing procedures – the range of locations representative of conditions across the entire eco-region. This leads to the conclusion that, already during the model development process, scientists from various disciplines and at various sites need to be involved in order to arrive at reliable products. Research networks enable the necessary development of a common language of concepts, models and data bases, and allow frequent interaction among actors. Networks are thus a logical format to execute the tasks of development, testing and application of models for eco-regional studies.

A systems research network has been established in the Asian rice producing countries through the SARP (Simulation and Systems Analysis for Rice Production) project. Extensive descriptions of the project have been given earlier (Ten Berge 1993; Kropff *et al.* 1994). This paper presents an overview and discusses the main project features in retrospect. It also includes summaries of three cases of eco-regional work, executed within the framework of SARP.

The SARP project and its development phases

SARP is a collaborative project among agricultural research centres. It involves a number of national agricultural research centres (NARCs) in the Asian region, the International Rice Research Institute (IRRI), the DLO Research Institute for Agrobiology and Soil Fertility (AB-DLO, formerly CABO), and the Department of Theoretical Production Ecology of the Wageningen Agricultural University (TPE-WAU). Among the participating NARCs are a number of universities. The NARCs altogether represent eight countries: India, China, Bangladesh, Korea, Thailand, Malaysia, Indonesia and The Philippines. The locations of the centres are depicted in figure 1, the corresponding acronyms are given in list of acronyms. Each of the centres is represented by an interdisciplinary team of scientists.

While the longer term objective of the project is to improve sustainable productivity of rice based systems, the immediate goal is to build research capacity in systems analysis and crop simulation. Three phases are distinguished from an administrative viewpoint: Phase I (1984–1987), Phase II (1988–1991), and Phase III (1992–1995). During the first two phases, the international training programs were the core of project activities. Three consecutive groups of scientists were trained during the 1986–1987, 1988–1989 and 1990–1991 international training programs. A total of 91 NARCs based scientists received the formal training. Their disciplinary backgrounds are listed in table 1.

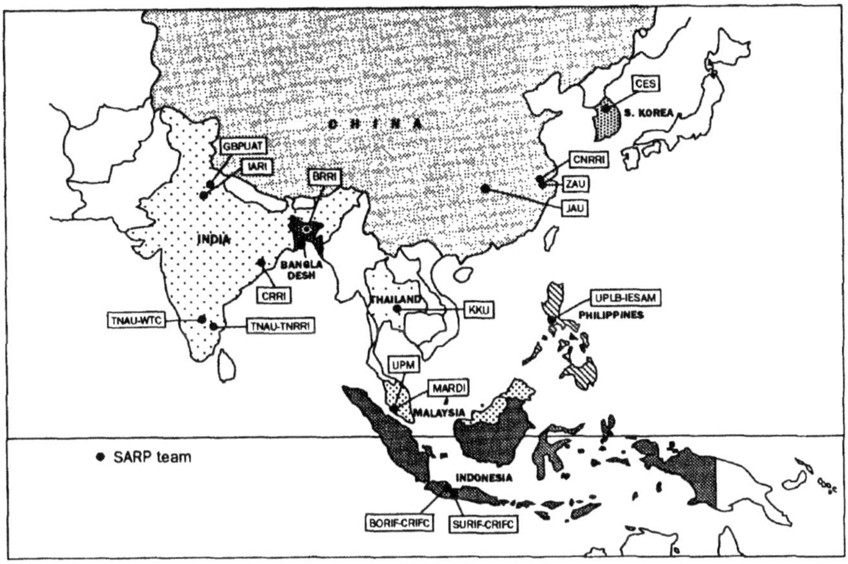

Fig. 1. Countries and NARCs participating in the SARP project. Only primary teams are indicated. (Secondary teams were established via in-country training programs).

Table 1. Number of scientists that received SARP training, per year, and discipline.

	1986–1987	1988–1989	1990–1991	Total
Agronomy	8	4	5	17
Plant/crop physiology	3	7	4	14
Entomology	5	3	5	13
Plant breeding	2	4	2	8
Soil chemistry/fertility	3	4	1	8
Plant pathology	3	1	3	7
Soil physics	2	2	4	8
Statistics/computer	3	–	3	6
(Micro)meteorology	2	1	1	4
Forestry	–	–	1	1
Irrigation/water management	–	–	1	1
Food engineering/aquaculture	–	–	1	1
Weed science	–	–	1	1
Horticulture	–	–	1	1
Nematology	1	–	–	1
Total	32	26	33	91

Building upon case studies initiated as part of the training programs, activities gradually transformed during Phase II into a collaborative international research program. Phase III aims at further developing this research program, thereby simultaneously consolidating the established systems research capacity in the form of a research network, and elaborating applied research issues towards tangible outputs. Thus, alternative to the formal Phases I–III, the project's history can be divided into a 'team establishment phase' (up to 1990) and an 'application phase' (from 1990 onward). This distinction is maintained in the following, although developments were obviously not identical at all centres, nor did occur simultaneously.

Key elements of the team establishment phase

Institutional aspects

NARCs were asked to establish small teams (originally 4 members) to participate in the training program. A prerequisite for participation of NARCs in SARP was that their team would be enabled to continue to work as a unit, and that they would be full time available to systems research work, at least for the duration of the one-year training program. Participation requirements for individuals were a good command of the English language, and some research experience preferably equivalent to MSc or PhD level. Teams were usually grouped around an agronomist, plant breeder or crop physiologist, and included scientists from other disciplines such as soil science, agrometeorology, entomology and phytopathology. Two groups of teams were established via the first two international programs (1986–1987 and 1988–1989). The last training program (1990–1991) was used to reinforce these teams, rather than admitting new centres to the network.

Each team was headed by a team leader, responsible for contacts with SARP staff and for integrating research plans. A senior scientist from each centre was assigned as team supervisor, whose main task was to help develop case studies in line with institute research priorities, and to support team activities at the home base by assuring access to research facilities.

Case studies undertaken by teams were directly derived from the institute's current research agenda, thus ensuring a continued interest also after the training phase.

Hardware was donated to each centre for easy access by all team members, and one team member was responsible for maintenance. At a later stage, with the popularization of personal computers, hardware became less of a constraint, although it still is in some centres. The maintenance of models and other software, on the other hand, has gradually become more problematic, one of the causes being the – encouraged – free proliferation of models and modules.

Training aspects

The goal of the training phase was to transfer the skills of model building, not to teach the use of particular models. Where existing models were used, they were fully open and were explained line by line. They served as an example or starting point for further elaboration, and their source codes were freely available. This philosophy proved useful for most participants. Younger participants, however, generally felt more at ease developing programming skills than the more senior scientists.

The training program consisted of three sections: the formal course (originally eight, later six weeks), the case study, conducted at the home institute, and a study workshop, after roughly eight months, to report results of the cases studies. During the formal course, participants were guided through all steps from word processing, the use of operating systems, principles of modelling, simulation languages, to model building and evaluation. The scientific contents covered the principles of potential production (crop physiology, agrometeorology), water limited production (soil physics, plant-water relations), nutrient limited production (plant-nitrogen relations; 3-quadrant analysis), and yield reducing factors (crop-pest interactions; principles of population dynamics). All participants attended the entire program, irrespective of their disciplinary background. To some extent this was seen as an advantage, enabling more cross-disciplinary interactions among team members. Many participants, however, felt that a flash course across so wide a range of topics was too much, and in later national courses, organized by NARCs themselves, course contents were much reduced.

One or more case studies on a given rice cropping system were formulated by each team at the conclusion of the training course. Bottlenecks within the system were analyzed, for example problems related to transplanting, timing, and the choice of *pre-* and *post* rice crops. In some cases, especially in the field of rice pests, more discipline-oriented cases were formulated and conducted. Case studies initially emphasized the development of simulation skills, but later included experimental work to develop, test and validate models.

Teams were visited by SARP staff during the execution of the case studies. The duration of these visits was one to two weeks per team during the establishment phase. The main purpose was to provide direct hands-on support to the team's simulation activities and to interact with the team supervisor and institute's management. Lateron, team visits were shorter and more devoted to the planning, execution, and analysis of field experiments.

Key elements of the application phase

Scientific organization

While emphasis in the first SARP years was more on developing modelling skills, participants developed a stronger interest in combining experimental and modelling work during the early application phase. This allowed the improvement of model quality and broadening of application ranges. So-called 'common experiments' became the basis for coordinated network-wide joint experimentation. More recently a return to 'desk activity' was observed, associated with the compilation and analysis of vast sets of collected data in view of the project's ending.

To attain more focus in research efforts, 'research themes' were formalized in 1990. Workshops became centred around these themes: agro-ecosytems; potential production; water, nutrients and roots; and pests, diseases and weeds. Although SARP staff at IRRI and in AB-DLO/TPE-WAU were not assigned to particular themes, their activities gradually differentiated too. This was institutionalized with the start of the third project phase in 1992. From then on, a 'theme coordinator' was assigned to each of the four research themes, to support the research activities at the NARCs. Two coordinators were based at IRRI, and two at AB-DLO/TPE-WAU. An additional task of the coordinators was to address gaps in the understanding of rice production systems and in the models, by conducting their own research. They played a central role in developing various types of specialized software, e.g. models, data entry sheets, and 'user shells'. The two IRRI based coordinators took a relatively large share of logistic support to the project – including the organization of workshops and the editing of the SARP newsletter – in addition to their scientific work.

The coordinators were, and still are, supported by so-called 'theme leaders', senior scientists at IRRI and AB-DLO/TPE-WAU providing scientific guidance in research planning. While the costs of coordinator positions were borne entirely by the project, time inputs by the theme leaders were core inputs from IRRI, AB-DLO and WAU-TPE.

The differentiation of research has been instrumental in providing a better structured support to thematic activities. It counteracted however, to some extent, the interdisciplinary work originally envisaged. Another drawback was that coordinators found it increasingly difficult to support all team activities during their visits. The role of workshops became therefore more prominent.

While a relatively broad structure of four themes was sufficient upto 1993, research activities are now (1994 onward) grouped into six so-called 'application programs', to strengthen the focus on practical research in the network. These programs are:

a. Agro-ecological zonation and characterization
b. Optimization of crop rotation and water use
c. Application of models in plant breeding programs
d. Evaluation of climate change impact on rice production
e. Optimization of nitrogen management
f. Optimization of pest management

Much of the work conducted in these application programs can be classified as 'eco-regional', in the sense that it builds on a common scientific basis integrating 'eco-region wide' contributions from many scientists, and using models and approaches that were co-developed and tested across the wide range of conditions found in the major rice production areas of south, south-east and east Asia. Characteristic is, in our view, the mutual (cross-locational) enhancement of research efforts, through common model development, common experimentation and model parametrization, and the development and application of common methodologies in data selection and analysis. Some results of three such eco-regional application programs are presented in the following paragraphs.

Case I. Improved nitrogen management

This program aims at improving nitrogen management in rice, by developing tools for generating localized recommendations. Optimization of nitrogen (N) management in rice is high on the list of priorities of all research centres participating in SARP. While N-fertilizer recommendations are still mostly issued as 'blanket recommendations', it is recognized that substantial gains can be made by site tailored management. This requires a general description of the key processes, combined with site specific parameter values. Suitable procedures must be designed such that model inputs can easily be derived from field experiments at research stations.

A series of three 'common experiments' was conducted during 1990–1994 (Ten Berge *et al.* 1994). These experiments were directed at three key processes: fertilizer N recovery, crop N uptake capacity, and utilization of N after absorption by the crop. Teams in India (TNAU-TNRRI, SWMRI, CRRI), China (ZAU, JAU), Korea (CES), Indonesia (BORIF, SURIF), Malaysia (UPM, MARDI), and The Philippines (IRRI) participated in various phases of the experiments. This resulted in the formulation of a simple model, based on empirical soil and crop parameters. Soil parameters are the native soil N supply rate, directly estimated from N uptake in non-fertilized plots, and the fertilizer N recovery *vs.* time after transplanting. The crop parameters are maximum N uptake rates (relative to growth; and as an absolute rate), maximum crop N concentration, grain N content, and an empirical calibration factor (FSV) expressing the efficiency at which radiation and leaf nitrogen are used. After parametrization for a given combination of site, variety and

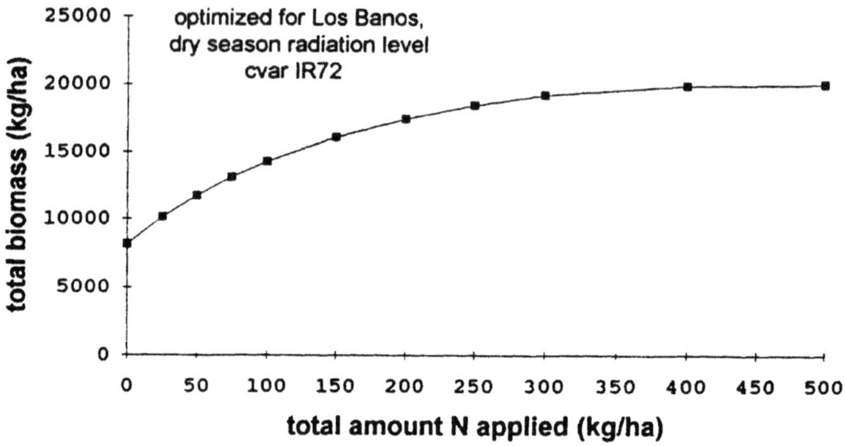

Fig. 2. Maximum attainable crop biomass for cultivar IR72 at selected total N input levels, for dry season conditions at Los Baños (Philippines), as generated by MANAGE-N.

weather conditions, the model is run with an optimization procedure which generates a 'best attainable' N response curve, and the optimal timing of fertilizer dressing for each N-application level (figures 2 and 3). Some conclusions:

- The efficiency factor FSV often steeply decreases after flowering. This was found frequently at two sites in India and in China, but not in The Philippines.
- Wet season efficiencies are lower than dry season values under Indian conditions, while the reverse is true at the IRRI farm.
- At several locations the N uptake capacity is strongly reduced from first flowering onward – or even earlier, as in some soils of West Java.

Although these contrasts are not fully understood, the observations can be used to derive numerically optimized local N management recommendations. A number of SARP teams are now comparing computer based recommendations with current 'standard' recommended practices. It was concluded from these comparisons that, in sufficiently fertile soils, basal N application should be reduced or entirely skipped and saved for later application, to increase yields at the same input level. So far, basal application has been widely recommended.

The package is now polished and wrapped as a user-friendly tool (MANAGE-N), for use by non-modellers at experiment stations. A further step could be to involve extension workers. Tamil Nadu Rice Research Institute has organized a course for scientists from 10 substations, guiding them in collecting the relevant data and using the software to generate specific rec-

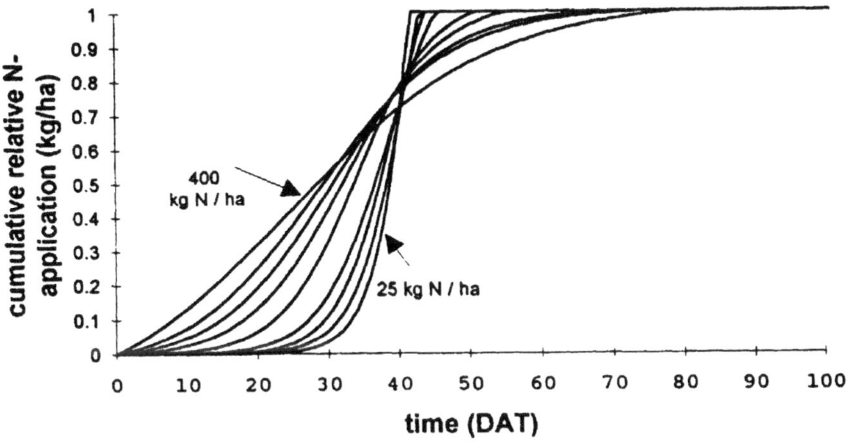

Fig. 3. Nitrogen fertilizer recommendation curves for maximum biomass production in rice cultivar IR72, for dry season conditions at Los Baños (Philippines), as generated by MANAGE-N.

ommendations. Large inter-site contrasts are found in this region, notably with respect to soil N supply and pest-N interactions. MANAGE-N will be further developed to allow the assessment of management options in contingency situations, i.e. for aged seedlings, after flood damage, and to evaluate the consequences of withholding further N dressings when pests occur.

Case II. Impacts of climate change on rice production

This program is centred around the Asia-wide evaluation of the effects of climate change on rice production. These have been studied by four SARP teams within the context of a larger IRRI-EPA collaborative project (Matthews *et al.* 1994). Research teams from Korea (CES), Malaysia (UPM), China (CNRRI) and India (TNAU-TNRRI) participated in the simulation studies, evaluating the effects of increased temperatures and doubled CO_2 concentrations. One aspect of the simulation models that needed refinement for this study was the relation between temperature extremes and spikelet formation and sterility. The results obtained at all sites have benefited from the regional approach, mobilizing expertise in this specialized field.

Common elements, apart from the use of common models, were the joint collection of weather data, varietal parametrization, and the use of common climate change scenarios. The latter were obtained from the General Fluid Dynamics Laboratory (GFDL), the Goddard Institute of Space Studies (GISS), and the United Kingdom Meteorological Office (UKMO), and were generated for the entire Asian rice region with the help of general circulation models. These models predicted a rise in temperature in all major rice zones, and increased precipitation in 60% of the area. CO_2 levels were assumed to either double or remain constant.

Rice responses to these changes were simulated with the help of the SIMRIW and ORYZA-1 models. Both models simulate the development and potential growth of rice in relation to temperature, solar radiation and varietal characteristics. Crop models can quantify the trade-off between yield decrease, arising from the predicted warming, and yield increase due to increased CO_2 levels. The uncertainty in these evaluations is mainly associated with uncertainties about the climate shifts which are used as inputs. Yield declines were predicted under all scenarios for Thailand, Bangladesh, and South East China, while increases were predicted for Malaysia, Indonesia and western India (table 2). The overall production in the region changed by −12% to + 7%, depending on the climate model used. For the Philippines, wet season yields would increase, but dry season yields would decline in some regions and rise in others.

Overall rice production in China would decline, notably in south-east China, although there was a considerable differentiation between early and late rice seasons. Korean rice yields would generally increase. In Japan, net production changes were predicted to be negligible, although the geographic distribution would change, the northern areas benefiting from longer growing season and reduced cold damage, the south suffering from the damaging effects of high summer temperatures on spikelet fertility.

The study points at the importance of breeding efforts towards higher temperature tolerance during flowering. The gains of such investments can be quantified with the help of these process-based simulation models, and can be weighed against other investments. Likewise, management options such as earlier planting and the use of longer duration varieties can be evaluated.

Case III. Agroecological zonation studies

A common methodology for agroecological zonation and characterization studies was developed (Bouman 1994; Bouman *et al.* 1993a) and is now being applied at several centres participating in the network. The approach has developed from the recognition of common problems experienced by the different teams. Such problems stem, in part, from the fact that simulation models are often developed and validated at crop-field-season scale, while their regional application involves the use of less detailed data sets. The user is then faced with issues like parameter uncertainty, gaps in basic input data, and year to year variations in weather. The choices made in solving such problems may largely determine the outcome of such simulation studies. The following framework was accepted as a step towards minimizing the bias arising from these subjective choices.

First, the overall objective of the study is stated (e.g. optimizing rice cropping systems), which is then further refined in terms of specific objectives

Table 2. Changes in total rice production predicted by the ORYZA1 model for each country and region, under three climate change scenarios. AEZ codes refer to agroecological zones. For other acronyms see text. (Source: Matthews *et al.* 1994)

		Current	GFDL		GISS		UKMO	
Country	AEZ	'000 t	%change	'000 t	%change	'000 t	%change	'000 t
Bangladesh	3	27691	14.2	31621	–5.0	26298	–2.8	26919
China	5	8854	–7.4	8201	0.3	8881	–25.2	6619
	6	79872	0.8	80484	–21.7	62514	–19.5	64334
	7	91828	5.8	97196	5.8	97135	3.1	94695
	8	2361	–6.4	2209	–14.2	2026	–27.6	1710
India	1	32807	4.6	34305	–10.8	29272	–5.5	31017
	2	49949	1.8	50849	–2.9	48493	–7.9	46002
	5	227	–7.4	210	0.3	228	–25.2	170
	6	26628	5.4	28069	3.2	27480	–1.3	26287
	8	1011	–6.4	946	–14.2	867	–27.6	732
Indonesia	3	44726	23.3	55155	9.0	48748	5.9	47387
Japan	8	12005	–6.4	11231	–14.2	10300	–27.6	8696
Malaysia	3	1744	24.6	2173	17.6	2050	26.8	2211
Myanmar	2	13807	21.5	16776	–10.5	12356	1.2	13974
Philippines	3	9459	14.1	10797	–11.8	8340	–4.7	9018
South Korea	6	8192	–13.6	7078	–5.3	7755	–21.9	6401
Taiwan	7	2798	11.8	3128	12.8	3156	28.0	3583
Thailand	2	20177	9.3	22044	–4.7	19230	–0.9	19989
Total		434136		462472		415129		409743
%change				6.5		–4.4		–5.6

that can be addressed with the help of simulation models, such as the quantification of potential yields, of actual yield gaps, of water requirements, of optimum planting times for given crops. As a next step, the appropriate model is selected, depending not only on objectives but also on the availability of input data. Along with the model, other tools can be selected such as a weather generator, a parameter distribution generator, and weather and soil data bases.

For each land unit, calculations for many years of weather data are made first, using single (e.g. mean) values for crop and soil parameters. The variation in predicted outputs quantifies the effect of weather variation. Depending on the importance of these effects, one may choose to define worst, best and average years, e.g. in terms of yield level. Subsequently, for each of these extreme cases, the analysis is repeated, now using probability distributions rather than single values to characterize crops and soils. This leads to probability distributions of crop performance in the distinguished representative weather types. These are graphed as maps or tabulated, or they are used as a basis for further risk assessment studies. It is often useful to compare the results with the objectives specified at the outset of the study, and to collect further data where necessary, or to introduce adaptations in the model.

The software developed for this framework includes, apart from models for potential, water limited and nitrogen limited rice growth, also tools for weather generation (IMSP, SIMMETEO), for generation of parameter distributions (RIGAUS), a weather data system (WEATHER) and a user friendly shell (SARP-Shell). This shell facilitates the use of the above tools, the selection of input data sets and the linkage with models, the performance of reruns, and the analysis and presentation of outputs.

A number of zonation studies have been reported (Bouman *et al.* 1993b), with emphases on linkage with GIS (Wopereis *et al.* 1993; Pannangpetch 1993), uncertainty and risk analysis (Aggarwal 1991; Lansigan and Orno 1993), and economic analysis (Pan Jun 1994; Labios *et al.* 1994). An example is given in figure 4. In 1994–1995, SARP teams will proceed with the zonation work along these lines in India (TNRRI, CRRI), Indonesia (BORIF, SURIF), China (ZAU, CNRRI), The Philippines (PhilRice, UPLB) and Bangladesh (BRRI). Some of these groups will address more specifically yield potentials, yield gap analysis and annual yield variation at given sites; others will perform economic analysis, along the lines suggested by Pandey (1994).

National training courses

Since the aim of the project was to build capacity in systems approaches in the region, existing SARP teams and NARCs were encouraged to organize their own, national, training programs after the formal SARP training phase which ended in 1991. Given the strong demand by the participating countries for more training in this field, six such courses have now been successfully organized and conducted by SARP scientists at the NARCs in India (2), China (2), Indonesia, and The Philippines. A few more are in preparation. In all cases, most of the participants were from sister institutes of the organizing NARCs, in addition to direct colleagues of the organizing teams. SARP staff support to these courses consisted of guest lectures, documentation, software, and sometimes small financial support for local travel and accomodation. Several courses were completely independent of SARP staff and financial support. The course format was often a concise form of the SARP blueprint, including a case study and a final reporting workshop, covering potential and water limited production processes. Two specialized training courses were held by TNAU (India), one on nitrogen management, the other on water management in rice.

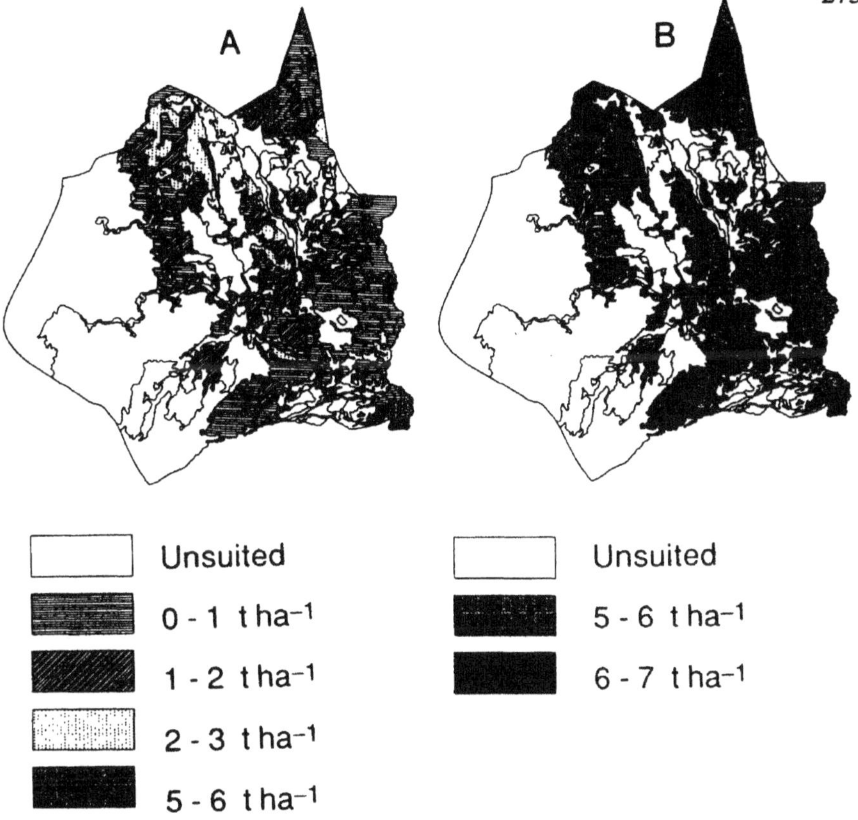

Fig. 4. Map of simulated rainfed rice yield (t ha^{-1}) for Tarlac Province in The Philippines, at 10% (a) and 90% (b) cumulative probability. Simulations were based on observed soil hydraulic characteristics in the area, and on historic (5 years) and generated long-term (25 years) weather data. (Source: Wopereis *et al.* 1993)

Fellowships

The project provides fellowships to support long (1 year) and short (3 months) duration visits of network members to IRRI and AB-DLO/TPE-WAU. Visits with clearly defined goals – such as developing specific (sub)models, analyzing data sets, and finalizing workshop proceedings – are an effective mechanism for interaction and simultaneously serve the network as a whole via their outputs. To some extent these visits reduce the need for extended visits by SARP staff to teams, as conducted in the earlier project phases, although obviously only a limited number of participants can be addressed directly by this mechanism.

In addition to the above, SARP offers five PhD sandwich positions. A 'sandwich' consists of an initial visit (to AB-DLO/TPE-WAU or IRRI) followed by field research at the home institute, again followed by a final phase in Wageningen during which the thesis is completed and presented. These fellowships were introduced to create opportunities for network members while avoiding the long term absence from the home institute usually associated with PhD programs. In this respect the mechanism works well; another advantage is that certain topics can be developed 'in depth', which would otherwise leave gaps. In terms of absolute numbers, the contribution of the 'PhD fellowships mechanism' to reinforcement of the research network is limited. Its main overall contribution lies in the improvement of insights and models, made available to other network participants.

SARP Research Proceedings

A series of publications was initiated for team members and SARP staff to publish and report developments in modelling and experimental work related to model development. Issues often take the shape of workshop proceedings. They may also be devoted entirely to a particular model or group of modules, or may compile the results of network-wide experiments. It is anticipated that 10–15 issues will be completed by the end of 1995. The currently available titles are listed in Appendix 1. The issues are officially registered and are sent to all 'rice libraries'. The presented papers may serve to encourage more formal, internationally refereed publications.

Transfer of coordination and network continuity

The research coordination (see section 'Key elements of the Application Phase') was, upto 1994, in the hands of full time SARP staff members. During the last two project years (1994–1995), coordination activities are transferred to four members of the participating NARCs. They were assigned by the SARP Steering Committee, a group of NARCs' research leaders.

The formal and logistic aspects of these assignments are complex. On the one hand a long term commitment of NARCs to research coordination is sought; on the other hand the capacity per centre is narrow (single individuals) and the time frame covered by granted funding is short (two years). The issue is further complicated by the recognition that international coordination is less effective if actuated from a national platform. The consequence of taking an international centre (IRRI) as a basis (as in SARP) precludes, however, the development of a stronger institutional commitment of NARCs to international network coordination.

Reducing the international coherence for the benefit of more regional or national 'sub-networks' would seem a valid option, and could be effectuated via national training programs. This approach could meet with new difficulties, including those associated with the acceptance of leadership roles. The necessarily smaller pool of scientific expertise might also be insufficient to effectively energize a network in the long term. For the near future, SARP chose for theme oriented coordination by scientists from NARCs, based at IRRI and supported through 'visiting scientist' fellowships.

Implications for NARCs

In most of the NARCs, teams work effectively across the disciplinary department boundaries. Instrumentation is shared, and often members from different departments join in common experimental work. Team supervisors play a very important role here. In groups lacking an active and involved supervisor, a senior team member can sometimes enhance the interaction among team members. It appeared easier to develop interdisciplinary work involving agronomy, soil science, crop physiology and plant breeding than to involve entomologists and phytopathologists. One of the reasons may be that these disciplines are traditionally more focused on the biology of the plague organisms than on their interaction with the crop and that, as a consequence, field experimentation is organized differently.

The shift from traditional empirical work towards the investigation of causal relations, as implied by the systems approach, brings about some operational changes at the institutes. In the long run, research output generated via a systems approach has more general validity than purely empirical work, thus allowing broader extrapolation to other conditions (weather, season, soil type, cultivars, crops). Such an approach is therefore more cost effective. Within a short time frame, however, experimentation with a systems view requires more time input by scientists and by field and laboratory staff. In addition, the costs of individual experiments may increase, due to the costs of extra chemical analyses of plant and soil samples, as more in-depth analysis is often required. This is felt as a serious drawback, the more so because SARP provides virtually no financial support to cover these expenses.

The absence of direct operational financial support was part of the project philosophy. It seems now, however, that the advantages of this approach (no funding-induced deviation from the institute's identified research priorities) are outweighed by the disadvantages: assignment of capable staff to other (paying) projects; and insufficient access to experimental facilities, including support staff to assist in field and analytical work. Various SARP teams have drawn attention to these limitations during the more recent years. Core funding at many of the centres is probably insufficient to sustain the specific operational expenses of systems research for any prolonged period without outside support. Team members are creative and go to great lengths in solving such difficulties internally, but the sustainability of this format is at least questionable.

Modelling activities are sometimes not very accessible to those not skilled in computer use. Frequent participation, by the same individuals, in overseas workshops may also generate an imbalance among scientific staff. Some of the centres expanded their teams in response to such inequities, thus exposing other scientists to the new methodologies. This has not always proved effective from a scientific viewpoint, because new members often lack a basis in modelling.

Other teams have successfully organized their own training programs in order to extend the number of colleagues acquainted with systems approaches (see section 'National training courses'). SARP organized a short course in modelling for new members of exisiting teams in 1994. Other attempts to reinforce existing teams include the development and distribution of self-instructive courseware and user-friendly software, to facilititate the operation of models (see section 'Software developments'). It appears that the self-instructive courseware, in the currently available form, requires hands-on support to fully express its potential.

As in Europe, there is a strong tendency also in Asia for research centres to organize their research in programs and projects which are to be acquired in internal (national) competition. Several SARP teams have been successful in obtaining such grants (e.g. in India and Indonesia).

Software developments

SARP's main aim being the transfer of systems research skills, emphasis has been on the underlying principles of modelling and systems analysis, rather than propagating particular ready-made models. After all, each real world problem requires its own approach, and claims of universal models are generally illusive. Nevertheless, a demand for user-friendly ready-made models has developed for two reasons. One is the desire of many institutes to involve more novice team members without guiding them through lengthy training courses; the other is the s demand for models that execute specific tasks, e.g. optimization of management, agro-ecological zonation, and exploration studies. To execute these tasks, models are conveniently run under a user friendly 'shell structure', which was developed to facilitate selection of models and model options, selection of input data sets, and the editing of data files. Examples of 'shell' applications in the SARP context are zonation studies for potential and water limited production, with smooth execution of multiple runs; various sensitivity analyses; and optimization of nitrogen use (see sections 'Cases I–III').

The introduction of user interfaces in turn sets its own demands of standardization of models and data formats. This curtails users' initiatives to 'play' with the model source codes. The dilemma between 'models easy to run' and 'models easy to adapt' reflects the existence of different user groups: those who wish to apply models in straightforward analyses, and those more involved with testing of hypotheses and model development. The solution chosen by SARP is to maintain two 'tracks', one being the free development of 'easy to change' models in a real simulation language (FST), offering programming versatility after only little initial training, the other being the development of more formalized model codes in FSE standard FORTRAN for running under the user interface. FSE (FORTRAN Simulation Environment; Van Kraalingen 1991) is a design structure for dynamic models. FST (FORTRAN Simulation Translator; Van Kraalingen *et al.* 1994) converts an FST model code into FSE; this translator has replaced the earlier used PCSMP language. Although FSE models still require slight adaptations before they can be run under the user interface, this two-track approach offers sufficient continuity from 'model development' to 'model application' to serve various user groups.

One principle maintained throughout these developments is that all models and modules are fully documented, both in terms of their scientific contents and their technical form, and that all source codes are available to all users.

Conclusions

Systems research networks such as SARP are useful to establish a common basis of concepts and models, and to design experiments to improve, validate and parameterize models for various applications. The latter include local in-depth studies on production constraints, but also eco-region-wide analyses for increased research efficiency and more adequate research prioritization. Both types of applications benefit from the specialist expertise mobilized through a network format, if combined with on-site interdisciplinary team approaches. The variety of environmental conditions and germplasm covered by a sufficiently widespread network allows the rigorous testing and adaptation of systems tools, including models, before their application. We believe that this should drastically improve the predictive capacity of models. For eco-regional analysis, i.e. the characterization of production systems and production conditions across eco-regions, and their expected responses to proposed changes, models are combined with data bases. The development of these along universal lines, in association with certain modelling standards, requires further attention in the near future. Agricultural research world-wide will benefit from attempts to standardize models and data formats.

Establishing systems research networks requires a long term investment in training, coordination and support. It involved, in the case of SARP, six years of formal training programs, followed by a joint research phase on applied topics. Frequent team visits by research coordinators, exchange fellowships, thematic workshops, joint research planning and a collective 'research proceedings' series all proved useful mechanisms for strengthening research capacity in the network. Simultaneously, these elements allowed the development of a common knowledge base expressed in validated models, quantified parameters, and research reports.

Appendix 1. SARP Research Proceedings

Current (January 1995) titles available in the SARP Research Proceedings series.

Title	Topic	Description
The SARP-project. Phase III (1992–1995). Overviews, Goals, Plans.	Project overview Systems Analysis	Workshop proceedings
Agro-ecology of rice-based cropping systems.	Agro-ecology	Workshop proceedings
Mechanisms of damage by stem borer, bacterial leaf blight and sheath blight and their effects on rice yield.	Crop protection	Workshop proceedings
Nitrogen economy of irrigated rice: field and simulation study.	Rice nitrogen	Workshop proceedings
ORYZA1 An Ecophysiological model for irrigated rice production.	ORYZA1	Model documentation
The development, testing and application of crop models simulating the potential production of rice.	Potential production	Workshop proceedings
Analysis of damage mechanisms by pests and diseases and their effects on rice yield.	Pest and disease management	Workshop proceedings; Methodology
The use of crop growth models in agro-ecological zonation of rice.	Zonation	Methodogy
Agro-ecological zonation, characterization and optimization of rice-based cropping systems	Agro-ecology	Workshop proceedings
ORYZA simulation modules for potential and nitrogen limited rice production.	ORYZA modules	Model documentation

Acronyms

AB-DLO	Research Institute for Agrobiology and Soil Fertility (Wageningen, The Netherlands)
BORIF-CRIFC	Bogor Research Institute for Food Crops, Central Research Institute for Food Crops (Bogor, Indonesia)
BRRI	Bangladesh Rice Research Institute (Joydebpur, Bangladesh)
CES	Crop Experimental Station (Suweon, South Korea)
CNRRI	China National Rice Research Institute (Hangzhou, China)
CRRI	Central Rice Research Institute (Cuttack, India)
DGIS	Directoraat Generaal voor Internationale Samenwerking (Directorate General for International Cooperation)
DLO	Dienst Landbouwkundig Onderzoek (Agricultural Research Department)
EPA	Environmental Protection Agency
FSE	FORTRAN Simulation Environment
FST	FORTRAN Simulation Translator

GIS	Geographic Information System
IARI	Indian Agricultural Research Institute (New Delhi, India)
IESAM-UPLB	Institute for Environmental Studies and Management, University of the Philippines (Los Banos, Philippines)
IRRI	International Rice Research Institute (Los Banos, Philippines)
JAU	Jiangxi Agricultural University (Nanchang, China)
KKU	Khon Kaen University (Khon Kaen, Thailand)
MARDI	Malaysian Agricultural Research and Development Institute (Serdang, Malaysia)
NARC	National Agricultural Research Center
PUAT	Pantnagar University of Agriculture and Technology (Pantnagar, India)
SARP	Simulation and systems Analysis for Rice Production
SURIF-CRIFC	Sukamandi Research Institute for Food Crops, Central Research Institute for Food Crops (Sukamandi, Indonesia)
TNAU-SWMRI	Tamil Nadu Agricultural University, Soil and Water Management Research Institute (Thanjavur, India)
TNAU-TNRRI	Tamil Nadu Agricultural University – Tamil Nadu Rice Research Institute (Aduthurai, India)
TNAU-WTC	Tamil Nadu Agricultural University – Water Technology Center (Coimbatore, India)
TPE	Department of Theoretical Production Ecology, WAU
UPM	Universiti Pertanian Malaysia (Serdang, Malaysia)
WAU	Wageningen Agricultural University (Wageningen, The Netherlands)
ZAU	Zhejiang Agricultural University (Hangzhou, China)

References

Aggarwal P K (1991) Estimation of the optimal duration of wheat crops in rice-wheat cropping systems by crop growth simulation. Pages 3–10 in Penning de Vries F W T, Van Laar H H, Kropff M J (Eds.) Simulation and systems analysis for rice production (SARP). Pudoc, Wageningen, The Netherlands. ISBN 90-220-1059-7.

Ten Berge H F M (1993) Building capacity for systems research at national agricultural research centres: SARP's experience. Pages 515–538 in Penning de Vries F W T, Teng P S, Metselaar K (Eds.) Systems Approaches for Agricultural Development. Kluwer Academic Publishers, Dordrecht, The Netherlands.

Ten Berge H F M, Wopereis M C S, Shin J C (Eds.) (1994) Nitrogen economy of irrigated rice: field and simulation studies. SARP Research Proceedings, AB-DLO, Wageningen, The Netherlands. ISBN 90-73384-22-2.

Bouman B A M, Penning de Vries F W T, Riethoven J J M, Kropff M J, Wopereis M C S (1993a) Application of simulation and systems analysis in rice cropping optimization. Pages 1–15 in Bouman B A M, Van Laar H H, Wang Zhaoqian (Eds.) Agroecology of rice based cropping systems. SARP Research Proceedings, AB-DLO, Wageningen, The Netherlands. ISBN 90-73384-18-4.

Bouman B A M, Van Laar H H, Wang Zhaoqian (Eds.) (1993b) Agroecology of rice based cropping systems. SARP Research Proceedings. AB-DLO, Wageningen, The Netherlands. ISBN 90-73384-18-4.

Bouman B A M (1994) A framework to deal with uncertainty in soil and management parameters in crop yield simulation; a case study for rice. Agricultural Systems 46:1–17.

Van Kraalingen D W G (1991) The FSE system for crop simulation. Simulation Report CABO-TT nr 23. CABO-DLO, Wageningen, The Netherlands.

Van Kraalingen D W G, Rappoldt C, Van Laar H H (1994) The FORTRAN Simulation Translator (FST), a simulation language. Pages 219–230 in Goudriaan J, Van Laar H H (Eds.) Modelling crop growth processes. Kluwer Academic Publishers, Dordrecht, The Netherlands.

Kropff M J, Penning de Vries F W T, Teng P S (1994) Capacity building and human resource development for applying systems analysis in rice research. Pages 323–339 in Goldsworthy P, Penning de Vries F W T (Eds.) Opportunities, use and transfer of systems research methods in agriculture to developing countries. Kluwer Academic Publishers, Dordrecht, The Netherlands.

Labios R V, Delos Santos R E, Salazar A M, Villancio V T (1994) Rice-upland crop rotations in rainfed lowland rice areas in Bulacan, Philippines. In Lansigan F P, Bouman B A M, Van Laar H H (Eds.) Agro-ecological zonation, characterization and optimization of rice-based cropping systems. SARP Research Proceedings. AB-DLO, Wageningen, The Netherlands (in prep.).

Lansigan F P, Orno J L (1993) Simulation analysis of risk and uncertainty in crop yield due to climatic variations and change. Pages 115–125 in Bouman B A M, Van Laar H H, Wang Zhaoqian (Eds.) Agroecology of rice based cropping systems. SARP Research Proceedings. AB-DLO, Wageningen, The Netherlands. ISBN 90-73384-18-4.

Matthews R B, Kropff M J, Bachelet D, Van Laar H H (Eds.) (1994) The impact of global climate change on rice production in Asia: a simulation study (in press).

Pan Jun (1994) Systems analysis and simulation applied to the 'Central China Double and Single Rice Cropping Region'. In Lansigan F P, Bouman B A M, Van Laar H H (Eds.) Agro-ecological zonation, characterization and optimization of rice-based cropping systems. SARP Research Proceedings. AB-DLO, Wageningen, The Netherlands (in prep.).

Pandey S (1994) Risk analysis and crop growth modelling. In Lansigan F P, Bouman B A M, Van Laar H H (Eds.) Agro-ecological zonation, characterization and optimization of rice-based cropping systems. SARP Research Proceedings. AB-DLO, Wageningen, The Netherlands (in prep.).

Pannangpetch K (1993) Application of model simulation to evaluate rice production at the district level. Pages 16–26 in Bouman B A M, Van Laar H H, Wang Zhaoqian (Eds.) Agroecology of rice based cropping systems. SARP Research Proceedings, AB-DLO, Wageningen, The Netherlands. ISBN 90-73384-18-4.

Wopereis M C S, Kropff M J, Hunt E D, Sanidad W, Bouma J (1993) Case study on regional application of crop growth simulation models to predict rainfed rice yields: Tarlac Province, Philippines. Pages 27–46 in Bouman B A M, Van Laar H H, Wang Zhaoqian (Eds.) Agroecology of rice based cropping systems. SARP Research Proceedings, AB-DLO, Wageningen, The Netherlands. ISBN 90-73384-18-4.

Collaboration between national, international, and advanced research institutes for eco-regional research

P.R. GOLDSWORTHY, P.B. EYZAGUIRRE and S.W. DUIKER

International Service for National Agricultural Research (ISNAR), P.O. Box 93375, 2509 AJ The Hague, The Netherlands

Key words: agricultural research, eco-regional, institutions, natural resources management, systems research

Abstract. Global concerns about environmental degradation, loss of biodiversity, and the increasing demands that the expected large increases in population will make on the world's natural resources, have combined to alter the environment in which agricultural research is being conducted. This has led to a shift in the focus of research from commodities to a land-use systems perspective, and the integration of natural resource management concerns into the agricultural research agenda. The organizational changes which are taking place in the international agricultural research system in response to this broader research agenda are described. The management and institutional implications of these changes for research conducted at a national and regional level, and for the collaboration between international and national research organizations, are discussed. Examples are given to illustrate the diverse characteristics of recent initiatives which address natural resource management problems through partnership arrangements among national, regional and international research institutions. The paper concludes there are unlikely to be simple prescriptions for successful partnership arrangements, but more probably, general guiding principals that could be captured by sharing experiences among the different partnership groups.

Introduction

The paper looks at changes in the institutional environment for agricultural research arising from the global concerns about sustainable uses of natural resources. Our primary concern is the nature of the collaboration between national agricultural research institutions and the international agricultural research centers (IARCs) of the Consultative Group on International Agricultural Research (CGIAR). First we consider how the CGIAR as part of a wider global research community is refocusing its research agenda to derive greater complementarity with components of the global research system. The paper describes the sharper distinction that will be made between the global and regional research activities of the CGIAR, in order to increase the participation of national institutions, in the planning, and conduct of research programs. This should result in a sharing of responsibility that contributes more effectively to both sustained growth in agricultural production, and conservation of the natural resource base.

J. Bouma et al. (eds.), Eco-regional approaches for sustainable land use and food production, 283–303.

Global context of sustainable agriculture

Although global population growth rates have begun to decline, the current increases in absolute numbers are unprecedented and will continue well into the next century. Conservative estimates project the need to double supplies of food grains within the next 30–40 years to meet the growing demand. The need will be greatest in South Asia and Africa because of their continuing high rates of population growth (Crosson and Anderson 1992).

While there is growing concern globally about the impact of agricultural activities on the Earth's natural resources, the highest priority for governments in developing countries will continue to be the need to meet the growing demand for food. The dramatic increases in the production of food, fibre, and fuel that took place in the last three decades were a combined result of major productivity increases in the principal food crops and of bringing more land and other resources into production. Future increases in production to meet the growing demand will have to depend to a larger extent on improving the productivity of land and other resources that are already in use. However, indications that the earlier rapid rate of growth of food-grain production is slowing down is now a cause for concern. There are doubts for example about the sustainability of the increasingly intensive cropping systems in South Asia (Byerlee 1992; Fujisaka *et al.* 1994). While agriculture, livestock, and forestry will continue to be among the largest users of these resources (Antle 1993), there is a growing awareness that they are not the only or even the main users, and that there is increasing competition for these resources.

Changing research environment

In the 30 years since the first CGIAR International Agricultural Research Centers were established there has been a major change in the environment in which research is conducted (Crosson and Anderson 1993). There are now more and diverse set of institutions that are contributing. The CGIAR accounts for only about 5 percent of the global financial commitment to agricultural research for the developing world. The other main actors are national agricultural research systems (NARS),* bilateral and multilateral donors, non-government organizations (NGOs), universities, and the private sector.

The goals have also evolved. There is now wider recognition that all agricultural technologies affect the resource base, and that they affect other users of resources. There is a realization therefore that a longer view, more spatial integration and an intersectoral perspective are required in research to assess the consequences of introducing new technologies. This has led to a shift

* NARS as referred to in this paper is taken to include all the institutions in the public sector, including universities, that are capable of contributing to research related to the development of agriculture, forestry and fisheries).

in the focus of research from commodities to a land-use systems perspective, and the integration of natural resource management (NRM) concerns into agricultural research. The result is a much broader research agenda for national and international agricultural research.

Although there are substantial differences among NARS, and some have become stronger in the past 20 years, there is still a lack of institutional and research capacity in many countries to do the research that is needed. To strengthen their capacity and address the new broader research agenda, stronger linkages will be needed between NARS, policy makers, local institutions and non-government organizations (NGOs) involved in the use and conservation of natural resources. Research institutions in industrialized countries and those in the developing countries will need to be linked in new partnerships that reflect opportunities created by developments in science (CGIAR 1994a). Thus, further improvements in agricultural productivity will depend not only on improved methods of production, but also on institutional reforms (TAC 1989; Ruttan 1991).

There are four key issues related to the organizational and institutional changes that the integration of environmental and development goals will involve: cross-sectoral planning; information, as a key requirement for the planning process; a participatory decentralized approach in both planning and implementation; and institution building as an essential condition for the sustainable development of agriculture and a more rational use of resources (ISNAR 1993; Goldsworthy *et al.* 1994a).

The response by international research to changing needs

Changes in CGIAR research priorities and activities

The focus of the IARCs has moved progressively from research on increasing food production to one that aims for a balance between production-oriented objectives and research to improve the efficient resource use and ensure the sustainability of agricultural systems. A review of CGIAR priorities and strategies, conducted in 1985 by the Technical Advisory Committee (TAC) of the CGIAR, identified sustainability, resource management and environmental degradation, and evolving partnerships with NARS as issues for particular attention. Following the Bruntland report in 1987, the CGIAR began to be more explicit about the incorporation of sustainability concepts into its plans (TAC 1989).

The long term nature of resource management research, the difficulty of defining priorities between its multiple objectives, and of measuring and attributing its impact, has meant it has often proved easier for the IARCs to develop more productive, disease-, and pest-tolerant crop plants than to devise new or improved production systems. The integration of resource management into the research agenda changes the relative importance of different research activities. Methods to increase agricultural production are not enough on their own; there are also social and environmental goals, which require simultaneous action at a national policy level. As a consequence the relative importance of policy formulation and analysis increases (Janssen 1994).

An agro-ecological basis for research

The CGIAR remains committed to strategic research consisting of global activities throughout the developing world. From its examination of CGIAR priorities TAC identified a continuing long term need for international research in the following five main sets of activities (TAC 1992; and CGIAR 1994b):

- conservation and management of natural resources;
- germplasm enhancement and breeding;
- production systems;
- socioeconomic, public policy and public management research;
- capacity and institution building.

These global research activities are to be complemented by others that are focused on the needs of specific regionally defined agro-ecological zones which TAC refers to as 'eco-regions' (TAC 1990; CGIAR 1993). The combination of the two kinds of activities will enable the CGIAR to organize and implement research on resource management while continuing the emphasis on agricultural productivity. One of the objectives of adopting an eco-regional approach to international agricultural research is to enable NARS to assume greater responsibility for research activities within these eco-regions, particularly activities concerned with the improvement of production systems.

TAC argued that two key issues, the strengthening of the capacity of NARS to conduct research, and the strengthening of research on resource management and conservation, could be better addressed through an eco-regional approach. It is expected that in time most NARS will have developed adequate capacity to meet their own research needs, and that by then there will be regional and eco-regional mechanisms in place for transnational research on NRM. The CGIAR would then focus on strategic global research activities as its long term commitment (CGIAR 1994a). Meantime, managing this mix of two kinds of activities and maintaining a balance will not be easy.

An eco-regional approach will also require strong political support in participating countries. Unless policy makers as well as scientists are involved in the planning and implementation of eco-regional activities the results of eco-regional research may have little influence on policy decisions concerning the use of natural resources in individual countries. For this reason a key element in the strategy will be to link policy formulation to technology development and diffusion, and make collaborative programs with NARS the mode of implementation.

A systems approach to research

With time the centers have had to pay more attention to less favored environments in which the rural populations are ecologically and economically disadvantaged. Yields in these environments are characteristically very variable, largely because weather conditions, particularly rainfall, are also variable from year to year. The centers have therefore become more aware of the need to quantify the factors in these environments that make the production of food crops and livestock uncertain.

Until recently the methods available could only indicate environmental classification which was too broad to meet their needs. The intercentre conference on the characterization, classification and mapping of agricultural environments in Rome in 1986 was one of the first significant steps taken by the centers to promote an exchange of ideas amongst themselves, the developing nations with whom they cooperate, and other international agencies on ways to develop closer collaboration in the collection and sharing of data for the characterization of agricultural environments (Bunting 1987).

Systems approaches will feature prominently in eco-regional research. They are essential for addressing sustainability and resource management issues in agriculture (Lynam and Herdt 1989). First, these approaches serve to identify the agro-ecological production systems that are characteristic of different environments, including the social, cultural, and economic components of those systems. By distinguishing a hierarchy of system levels, they serve to clearly define the geographic scale and the temporal dimension of the problem to be solved. Second, systems approaches call for greater attention to the relationships between production and the environment, with an inventory of the environmental resources and better understanding of how they are being used. The purpose of an agro-ecological approach is also to increase knowledge of the potential uses of resources and of how they can be developed. Thirdly, research with a systems perspective applied to natural resource management research, reveals the different sets of institutional issues that arise at the various systems levels. These issues include, for example, the need for interinstitutional linkages, and which institutions are likely to have a comparative advantage in working at a given scale or to achieve a particular objective (Goldsworthy and Penning de Vries 1994).

All the centers and many national research institutions are using systems methods of one kind or another to characterize environments and agro-ecological systems, to analyze land-use patterns and options, to understand soil, water, and plant-nutrition relations, and to study the distribution of crops and cropping systems. Table 1 summarizes some of the systems methods used

Table 1. Examples of systems approaches employed by IARCs.

IARC	Use of systems approaches	Methods
CIAT	select and describe sites for crop improvement in cassava, beans, rice; select sites for NRM research, soil chemistry – long-term trends studies of crop growth; whole-farm models agrosystems analysis/study of pesticide residues	GIS, expert-decision-support systems, stochastic rainfall models
CIMMYT	characterization of production environments characterization genotype responses to environment conservation tillage methods in mixed maize-legume cropping systems	combination of GIS and simulation models; cluster analysis
CIMMYT/IRRI	systems methods to investigate probable causes of declining yields in wheat-rice based systems	combination of GIS, soil + crop simulation models, multiple regression and cluster analysis
CIP	choose sites that represent diversity of Andean highlands. studies on environmental + policy aspects of pesticide use.	stochastic simulation models. economic models.
ICARDA	study of crop/livestock production systems in variable environments; assessment of risk	crop simulation; spatial weather generator; hydrological models; stochastic linear programming models
ICRAF	biological modeling of agroforestry characterize land-use patterns in which agroforestry can be used	GIS. Tree-, crop-, and soil simulation models: expert-decision-support systems.
IITA	characterize land use and agricultural production systems	GIS decision support systems, crop + pest simulation models
IRRI	sustainable irrigated + rainfed rice production systems; risk assessment cross-ecosystem studies	GIS and modular system of simulation models
WARDA	ecosystems studies. explore feasibility of improving the adaptation of West-African rice production systems	GIS, simulation models, modified linear programming procedures

by the CGIAR centers. Though some of these methods are still experimental, the use of others is increasing (Goldsworthy and Penning de Vries 1994).

System levels. Agro-ecological systems consist of a hierarchy of different systems levels. An ascending agricultural hierarchy and a corresponding spatial hierarchy are shown in table 2. There is no agreed taxonomy for the purposes of system analysis. However, it is clear that a geographical

Table 2. Examples of hierarchy of systems levels and of spatial scales in agriculture.

Agricultural hierarchy	gene	plant	plant community (crop)	cropping system	farming system	land-use system	agroecological system
Agricultural spatial scales		soil unit	field	farm unit	farming landscape	catchment area	agroecological region

organization will be one key dimension in planning eco-regional and resource management research activities. The different levels of decision making that determine the use of natural resources, will be another dimension (Janssen 1994).

Combinations of systems approaches. Most of the centers are now using geographic information systems (GIS). It is a powerful tool for dealing with the spatial dimension of complex agro-ecological problems and a useful complement to simulation methods for geographic extrapolation of research findings. CIAT, CIMMYT, ICARDA, IRRI, and WARDA provide examples (see Dingkuhn, Harris *et al.*, Harrington *et al.*, Kropff *et al.*, and Torres and Gallopin, in Goldsworthy and Penning de Vries 1994). This combination of methods will be particularly useful for resource management research because of the location-specific nature of many methods of resource management. The specific adaptation of the rice cropping systems that WARDA is helping to develop for different areas in West Africa illustrates this point (Dingkuhn 1994). The combination of GIS and simulation methods brings more information to bear on a problem, which now makes it easier to target research, whether for varietal improvement or resource management.

The development of systems methods and models. The CGIAR centers, like NARS, are users rather than the developers of the various methods and models. CIMMYT, IRRI, and ICARDA, among others, use systems methods as tools to make predictions about and support decisions in their own research. Their experience and that of other users has helped those outside the CGIAR who are developing models, to focus further research on critical areas, and to refine the use of systems methods for various applications (Goldsworthy and Penning de Vries 1994).

The interdisciplinary nature of systems research. The new resource management focus and an eco-regional approach increases the complexity of the research task. The changing demands and opportunities put pressure on both NARS and the IARCs for institutional change. In general the centers have been able to respond quickly to these changes. CIAT, IRRI and IITA for example, have altered their internal structure as part of the move to integrate production and resource management research (Kropff *et al.* 1994, Jagtap 1994, Torres and Gallopin 1994).

Institutional organization and management

Organizational principles

Clearly, it will be necessary to establish operational mechanisms that will facilitate the collaboration between the IARCs and their partners in the same eco-region, but also to facilitate the exchange of information and ideas concerning research at a global level. Table 3 shows some of the coordinating mechanisms that have been used in the past, and the options that have been suggested for further development of an eco-regional approach, with their corresponding advantages and disadvantages.

The CGIAR needs mechanisms (rather than new Centers) that will encourage new, effective partnerships and new research approaches to research on sustainable growth in agriculture. As yet there is no organizational model which takes adequate account of the physical, biological and human factors that influence sustainability. And there is no single organizational model that will serve the needs of all eco-regions, given the diversity of NARS capabilities, the different centers' mandates, and the diversity of resource management issues. But from the discussions in TAC and the CGIAR a set of organizational principles were identified for conducting research on agriculture and natural resource management at an international level (TAC 1993; see box 1).

[Box 1] Organizational principles for conducting research on agriculture and natural resource management at an international level

- operate on a regional basis;
- focus on important agro-ecological areas with clearly identifiable problems;
- integrate production and resource management research;
- employ an interdisciplinary approach, with a balance of natural and social sciences;
- involve NARS and other stakeholders actively; seek to identify complementary roles and research capabilities;
- adopt a flexible procedure for determining research priorities and flexible systems of governance and funding;
- monitor coherence of activities against global objectives.

Table 3. CGIAR coordinating mechanisms to avoid duplication of effort. (Source: TAC 1991; CGIAR 1993)

	Mechanism	Focus	Examples	Advantages	Disadvantages
1.	Global mandate	Commodity improvement	CIMMYT, IRRI	Clear leadership; has worked well for commmodities	Little attention to resource management
2.	Regional mandate	Ecoregional/ commodity	ICARDA, IITA, ICRISAT	Knowledge of region and main agroecologies	Responsibilities within CGIAR not always clear: difficult to strike balance commodity/ resource management
3.	Intercenter network	Thematic	Agroecological characterization group	Informal grouping: common interest: avoid duplication of effort; can evolve to become formal.	Additional responsibility and load on convener center
4.	Liaison entity	Specialized field	IPGRI/Intercenter Working Group on genetic resources (ICWG-GR)	coordination at CGIAR-system level IPGRI unites efforts of other CGIAR centers. Clear leadership + identity.	
5.	Center-initiated ecoregional models	Ecoregional research	CONDESAN, Sustainable Mountain Agriculture Development Initiative, African Highlands Initiative, Alternatives to Slash-and-Burn Initiative, etc.	Clear sense of ownership and identity. No need to redefine responsibilities.	No coordination at CGIAR system level to ensure system-wide coherence.
6.	Coordinated intercenter ecoregional models	Ecoregional research	(none yet)	DGs can develop mechanisms to avoid duplication. No new governance needed. Clear sense of ownership and identity.	Workload for centers increases. No mechanism to resolve conflicts.
7.	Program funded-matrix ecoregional activities	Ecoregional with some strategic research	(none yet)	More systematic coverage of ecoregional research. Duplication avoided (provided new mechanisms for coordination are developed). Is more cost-effective than 5 and 6. Could begin to be implemented as part of MTP.	Restricts freedom of individual centers.
8.	System-wide model		(none yet)	More structured and transparent approach to ecoregional research. More coherent across whole system.	No short term resolution of how to introduce ecoregional activities. Serious funding constraints of present would be difficult.

1. Mechanisms 1–4 have been used in the past.
2. Mechanisms 5–6 are mutually exclusive options for implementing an ecoregional approach in the CGIAR system. Only option 5, the center-initiated model, has been used so far.

An organizational matrix

An organizational matrix has been proposed as a basis for allocating responsibilities for doing CGIAR sponsored research. A schematic illustration of the matrix is shown in figure 1. The five sets of activities mentioned previously (see section 'An agro-ecological base for research') of the global program are shown as themes or columns. Centers are represented by the rows of the matrix, and the cells are related sets of time bound research activities with scheduled objectives and outputs, and resource requirements. Such a matrix would show the links between CGIAR centers and other participants in the international agricultural research system. It also shows the relation between CGIAR centers and programs. However, the basis for determining the balance between system wide activities and center priorities is not yet clear (CGIAR 1994a).

Financing the programs. One of the aims of the organizational changes within the CGIAR system is to achieve a closer integration between the agreed research agenda and the budget, so that the agenda drives the budget and not vice-versa. It has been suggested (CGIAR 1994a) that the organizational matrix would facilitate the development of a balanced funding plan for the agreed agenda of activities. The matrix leaves individual donors free to contribute to CGIAR system wide programs (columns), to centers (rows), or to individual center-activity combinations (cells). The aggregation of contributions would be the basis of the funding plan. Concern has been expressed that this may favor restricted rather than unrestricted core, with donors contributing to individual activities rather than to general system support. Operational procedures to ensure balanced funding of the agreed agenda will therefore need to be developed.

Assigning management responsibilities. The organizational principles that are being followed will need to be accompanied by operational mechanisms to ensure that structure follows function. Figure 1 illustrates some of the many possible forms of project management that might be required. The matrix shows that the allocation of management responsibilities would be complex. Natural resource management research has to be done *in situ* and it needs to be decentralized. It is not an appropriate responsibility for a global entity. In contrast central leadership has clear advantages in the case of germplasm collection and conservation, as described in the previous section concerning the role at a global level of IPGRI and ICWG-GR, but illustrated here at a commodity level for rice (figure 1, column C). Several CGIAR centers, working in different rice growing environments are involved in rice germplasm collection, but IRRI provides leadership.

There are other areas of research illustrated in figure 1 in which similar kinds of activity may be conducted in different environments, but where

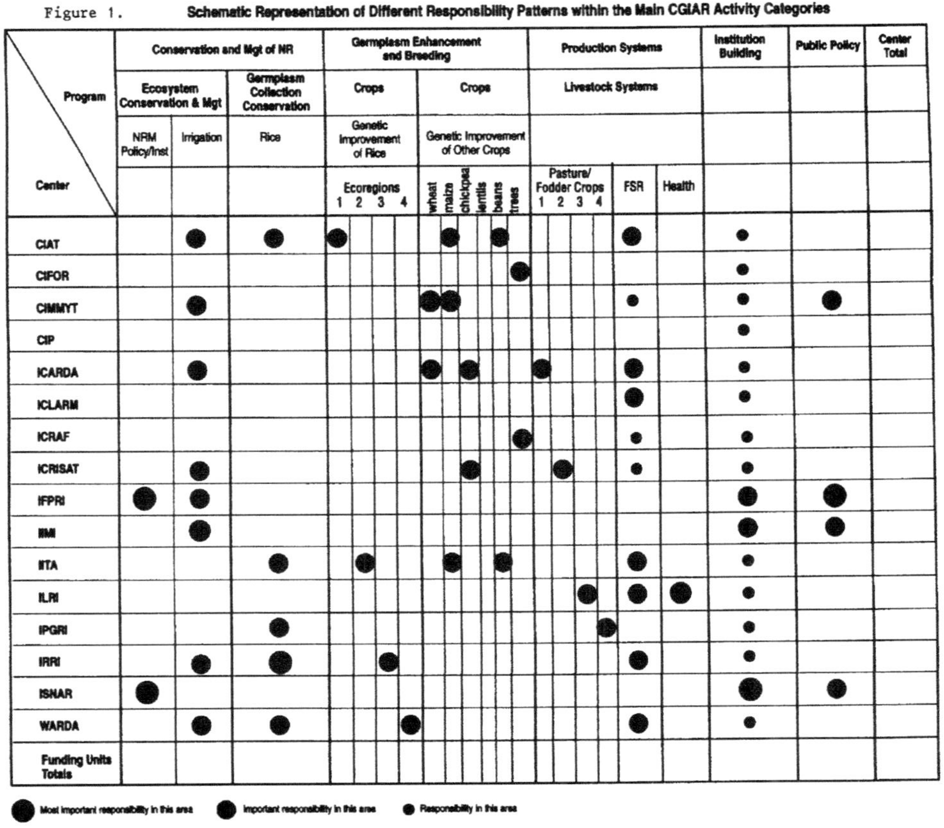

Fig. 1. Schematic representation of different responsibility patterns within the CGIAR activity categories.

there is no sufficiently common thread to make a case for central management responsibility. An examples is rice breeding activities at CIAT, IITA, IRRI, and WARDA. The needs of rice growers served by each center are different, although there may be some opportunities for combined efforts (e.g., at IITA and WARDA). Work at different CGIAR centers on fodder crops as a component of very diverse farming systems illustrates a similar situation that requires decentralized responsibility within the CGIAR.

Eco-regional activities may give rise to combinations of operational mechanisms in addition to those shown in figure 1, so that eco-regions become major axes for management responsibility, through the eco-regional entities that are described later. The new strategies and the changes in organization which follow from them are intended to improve the accountability of the research system. However, the added complexity raises questions of diffused responsibility for the management of activities; questions that will have to be resolved. The eco-regional concept therefore still needs to be elaborated to make clear principles and criteria that will be used to determine the allocation of global and eco-regional responsibilities. Clearly mechanisms will also be required to determine how the funds are to be allocated among participants.

Eco-regional entities

TAC (TAC 1991) refers to 'eco-regional entities' as an institutional mechanism given responsibility for CGIAR sponsored research in a clearly defined eco-region. The entity could be a center or a sub-center (e.g., the ICRISAT Sahelian Center) or a smaller organization. An eco-regional entity would have a clear international role to perform, and it would take responsibility for most of the research on NRM. Accountability will be vital. As a consequence, a leading responsibility will often have to rest initially with a CGIAR center, since they are usually the only entities with legal existence in a transnational context.

TAC makes a distinction between a lead center and a convener center. A lead center is responsible to TAC and donors and therefore requires the commensurate authority. However this can make it difficult for others to feel ownership, and for the lead center to establish the kind of collaboration required with national and regional organizations for a successful participatory decentralized approach. It raises the question of how best NARS can be brought fully into the eco-regional activities, and how as system-wide collaboration is being sought, a system that is open to collaboration with multiple actors outside the CGIAR can be maintained most effectively.

A convening center on the other hand, is a facilitator, and it would normally be located within the eco-region. Subsequently it may or may not become research leader.

The role of eco-regional centers. With their regional basis of operation and interdisciplinary approach eco-regional centers will be able to collaborate with national institutions to integrate production and resource management research and to focus on important agro-ecological areas and problems. They will seek to identify complementary roles for their research capabilities. Some of the activities by which they could do this are shown in the Box 2.

[Box 2] Activities of eco-regional centers to complement the research of NARS

- assisting national institutions in the development of resource inventories;
- demonstrating the potential and the future consequences of NRM research;
- assisting NARS in determining the research domains of highest priority;
- collecting and disseminating information relevant to national policy options;
- contributing to the development of methodology for evaluating different strategy options in NRM; and,
- by organizing many of the less specialized training activities previously undertaken independently by individual centers (e.g., the use of farming systems research diagnostic methods, experiment station management, and some of the widely used applications for GIS).

Interinstitutional collaboration

Collaboration between national systems and CGIAR centers

Given the complex NRM research the different centers working in a country or region will need in future to cooperate more closely among themselves, with their donors who are providing support, and above all with the appropriate national and regional agencies. An eco-regional approach will be one means to rationalize training and capacity building activities among the centers, and avoid the kind of overlap of independent initiatives that have been a burden to the NARS in the past (TAC 1991).

Strengthening national systems. The success of international agricultural research depends on the strengths of the participating institutions, whether national systems, regional organizations, international centers, or advanced research institutions in industrialized countries. Part of the rationale for the eco-regional strategy is the increasing scope for collaboration among the stronger national systems, and between them and international centers, as a means to contribute to the greater public good derived from agricultural research. Eco-regional entities will work with national research systems and regional organizations, helping them to coordinate the work they do with the global centers of the CGIAR. National and regional organizations will participate actively in determining the research agenda, and in the design, conduct, and interpretation of research at key locations. Ideally all adaptive and applied research would be done by the national and regional programs.

Scale and scope of NARS participation. The eco-regional work that the centers are planning includes broad geographic areas of considerable diversity – broader than the scope of individual national systems (e.g., African Highlands Initiative; and Agricultural Sustainability on Central-American Hillsides [CIAT, CIMMYT, CATIE, IICA]), the Alternatives to Slash and Burn Initiative [ICRAF], and the Desert Margins Initiative [ICRISAT]). Even at a smaller scale, such as major watersheds and river basins, there are issues that cut across national boundaries. NARS have expressed concern at the prospect that they may be called on to divert part of their capacity and financial resources to work on their neighbors problems, and that this would detract from their ability to address their own priority national concerns (Goldsworthy *et al.* 1994b, 1994c). Clearly, a balanced approach to national and international dimensions is required. Establishing specific benchmark sites that represent the agro-ecological regions of importance to national programs could serve as a way to span differences in spatial scale. It would enable individual NARS to contribute to a larger eco-regional objective while focussing on issues of direct relevance to their national interests.

Implications for NARS of the eco-regional approach

Interdisciplinary research in NARS

Few national research institutes have the resources to gather all the disciplinary skills required to make up interdisciplinary teams for NRM research. The solution is not to create new institutions, but through new institutional arrangements to strengthen linkages between existing institutions. These new arrangements might include reorganization and merging of institutes, new partnerships with NGOs and universities, collaboration with private sector research organizations, and opportunities for cooperation between nations with similar natural resources and environmental problems. More effective mechanisms for coordinating NRM research among the institutions involved, across spatial levels and administrative divisions, is also essential. The eco-regional consortia are one form of interinstitutional collaboration that is being tested.

Use of systems approaches in NARS

While systems methods are now used widely in research in the IARCs, the institutional difficulties of introducing systems methods are likely to be more marked in NARS. Most of them are part of the public civil service, organized along disciplinary or departmental lines, and they do not often have the autonomy of the IARCs to make the organizational changes needed (Bengtsson 1987).

The shortage of trained staff in particular disciplines is another major bottleneck (Oram 1991). Training programs in systems research are badly needed, but few are available (Penning de Vries *et al.* 1991, Singh *et al.* 1994). Although the IARCs offer some training they have no comparative advantage in this specialized subject. Two international networks, SARP and IBSNAT, have run courses, and as a result of their work the Asian Institute of Technology now offers a masters degree course in agricultural systems research, and some universities in Asia have begun to include long and short courses in their curricula.

In spite of these still limited training opportunities, the capacity of some NARS to use systems approaches is developing rapidly (see Cheng Xu, Sankaran, and Sarr, in Goldsworthy and Penning de Vries 1994). The different experiences of the IARCs may therefore serve as useful indicators to help NARS identify the factors that influence institutional integration of systems methods, and thus enable them to avoid some of the problems.

Transfer of research results

Much of the NRM technology will be location specific and will be based on an understanding of the way existing agricultural systems work. There are unlikely to be any universal technical prescriptions. Therefore an issue for NARS is how to apply the knowledge acquired in one situation in another. It is here that systems methods are particularly valuable. The combination of GIS and simulation methods, with the facility this offers to bring more information to bear on a problem, makes it possible to target research, whether it is for varietal improvement or resource management, in a way that it was not possible to do until very recently. This facility, as noted earlier, helps in the transfer of research results from one location to other similar locations, or to locations that differ from it in known ways. Information is the key to success for this kind of research, and information collection and management therefore becomes particularly important.

CGIAR eco-regional initiatives

Current eco-regional initiatives (TAC 1993)

Eco-regional research is intended to address issues of global concern, such as poverty, food security, loss of biodiversity, climate change, migration from the countryside to the towns, and political and social stability. A number of eco-regional research initiatives have been or are being established to address some of the aspects of these global concerns related to agricultural and resource management. They have been established by CGIAR centers, or partnerships of centers in consultation and in partnership with developing country institutions. The characteristics of these initiatives vary considerably. There are perhaps about ten of them, but we have summarized the features of four of them in table 4 to illustrate the kind of variation which occurs.

System levels. First, the individual initiatives are concerned with different levels of system complexity, ranging from an eco-regional zone to a cropping system. For example, CONDESAN (Consorcio para el Desarollo Sostenible de la Ecorregion Andina) is concerned with a diverse mountainous eco-region, but is within a geographically entire area with a common culture and history. The African Highlands initiative (not shown in table 4) is similar in this respect. The initiative to examine alternatives to 'slash-and-burn' (ASB) in tropical rainforest addresses a land-use system which is practiced in many different parts of the world, while the lowland rainfed rice and rice-wheat systems consortia (LRRC and RWSC) are concerned with problems affecting particular cropping systems in South and Southeast Asia.

Research problem. The problems that have to be addressed in CONDESAN include the most diverse and complex set of biophysical, socioeconomic and institutional issues, and amount to complete ecosystem analysis on a very large scale. Similarly, the land-use and land tenure issues which the ASB initiative has to address in Indonesia are likely to be very different from those in Cameroon. At the other end of the range the RWSC is addressing an equally important combination of biophysical, socioeconomic and institutional issues, but the context of the problems is much more circumscribed, and some of the main biophysical issues will involve research on soil processes at a very detailed level.

Number and diversity of stakeholders. There is a very large number of institutions participating in CONDESAN; seven CGIAR centers, seven other international organizations, and 69 institutions from the seven participating countries. There are fewer participants in the ASB initiative, but they represent countries which are geographically widely separated and culturally diverse. Participation in the LRRC and the RWSC is more restricted; in each case it

Table 4. A comparison of four ecoregional initiatives led by CGIAR centers.

	CONDESAN	Alternatives to Slash and Burn initiative (ASB)	Lowland Rainfed Rice Consortium (LRRC)	Rice-Wheat Systems Consortium (RWSC)
System level	Ecoregion	Land use system	Land use system/Production system	Production system
Problem identified	Degradation of mountain ecologies in the Andean Region	Destruction of tropical rainforest	Rice production in variable, high risk rainfed lowland environments	Declining factor productivity in rice-wheat production systems of the Indogangetic Plains
Strategic issue	Ecosystem conservation and management: ecosystem analysis; land conservation and management; conservation of biodiversity; policy research.	Conservation of forest, particularly tropical rainforest ecologies	Characterise agroecology and understand physical and socioeconomic diversity and variability of rainfed lowland rice production systems: crop germplasm improvement, efficient and sustainable use of resources.	Sustainability of production: strategic research on soil and soil/water management; water use policies. Analysis of technical, socioeconomic and institutional constraints in rice-wheat cropping systems.
Countries involved	Argentina, Bolivia, Chile, Colombia, Ecuador, Peru, Venezuela	Cameroon, Zambia, Brazil, Peru, Mexico, Indonesia, Philippines, Thailand	Bangladesh, India, Indonesia, Philippines, Thailand	Bangladesh, India, Nepal, Pakistan
Lead center	CIP	ICRAF	IRRI	ICRISAT
Participating institutions: CGIAR	CIAT, ICRAF, IFPRI, ILCA, IPGRI, ISNAR	CIFOR, CIAT, IFPRI, IITA, IRRI		CIMMYT, IFPRI, IIMI, IRRI
Other Intern. Inst.	7	3		
Government and parastatal research org.	8	9	8	4
Universities	18		4	
NGOs	24			
Networks	12		2	
Total number of participants	76	18	15	9
Organization	*Consortium:* 4 Program components. 7 countries, many institutions. Complex interinstitutional links. Therefore separation between policy and strategic planning and implementation at level of program and benchmark sites.	*Consortium:* Thematic Working Groups. 8 countries on 3 continents. Complex inter-institutional links between and within continents. Therefore 4 levels of decision-making.	*Consortium:* of participating countries with IRRI. Demand driven; mutually agreed objectives. Formal request and agreement from senior government level for committment to land-use objectives required by consortium. Implementing institutions also decide on policy and planning.	*Consortium:* Informal association of collaborating country institutions and IARCs: organised to collaborate in research. Implementing institutions/scientists also decide on policy and planning.
Management	Elected *Advisory Council* for policy and strategic planning by 8 principal institutions and donors. An *Executive Committee* for management of institutional linkages. *Coordinator:* overall management. *Steering Committees* (4) and *Coordinators* for each program component: prepare project proposals.	*Steering Committees:* 1) *Global:* for policy and strategic planning (12 of 17 participating institutions represented). 2) *Regional:* coordination by CIAT (LAC), IRRI (Asia), IITA (SSA). 3) *National:* Coordination within participating countries responsible for benchmark sites: NARS chair committees. 4) *Local:* includes local communities and other implementing projects. *Global Coordinator:* overall management.	*Steering Committee:* advisory body to consortium nominated by governments of participating countries. *Coordinator:* director of IRRI Lowland Rice Ecosystem: overall management. *Site Committee:* multidisciplinary team of scientists. A national guiding committee to ensure consortium activities consistent with national priorities. IRRI back-up.	*Steering Committee:* DGs of IARCs and senior representatives of the Agricultural Research Councils/Min. of Agric. of 4 countries. ICRISAT serves as convener, CIMMYT and IRRI provide technical leadership.
Funding	Shared cost	Shared cost	Shared costs	Shared costs
Additional resources	from CIDA/ IDRC, and possibly Switzerland, Germany, Netherlands	Global Environment Fund (UNDP) (56% to CG institutes, 44% to other participants).	Asian Development Bank (Phase I), Rockefeller Foundation, DGIS (Phase II)	Australia: Bangladesh USAID: Nepal IBRD: India IFAD: ICRISAT ADB: IRRI & NARS
Improved efficiency	Provides coordination; avoids duplication; information dissemination	Combined efforts greater than sum of parts: already standardization of methods enabling global comparisons.	Establishment of multidisciplinary teams; coordination of efforts of participants; establishment of key research sites serving as nuclei around which other international efforts in less favoured environments are being built. Human resource development. Information dissemination.	

includes only those countries and institutions which are directly interested in the specific topic which the consortium is addressing. In the LRRC a formal commitment to the objectives of the initiative is required from would-be participant countries before they join; a measure which underlines that it is for a specialized interest group.

Management structure required. Because of the large number of participants in CONDESAN and the complexity of the issues it addresses, a management structure has been adopted in which there is a clear hierarchy of decision-making levels which separates the management from the implementation functions. Thus, while there are many participants at the implementation level, the actual partnership in consortium management is more restricted.

The management structure in the ASB initiative shows a similar hierarchy of decision levels, but one which reflects the geographically dispersed nature of the operation, with global, regional, national and local levels of decision making.

The LRRC addresses issues which though perhaps less diverse than those of the CONDESAN or ASB initiatives, concern variable rainfed environments. Therefore it too has adopted a decentralized form of management structure which can accommodate variations between and within the partner countries, of the issues to be tackled. However there is less separation of management and implementation functions than there is in CONDESAN or ASB. In the LRRC, implementing institutions also decide on policy and planning.

In the RWSC, the problems are even more specific, and there is little or no separation of the management and implementation functions. There is a senior policy level steering committee, but there is also a direct collaborative relationship between CIMMYT, IRRI and individual scientists in the partner countries. The consortium is also an example of the role of an eco-regional entity (in this instance ICRISAT) as convener; ICRISAT has no mandated responsibility for research on rice or wheat, but because of its location in the region where the work is done, it is well placed to act as facilitator to the consortium.

Conclusions

The medium term vision of the CGIAR involves a move towards an eco-regional approach to research, integrating agricultural and NRM concerns. In this paper we have reviewed the rationale behind this move and some of the institutional and management issues involved. It is anticipated, that, as national systems become stronger, the CGIAR will remain as a set of global research activities, mainly on issues such as conservation of genetic resources, global climate change, and global policy issues.

To the extent that the examples of eco-regional initiatives described represent the variation in the mode of implementation of the eco-regional approach, they pose the question whether there is or ever can be an 'eco-regional paradigm' as such. The diversity of regional environments and the many different manifestations of the central global issues to which we have referred suggest that a correspondingly diverse set of institutional and management arrangements will be required, each tailored to the particular task. This view should not be seen as a council of despair, but both as a prompt for more vigorous efforts to identify what common lessons can be learned about successful institutional and management arrangements, and investigational methodologies, and as a caution against expectations that there will be any simple prescriptions or models for the design and management of eco-regional approaches.

Acronyms*

CGIAR	Consultative Group on International Agricultural Research
CONDESAN	Consorcio para el Desarollo Sostenible de la Ecorregion Andina (Consortium for the Sustainable Development of the Andean Eco-region)
GIS	Geographical Information System
IARC	International Agricultural Research Center
NARC	National Agricultural Research Center
NARS	National Agricultural Research System
NGO	Non-Governmental Organization
NRM	Natural Resource Management
TAC	Technical Advisory Committee

References

Antle J M (1993) Environment, development, and trade between high- and low income countries. American Journal of Agricultural Economics 75:784–788.

Bengtsson B (1987) Research and development priorities and systems for increasing food production and rural income of small holding sectors in Africa. Pages 294–307 in Proceedings of the FAO/SIDA seminar on increased food production through low-cost food crops technology. March 2–17, 1987, Harare, Zimbabwe.

Bunting A H (Ed.) (1987) Agricultural environments: characterization, classification and mapping. Proceedings of the Rome workshop on agro-ecological chacterization, classification and mapping. April 14–18, 1986. CAB International, Wallingford, UK.

Byerlee D (1992) Technical change, productivity, and sustainability in irrigated cropping systems of South Asia. Journal of international Development 4(5):477–496.

Cheng Xu (1994) Needs and priorities for the management of natural resources: a large country's perspective. Pages 199–212 in Goldsworthy P R, Penning de Vries F W T (Eds.) Opportunities, use, and transfer of systems research methods in agriculture to developing countries. Kluwer Academic Publishers, Dordrecht, The Netherlands.

* The acronyms of the CG-centers and IARCs are given in the acronyms list in this book.

CGIAR (1993) CGIAR response to UNCED's Agenda 21. Report of the CGIAR Task Force. September 1993. CGIAR, Washington, USA.

CGIAR (1994a) A Research Agenda for the Future. ICW/94/13, October 1994. CGIAR, Washington, USA.

CGIAR (1994b) Research for a Food Secure World. ICW/94/11, October 1994. CGIAR, Washington, USA.

Crosson P, Anderson J R (1992) Resources and global food prospects: supply and demand for cereals to 2030. World Bank Technical Paper No. 184. World Bank, Washington DC, USA.

Crosson P, Anderson J R (1993) Concerns for sustainability: integration of natural resource and environmental issues in the research agendas of NARS. ISNAR research Report No. 4. International Service for National Agricultural Research, The Hague, The Netherlands.

Dingkuhn (1994) Systems research at WARDA. Pages 341–150 in Goldsworthy P R, Penning de Vries F W T (Eds.) Opportunities, use, and transfer of systems research methods in agriculture to developing countries. Kluwer Academic Publishers, Dordrecht, The Netherlands.

Fujisaka S, Harrington L, Hobbs P (1994) Rice-wheat in South Asia: systems and long-term priorities established through diagnostic research. Agricultural Systems 46:169–187.

Goldsworthy P R, Penning de Vries F W T (Eds.) (1994) Opportunities, use and transfer of systems research methods in agriculture to developing countries. Kluwer Academic Publishers, Dordrecht, The Netherlands.

Goldsworthy P R, Eyzaguirre P, Boerboom L (1994a) The institutional and organisational implications of integrating natural resource management and production-oriented research. Pages 139–152 in Goldsworthy P R, Penning de Vries F W T (Eds.) Opportunities, use, and transfer of systems research methods in agriculture to developing countries. Kluwer Academic Publishers, Dordrecht, The Netherlands.

Goldsworthy P R, Penning de Vries F W T, Van Dongen J (1994b) The use of systems methods by NARS for addressing natural resource issues. Briefing Paper No. 16, December 1994. ISNAR, The Hague, The Netherlands.

Goldsworthy P R, Penning de Vries F W T, Van Dongen J (1994c) The use of systems methods in international agricultural research centres. Briefing Paper No. 17, December 1994. ISNAR, The Hague, The Netherlands.

Harris H C, Nordblum T L, Roriguez A, Smith P (1994) Experience of the use of systems analysis in ICARDA. Pages 295–302 in Goldsworthy P R, Penning de Vries F W T (Eds.) Opportunities, use, and transfer of systems research methods in agriculture to developing countries. Kluwer Academic Publishers, Dordrecht, The Netherlands.

Jagtap S S (1994) The use of systems analysis at international and program levels: IITA's experience. Pages 313–322 in Goldsworthy P R, Penning de Vries F W T (Eds.) Opportunities, use, and transfer of systems research methods in agriculture to developing countries. Kluwer Academic Publishers, Dordrecht, The Netherlands.

Janssen W (1994) Characteristics of NRM research: institutional and management implications. Paper presented at ISNAR/DSE workshop on 'Research policy and management for agricultural growth and sustainable use of natural resources', December 1994. ISNAR, The Hague, The Netherlands.

Kropff M J, Penning de Vries F W T, Teng P S (1994) Capacity building and human resource development for applying systems analysis in rice research. Pages 323–339 in Goldsworthy P R, Penning de Vries F W T (Eds.) Opportunities, use, and transfer of systems research methods in agriculture to developing countries. Kluwer Academic Publishers, Dordrecht, The Netherlands.

ISNAR (1993) Agenda 21: issues for national agricultural research. Briefing Paper No. 4. December 1994. ISNAR, The Hague, The Netherlands.

Lynam J K, Herdt R (1989) Sense and sustainability: sustainability as an objective in national agricultural research. Agricultural Economics 3:381–398.

Oram P (1991) Institutions and technological change. In Agricultural sustainability, growth, and poverty alleviation: issues and policies. Proceedings of a Deutsche Stiftung für Internationale Entwicklung (DSE) & International Food Policy Research Institute (IFPRI) conference. September 3–27, 1991. Feldafing, Germany.

Penning de Vries F W T, Van Laar H H, Kropff M J (1991) Systems simulation at IRRI. IRRI Res. Pap. Ser. 151. IRRI, Los Baños, Philippines.

Ruttan V W (1991) Sustainable growth in agricultural production: Poetry, policy and science. Page 13 in Agricultural sustainability, growth, and poverty alleviation: Issues and policies.

Sankaran S (1994) The use of systems analysis methods – the experience at a national level (India). Pages 213–226 in Goldsworthy P R, Penning de Vries F W T (Eds.) Opportunities, use, and transfer of systems research methods in agriculture to developing countries. Kluwer Academic Publishers, Dordrecht, The Netherlands.

Sarr D Y (1994) Opportunities for use of systems approaches in agricultural research in developing countries – the Senegal experience in research systems. Pages 227–232 in Goldsworthy P R, Penning de Vries F W T (Eds.) Opportunities, use, and transfer of systems research methods in agriculture to developing countries. Kluwer Academic Publishers, Dordrecht, The Netherlands.

Singh G, Pathak B K, Penning de Vries F W T (1994) Requirements for the use of systems research in agricultural and environmental sciences. Pages 255–265 in Goldsworthy P R, Penning de Vries F W T (Eds.) Opportunities, use, and transfer of systems research methods in agriculture to developing countries. Kluwer Academic Publishers, Dordrecht, The Netherlands.

TAC (1989) Sustainable agricultural production: implications for international agricultural research. TAC Secretariat, FAO, Rome, Italy.

TAC (1990) Towards a Review of CGIAR Priorities and Strategies. TAC Secretariat, FAO, Rome, Italy.

TAC (1991) An eco-regional approach to research in the CGIAR. TAC Secretariat, FAO, Rome, Italy.

TAC (1992) Review of CGIAR research priorities and strategies part II. TAC Secretariat, FAO, Rome, Italy.

TAC (1993) An eco-regional approach to research in the CGIAR. Report of a TAC/Centre Director's Working Group. CGIAR, Washington, USA.

Torres F, Gallopin G (1994) Systems research methods and approaches at CIAT-current and planned involvement. Pages 271–280 in Goldsworthy P R, Penning de Vries F W T (Eds.) Opportunities, use, and transfer of systems research methods in agriculture to developing countries. Kluwer Academic Publishers, Dordrecht, The Netherlands.

Developing an R & D model for the humid tropical eco-region in Asia

P.S. TENG, M. HOSSAIN and K.S. FISCHER

International Rice Research Institute (IRRI), P.O. Box 933, Manila, Philippines

Key words: agro-ecological zone, CG-centers, eco-regional program, inter- and intra-institutional collaboration, NARS, NGO, research priority

Abstract. Of most significance for an eco-regional initiative in the tropics are strong consortia and networks that are now managed by the national agricultural research systems of the region. The R & D model intends to further refine these by extending the scope of the collaborative mechanism to include issues that cut across and within the four agro-ecological zones (AEZs) in a geographic mode, i.e. eco-regional mode. A framework will be created to determine the common issues between contiguous eco-regional zones, called domains, within which ecosystems will be categorized for prioritization of issues. Initially, the effort will focus on domains in AEZs 2, 3, 6, and 7, in which rice predominates but other crops play an important role. The framework further recognizes that contiguous rice-based ecosystems transcend AEZ boundaries and that rice-based ecosystems are key domain determinants. An eco-regional R & D Model, which includes four steps, is proposed:

Step 1: *Ex ante* analysis of eco-regional issues and knowledge gaps, and Strengthening of collaborative mechanisms (e.g. consortia, networks, inter-Center collaboration) for eco-regional work;

Step 2: Specification of R & D priorities, Development of an Eco-regional Action Plan to respond to eco-regional issues; identification of additional collaborators;

Step 3: Participatory implementation of action plan to meet research priorities; synthesis of R & D results; creation of a supportive policy environment for eco-regional research; and

Step 4: Development and implementation of monitoring mechanisms to evaluate improvements in the eco-region.

Steps (1) and (2) are proposed for the first 2 years. During this period, application of geographic information systems for improving AEZ-ecosystem characterization will be done in conjunction with systems analysis of key issues. A Mega workshop is proposed toward the end of the second year, to include key stakeholders from IARCs and NARS in the eco-region, to prioritize eco-regional issues and develop an Eco-regional Action Plan (EAP). Issues that cut across AEZs or ecosystems include sustainable improvement of rice-legume and other lowland systems, integrated pest management through managing habitat diversity, and enhancing the sustainability of upland systems.

Rationale for research at the eco-regional level

The Asian region is generally acknowledged as the most dynamic growth area of the 1990s, and forecasts are that this region will continue to be a key influence on the global economy in the next century (Naisbett 1994). However, such growth has not seen significant decline in the proportion of

J. Bouma et al. (eds.), Eco-regional approaches for sustainable land use and food production, 305–330.

poor people or significant improvement in the quality of the environment and life for many of the target beneficiaries of such growth.

Major social issues in most of the region remain – underemployment, malnutrition, illiteracy, and widespread poverty. There is only a remote possibility that food demand will decline in the near future since population continues to grow at over 2.0% per year in the region, and unmet demand is still large due to lack of purchasing capacity (Zeigler *et al.* 1994). Poverty induces expansion of cultivation to marginal land and exploitation of women and child labor. Abiotic stresses – droughts, temporary submergence, waterlogging and salinity – remain a major challenge to germplasm improvement research. A large proportion of arable land is still cultivated at the vagaries of nature and suffer from phosphate and zinc deficiency, while soil erosion is the primary constraint in the sloping uplands of many southeast Asian countries.

Thus, in spite of this region being a success story for the 'Green Revolution' of the 1960s, there exists today a challenge to maintain the natural resource base and concurrently utilize technologies that will improve social justice as well as grow enough food for the expected increase in population.

Centers belonging to the Consultative Group for International Agricultural Research (CGIAR) in the Asian region have met past challenges admirably through germplasm improvement, in which scientific knowledge is 'captured' in the form of seed. Indeed this success has in the past lead to misunderstandings of the role of the commodity-based centers such as IRRI and ICRISAT as solely germplasm centers with little to no work on natural resource conservation and management (TAC 1992). TAC, in a comprehensive review of CGIAR priorities and strategies noted that for the longer term, a new paradigm for global efforts would need to be adopted if sustainable development goals were not to conflict with those of increased food production. The paradigm would explicitly recognize the pivotal role of national agricultural research systems in natural resource management research, with CGIAR centers contributing as a partner on strategic issues. Painful as it was, TAC correctly noted the need for international centers to work closer together than presently in common geographic domains with harmonized use of resources and toward common goals (Gryseels *et al.* 1992). It also recommended that global research activities of the CGIAR be complemented by an increase in research focused on agro-ecological zones (AEZs). The above furthermore appears to be coincident with thinking and action in the Asian region to incorporate more "ecological" approaches into agricultural practices; a notable example is the strong development of integrated pest management (IPM) in this region (Teng 1994).

Research that is eco-regional in scope requires concepts and techniques that allow integration and interaction of different levels of biological, physical or social organization. For a single commodity, this would imply integration of knowledge at the cellular, tissue, plant, crop, cropping system, community and landscape levels. Until recently, it was difficult if not impossible to achieve such integration or interaction, especially with knowledge on the natural resource base. Furthermore, research on natural resource management tends to be site-specific and opportunities for generalization to wider domains and spill-over effects are often limited (TAC 1992). To minimize this limitation, TAC suggested the conceptualization, conduct, and interpretation of such research in the context of homogeneous agro-ecological zones that are regionally defined, i.e. eco-regions. All these reflect a trend of strong locality-specific knowledge that is related to form higher level associations. This reliance on local knowledge is a strong basis for launching larger scale (eco-regional) activities, and is made possible by the rapid and significant developments in electronic networking, communication, data analysis, system modelling and geographic information systems in the region (Kropff *et al.* 1993). Indeed, a new age has arrived, where it is not only possible, but essential, to "*Think Locally, Act Globally*" (Naisbett 1994) to maintain a competitive advantage.

TAC (1992) defined eco-regional research activities as those concerned with ecosystem conservation and management, production system development and management, socioeconomic, public policy and public management research, and institution building. It considered germplasm collection, conservation, characterization and evaluation, enhancement and breeding as global commodity improvement activities. The guiding principles for the adoption of the eco-regional approach to research, as stated by TAC, are:

- fill gaps in the coverage of research relating to natural resource conservation and management;
- rationalize overlapping commodity mandates and minimize overlaps in research on natural resource conservation and management, by clearly delineating responsibilities for different research activities;
- provide focal points within an organized agro-ecological framework for coordinating decentralized research activities; and
- streamline interactions between NARS and CGIAR centers to avoid confusion at the national level, by coordinating institution building efforts and other activities (TAC 1992).

TAC recognized the difficulty of applying these principles and had advised Centers to adopt a pragmatic approach in applying them (TAC 1992).

In the long run, TAC (1992) expects the following outputs from comprehensive eco-regional research:

a) to determine an effective approach to natural resource management research that bring sustainable improvements in productivity to agriculturally dependent rural communities;
b) to understand the principles of management of soil, water and biological processes and their interactions in different environments;

c) to determine an effective mechanism to link policy formulation with technological opportunities at different levels of population pressure, social organizations and employment opportunities;
d) to understand the principles of farmer and community decision making in relation to the trade-off between short-term gain and long-term sustainability of production;
e) to build a human resource capacity to help NARS implement an effective research approach to natural resource management.

TAC also emphasized that research directed towards the improvement of sustainable production systems would have to be multi-commodity in its coverage and move into areas of research on land use and natural resource conservation.

To the extent that it is possible to judge before a detailed analysis, many of these considerations are in direct congruence with issues articulated repeatedly by the NARS in the Asian humid tropical eco-region.

IRRI, as the oldest CGIAR center in Asia, has recognized the strong developmental trends in the region, and has incorporated much of TAC's suggestions for a changed paradigm into its current medium term plan for 1994–1998 (IRRI 1994a). Within its core research, IRRI programs are based on rice-agro-ecosystems – a move toward an eco-regional approach to research that was made in 1988. There are four ecosystem-specific programs, respectively addressing irrigated rice, rainfed lowland, rainfed upland and flood-prone systems; in addition, a fifth program, the cross-ecosystems, addresses issues which cut across the ecosystems or are ecosystem neutral. The institute therefore feels it is well-positioned to play a lead convening role in the Asian region to bring together all concerned 'stakeholders' to implement an eco-regional program.

In this paper, we provide background information that was used in formulating the approach to derive an eco-regional program. We have called the approach a Research and Development (R & D) model and the program an Eco-regional Action Plan. This paper and IRRI's eco-regional initiative does not propose research *per se* in the initial period, but rather proposes conducting a joint systems analysis to set the scope of eco-regionalism, to identify issues, to set priorities and to jointly arrive at mechanisms to implement an eco-regional program with clear goals and measurable achievements. The initiative in no way excludes ongoing work at different centers which have eco-regional relevance. It serves to arrive at a conceptual framework to determine how current and new activities can best fit into a preferred eco-regional scenario. To do so it proposes a two year period of rigorous interaction between stakeholders using conceptual and empirical approaches.

Essential considerations for designing an eco-regional program

The term 'eco-region' connotes a geographic domain within Asia based on the concept of AEZs. Development of an eco-regional program will therefore have to recognize the key elements of the eco-region, which include the AEZs, land use within the AEZs, the features of the natural resource base, cross-cutting issues, institutional aspects, and mechanisms for inter-institutional collaboration.

Agro-ecological zones (AEZs)

All AEZs were estimated in the TAC study using simple crop growth criteria, namely, (a) a moisture regime based on rainfed length of growing period (RLGP) and (b) the thermal regime based on mean daily temperature (MDT). The humid AEZs have RLGPs of > 270 days and the warm AEZs have MDTs > 20 °C. Using these criteria, the TAC study resulted in 9 AEZs located in four regions – Sub-Saharan Africa (SSA), West Asia–North Africa (WANA), Asia, and Latin America and Caribbean (LAC) (TAC 1992). Through a multi-stage process, the 36 region * AEZ combinations were reduced eventually to six eco-regional programs.

In this paper, we recognize TAC's consideration that an eco-region encompasses a set of AEZs that are geographically bound. The warm tropical eco-region for which IRRI was assigned by TAC to convene an eco-regional program is considered to be the region covered by four AEZs, respectively, AEZs 2 (warm, sub-humid tropics), 3 (warm, humid tropics), 6 (warm, sub-humid subtropics with summer rainfall) and 7 (warm, cool, humid subtropics) – figure 1. The Asian warm humid eco-region therefore comprises the following:

AEZ 2 (warm, sub-humid tropics) – includes parts of India (Assam, Andhra Pradesh, Orissa, West Bengal, Bihar, Kerala), and the entire areas of Myanmar, Sri Lanka and Thailand;
AEZ 3 (warm, humid tropics) – includes the entire areas of Bangladesh, Indonesia, Malaysia, Philippines, Laos, Cambodia and Vietnam;
AEZ 6 (warm, sub-humid subtropics with summer rainfall) – includes part of India (Uttar Pradesh) and parts of China (Shandong, Henan, Shanghai); and
AEZ 7 (warm, cool, humid subtropics) – includes the entire island of Taiwan, and parts of China (Jiansu, Zhejiang, Anhui, Fujian, Jiangxi, Hubei, Hunan, Guangdong, Guanxi, Sichuan).

We will use the AEZs delineated by TAC in their priorities document (TAC 1992), but recognize that further refinements are possible. These refinements will have to be done as part of the proposed R & D model to ensure a more accurate delineation of geographic domains at the sub-national level.

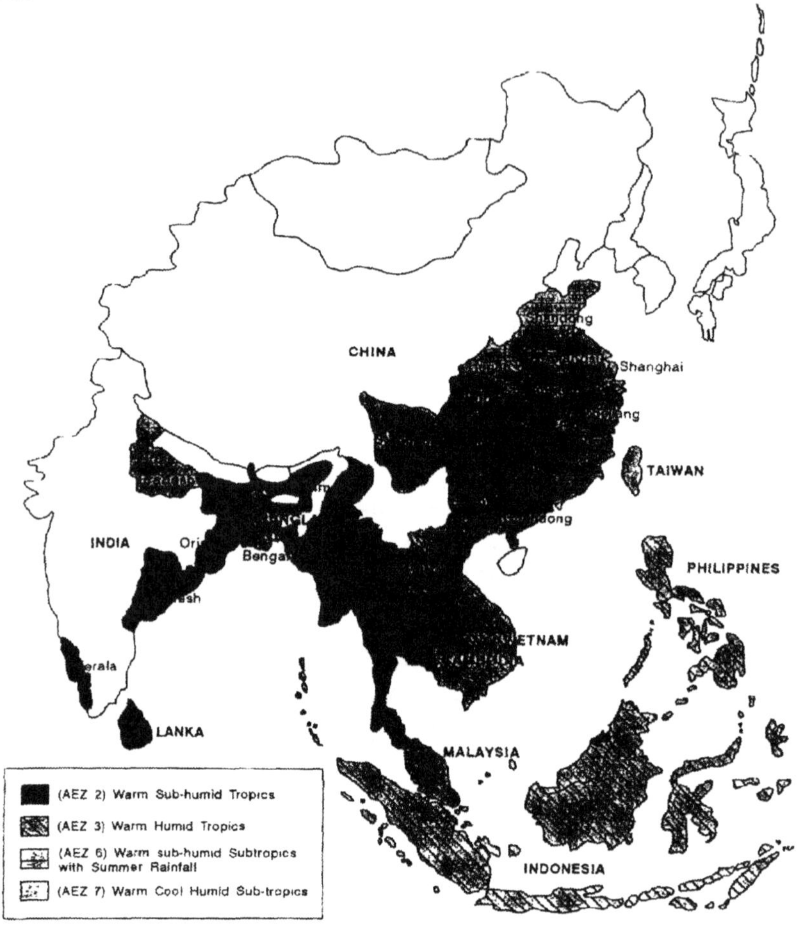

Fig. 1. Map showing the geographic distribution of the Asian humid tropical eco-region encompassing four agroecological zones.

Land use patterns in the AEZs

Land use patterns in the four AEZs of the tropics and subtropics, vis-à-vis food crops, is summarized in figure 2 for the important food crops (IRRI 1993). Other AEZs in Asia (1, 5 and 8) are discussed in relation to the relative importance of different food crops, and implicitly, to the potential role that international and national research institutes can play in eco-regional research. Rice is the dominant food crop in AEZ 7 (78% of total arable area), AEZ 3 (74%) and AEZ 2 (70%). It is a major crop in AEZ 6 (21% of total arable area, after wheat and maize which occupy 46%) and AEZ 1 (20% of total area) but is insignificant in AEZ 5 and 8. The mix of crops and other agricultural

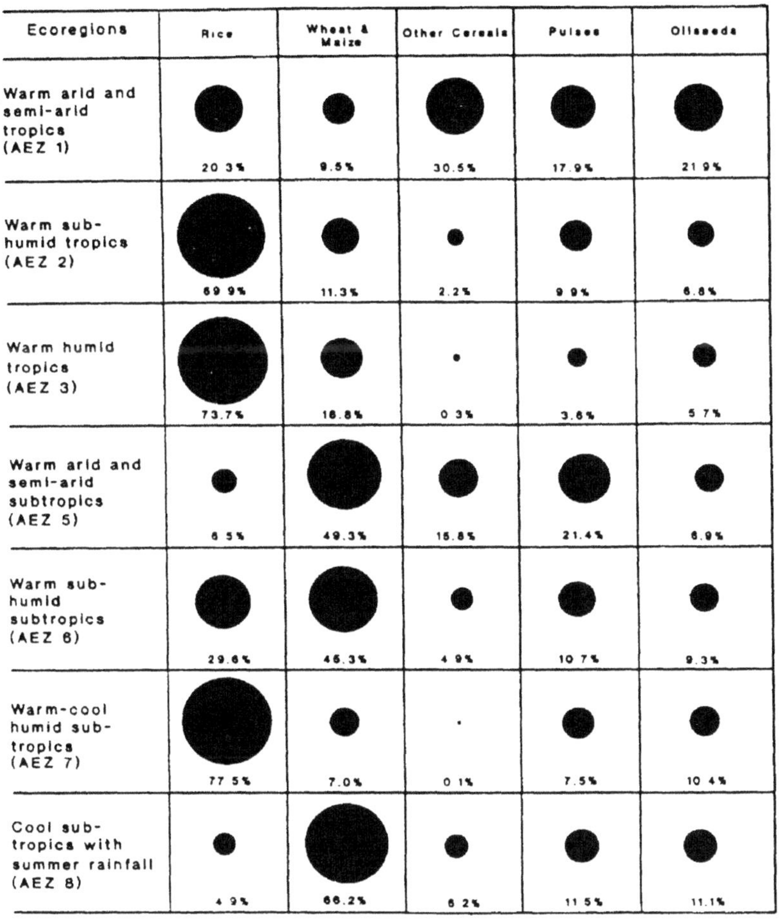

Fig. 2. Land use patterns for food crops in the agroecological zones in which rice is grown. (Figures indicate percent of area in each crop. (Source: IRRI 1993)

activities such as livestock and aquaculture in each AEZ suggests a basis for determining which institutions are potential stakeholders in that AEZ.

Because rice is such a dominant component of the landscape in the Asian eco-region of concern, and features strongly in the economies of many countries in this region (Hossain and Laborte 1994), any eco-regional program will have to explicitly recognize this. However, the role of rice is influenced by the topography in which it is grown, which in turn is reflected in the type of rice ecosystem. The interface between rice ecosystems (as defined by IRRI) and TAC's AEZs is shown in figure 3.

The irrigated rice ecosystem, which accounts for 57% of global rice area, is the dominant rice ecosystem in AEZ 3, 5, 6 and 7 of the eco-region. Irrigated

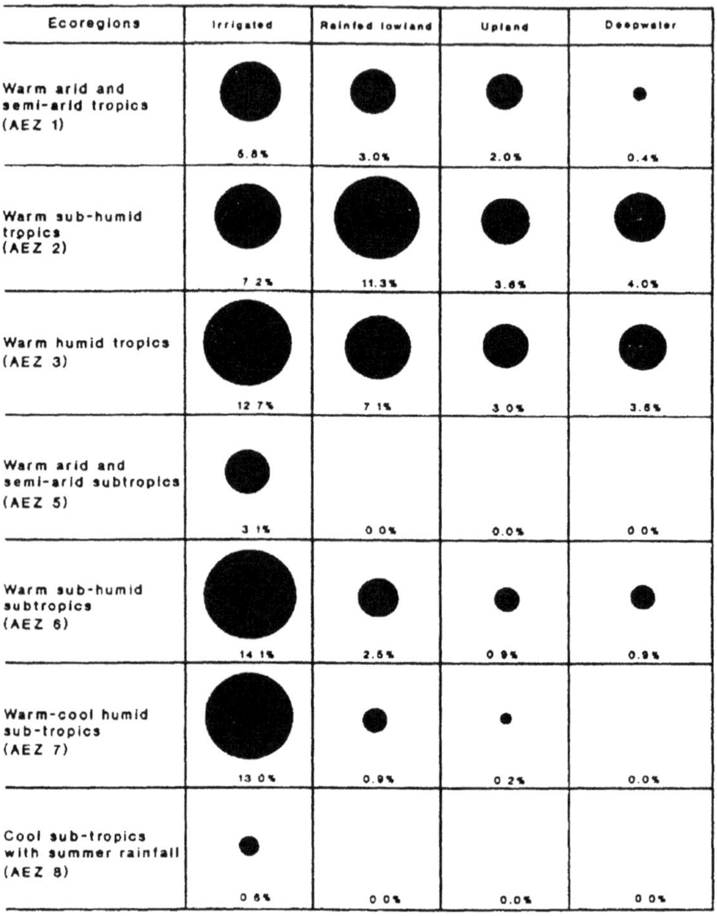

Fig. 3. Relative occurrence of rice ecosystems in different agroecological zones (figures indicate percent of total rice area for each AEZ and each rice ecosystem). (Source: IRRI 1993)

rice is also the major ecosystem in AEZ 1 (the warm arid and semi-arid tropical areas of India) and 8 (the cool sub-tropical areas in China, Nepal, Japan, the Korean peninsula and parts of India (Jammu and Kashmir).

The unfavorable rice ecosystems dominate in AEZ 2 (Eastern India, Myanmar, Thailand) and occur in significant areas in AEZ 3 (Bangladesh, Cambodia, Laos, Vietnam).

Non-rice crops which are of some importance in AEZ 2, 3, 6 and 7 are as follows:

- Pulses and oilseeds in AEZs 2, 6 and 7;
- Maize in Philippines and Indonesia (AEZ 3)

- Wheat in AEZ 6 (parts of India, China)
- Agroforestry in upland areas in Southeast Asia (AEZ 2,3)
- Cassava in Thailand (AEZ 2)
- Vegetables in South and Central China, and Taiwan (AEZ 7).

Several important rice growing areas lie outside the AEZs under consideration. The whole of Japan, the Korean peninsula, Nepal,and parts of India (Jammu, Kashmir) and China, are included in AEZ 8, the cool sub-tropics with summer rainfall, and are not in either of the two Asian eco-regional programs proposed by TAC.

The availability of current and complete databases on land use at the sub-national level is prerequisite for analyzing the contributions of various commodities to the eco-region's food needs. An integrated database does not appear to exist and would be a high priority for an eco-regional plan. Furthermore, the application of a common methodology for land use evaluation (Stomph *et al.* 1994) in this eco-region has yet to be done to allow comparative analysis of resource utilization.

Elements of the natural resource base and sustainability

The natural resource base includes soils, water, microbes, wildlife, forests, plant genetic resources, crops, fish, and livestock. The eco-regional approach was proposed by TAC as a vehicle for increasing focus on the conservation and management of natural resources with the goal of developing sustainable food production systems.

For food production and economic development to be sustainable, the link between the rural and urban sectors must be recognized, and modern technology must be blended with indigenous knowledge systems. One of the challenges facing scientists in the Asian eco-region is how to preserve some of the traditional techniques that work under less intensive systems in the face of increasing pressure to intensify so that more food can be produced. In Asia, proponents of sustainable agriculture (SA) often cite the Fukuoka farm in Japan (Fukuoka 1983) and the Fantilanan farm in the Philippines (Pelegrina *et al.* 1992) as proof that SA can be practiced economically. However, to what extent these farms can be used as a model for extrapolation or indeed provide a model for study is not certain. In the two countries, SA has not multiplied significantly, yet non-government organizations (NGOs) consistently use low yield systems as examples of sustainable technology for managing natural resources.

Conservation and management of the natural resource base are particularly germane to the achievement of sustainable agricultural systems. "Sustainable agriculture should involve the successful management of resources for agriculture, to satisfy changing human needs while maintaining or enhancing the quality of the environment and conserving natural resources" (TAC

1992). The U.S. Board for International Food, Agriculture and Development proposed that "Sustainable agriculture should conserve and protect natural resources and allow for long-term economic growth by managing all exploited resources for sustainable yields" (BIFAD 1988). The common themes are to conserve, if not enhance, the natural resource base while satisfying the dynamic demands of society.

The social and policy aspects of development are further recognized by workers in SA. Gips (1987) described SA as being "ecologically sound, economically viable, socially just and humane". Harwood (1987) stated that "sustainable agriculture must make optimal use of the resources available to it to produce an adequate supply of goods at reasonable cost: it must meet social expectations, and it must not overly expend irreplaceable production resources". Thus, in planning for an eco-regional program for humid tropical Asia, it would be critical to include data on the social, as well as as institutional and policy aspects of the system.

Cross-cutting issues

There is as yet no general agreement among scientists as to what constitutes total system sustainability. However, there is agreement on the major causes of unsustainability, from which can be identified a corresponding, antithetic topic offering opportunity for research (table 1).

This framework is presented to illustrate some potential areas for eco-regional collaboration which could directly contribute to arresting the decline in the natural resource base in some cases, and in others to enhancing it. All the topics cut across the mandates of single institutions for their research and resolution. In the eco-regional initiative proposed for the humid/subhumid tropics/subtropics of Asia, no presupposition is made on the priority, and current work on any of the above issues will be factored into the analysis process to design the eco-regional action plan.

Leveraging the comparative advantage of institutions

CGIAR international centers

In addition to IRRI, many of the CGIAR centers have research activities in the Asian tropical eco-region – ICRISAT, CIMMYT, CIP, ICLARM, ICRAF, IIMI, IBSRAM, IFPRI and ISNAR. In the proposed initiative, centers would be expected to contribute the skills and knowledge commensurate with their commodity or specialty mandates, strategies and workplans. Since all centers have to some extent incorporated elements of eco-regional thinking into their workplans, it is further anticipated that there will be a need to rationalize these. For example, in resource management, research for areas where non-rice crops are important components of the agro-ecosystem, IRRI will seek research collaboration with other Centers such as ICRISAT and CIP in AEZ 2, ICRAF, IBSRAM and CIMMYT in AEZ 3, and AVRDC in AEZ 7.

Table 1. List of cross-cutting issues on sustainability that have eco-regional relevance.

Cause of unsustainability	Opportunities for research
Reduced, unstable yield caused by pest outbreaks	IPM implementation tools Managing biodiversity for pest management
Degradation of soil-nutrient resource base	Causes of yield decline in intensive systems Improving input use efficiency
Destabilization of watershed and related erosion/salinization	Designing new upland life support systems Improved water conservation and utilization
Global climate change	Effects of flooded rice on global warming Effects of climate change on rice
Loss of genetic diversity in agronomic species and ecosystems	Conservation of biodiversity and habitat diversity
Environmental pollution	Rational and safe use of synthetic inputs
Inequitable technologies and social systems, and poverty	Profit-generating scale-neutral technologies Assessing impact of technology on rural households Diversified cash-generating enterprises
Inappropriate policies	Generate knowledge to support "sustainability-friendly" policies, assessment of potential impact

IRRI's contact with NARS is mostly concentrated in the four AEZs. Rice accounts for 90% of total foodgrain production in AEZ 7, 77% in AEZ 2 and 71% in AEZ 3: increased rice production with greater input use efficiency would make a substantial impact on the well-being of farmers in these eco-regions, and natural resource management research on rice-based cropping system would contribute to enhancing environmental quality. Thus, it makes sense for IRRI to take a leadership role in eco-regional research in AEZ 2, 3, 6 and 7, and to initiate dialogues with relevant research centers to coordinate collaboration on natural resource management research. For AEZ 1 and 5, IRRI will need to cooperate with ICRISAT and CIMMYT whose mandated crops are dominant in those AEZs.

NARS

NARS, which in this context include agricultural research centers and agricultural universities, have the strong location advantage for natural resource work. In some instances, capacity for conducting aspects of systems research related to eco-regional topics already exist in many national programs (Kropff *et al.* 1993). However, in general, capacity for newer technologies such as GIS is still relatively undeveloped.

NGOs

The comparative advantage of NGOs are: a better understanding of the needs of resource poor farmers derived through close association for long periods, realistic judgement of the limitation to adoption of specific technologies, and access to non-formal organizations. NGOs will be able to provide a 'reality check' during the planning of the eco-regional program as well as participate in its implementation.

Other institutions

The most important non-CGIAR international center in the region is AVRDC, for which active collaboration will have to be sought on all issues related to vegetable improvement and production. The concepts and techniques for research at the eco-regional level have received attention at several advanced institutions, notably the Wageningen Agricultural University and DLO in Wageningen, The Netherlands, and the U.S. universities belonging to the recently-formed International Consortium for Agricultural Systems Applications (ICASA). The experiences of these will be tapped to assist in the analysis.

Mechanisms for cooperation

While TAC called for eco-regional research that is directed towards solution of problems within the framework of a broadly defined agro-ecosystem, it also stressed that such research be strongly linked with NARS research focused on increasing food production and be implemented through collaboration with other CGIAR centers, NARS and local development organizations including NGOs. Some of these collaborative mechanisms are discussed here.

Consortia: partnership with NARS

Consortia are groupings of research institutions who share facilities and capacity to solve specific problems that have been mutually identified. TAC (1992) felt that the research consortia have potential to be a mechanism for addressing eco-regional concerns in the heterogeneous rainfed ecosystems. In IRRI's Medium Term Plan, three consortia are envisaged. Of these, two were established in 1991 – one for the upland rice ecosystem and the other for the rainfed lowland rice ecosystem. The third, on flood-prone rice ecosystems, is being planned. Also, the development of a fourth consortium on irrigated rice, with emphasis on research issues in intensive systems, is underway.

In the less favorable environments of the rainfed lowlands (including the flood-prone ecosystems), and the rainfed uplands, research consortia aim at an effective partnership with NARS to address strategic research issues related to abiotic and biotic stresses. The consortia mechanism further improves coordination between research on productivity/germplasm improvement and natural resource management. The strategic research themes are explicitly directed towards eco-regional issues like sustainable productivity increases and maintenance of the soil and water resources base in AEZs 2 and 3. Research is also directed at improving the long-term health of the soil through modification of cropping practices and rice germplasm characteristics. The research consortia are particularly attractive in that eco-regional as well as equity issues can be addressed.

Networks: exchanging tools and technology

The Crop Resource Management Network (CREMNET) is being established by IRRI as a companion network to the highly successful INGER (International Network for Genetic Evaluation of Rice). Building on the concepts of INGER, (for seed technology), and on the past experiences of Asian Rice Farming Systems Networks, the new CREMNET will focus on the exchange of non-seed technologies. Other centers have similar networks and it is envisaged that network cooperators will play an active role in the systems analysis, knowledge/data collection and in determining priorities.

IRRI and FAO have developed collaborative initiatives on their work on IPM in the region, such as with IPM-NET. Other research networks include the SARP and the Asian Rice Biotechnology Network.

Management mechanisms for effective eco-regional partnerships

Discussions with leaders in Asia have led IRRI to initiate establishment of a Council for Collaborative Research in Asia. Opinions and inputs of Asian leaders are being canvassed. One of the important objectives of the Council, which will meet annually, would be to develop concepts for an eco-regional approach of relevance to the Asian Region. Because rice is the major crop in AEZs 2, 3 and 7 and significant in AEZs 5, 6 and 8, this Council will include in its membership NARS leaders with responsibilities not only for rice research in their countries but for other agricultural commodities. The Council will assist in securing the necessary political and administrative support for NARS participation in eco-regional activities.

Inter-center collaboration

Aside from the humid tropical Asian area for which IRRI will convene an eco-regional program, it will also involve itself in eco-regional activities in AEZs 1 and 5 by playing a cooperative role in conjunction with the lead center, ICRISAT.

IRRI has discussed with ICRISAT the sharing of responsibilities for resource management issues in the major agro-ecological zones of Asia and has initiated activities to develop and share common databases. Three agro-ecological zones – AEZs 1, 2 and 5 – have been identified as areas of common interest to both institutes. Options on other agro-ecological zones in the region are being kept open. Joint activities would include the development and maintenance of common databases using common software for GIS applications for all eco-regions in Asia. There has been an exchange of scientists to develop the common approach. Collaborative research will focus on factors related to degradation of production systems in which mandated crops for the centers are involved. Emphasis is on nutrient cycling, and potential for enhancement in rice-legume systems in rainfed and irrigated systems.

IRRI and IFPRI are collaborating on research to define policy issues in the Asian region. A detailed study is defining the demand for rice and other foods in the Asian Region and is linking that with supply issues including the balance between production and the use of natural resources.

In addition, CIAT has a collaborative project scientist at IRRI.

IRRI is also developing a collaborative project/activities with AVRDC on diversifying cropping systems to increase farmer income. AVRDC and IRRI have outlined and initiated in 1994 three collaborative projects, viz., Diversification and income improvement in Asian rice systems, Vegetables and legumes in rice systems in which factor productivity is declining, and integrated pest management for rice and vegetable sequences. Also they will exchange information and participate in planning meetings in the Uplands for Sustainable Systems.

With ICRAF, IRRI is cooperating in areas where agroforestry systems are an important component of a sustainable system in the Uplands.

Rice-fish research has been ongoing for many years in IRRI-ICLARM collaboration.

The collaborative project on rice-wheat systems by CIMMYT and IRRI to improve productivity and sustainability, began in June 1991, and linked nine NARS institutes (Lead centers) in Bangladesh, India, Nepal and Pakistan in consortium-style research. IRRI and CIMMYT provide strategic research support. The initial support by the ADB is augmented with substantial complementary funding being made available by the National Governments with the assistance of World Bank loan support for research and technology transfer. In addition, IRRI and CIMMYT are expected to contribute to the initiative through their core programmes. ICRISAT is now providing logistical support for the rice-wheat collaborative project between CIMMYT, IRRI and NARS.

Collaborative relationships with regional and global projects

IRRI is an Asian member of three global projects dealing with methodologies at the systems and watershed levels – the Slash and Burn project (UNDP), the IPM Collaborative CRSP (US AID) and the SANREM (USAID) project. In addition, IRRI is an active node in terrestrial ecosystem projects sponsored by the International Geosphere Biosphere Program (IGBP) through participation in the International Global Atmospheric Chemistry Project (IGAC) and the Global Change in Terrestrial Ecosystems Project (GCTE). These projects enable a detailed understanding of ecosystem processes, such as carbon fluxes, at global, eco-regional and landscape scales.

Developing a conceptual model to guide eco-regionalism in Asia

Building on a solid capacity for eco-regional activities

Any eco-regional program must have heavy emphasis on conservation and management of the natural resource base. In 1992, IRRI allocated 27% of its core resources and 19% of its complementary resources to TAC activity category 1 (conservation and management of natural resources). The corresponding figures for 1994 are 28% and 32%, respectively.

IRRI, in collaboration with Wageningen Agricultural University (WAU), and the Research Institute for Agrobiology and Soil Fertility (AB-DLO), The Netherlands, leads a seven-country collaboration to apply systems analysis and simulation to rice production problems (SARP project). Teams of well trained NARS scientists, with their new skills in systems modelling, GIS and data bases, will be the basis for *ex ante* assessment and extrapolation of systemwide issues (Ten Berge *et al.* 1992). Coordination of the four research themes (agro-ecology, potential production, soils & water, pests) are now undertaken by NARS scientists and clearly demonstrates the maturity in the region for such work (Kropff *et al.* 1994).

In the 1994–1998 Medium Term Plan, IRRI developed three MegaProjects to address strategic research issues with an eco-regional dimension: reversing trends of declining productivity in intensive irrigated rice, improving rice-wheat cropping systems, and exploiting biodiversity for sustainable pest management. All these projects actively involve NARS partners and have been the subject of much previous discussion (IRRI 1994a).

Due to the diversity of rainfed farming systems, IRRI considers that integrated systems approaches are necessary. IRRI's upland rice programme emphasizes the rice-based component of upland farming systems within a holistic upland agro-ecology.

IRRI is the convenor of a project conducted by the National Research Systems to develop methodologies for setting research priorities, with a focus on the use of biotechnology in rice. The large country program effort in characterizing constraints at the farm level in each of the rice ecosystem allows for identification of research needs to set priorities at the ecosystem, and eco-regional level.

Conceptual system model: toposequence, and strong link between ecology and social economics

Rice cultivation is largely dictated by the availability of water. The dominance of rice in the eco-region and its relationship to other crops and enterprises may be schematically represented in a toposequence (figure 4). An ecological perspective of the eco-regional approach is further proposed in this paper to guide future research on the biophysical aspects of the eco-region. In this toposequence, the major rice ecosystems may be seen to occupy specific niches along a landscape strongly influenced by hydrology. The various rice ecosystems exist with other agricultural and non-agricultural plant and animal species to each constitute a biome (Aber and Memillo 1991), and collectively make up the terrestrial (and semi-aquatic) ecosystems in the region. In each biome, such as the upland rice ecosystem (figure 4), system components and diversity are characterized by dominant man-managed crops or natural communities, depending on the particular AEZ. In large parts of AEZ 2, rice is only a minor component of upland systems, with forest species and other cereals occupying majority niches (figures 2 and 3). A third dimension may be added to figure 4 to denote the spatial element implied by AEZs. Added to the schematic of figure 4 are ecological concepts concerning ecosystem function, such as the processes involved in carbon, water and nutrient cycling, all of which underpin sustainability of the natural resource base. A further operational concept is that of ecological gradients and their influence on climax communities.

In agricultural systems, it is particularly necessary to ask what optimal mix of plant and animal species can be deployed in a particular biome so that there is renewal of the resource base. It is also necessary to determine the extent of human intervention. The irrigated rice ecosystem (dominant in AEZs 2, 3, 7) is gradually loosing land to accommodate urbanization and industrialization, and approaching the ceiling of rice yield. It is characterized by intensive use of agro-chemicals with associated potential effects on the environment and human health. Results from long-term experiments on continuous, irrigated rice cropping have shown that sustaining high yield and total factor productivity in these systems is a major concern, and a threat to income of irrigated rice farmers.

A major portion of irrigated rice is in East Asia where rapid economic progress, urbanization and industrialization have already absorbed surplus labor from rural areas and put an upward pressure on the wage rate. These economic forces are gradually eroding the comparative advantage of rice as an economic activity even at high levels of yield and the government has to protect farmers incentives by providing subsidies. Decline in population growth and changes in food habits associated with high income levels have reduced the demand for rice. But the demand for better quality rice, which benefited little from research, has been growing. In South and Southeast Asia, however,

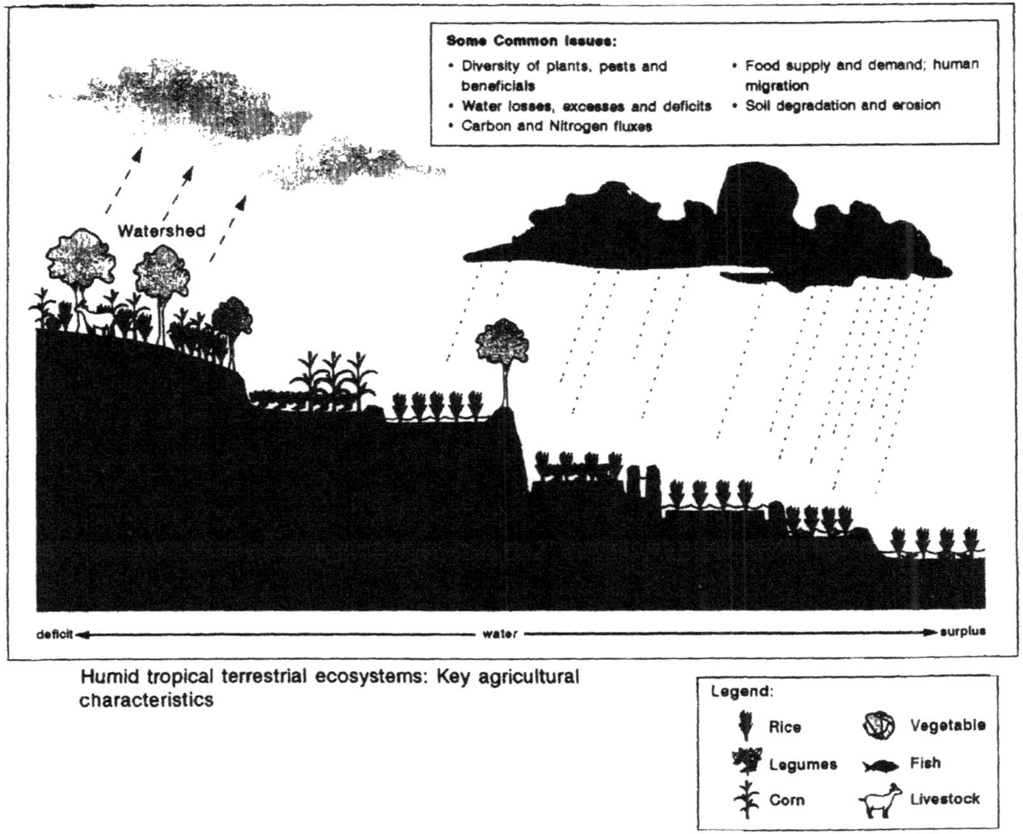

Fig. 4. Schematic showing an idealized, typical toposequence for the humid tropics of Asia, with the major rice biomes.

the irrigated rice ecosystem still has large potential of increasing production, which needs to be exploited for keeping food prices low, for improving the well-being of the urban poor. These areas need strategic research support for shifting the yield frontier and increasing labor productivity.

The unfavorable environments (AEZs 2 and 3) are regions where CGIAR Centers' technologies have yet to make an impressive impact and area expansion has been an important source of growth of food crop production. Progress in non-agricultural activities has been inadequate to generate employment for the new entrants to the labor force, creating more pressure on agriculture to absorb the surplus labor, constraining expansion of labor-saving technology and putting downward pressure on the real wage rate (Zeigler *et al.* 1994).

This interplay between ecology and socio-economics is explicitly recognized in the systems analysis proposed in this paper.

Basic elements of the conceptual R & D Model

The eco-regional approach is a new way of thinking about and doing research. It is important that a thorough and systematic effort be conducted to apply this approach to the Asian humid tropics. Therefore, in this proposal for a system wide initiative for the humid tropics, we have emphasized the importance of doing an exhaustive analysis of the eco-region, to involve all the major stakeholders, before embarking on the research itself.

The general methodology is to conduct a systems analysis of the eco-region (number of agro-ecological zones geographically bound). Objectives include to identify research issues that are eco-regional in scope; to prioritize these issues for further R & D into sustainable agricultural strategy and technology; to identify and develop the mechanisms for a collaborative program on the priorities, to involve national and international research organizations and NGOs; to develop a methodology and system for synthesizing eco-region research into application tools for specific geographic domains within the eco-region; and to develop and implement a system for monitoring improvements in the eco-region according to predetermined sustainability criteria.

The systems analysis will be done by a team made up of representative experts from IARCs, NARS, NGOs and advanced institutions, and will utilize up-to-date data collected in the project on physical (including weather), biological, chemical,and socio-economic attributes of the different ecosystems in the eco-region. Simulation modelling with the relevant crop models, and associated techniques such as GIS will be used to develop realistic geographic domains within AEZs with land use choices. We recognize that prioritization of issues into an Eco-regional Action Plan (EAP) will have to recognize both the supply (e.g. research capacity and interests) and demand (needs, problems) aspects of research, and in this project will use both subjective and objective criteria, in a participatory manner, to set priorities and develop the EAP. In anticipation of EAP implementation, existing collaborative research sites of the relevant IARCs will be strengthened, e.g. the rainfed upland and lowland consortium sites of IRRI, and key sites of ICRISAT and CIMMYT, all of which are in reality, research sites belonging to national institutions.

Because of the magnitude and complexity of this project, we propose that a fixed-term position for an Eco-regional Systems Coordinator be funded from the project. The position description will be jointly developed by a search committee (with representatives from the IARCs, NARS and NGOs) and the person is to be based at IRRI to enable linkage with the active systems analysis group.

Conceptual process model: R & D steps

The goal of the R & D model is to develop an operational model for eco-regional research that addresses the link between agriculture and other sectors and stresses the role of all the major stakeholders in a geographic area of Asia covered by the warm humid/subhumid tropics and subtropics.

This operational model – the Eco-regional R & D model – emphasizes strong feedback between the different objectives (represented by "steps" in figure 5) and is based on a systems approach to agricultural development. It uses the convention initially discussed by Dent and Anderson (1971) in applying systems analysis for agricultural management. The term 'model' is used here to mean the representation of a process, and not a computer simulation or mathematical entity. Four steps are envisaged (figure 5). Steps (1) and (2) are anticipated to take 2 years (IRRI 1994b). Parts of steps (3) and (4) will proceed concurrently with steps (1) and (2) and will require the strong support of the donor community and NARS.

Implementation of the first phases of the R & D model

The first step in the R & D model is *ex ante* analysis of eco-regional issues and knowledge gaps, concurrent with the strengthening of collaborative mechanisms for eco-regional activities. It is recognized that AEZs are often contiguous in Asia and that ecosystems may also cut across the same or several AEZS; for example, the rice-wheat system is found spread across AEZs 2, 4, 5, and 6. We, therefore, have proposed using a matrix of AEZs versus ecosystems (cereal-legume; cereal-vegetables; cereal-cereal) to define the potential interaction areas that lend themselves to improvement. These overlap areas consist of the boxes in figure 5 and represent interactions, such as in AEZ 2 between irrigated rice and vegetable. The following activities are proposed.

Conceptualization of eco-regional issues
In this aspect of the approach, system understanding and problem specification will be done. Both 'soft' (conceptual, intellectual) and 'hard' (empirical, data-based) systems analysis will be conducted, utilizing, respectively,

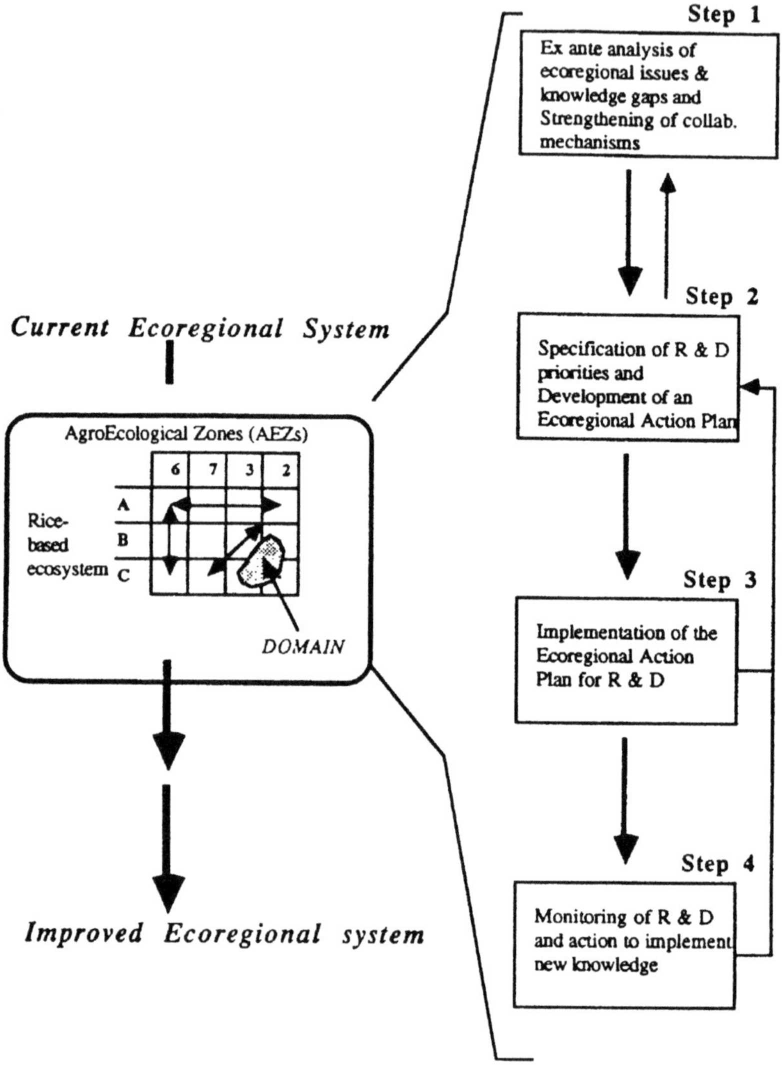

Fig. 5. Schematic showing the different steps in IRRI's proposed R & D Model for an eco-regional approach. (Source: IRRI 1994b)

brainstorming versus simulation analysis techniques. A workshop will be held at the outset, to conduct this conceptualization and to harmonize the data-gathering process and analytical procedures, including GIS procedures. The partners in this are anticipated to be international centers, NARS, and NGOs.

Knowledge/data collection and analysis
Included in the first step of the proposed Eco-regional R & D Model will be activities to build up the knowledge/data bases on items, such as land-use patterns, biophysical characteristics of the ecosystems, socioeconomic variables, policies, stakeholder properties, and R & D capacities in the major AEZs (figure 5). The subsequent analysis depends, in part, on the successful development of suitable data bases (natural resources, socio-economic data) and application of GIS tools. The analysis will take into consideration the ecological properties, socioeconomic parameters, natural resource base, and infrastructural capacities of each interaction point in the matrix. Teams of NARS scientists who have undergone training in systems analysis during the 10-year SARP project will be asked to participate in this analysis step, working closely with consortia and networks. IRRI will provide the leadership in this eco-regional initiative for this exercise and will actively involve sister IARCs in the region. The partners in this are anticipated to be international centers, NARs, NGOs, ICASA, selected advanced institutions, and FAO.

Knowledge synthesis
A mini workshop to synthesize data and knowledge for presentation to the Mega-workshop will be held. The partners in this are anticipated to be international centers, NARs, NGOs, ICASA, selected advanced institutions, FAO.

The first step in this R & D model is expected to result in (i) specification of cross-cutting issues that lend themselves to eco-regional research and resolution, (ii) knowledge/data bases and GIS maps that present alternative scenarios of land use under different constraints, and their potential for resolution by R & D, and (iii) identification of the key stakeholders for the future action plan. Stakeholders in the eco-region are further expected to be sensitized to the eco-regional concept.

Strengthening of eco-regional mechanisms
Strengthening of present collaborative research mechanisms for eco-regional approach – consortia, networks, NARS-NARS links, NARS-IARC links – will be initiated during this first step. In anticipation of a holistic eco-regional program in which current IRRI consortia and network sites will play an important role, the capacities at current sites for systems-level research will have to be strengthened, new sites will have to be added, and new partners (such as other IARCs) will have to be included. The partners in this are anticipated to be Consortia of IRRI, Key sites of ICRISAT, CIMMYT, and AVRDC, NARS institutions, and NGO sites.

The second step is specification of R & D priorities and development of an EAP to respond to eco-regional issues. An action plan that utilizes the comparative advantage and resources of each IARC stakeholder and NARS center in the eco-region will have to be developed. To the extent that is possible, implementation organizations such as FAO, NGOs, and national extension agencies, will be involved in on-farm evaluation and technology extension. The intervention objectives of the action plan for each zonal system will further have to be explicitly time- and resource-bound. The process of developing this action plan, and securing the necessary commitment from all major stakeholders in the eco-region, will be initiated during the Mega-workshop. Research priorities in the Asian eco-region will inevitably be influenced by the extensiveness of a topic, e.g. research on strategies to manipulate insect and pathogen habitats for sustainable pest management (see IRRI Mega Project) will affect several AEZs and ecosystems. However, political, social, and economic factors inevitably influence the potential usefulness of R & D efforts.

It is our intention to present the careful analyses from step 1 to an audience that has the ability to make an action plan work. A Mega-workshop of major stakeholders in the eco-region will use outputs from the analysis of AEZs and ecosystems. The workshop will include representatives of all the IARCs and NARS anticipated to collaborate on specific research issues, and will be used to secure political support at all levels for the R & D Eco-regional action plan. Implementation agencies (international, national, and local) will be invited to provide the relevance check. A main activity in the workshop will be to rationalize current activities and mechanisms of an eco-regional nature with the priorities determined in this more systematic and holistic initiative. The partners in this are anticipated to be international centers, NARS, NGOs, and FAO.

The EAP will be an important step forward for the humid tropics, especially if we can secure a majority buy-in to the concept of having a research strategy for the eco-region.

To monitor progress in the EAP, and to ensure relevance of the outputs, monitoring mechanisms will have to be put in place. One proposed mechanism is a Council for Collaborative Research, to consist of members who are currently leaders of agricultural institutions in countries of the eco-region. Working groups will be organized for each of the research themes in the EAP. It is also proposed that regular monitoring workshops be held to evaluate progress of research to meet the priorities identified in the EAP. We are not presumptuous at this point to specify the number of such working groups or workshops as these should be identified during the project.

Beyond 1997

Implementation and monitoring

Participatory implementation of the action plan will be needed to meet research priorities. A single milestone will not be feasible, but rather, a time series of outputs to be synthesized and applied is anticipated. Specific outputs are expected to include: knowledge generated by research teams to address priorities; a research-monitoring mechanism in place to evaluate work on prioritized eco-regional issues; and knowledge to support policy environments that favor natural resource management for sustainable agriculture. To obtain these outputs, proposed activities are:

(1) Coordinated research conducted on priority topics of eco-regional importance in the AEZ-ecosystem combinations selected by the Mega Workshop. A preliminary categorization of research themes, to help focus research, is proposed:
   - Genetic evaluation of crop and animal species for optimal ecosystem output
   - Diversifying habitats for sustainable pest management
   - Food supply and demand in the humid Asian eco-region
   - Nutrient and water cycling for sustaining agricultural and industrial growth
   - Optimal land use patterns to maximize potential agricultural productivity
   - Strategies for optimal crop-livestock combinations that enhance natural resources.

   These and other research themes could further be organized to directly address a specific cause of unsustainability or problem associated with the natural resource base (see section 'Cross-cutting issues'). Each research theme will involve cooperative research between different mixtures of IARCS, NARS and NGO partners, who will collectively form a working group. The partners for this are anticipated to be IRRI, ICRISAT, CIMMYT, AVRDC, ICLARM, ICRAF, IFPRI, NARS, NGOs, and FAO.
(2) Synthesis of results and pilot testing in cooperation with implementation agencies. The partners for this are anticipated to be IRRI, ICRISAT, CIMMYT, AVRDC, ICLARM, ICRAF, IFPRI, NARS, NGOs, FAO, ICASA, Advanced institutions, and National extension services.
(3) We propose preparing the details of activities for step 3 during the Mega-workshop, in consultation with all potential partners in each research theme.

IRRI recognizes that research and research mechanisms alone are not sufficient to engender change in an eco-region. Common policy frameworks, based on sound scientific knowledge for resource management, are needed to support change. As part of a functional eco-regional model, we anticipate cooperation with IFPRI and with universities in the region to identify policy changes needed and the support information of these changes, to meet certain intervention objectives. An example is the case of reducing insecticide use through IPM in all lowland rice-vegetable systems in AEZs 2, 3, 6 and 7. Such an extensive eco-regional issue requires that supportive policy be in place to implement a substantive information base that has been generated by research.

The magnitude and complexity of eco-regional issues suggest that longer time frames may be needed for their resolution. A monitoring and review mechanism will ensure continuing relevance of the action plan and will assess progress in relation to an eco-regional set of priorities. Research done under the umbrella of this eco-regional initiative will be critically reviewed, and changes in the eco-region will be reflected in changed activities. In step (4) of the Eco-regional R & D Model, anticipated outputs are a mechanism to ensure utility of eco-regional research outputs, and ultimately, improved systems of technology adopted in a conducive policy environment, for resolving specific eco-regional issues. The activities to achieve this include:

(i) Development of a Council for Collaborative Research in Asia. Opinions and inputs of Asian leaders will be canvassed. One of the important objectives of the Council, which will meet annually, would be to endorse concepts and outputs; and
(ii) Yearly working group meetings to evaluate technologies, strategies and their implementation for eco-regional improvement.

Anticipated impact

Diversified food sources are needed to feed the projected 10 Billion population by 2050, using strategies and technologies that are sustainable and environmentally-friendly. The proposed initiative is expected to make a significant contribution towards this by developing a systematic eco-regional plan that would then allow selective buy-in by donors. Both intermediary, short-term, and final, mid to long-term benefits are expected in the following areas:

(1) Improvement in total productivity of selected ecosystems through improved technologies for natural resource management in specific geographic domains of the eco-region; development and deployment of sustainable combinations of crops that optimize water and labor resources, reduce erosion of watersheds, and maximize total utilizable biomass from agricultural ecosystems; and reduction in pesticide use and concurrent increase in use of biodiversity conservation and cultural techniques for integrated pest management.
(2) More focussed and rationalized research on the foundations of sustainable production systems, through increased knowledge on nutrient cycling in the eco-region and the contributions of natural and synthetic sources of N and P; and increased knowledge of pest (insect, pathogens, weeds, rodents, molluscs) ecology and epidemiology with respect to

habitat diversity and exploitation of these for sustainable pest management.

(3) Strengthened cooperation between and within national and international partnerships for research on eco-regional issues through more efficient systems of collaborative research between IARCs, NARS, and NGOs; and enhancement of capacities at consortium and network sites for applying systems approaches.

In conclusion, an eco-regional initiative to develop sustainable agricultural systems based on sound policy and technology for managing the natural resource base is long overdue in the humid Asian tropics. We believe and endorse the CGIAR's efforts in fostering this scientific endeavor and IRRI will play its role to convene such an initiative in a region that grows over 90% of the world's rice, where rice forms over 70% of the food requirements, and where over 60% of the world's population resides.

Acronyms*

AEZ	Agro-Ecological Zone
CGIAR	Consultative Group on International Agricultural Research
EAP	Eco-regional Action Plan
FAO	Food and Agriculture Organization of the United Nations
GIS	Geographical Information System
IARC	International Agricultural Research Center
ICASA	International Consortium for Agricultural Systems Applications
NARC	National Agricultural Research Center
NARS	National Agricultural Research System
NGO	Non-Governmental Organization
SA	Sustainable Agriculture
TAC	Technical Advisory Committee

References

Aber J D, Melillo J M (1991) Terrestrial ecosystems. Saunders College Publishing, Philadelphia, USA.

BIFAD (Board for International Food and Agricultural Development) (1988) Environment and natural resources: strategies for sustainable agriculture. Occasional Paper no. 12. US AID, Washington DC, USA.

Dent J B, Anderson J R (1971) Systems analysis in agricultural management. John Wiley & Sons, New York, USA.

Fukuoka M (1983) The natural way of Farming. Principles and practices. Japan Publication Inc., Tokyo, Japan.

Gips T (1987) Breaking the pesticide Habit. International Organization of Consumer Unions, Penang, Malaysia.

* The acronyms of the CG-centers and IARCs are given in the acronyms list in this book.

Gryseels G, De Wit C T, McCalla A, Monyo J, Kassam A, Craswell E, Collinson M (1992) Setting agricultural research priorities for the CGIAR. Agricultural Systems 40:59–104.
Harwood R R (1987) Low-input technologies for sustainable agricultural systems. In Ruttan V W, Pray C E (Eds.) Policy for Agricultural Research. Westview Press, Boulder, USA.
Hossain M, Laborte A (1994) Asian rice economy: recent progress and emerging trends. Taipei: Food and Fertilizer Technology Center.
IRRI (1993) Rice almanac. International Rice Research Institute. Los Baños, Philippines.
IRRI (1994a) IRRI medium term plan 1994–98. International Rice Research Institute. Los Baños, Philippines.
IRRI (1994b) An eco-regional approach to research and development in the humid/sub-humid tropics and subtropics of Asia. International Rice Research Institute. Los Baños, Philippines.
Kropff M J, Penning de Vries F W T, Teng P S (1994) Capacity building and human resource development for applying systems analysis in rice research. Pages 323–341 in Goldsworthy P, Penning de Vries F W T, (Eds.) Opportunities, use and transfer of systems research methods in agriculture to developing countries. Kluwer Academic Publishers, Dordrecht, The Netherlands.
Naisbett J (1994) Global paradox. William Morrow & Co. Inc., New York, USA.
Pelegrina W R, Marges H E, Calinga Ma T E (1992) Sustainable agriculture as practised by farmers in the Philippines (case studies). SIBAT, Philippines.
Stomph T J, Fresco L O, Van Keulen H (1994) Land use systems evaluation: concepts and methodology. Agricultural Systems 44:243–255.
TAC (1992) Review of CGIAR priorities and strategies. Technical Advisory Committee of the Consultative Group on International Agricultural Research (TAC/CGIAR).
Ten Berge H F M (1992) Building capacity for systems research at national agricultural research centres: SARP's experience. Pages 515–538 in Penning de Vries F W T, Teng P S, Metselaar K (Eds.) Systems approaches for agricultural development. Kluwer Academic Publishers, Dordrecht, The Netherlands.
Teng P S (1994) Integrated pest management in rice. Experimental Agriculture 30:115–137.
Zeigler R S, Hossain M, Teng P S (1994) Sustainable agricultural development of Asian Tropics and Subtropics: emerging trends and potential. In Proceedings of IFPRI Eco-regional Workshop, November 7–11, 1994, Airlie House, USA (in press).

SECTION F

Case studies of eco-regional approaches at the farm/enterprise level

Designing optimal crop management strategies

P.K. THORNTON[1], J.W. HANSEN[2], E.B. KNAPP[3] and J.W. JONES[2]

[1] *International Fertilizer Development Center (IFDC), P.O. Box 2040, Muscle Shoals, Alabama 35662, USA;* [2] *Department of Agricultural Engineering, University of Florida, Gainesville, Florida 32611, USA;* [3] *Hillsides Program, Centro Internacional de Agricultura Tropical (CIAT), Apartado Aereo 6713, Cali, Colombia*

Key words: crop management, enterprise, farm, maize, Malawi, model, coffee, Columbia, season-specific management, resource balance, expected utility

Abstract. The identification of optimal crop management strategies at farm and enterprise levels presents special difficulties – in particular, the nature of the farm itself, operating in an environment defined by biophysical, socioeconomic, and politico-cultural variables, the often intractable effects of human agency, and, in an eco-regional context, the problems of defining appropriate spatial and temporal scales for analysis; these all combine to form a perplexing problem domain. We illustrate the potential use of simulation models in addressing some of these problems with respect to two case studies: season-specific enterprise management in the mid-altitude maize ecology of central Malawi and an analysis of management options facing a hillside smallholder induced to move out of coffee production in the lower Andes of central Colombia. The first case study shows that simple analyses can provide season- and region-specific management recommendations on input use that could improve productivity and stabilise economic returns. The second highlights the importance of price and weather risk in analysing management options and their impact on farm viability.

Introduction

The design of optimal crop management strategies in an eco-regional context for sustainable land use and food production is a formidable task. '*Designing*' connotes some well-controlled activity in problem solving, the solution being a functional and pleasing whole. Such a connotation is scarcely appropriate for agricultural technology. How should agricultural technology be designed? Even if a design is functional, how is it to be adopted? We are far from being able to predict the adoption of technology, however desirable this may be; too often we do not really know what farmers do and why they do it (Carter *et al.* 1994). Researchers' and farmers' mindsets rarely meet, it seems.

The connotation of '*optimal*' is equally problematic in agriculture. Human actions never take place in a vacuum, and what is optimal for one person may be quite deleterious for another. In any case, what is it that we should be optimizing? Such elementary considerations underline the need for an objective function that incorporates whatever is deemed important in any particular situation – some combination of farmer, social, and environmental goals, for example. We can say with certainty that such objective functions

J. Bouma et al. (eds.), Eco-regional approaches for sustainable land use and food production, 333–351.

have multiple attributes, may involve many players, and in the real world are always tempered by political expediency; it is much more difficult to delineate what they should comprise.

The '*crop*' connotation highlights an important bias. If we accept that models are necessary tools, if only because there are few other ways in which many long-term issues can be addressed, then we must admit that good management-orientated models exist only for some crops. Management-orientated animal models (and their feedbacks to crop and soil) are, in general, not so well developed. The implications for farm-level analyses are obvious – one may well ask, if the important biophysical process at the farm level cannot be satisfactorily simulated, how can we identify what is appropriate and viable?

Again, '*management*' connotes a well-planned, well-controlled activity whereby objectives are attained through intervention in the system. In agriculture in particular, control is often limited at best, and external factors may conspire to produce system behaviour that is hard to distinguish from chaos. In any case, management is dynamic and adaptive – this must be true *a priori* of farming systems that have survived and nurtured populations over long periods of time. Long-term viability of farming systems is not a desirable feature so much as a necessity, and the history of agriculture is largely the history of man's ingenuity in adapting farming systems to meet his needs in response to change. This has implications on the issue of time horizon for decisionmaking, which is touched on below.

There is then the notion of '*strategies*' as a set of predetermined actions assessed under conditions of imperfect and incomplete knowledge. Uncertainty and risk are inherent in agriculture, and at the farm and enterprise levels their importance is enormous, whatever the source – weather, markets, pests, or prices, for example? Indeed, almost everything that farmers, policy makers, and researchers do can be interpreted to some degree as a risk avoidance or risk amelioration measure. In identifying interventions that are supposed to address risk, we are obliged to make measurements of levels, trends, and variability. Without such measurements, it is difficult to say anything meaningful about system performance in response to a particular intervention or strategy.

Faced with such problems, how to proceed? We have to deal with a complicated biophysical environment, made more complex by the interactions at the economic and social level and by human interventions; we have to understand the exact purpose for which the agricultural system is to be operated; and we have to quantify changes in system performance? Models are clearly needed, but so is a realistic framework for analysis. In this paper we briefly examine some of the features that distinguish the farm and enterprise levels in the hierarchy of agricultural systems. The design of crop management strategies is considered with respect to two case studies: at the enterprise level, provision of season-specific information to help improve maize production in the mid-altitude maize ecosystem that covers much of southern Africa and provides the staple food for millions of people; and at the farm level, provision of information to help smallholder farmers in the lower Andes of Colombia increase household incomes and food security in the face of substantial variability in their economic and natural environments.

The enterprise and farm levels

An enterprise may be viewed as an activity that makes use of some of a farm's resources to produce an end-product of economic value. Two enterprises may be distinguished one from another not only by virtue of different end-products but also by differences in resource use to achieve what may be the same end-product. A farm is a household together with a collection of enterprises. Enterprise and farm are thus two levels in the hierarchy of an agricultural system. A traditional view of management is that decisions concerning the most appropriate enterprise organisation must precede those concerned with combining the enterprises into an overall farm (Barnard and Nix 1973). In practice, decisions are likely to be iterative between farm and enterprise level, because resource conflicts at the farm level cannot be resolved by a consideration of enterprises in isolation.

Agricultural systems have parallel ecological, social, and economic subsystems, however. The farm level is perhaps unique in that the boundaries of these subsystems are comparatively well-defined and consistent. It is more difficult to identify and delineate the next highest level in the hierarchy. A farm can be considered part of an ecological system (a watershed), a social system (a community), or an economic system (a regional commodity market or local industry) – how the farm is viewed depends on the purpose of the analyst. Both farm and enterprise, as entities that contribute to meeting the objectives of the farmer, comprise an ecological subsystem consisting of the farm landscape or fields and plots, a social subsystem consisting of the farm household, and an economic subsystem consisting of resources. Agriculture links people with resources such as land, forming managed ecosystems. Land defines the physical boundaries of the farm and, often, of each enterprise. Because the enterprise and the farm occupy different levels in the hierarchy of agricultural systems, there are spatial and temporal issues that impinge on their analysis. Broadly speaking, the data and analysis costs of working at higher hierarchical levels require a reduction in detail, so that increasing the time constants of ecological processes (i.e. the time step) usually results in a reduction of detail. The implications of ecological theory on increasing time constants with increasing system level or spatial scale do not always extend to the human subsystems that differentiate agricultural systems from other ecological systems, but often they do so.

The enterprise and farm levels differ in terms of the cost and complexity of the analysis and the type of information that is produced as a result. Enterprise-level analyses are often comparatively simple and need not be costly in terms of time and resources. On the other hand, the information provided may be of limited value because the linkages between the ecological, social and economic subsystems (at the enterprise level itself and between the enterprise and farm levels) are rarely considered. In contrast, farm-level analyses that deal with these linkages may be costly and complex, if indeed they can be carried out at all (in some respects we do not yet know precisely how to proceed with such analyses). Some of these considerations are discussed below in relation to two case studies.

The enterprise level – Maize in southern Africa

Problem and context

In much of the smallholder sector of southern Africa, yields of maize have stagnated at levels of about 1 t ha^{-1} (Blackie and Jones 1993). Food security concerns pervade the region, population growth rates are still in excess of 3% per year, there is little new land that can be brought into production, and the region suffers periodic drought that can devastate crop production. The impacts of population growth, food insecurity and inequity on human welfare are increasingly wide-ranging. In Malawi, for example, if population growth continues at its current rate, by the year 2050, achieving or maintaining self-sufficiency will require annual production of 4 tonnes of maize from each hectare of land currently used (Blackie 1994). Such productivity gains will require integrated approaches to soil fertility that make use of inorganic nutrients and locally available nutrients, with strategic inputs from biological nitrogen fixation and agroforestry wherever possible (van Reuler and Prins 1993). Breeding advances are unlikely to make much difference: there is no breeding-based green revolution waiting to happen in sub-Saharan Africa (Rohrbach 1994). Rather, the key lies in improved, adaptive crop management. This must be set in a broader framework still (World Bank 1989): an enabling environment of sound policies and efficient infrastructure and services to foster productive activities and private initiatives, and a much enhanced capacity, from the village to the highest echelons of government, to cope with change.

In southern Africa as a whole, some 90% of the maize area is grown by cultivators using either oxen or hand-hoe (Low and Waddington 1989). In Malawi, for example, plot sizes are generally small (less than 1 ha). Planting is often late, sometimes many weeks after the onset of the rains. Although many smallholders use hybrid seed, average yields per hectare are no better than they were 10 years ago, despite the fact that smallholder fertiliser purchases

Table 1. Characteristics of Five Sites in Central Malawi.

Site	Latitude 'S	Longitude 'E	Elevation m	Annual Rainfall Mean, mm	Annual Rainfall CV, %	Yearly Mean Temp °C	Years of Sample	n
Chitala Experiment Station	13 41	34 15	606	918	24	23.2	1964–1988	24
Chitedze Research Station	13 58	33 38	1149	897	19	20.1	1957–1992	35
Dedza Meteorological Station	14 19	34 16	1632	930	15	17.7	1958–1988	30
Kasungu Township	13 02	33 27	1036	812	22	21.2	1956–1989	29
Salima Township	13 45	34 35	512	1243	27	24.2	1957–1990	32

increased by 15% per year between 1982 and 1991 (Conroy 1993). Smallholders generally do not have the capital or labour resources to make optimum use of inputs, nor do they have adequate information on their use (Blackie 1994).

Approach and results

In the better environments of the mid-altitude zone, yields of 4 t ha^{-1} are quite achievable under smallholders' conditions, with appropriate hybrid seed/fertiliser technology and dependable supply of these inputs (Blackie 1994). Management information then becomes a critical resource. One response of research and extension services would be the provision of site- or region-specific management recommendations to replace current blanket recommendations of many years' standing. The temporal aspect could be addressed also – an input package that is appropriate in one year, even if the farmer can afford it, may not be appropriate in the next. With careful use, validated crop models offer real potential in the efficient derivation and provision of management recommendations suitable for a location and a season type. To illustrate the approach, we carried out an analysis of season type for a number of locations in central Malawi, in similar fashion to work carried out by Stewart (1991) and McCown *et al.* (1991) in Kenya and Sivakumar (1988, 1990) in the Sahel.

For five sites (table 1), we classified historical rainfall records by year according to a simple criterion of season type, after the method of Jones (1987). This involves calculating the ratio E_a/E_p (actual to potential evapotranspiration) using a simple water balance model and then calculating proxies for growing season characteristics. The 'start' of the growing season is defined as the day of year of the fifth consecutive day with E_a/E_p greater than 0.8. The 'end' of the growing season is defined as the day of year of the eighth successive day with E_a/E_p less than 0.5. The day number of the start of the growing season (values greater than 365 imply that the season did not start until the subsequent calendar year) was plotted against the number of growing days (figure 1). For each site, we arbitrarily divided the distribution of start of the growing season into three equi-probable ranges to give three season types: early, normal, and late rains, these arbitrary boundaries being superimposed as dashed lines in figure 1 (each season type has a probability of 1/3 of occurring in any year at any site). The relationship between start of the growing period and the number of growing days is variable but strong, particularly for Kasungu and Chitala. There is thus useful information to be found in just the date of the start of the growing period, essentially a function of the date of the start of the rains. Earlier rains imply a longer growing season, and visual inspection of figure 1 suggests that for each day's delay in start of the growing season there is about a one-day decrease in the length of the growing season (the implication is that the start of the rains is more variable than the end of the growing season).

The crop model CERES-Maize (Ritchie *et al.* 1989) was used to simulate maize growth and yield for each of the 29 seasons for Kasungu, using the same treatment for each season with the exception of planting date, which was set to the start of each growing season as defined above. (CERES-Maize was validated for three sites in central Malawi with field data from three seasons between 1989 and 1991; results were reported in detail in Singh *et al.* (1993) and Thornton *et al.* (1995). The variety grown was the hybrid MH–16 at 37000 plants ha^{-1}, and 30 kg N ha^{-1} as urea was applied. Simulated maize yields were then classified according to the three categories of season type. Yields for Kasungu are plotted in figure 1 as box plots, showing the 0th, 25th, 50th, 75th, and 100th percentile of the yield distribution for each season type and for all years in the sample. It was still possible to obtain a high yield even with late planting, but it was much less probable than if the crop were sown earlier (figure 1). The distribution of maize yields, all other things being equal, changes markedly depending on the start of the rains; the distribution that pertains in any particular season can be derived once the start of the season is known. As might be expected, the fertiliser schedule giving the greatest monetary returns to the maize-growing enterprise varies markedly depending on season type. Simulations showed that, for the late-starting season in Kasungu, the optimal rate of fertiliser application was about 30 kg ha^{-1} N as urea, using 1993 costs and prices, whereas in the early-starting season, the optimal rate was about 90 kg ha^{-1}.

Implications and limitations

In some environments, there is much information to be gleaned from simple analyses of historical weather records, and this can be complemented with information from other sources (such as field trials and a crop simulation model) to suggest site- and season-specific management recommendations that can subsequently be tested in experimental plots and farmers' fields. This type of analysis could also be used to help select maize cultivars for a region based on the length of the growing season. Because season length is

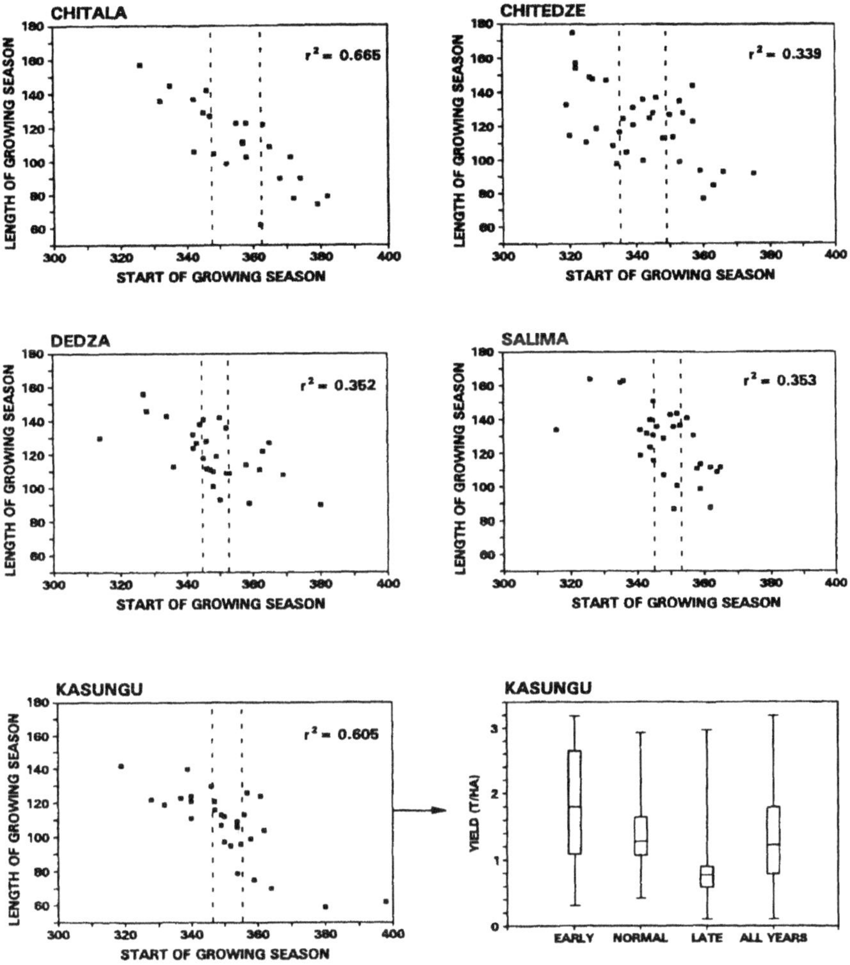

Fig. 1. Relationship between start of growing season and length of THE growing season for five locations in central Malawi, and simulated maize yield distributions by season type for one of these Locations, showing the 0th, 25th, 50th, 75th and 100th percentiles of each distribution.

influenced by the starting date of the rains, farmers could be advised, in a late-starting season, to grow short-duration maize cultivars or, alternatively, to grow other more drought-tolerant crops such as cassava or sorghum. This approach could also be used for decisions that must be made repeatedly during the growing season. Irrigation schedules that optimise economic returns could be identified on a regular basis, using historical weather records for the season to date and simulated records thereafter. If reliable 10-day weather forecasts were available, these could be incorporated into the analysis to improve the accuracy of the predictions. More work is required to estimate the economic value of such information and, if worthwhile, to refine the methodology, but the potential would appear to be considerable.

There are a number of limitations to this analysis. For example, there are some factors that impinge on maize production with late planting to which the crop model is not sensitive – an increased incidence of maize streak virus, an organism that can cause substantial reductions in yield, to name but one. Other limitations relate to the analytical framework itself and are harder to overcome. First, assuming that suitable recommendations can be delivered to the farmer, is the farmer in a position to be able to follow them? Can the farmer really plant the crop at the start of the rains and apply the optimal dose of fertiliser in a timely manner, as was assumed with the model runs made above? Probably not – lack of labour and other resources may well impede his progress. Figure 1 shows that the window for a 'normal' start to the rains is short (14 days or less). It takes little delay for a normal start to become a late start, as far as the farmer is concerned. Second, such recommendations are the best-bet in the long run. In any one particular year, it may be that the recommendation is not optimal (despite the fact that in the long run the farmer would be better off acting according to the recommendations, when faced with the same decision season after season). Farmers may not act according to the rational, 'best-bet-in-the-long-run' scenario, especially if their family's welfare is at stake. Third, the maize growing activity is but one enterprise on the farm. The analytical framework says nothing concerning resource conflicts and how these are to be resolved. It says little about the objectives of the farmer or the impact of decisions on the environment, unless these impacts are internalised and represented by a direct cost. Such problems point up the need for analysis at the farm level.

The farm level – A hillside farm in the lower Andes of Colombia

Problem and context

In the past, coffee provided a living for farmers in the marginal coffee-growing area in the Cauca River watershed of the lower Andes, while providing continuous canopy and ground cover to protect the steep andic soils from erosion. Declining world coffee prices and the spread of a new seed-boring insect pest, broca (*Hypothenemus hampei*), however, have reduced the ability of coffee to provide for farmers' livelihood. The Colombian Coffee Federation in the early 1990s offered one-off monetary incentives for farmers in marginal areas to abandon coffee production, to slow the spread of broca and to boost market prices. Farmers who participated found themselves with some financial capital but few good alternative enterprises and a highly uncertain future. Annual crops are now replacing coffee in many of the hillside farms in Colombia.

Which factors most seriously threaten the sustainability of these farms? What options do farmers and others have to increase the economic and food security of the family farm?

Farmers in the Cauca region are exposed to high levels of risk. Rainfall variability is high, both seasonally and annually. Market prices of agricultural products are also highly variable; prices for tomato and fresh maize follow seasonal patterns, while prices for beans, cassava and coffee are characterised by variability over longer cycles. The Colombian Coffee Federation has programs to moderate volatile world prices, but there is still considerable variability in coffee price received by the farmer. The shift from perennial to annual crops represents a shortening of farmers' planning horizons. Ashby (1985) argued that low profit margins and high price risk have forced farmers in the region to give short-term returns priority over investing in the long-term benefits of soil conservation. The annual crops that are replacing coffee leave steep lands exposed to erosive rains. Resulting soil erosion may irreversibly reduce crop productivity over time. Furthermore, the Cauca River watershed is strategic as a source of water and hydroelectric power for the city of Cali and for irrigated sugar production in the Cauca Valley. There is growing concern that the shift from perennial crops will result in increasing siltation and chemical pollution and poorer regulation of flow rates in the Cauca river. The important issue of natural resource degradation is not dealt with below, but is receiving attention in the case study.

The case-study farm (2 47′ N, 76 31′ W, 1650 m elevation) is of average size for the Cauca watershed (5.2 ha, of which 2.3 ha is cultivated with annual crops) and supports a family of six. Since removing his coffee bushes in 1993, the farmer has grown maize for the fresh market and beans. Because of the lack of family labour he routinely hires day-labourers. The family lives in a five-room stucco home of above-average quality for the area. Access to the farm is by a well-maintained gravel road 2 km from a small town and 4 km from the Panamerican highway. The farmer's level of technical knowledge is high, but he lacks capital – he has little savings, and credit is scarce and expensive. He has access to production inputs such as seed, fertiliser, and pesticides. The farm has neither capital equipment (such as backpack sprayer, truck, or motorcycle) nor animal-drawn implements. Field operations, including land preparation, are contracted out. The farmer is relatively innovative and regards himself as a conservationist.

Approach and results

We examined potential impacts of alternative crop management scenarios on farmer livelihood based on stochastic simulation of alternative farm scenarios using a whole-farm model. The farm model, which is still under development, uses simulation models of crop and soil processes, stochastic time-series models to simulate variability in weather and in prices, and a discrete-event model to simulate farm operations and use of resources for agricultural production and household consumption. The farm model executes process-level models to simulate soil water and nitrogen dynamics and the growth of bean, maize, cassava (Tsuji *et al.* 1994) and tomato (Scholberg *et al.* 1993). The crop models obtain stochastic daily weather inputs from the weather generator WGEN (Richardson 1985). The timing of field operations and material resource balances is based on information provided by the crop models. Operations are matched to labour and equipment requirements and stored in a queue until all fields on the farm have been simulated. Fixed costs, debt servicing, and household consumption are charged on a monthly basis; variable costs are charged at the time of each operation. Fixed and variable costs are represented as links with other resources, and stochastic prices determine exchange rates between resources. The farm model simulates monthly produce prices received by the farmer based on autoregressive moving average models (Box and Jenkins 1970) with deterministic trend and seasonal components; monthly prices are linearly interpolated to daily prices.

The simulation study required detailed information concerning the physical environment, prices, current and alternative management practices, and household consumption requirements. Weather generator coefficients were derived from historical weather records for a number of sites and spatially interpolated for the farm. Soils were sampled at eight locations and three depths to derive the soil variables required by the crop models. Laboratory and field measurements of soil water-holding capacities and contents were carried out. Topography and land use were characterised by a land survey. Historical price data were obtained from wholesaler records for cassava, fresh maize, and tomato (CAVASA 1994), from records of retail prices for beans (Castillo 1990, 1993), and from published sources for chemical inputs, coffee, and consumer price indices (Banco de la Republica, various dates). A price multiplier that accounts for market discounts, quality adjustments, moisture contents, and recovery was determined for each commodity. Crop management practices and input levels were obtained from unpublished enterprise budgets, farmer interviews, and scientists at CIAT. Information on savings, credit availability, and household expenditures was obtained from an interview with the farmer.

We examined the performance of alternative cropping systems by simulating the first six scenarios listed in table 2. Twelve-year sequences of each scenario were simulated, replicated twenty times with respect to weather and costs and prices (the last three scenarios in table 2 were simulated for six years in order to examine the relative contributions of weather and prices to risk). The results of the 12-year simulations, in terms of the trends and variation in wealth (i.e., monetary reserves plus the value of any harvested produce not yet marketed), are shown in figure 2 as percentile plots. Monocultures of coffee and fresh maize showed relatively stable wealth both through time and between replicates, with variability lower for maize than for coffee. Price

Table 2. Description of the scenarios simulated using the farm model.

Scenario	Description
Maize[a]-bean-bean-cassava	Three-year rotation with typical management and 1/3 of the cultivated area in each phase of the rotation.
Intensive management	Maize-bean-bean-cassava with higher planting densities and fertiliser levels.
Maize-bean-tomato-cassava	Three-year rotation with tomato, bean and maize irrigated as needed.
Coffee monoculture	All cultivated area in coffee.
Maize monoculture	All cultivated area double-cropped with maize.
Cassava monoculture	All cultivated area in cassava, with two crops every 3 years.
Weather and price risk	Maize-bean-bean-cassava with higher initial wealth.
No weather risk	Maize-bean-bean-cassava with the same (arbitrary 'average') year of weather used for all years of all replicates.
No price risk	Maize-bean-bean-cassava with prices based only on deterministic trend and seasonal components.

[a] All maize is assumed to be grown for the fresh market.

volatility contributed to the higher variability of the cassava monoculture. Variability was high for the diversified annual crop rotations. More than half of the replicates of maize-bean-bean-cassava failed within six years; intensifying management of this scenario increased both survivability and variability. Irrigated maize-bean-tomato-cassava showed both the greatest median accumulation of wealth and the highest risk. Bean prices fluctuate widely, and this fluctuation may explain the high variability of the diversified annual crop scenarios compared with the monocultures. Bean monoculture was not simulated because nematodes and soil-borne diseases prevent continuous bean production.

We investigated the effects of price and weather variability on whole-farm risk by comparing the last three scenarios in table 2. Results are shown in figure 3 as cumulative distribution function plots of final wealth after six years. Eliminating the variability of weather had little effect on wealth, whereas eliminating price variability had a large effect. The contribution of price variability to risk in this particular system is thus much greater than the contribution of weather variability. Part of this effect may be due to the inability of WGEN to reproduce the full variability of historical weather records (Jones and Thornton 1993).

The importance of risk in farmers' decisions can be examined using the theoretical framework of expected utility maximisation (Anderson *et al.* 1977). We ranked scenarios on the basis of certainty equivalent of income in each year calculated at different levels of risk aversion. Certainty equivalents were

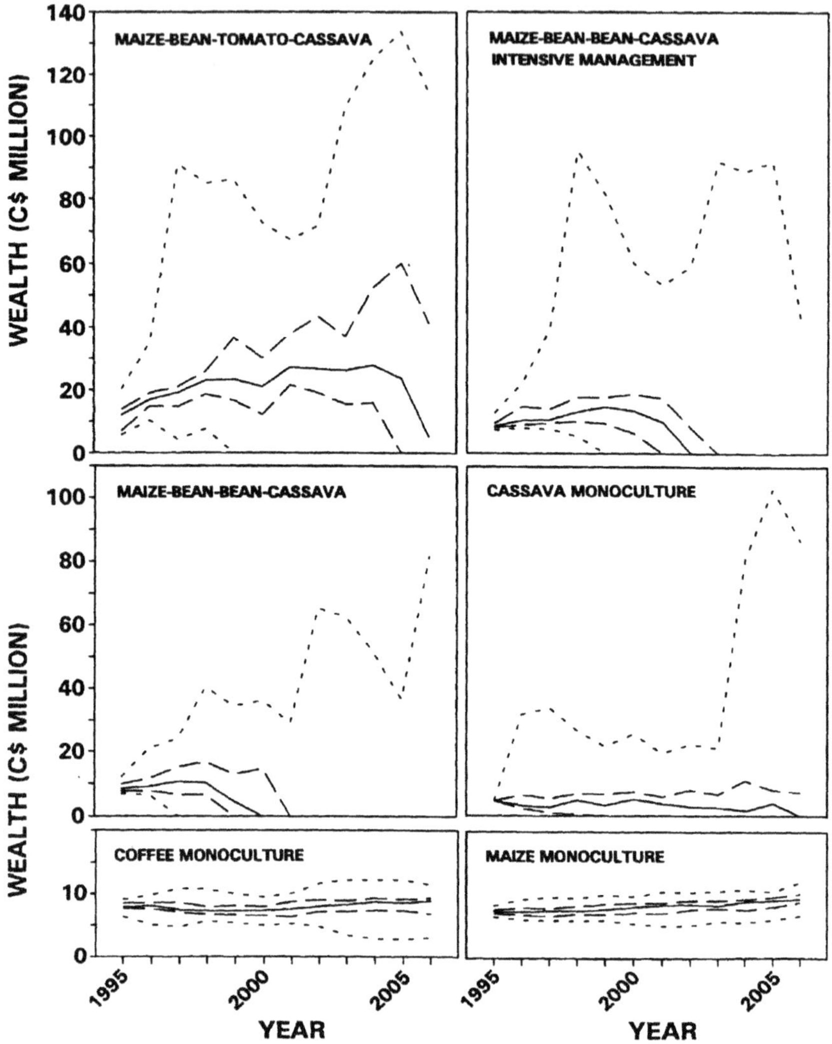

Fig. 2. Percentile plots of wealth (Colombian pesos × 10^6) by year for the first six scenarios listed in Table 2, showing movement of the 0th and 100th (- - - -), 25th and 75th (— — —) and 50th (——) percentiles of the simulated annual wealth distribution through time. The wealth scale for the coffee and maize monocultures is doubled for clarity.

discounted to net present value (NPV_{CE}) using the inflation-adjusted interest rate of 6.8% pertaining to 90-day certificates of deposit in Cali in April 1994. The form of utility function used to calculate the certainty equivalent of risky income is characterised by constant absolute risk aversion with respect to income. Maximising net present value is consistent with risk neutrality, while

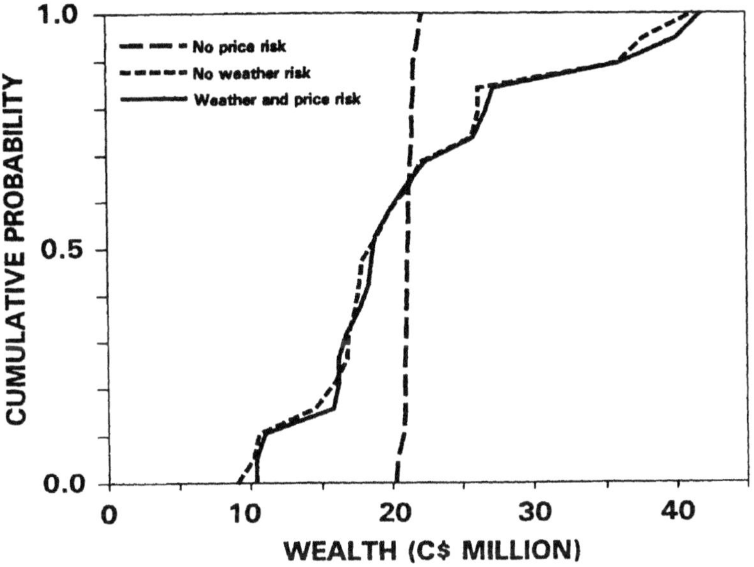

Fig. 3. Cumulative distribution functions of final wealth for the last three scenarios listed in table 2.

a safety-first rule such as maximising the probability of survival is consistent with a high level of risk aversion. To reduce the confounding effects of simulated farm failures, only the first six years of each scenario were considered. Results (table 3) show that the maize-bean-tomato-cassava scenario had the highest mean income and was ranked highest at low levels of risk aversion, whereas the maize monoculture was ranked highest at moderate to high levels of risk aversion. Preference for maize and coffee monocultures at high levels of risk aversion are consistent with the simulated survival rates of 100% for these scenarios.

Implications and limitations

Farmers are often forced to adapt to comparatively rapid changes in their physical and economic environments. By capturing the variability inherent in agriculture due to weather and prices, and by simulating long-term changes in soil and in other resources, simulation modelling offers a way to explore the potential impacts of management and external forces on the long-term viability of farming systems in a manner impossible to do experimentally. Such simulation is demanding in its information requirements and may be very sensitive to assumptions that are difficult to measure.

Although results are preliminary, the present study suggests that incorporating irrigated vegetables such as tomato into more traditional cropping systems increases household income and its variability. The results also sug-

Table 3. Preference ranking based on mean annual income, discounted certainty equivalent of income (NPV_{CE}), and survivability for the first six years of various simulated scenarios.

Scenario	Mean income	NPV_{CE}[a] $r = 0.00$	0.04	0.12	0.36	Survivability
Maize[b]-bean-bean-cassava	3	3	5	5	5	6
Maize-bean-tomato-cassava	1	1	1	6	6	4
Intensive management	2	2	2	2	4	5
Coffee monoculture	5	5	4	3	2	1[c]
Maize monoculture	4	4	3	1	1	1[c]
Cassava monoculture	6	6	6	4	3	3

[a]Net present value of the certainty equivalent of income for different values of the coefficient of risk aversion ($r \times 10^6$).
[b]All maize is assumed to be grown for the fresh market.
[c]No simulated failures were recorded.

gest that prices may have a much greater role than weather in determining risk in these production systems; policies to reduce price variability could thus improve the viability of farming systems in the study region.

The study helped to identify some of the weaknesses in the component crop simulation models, particularly related to soil dynamics. The models probably underpredict response to fertiliser management because they only simulate crop response to water and nitrogen, whereas production is believed to be limited more by phosphorus availability. Simulations of soil erosion and its impacts on crop productivity are also missing. Although the simulation results suggested that annual crop production can be profitable, they did not indicate whether productivity can be maintained on steep slopes in the presence of highly erosive rainfall after the removal of perennial coffee with its permanent groundcover. The farm model is still being developed, however, and the simulations outlined above will be repeated once sensitivity to phosphorus and soil erosion has been built into the process models.

Predicting farmers' decisions, such as technology adoption and borrowing and spending behaviour, is more difficult. Risk preferences are not easy to measure. People do not always make decisions consistent with the expected utility framework (Shoemaker 1982); this has been attributed to lack of ability to process optimal decisions (Day 1983) and biases in perceptions of outcome distributions (Musser and Musser 1984). Alternatively, the expected utility hypothesis may be an incomplete theory of decision making under risk; it is doubtful whether we really understand how risk is perceived by farmers. In addition, considerations other than risk and returns will often enter into farmers' decisions (Patrick and Blake 1981; Kliebenstein *et al.* 1980). As the above results show, the substitution of short-season annual crops for traditional perennial crops may be attractive in the short to medium term to individual hillside farmers. However, the consequences of such practices, if widely carried out within a watershed, may result in unacceptable social costs associated with the degradation of soil and water supplies.

Discussion and conclusions

In this paper we have considered how modelling tools might be used to provide decision makers with management information. Analysis at any one level in the hierarchy of agricultural systems is bound to have its limitations, if only because links exist to other levels that are not necessarily considered in the analytical framework. As both case studies have shown, these links may be of great importance. To conclude, we consider four questions.

What role does enterprise-level analysis have? Enterprise analysis is generally appropriate for investigating small or marginal changes in farming systems. For the farmer who will apply some fertiliser to the maize crop this year but is uncertain how much to apply, enterprise-level analysis will do very well. The limitations of enterprise analysis become important when more radical changes to the farming system are contemplated. For example, can resources be obtained for a particular enterprise? How will farm resources change over a long period of time? What variability is associated with a proposed diversification of enterprises, and how might this affect the farmer? Such questions cannot be answered at the enterprise level alone.

What role does farm-level analysis have? A major role of farm-level analysis is to make more explicit the consideration of the parallel ecological, social, and economic subsystems of agricultural systems. Enterprise resource constraints can be addressed and resolved; long-term variability can be assessed; and, in theory, impacts on farmer behaviour can be addressed. A fundamental weakness of current farm-level analysis is our inability to understand farmers' decision-making behaviour and the factors that affect it; this contrasts sharply with our ability to understand and predict many biophysical processes. Farm-level analyses should allow us to assess radical changes to farming systems. As for the enterprise level, however, some questions cannot be answered at this level alone. For example, what will happen to market prices if a number of farmers change production levels or enterprise mixes? What long-term effects might this have? How might off-farm effects of production practices such as pollution or resource depletion affect the sustainability of a community of farmers in a watershed?

How can enterprise- and farm-level analyses be extended to the regional level? Farm-level studies have only rarely been extended to the regional level, unlike enterprise-level studies, which have often been extended to the regional level, bypassing the farm level. The reasons for this, although not clear, presumably relate to complexity (which can be reduced either by moving up the hierarchy to reduce the level of detail or by moving down the hierarchy to avoid having to consider social, economic, and other inconvenient subsystems). In any case, extension to the regional level is difficult because the analyst has to come to terms with the issues of scale and spatial heterogeneity. Some ecological factors, such as soil and weather, tend to be spatially homogeneous (for any point chosen at random, the probability that a neighbouring point belongs to the same class is very high). For economic and social factors, this clustering is often much less pronounced. It is not clear how system characterisation can be done in a way that adequately treats the problems of spatial heterogeneity.

What is the status of our analytical tools? Five problems were listed in the introduction – how to design agricultural technology, what to optimise, limited biophysical models, even more limited behavioural models, and risk. In terms of the analytical tools currently available to deal with these problems, advances are clearly needed to bring methods of handling the socioeconomic environment more nearly into balance with methods of handling the biophysical environment. Four specific areas are listed where substantial collaborative progress is required if headway is to be made in addressing the enterprise and farm levels in an eco-regional context.

First, we need to develop models that ensure balance between data input and simulation output at multiple levels. It would make no sense to use (even if they could be obtained) tens of soil inputs at a scale of 1:1000000 if all that is required is an estimate of potential crop yield by soil mapping unit. It is equally clear that no one model (or even type of model) is capable of running the gamut of levels from enterprise to eco-region. Different types of models may be needed to ensure that the detail and scale of model input data match the detail and scale at which they are available or able to be manipulated. The relationship between model type and level of analysis in agricultural systems is largely unexplored territory.

Second, it is essential that we be able to carry out analyses of the same problem at different scales, because this is precisely the root of many conflicts (social costs versus individual costs, for example). Farm planning cannot be isolated from regional planning. The links between farm and regional levels are not well-developed. One critical area of work is to improve methodologies for classification and typology of systems and subsystems (ecological, economic, and social), while recognising that methods of classification depend on the purpose of the analysis (Dent 1994). The prospects of a useful Minimum Data Set for economic and social factors are not good because of spatial heterogeneity – if one is eventually defined, it is unlikely to be based on notions of Euclidean distance.

Third, a major deficiency in current computer-based decision support systems is that they are heavily biased towards simulating production for planning purposes. As noted above, the perspective of eco-regional research connotes multiple stakeholders with multiple and often competing objectives. Assuming that adequate models exist to deal with the range of temporal and spatial scales necessary, current methods for quantifying and dealing with multiple objectives and multiple stakeholders are inadequate in terms of behavioural assumptions and/or practicability. Such methodologies are likely to arise as a knock-on effect of the development of improved behavioural models, so the time lag involved may be substantial, unfortunately.

Fourth, we have little idea of the value of the information that is lacking and the likely cost of generating it. Data drive models; models can, with care, produce useful information. Estimates of the costs and benefits of more data, and the costs and benefits of generating information, are essential if resources are not to be wasted and if some sort of research prioritization is to be feasible. Such estimates necessitate simple but efficient assessment methodologies, and these have yet to be developed and applied.

Acronyms

CIAT	Centro Internacional de Agricultura Tropical (International Center for Tropical Agriculture)

Acknowledgements

We thank, without implicating them in any way, J. Barry Dent and Carlos Baanante for critical comments on an earlier draft of this paper. Errors and omissions remain our responsibility.

References

Anderson J R, Dillon J L, Hardaker J B (1977) Agricultural Decision Analysis. Iowa State University Press, Ames, Iowa, USA.

Ashby J A (1985) The social ecology of soil erosion in a Colombian farming system. Rural Sociology 50:377–396.

Banco de la Republica (various dates) Revista del Banco de la Republica, various issues. Bogotá, Colombia.

Barnard C S, Nix J S (1973) Farm Planning and Control. Cambridge University Press, Cambridge, UK.

Blackie M J (1994) Maize production for the 21st century: the African challenge. Proceedings, Fourth Eastern and Southern Africa Regional Maize Conference on Maize Research for Stress Environments, Harare, Zimbabwe, 28 March–1 April, 1994.

Blackie M J, Jones R B (1993) Agronomy and increased maize productivity in eastern and southern Africa. Biological Agriculture and Horticulture 9:147–160.

Box G E P, Jenkins G M (1970). Time series analysis: forecasting and control. Holden-Day, San Francisco, California, USA.

Carter S E, Bradley P N, Franzel S, Lynam J K (1994) Dealing with spatial variation in research for natural resource management. Proceedings of the workshop on Sustainable Mountain Agriculture, Nairobi, Kenya, December 1993, IDRC.

Castillo W C (1990, 1993) Series estadisticas de precios promedios corrientes de frijol a nivel de consumidor en Cali y Palmira Valle del Cauca. CIAT, Cali, Columbia (unpublished).

CAVASA (1994) Series Historicas de precios por kilogramo registradas en el mercado mayorista de CAVASA. Corporacion de Abastecimentos del Valle del Cauca S.A., Cali, Colombia.

Conroy A (1993) The economics of smallholder maize production in Malawi with reference to the market for hybrid seed and fertiliser. Unpublished PhD thesis, University of Manchester, Institute of Development Policy and Management, Manchester, UK.

Day R H (1983) Farm decisions, adaptive economics, and complex behavior in agriculture. In Baum K H, Schertz L P (Eds.) Modeling farm decisions for policy analysis. Westview Press, Boulder, Colorado, USA.

Dent J B (1994) Theory and practice in FSR/E: consideration of the role of modelling. Paper presented to the International Farming Systems Symposium, Montpellier, France, November, 1994.

Jones P G (1987) Current availability and deficiencies in data relevant to agro-ecological studies in the geographic area covered by the IARCs. Pages 69–83 in Bunting A H (Ed.) Agricultural Environments. CAB International, Wallingford, UK.

Jones P G, Thornton P K (1993) A rainfall generator for agricultural applications in the tropics. Agricultural and Forest Meteorology 63:1–19.

Kliebenstein J B, Barrett D A, Heffernan W D, Kirtley C L (1980) An analysis of farmers' perceptions of benefits received from farming. North Central Journal of Agricultural Economics 2:131–136.

Low A C, Waddington S R (1989) On-farm research on maize production technologies for smallholder farmers in southern Africa: current achievements and future prospects. Pages 491–511 in Gebrekidan B (Ed.) Maize improvement, production and protection in eastern and southern Africa. CIMMYT, Mexico City, Mexico.

McCown R L, Wafula B, Mohammed L, Ryan J G, Hargreaves J N G (1991). Assessing the value of a seasonal rainfall predictor to agronomic decisions: the case of response farming in Kenya. Pages 383–410 in Muchow R C, Bellamy J A (Eds.) Climatic risk in crop production: models and management for the semiarid tropics and subtropics. CAB International, Wallingford, Oxon, UK.

Musser W N, Musser L M (1984) Psychological perspectives on risk analysis. Pages 82–92 in Barry P J (Ed.) Risk management in agriculture. Iowa State University Press, Ames, Iowa, USA.

Patrick G F, Blake B F (1981) Measurement and modeling of farmers' goals. Southern Journal of Agricultural Economics 12:199–204.

Richardson C W (1985) Weather simulation for crop management models. Transactions of the ASAE 18(5):1602–1606.

Ritchie J T, Singh U, Godwin D C, Hunt L A (1989) A User's Guide to CERES Maize V2.10. IFDC, Muscle Shoals, Alabama, USA.

Rohrbach D (1994) Maize research in stressed environments in perspective: the broader challenge of improving farmer well-being in semiarid areas. Proceedings, Fourth Eastern and Southern Africa Regional Maize Conference on Maize Research for Stress Environments, Harare, Zimbabwe, 28 March–1 April, 1994.

Scholberg J M S, McNeal B L, Jones J W, Stanley C D (1993) Adaptation of a generic crop-growth model (CROPGRO) for field-grown tomatoes and peppers. Agronomy Abstracts.

Shoemaker P J H (1982) The expected utility model: its variants, purposes, evidence and limitations. Journal of Economic Literature 20:529–563.
Singh U, Thornton P K, Saka A R, Dent J B (1993) Maize modelling in Malawi: a tool for soil fertility research and development. Pages 253–273 in Penning de Vries F W T, Teng P S, Metselaar K (Eds.) Systems Approaches for Agricultural Development. Kluwer Academic Publishers, Dordrecht, The Netherlands.
Sivakumar M V K (1988) Predicting rainy season potential from the onset of rains in southern Sahelian and Sudanian climatic zones of West Africa. Agricultural and Forest Meteorology 42:295–305.
Sivakumar M V K (1990) Exploiting rainy season potential from the onset of rains in the Sahelian zone of West Africa. Agricultural and Forest Meteorology 51:321–332.
Stewart J I (1991) Principles and performance of response farming. Pages 361–382 in Muchow R C, Bellamy J A (Eds.) Climatic Risk in Crop Production: Models and Management for the Semiarid Tropics and Subtropics. CAB International, Wallingford, UK.
Thornton P K, Saka A R, Singh U, Kumwenda J D T, Brink J E, Dent J B (1995) Application of a maize crop simulation model in the central region of Malawi. Experimental Agriculture (in press).
Van Reuler H, Prins W H (Eds.) (1993) The role of plant nutrients for sustainable food crop production in sub-Saharan Africa. VKP, Leidschendam, The Netherlands.
Tsuji G Y, Uehara G, Balas S (Eds.) (1994) DSSAT Version 3. University of Hawaii, Honolulu, Hawaii, USA.
World Bank (1989) Sub-Saharan Africa: From crisis to sustainable growth. Washington DC, USA.

Sustainability and long-term dynamics of soil organic matter and nutrients under alternative management strategies

H. VAN KEULEN

DLO-Research Institute for Agrobiology and Soil Fertility, P.O. Box 14, 6700 AA Wageningen, and Section Animal Productions Systems, Department of Animal Husbandry, Wageningen Agricultural University, Marijkeweg 40, 6709 PG Wageningen, The Netherlands

Key words: sustainability, soil organic matter, simulation, long-term model

Abstract. Sustainable agricultural production systems should meet the requirements of the farm household in terms of food, income and leisure, without endangering the productive capacity of the natural resource base. Land use systems are characterized by the crops grown and their order, the production techniques applied and soil and crop management. To judge their degree of sustainability, the processes determining production and soil characteristics and their interactions, as influenced by environmental conditions and management practices should be understood quantitatively. An experimental approach to this problem is problematic: it requires long-term experimentation, which is laborious and time-consuming and takes too long to yield relevant results within the timeframe of the current priorities. Moreover, the erratic and unpredictable nature of environmental conditions makes it difficult to use time- and location-specific information for extrapolation and prediction. A modelling approach to the exploration of the dynamics of sustainability indicators seems promising for formulating criteria for different management strategies. In this paper a model is presented that harnasses existing knowledge into quantitative, descriptive relations. An example is given of calibration of this model on the basis of results of a long-term agronomic experiment. The performance of the model in terms of organic matter dynamics appears 'reasonable', and reproduces effects of soil and crop management in a recognizable form. The apparent discrepancy between organic nitrogen and organic carbon behaviour must be analysed in more detail, to establish its possible causes. It may well be that contrary to model predictions, additional losses of nitrogen through volatilization, denitrification or leaching have taken place. Further testing of the model is necessary, before confidence can be placed in its predictive capacity.

Introduction

Sustainable agricultural production systems should meet the requirements of the farm household in terms of food, income and leisure (Kruseman *et al.* 1993), without endangering the productive capacity of the natural resource base, to guarantee possible production in the future (Meerman *et al.* 1992). These two aspects are directly related, as the quality of the resource base determines its production capacity, and the amounts of external growth factors needed to attain target production levels.

J. Bouma et al. (eds.), Eco-regional approaches for sustainable land use and food production, 353–375.

However, a first requirement in defining sustainable systems is explicitizing and operationalizing the concept of sustainability. For this purpose, 'indicators' for sustainability may be defined, that can be measured objectively. This may not be easy for all aspects that sustainability encompasses, but when restricted to the agro-technical component, such indicators can be identified (Kruseman *et al.* 1993).

Agricultural production systems (or land use systems) are characterized by the crops (or varieties) grown and their order, the production techniques applied and soil and crop management (Stomph *et al.* 1994). To judge their degree of sustainability, the processes determining production and soil characteristics and their interactions, as influenced by environmental conditions and management practices should be understood quantitatively. An experimental approach to this problem is problematic: it requires long-term experimentation, which is laborious and time-consuming and takes too long to yield relevant results within the timeframe of the present priorities. Moreover, the erratic and unpredictable nature of environmental conditions makes it difficult to use time- and location-specific information for extrapolation and prediction. Hence, a simulation approach, in which existing knowledge is harnassed into quantitative, explanatory models appears attractive. Such models should be calibrated and validated on the basis of results of existing long-term experiments, and may then be used to explore the options and constraints for the development and implementation of sustainable cropping systems.

Considerations for model development

The quality of the natural resources, particularly the soil, and crop performance are affected by many interacting processes, that are largely determined by environmental conditions and soil and crop management. Detailed quantitative description of these processes requires comprehensive simulation models (Penning de Vries 1982), and such models, containing many of these processes, have been developed (e.g. PERFECT (Littleboy *et al.* 1989), EPIC (Sharpley and Williams 1990), CREAMS (Knisel 1980), CENTURY (Parton *et al.* 1989)). Major disadvantages of such models are (i) their extensive data requirements that often cannot be satisfied, (ii) the difficulty in validation, as many variables have not been measured, and certainly not over the time-span necessary to judge their long-term behaviour and (iii) the partial knowledge of many of the underlying processes that often leads to unbalanced descriptions, i.e. many details of well-known processes and gross generalizations of those processes that are poorly understood.

For exploration of the long-term effects of environmental conditions and crop and soil management on sustainability indicators, a summary model (Penning de Vries, op. cit.), in which processes and interactions are incorporated in a descriptive, rather than an explanatory fashion, and whose data requirements can be readily satisfied, often suffices (cf. Wolf *et al.* 1987, 1989; Janssen *et al.* 1989; Wolf and Van Keulen 1989). Such a model may operate with time steps of one year, disregarding intra-seasonal dynamics, and focussing on effects of strategic, rather than tactical or operational decisions.

In this paper, some of the main principles of such a model are discussed, and some preliminary results are presented.

Summary description of the model

Environmental conditions

The location is defined by its geographical coordinates, and is further characterized by the first and the last month of the growing season. Options for multiple cropping, although essential, have not yet been included. The plot is characterized by its slope and slope length.

The model uses mean monthly values of temperature, rainfall, vapour pressure and potential evapotranspiration (Penman-type) as weather inputs. Although time intervals of one year are used, introduction of this temporal variability in the model allows distinction between weather conditions during the growing period and in periods without a crop. From this base information, other weather characteristics are derived, such as average annual temperature, annual precipitation surplus, and average temperature, precipitation surplus and average vapour pressure deficit during the growing period.

Essential soil information comprises textural composition (sand, silt, clay); organic matter content (or organic carbon content), subdivided in a 'stable pool' and a 'labile pool'; inorganic phosphorus content, organic phosphorus content, again subdivided into a 'stable' and a 'labile' pool; initial settled soil bulk density; initial carbonate content; initial soil pH; and rooting zone depth. To derive the size of the stable and labile pools from total organic matter content, the method described by Wolf *et al.* (1989, 1987) for organic P and N, respectively, may be used.

The crops in the model are characterized by the minimum concentrations of nitrogen and phosphorus in grain, straw and roots, respectively (van Keulen and Van Heemst 1982), the harvest index (i.e. the ratio of economic yield to total aboveground dry matter production), the shoot/root ratio and their water use efficiency.

Soil organic matter

Soil organic matter dynamics are described on the basis of two pools, i.e. *stable* and *labile*; in each of the pools carbon, nitrogen, and phosphorus are treated separately. The pools are defined on the basis of their turnover rates (annual fractions of conversion) and cannot be identified with specific physical components in the soil, although it has been suggested that the 'stable' and 'labile' organic pools might be associated with the fine and coarse fraction of the soil, respectively (Dalal and Mayer 1986). It has been assumed that the microbial pool, although playing a key role in the transformations of organic components in the soil, and showing considerable intra-seasonal dynamics, only comprises a small proportion of the total organic pool (Anderson and Domsch 1989; Ladd and Foster 1988) and may be considered more or less constant when looking at inter-seasonal dynamics (Van Veen *et al.* 1984; Seligman and Van Keulen 1981). Hence, this pool is not explicitly considered in the model.

The rate of change of *carbon* in the *stable pool* comprises three terms: (i) transformation of stable organic matter into labile organic matter, depending on the decomposition rate of stable organic material, governed by its relative rate of decomposition defined as a function of temperature, moisture conditions and soil texture, (ii) transfer of resistant material from the labile pool to the stable pool and (iii) loss of stable organic material through erosion.

The relative rates of decompostion under optimum soil moisture conditions and at 50°C have been set at 0.15 and 1.5 yr^{-1} for the labile and stable pool, respectively. The effects of soil moisture and temperature are derived as weighted average values for the various months, using relations as illustrated by Van Keulen and Seligman (1987). The effect of soil texture, reported frequently (cf. Verberne *et al.* 1990; Ladd *et al.* 1981), is derived from the clay and silt content of the soil (Parton *et al.* 1989), such that the rate of decomposition decreases with increasing fraction of fine material. It has been assumed that 20% of the decomposing material in the labile pool resists attack (Seligman and Van Keulen 1981) and is transferred to the stable pool. The rate of loss through erosion is derived from the soil loss, taking into account the phenomenon of enrichment (Knisel 1980).

The rate of change of *carbon* in the *labile pool* comprises four terms: (i) carbon transferred from the stable pool; (ii) carbon originating from the labile pool itself; (iii) carbon originating from organic amendments; and (iv) loss through erosion.

Carbon transferred from the stable pool is assumed to go through microbial biomass, hence, a fraction of that carbon is released as CO_2 in microbial respiration. The fraction incorporated in microbial biomass, referred to as the growth efficiency, is set at 0.5.

The carbon originating from the labile pool itself (hence the complementary fraction of that transferred to the stable pool) is, after having gone through microbial tissue, re-incorporated into that pool. As this material undergoes microbial transformation twice, respiratory losses have been taken into account for both transformations. The relative rate of decomposition of the labile carbon pool is dependent on temperature, moisture conditions and soil texture, as for the stable pool, and in addition on the carbon/nitrogen or carbon/phosphorus ratio of the material, using the equations developed by Van Keulen and Seligman (1987).

The organic amendments comprise roots and stubble and possibly crop residues of last year's crop, but may include also green manure or animal manure incorporated in the soil. The rate of decomposition of this material depends again on environmental conditions, soil texture and its carbon/nitrogen or carbon/phosphorus ratio.

The rate of change of *nitrogen* in the *stable pool* is directly linked to the change in carbon. All nitrogen associated with the decomposing stable organic material (derived from the rate of loss of carbon, by dividing by its C/N ratio) is assumed to be transformed into inorganic nitrogen and leaves the stable pool. The carbon originating from the labile pool and from the organic amendments is incorporated in the stable pool at its current C/N ratio, provided that sufficient nitrogen is 'available' to maintain that ratio. If insufficient nitrogen is available, the C/N ratio of the added material is assumed to increase in proportion to the ratio mineral nitrogen 'available' to mineral nitrogen 'required'(see later). Losses through erosion are also taken into account.

The rate of change of *nitrogen* in the *labile pool* again closely follows that of the carbon: nitrogen leaving the labile pool through decomposition, is calculated from its current C/N ratio; nitrogen added to the labile pool originating from the stable pool is calculated from the C/N ratio of the stable pool; nitrogen originating from the labile pool and organic amendments is calculated from the carbon addition, assuming a fixed (currently set at 8) C/N ratio of microbial biomass, again under the restriction that sufficient N is 'available'. In addition, erosion losses are accounted for.

Nitrogen 'available' for incorporation in the organic pools is calculated from the total amount associated with the decomposing organic material, that originating from other sources (rain) and the nitrogen added as chemical fertilizer, taking into account losses due to leaching and denitrification, and competition by the crop.

The total nitrogen 'requirement' for incorporation in the organic components follows from the rate of change of carbon in the stable and labile pool and the carbon to nitrogen ratio aimed at, as explained before. If 'requirement' exceeds 'availability', the actual transfer for each of the components is multiplied by the ratio of 'availability' and 'requirement'. In this way the C/N-ratio of the organic pools can change in the course of time.

Although the cycles of organic nitrogen and phosphorus are not fully 'identical' (cf. Coleman *et al.* 1983), as a first approximation the description of the dynamics of the stable and labile organic phosphorus pools is very similar to that of the nitrogen pools.

The rate of change of *organic phosphorus* in the *stable pool* is directly tied to the change in carbon. All phosphorus associated with the decomposing stable organic material (derived from the loss in carbon divided by its C/P ratio) is assumed to be potentially available for transformation and is removed from the stable pool. The carbon originating from the labile pool and from the organic amendments is incorporated at the current C/P ratio of the stable material, provided sufficient phosphorus 'availability'. If insufficient phosphorus is available, the C/P ratio of the added material is assumed to be reduced in proportion to the 'availability'/'requirement' ratio.

The rate of change of *organic phosphorus* in the *labile pool* closely follows that of carbon: phosphorus leaving the labile pool through decomposition, is calculated from its current C/P ratio; phosphorus added to the labile pool originating from the stable pool is calculated from the C/P ratio of the stable pool; phosphorus originating from the labile pool or from organic amendments is calculated from the carbon addition, assuming that it is incorporated at the fixed (currently set at 40) C/P ratio of microbial biomass, again under the restriction that sufficient P is 'available'. In addition, it is assumed that part of the phosphorus 'cycling' in the labile pool (i.e. that originating from the labile organic pool and that originating from organic amendments) is transferred to the *labile inorganic pool.*

Phosphorus 'available' for incorporation in the organic pools is calculated as the total amount associated with the decomposing organic material, minus that transferred to the inorganic pool and the labile fraction of the phosphorus added as chemical fertilizer, taking into account competition by the crop.

The total phosphorus 'requirement' for incorporation in the organic components is calculated similarly to that for nitrogen. If 'requirement' exceeds 'availability', the actual transfer for each of the components is multiplied by the ratio of 'availability' and 'requirement'. In this way the C/P-ratio of the organic pools may change in the course of time.

Inorganic phosphorus

The inorganic component of soil phosphorus, that may comprise a substantial part of the total store, depending on soil type and management (cf. Paul and Clark 1988), is subdivided in a *labile pool* and a *stable pool.* The *labile pool* does not represent a specific physical or chemical phosphorus component, but may be considered comparable to 'potentially available' phosphorus, as extracted with such methods as 'Olsen', 'Bray', etc., and is assumed to have a turnover time of five years. The stable pool comprises components characterized by a turnover time of about 33 years, i.e. the relative rate of transfer amounts to 0.03 yr^{-1} (Wolf *et al.* 1987).

The rate of change of the *stable pool* thus is calculated as the sum of the contribution of the labile pool (relative turnover rate times the size of that pool), that of the stable fraction of inorganic P-fertilizer (dependent on fertilizer type), and that from weathering soil minerals, that has to be derived in principle from long-term experiments where approximately equilibrium conditions may be assumed, while the product of relative rate of transfer from the stable pool times its size is subtracted.

The rate of change of the *labile pool* consists of the contribution from the organic sources and that from the stable pool, while the transfer to the stable pool is subtracted.

Soil acidity

One of the factors that has (and is being) blamed for the steady decline in crop yields under tropical conditions is the decrease in soil pH as a result of the acidifying effect of nitrogenous fertilizers, nitrogen fixation by legumes and the high degree of leaching under many circumstances (cf. Ridley *et al.* 1990a; De Ridder and Van Keulen 1990; Brams 1971; Pichot 1971). An estimate of the changes in soil pH over time has therefore been included.

The basis for the calculation is the balance in acidity, resulting from acid production in the carbon and nitrogen cycles associated with organic matter dynamics (cf. Ridley *et al.* 1990b; Helyar 1976), and alkalinity production through plant nitrogen (nitrate) uptake and the possible addition of green or animal manures. Acid that is not neutralized originates from the build-up of organic carbon in the soil, from leaching of nitrate (produced through nitrification, either from inorganic fertilizers or from ammonium mineralized during organic matter decomposition), from the export of alkalinity in crop products or from leaching of carbonate. Acidity may be neutralized, either by exchange through plant roots or by free $CaCO_3$ in calcareous soils. The model takes into account that nitrification may be hampered at low soil pH (Van Cleemput and Baert 1984), so that part of the nitrogen remains in ammoniacal form, and does not contribute to acidification.

In case of a non-zero balance in acidity, soil pH may change, the degree of change being co-determined by the soil buffering capacity. The latter is derived from soil organic matter and clay content (Helyar *et al.* 1990), while effects of soil pH are taken into account. The latter are related to the fact that at low pH other buffering materials (amongst others silica released from the clay minerals), start playing a role (Bache 1988).

At low pH, bases at the exchange complex are exchanged against protons, resulting in reduced base saturation, which eventually leads to release of Al from the clay lattice (Bache 1988). Hence base saturation is derived from soil pH (Peech 1965; Pichot 1971).

The cation exchange capacity (CEC) is estimated on the basis of clay content (plus an indication of the predominant clay mineral), organic matter content and soil pH (Helling *et al.* 1964).

Crop nutrient availability

Nitrogen

The total amount of mineral nitrogen in the soil on an annual basis consists of the nitrogen mineralized from the stable and labile organic matter pools, that mineralized from the added organic amendments, the contribution from other natural sources (rain) and that added as chemical fertilizer. Part of the nitrogen is lost in the course of the year through volatilization, leaching and denitrification, part is immobilized again in the soil organic matter (through microbial action) and part is available for uptake by the crop. To calculate the losses, the total source is sub-divided into two pools: (i) that originating from natural sources (i.e. organic matter, rain and crop residues) and (ii) that originating from organic amendments and chemical fertilizer, assumed to be applied during the cropping season. For each of the two pools, the fractional losses through denitrification and leaching are derived from: a base value, defined as a function of precipitation surplus, for the natural sources on an annual basis, for amendments that during the growing season, and correction factors for CEC and clay content of the soil, assuming that higher CEC reduces leaching due to cation adsorption, and higher clay contents reduce leaching because of higher water storage capacity, and increases denitrification due to greater risks for anaerobiosis.

Mineral nitrogen remaining after the losses have been subtracted, is partitioned between immobilization and crop uptake, as a function of the ratio of total nitrogen 'requirement' of the organic matter (as defined above) and availability. This partitioning results in a situation, where with increasing quality (i.e. decreasing C/N-ratio) of the organic material, the proportion of the available nitrogen incorporated in the organic material is reduced. The exact quantitative relationship incorporated at the moment is based on 'intelligent guesswork' and needs further reconsideration.

Phosphorus

The total amount of available phosphorus in the soil on an annual basis consists of the phosphorus mineralized from the stable and labile organic matter pools, that mineralized from the added organic amendments, the contribution from the labile inorganic pool and the labile fraction of the applied chemical fertilizer. Part of the phosphorus is immobilized again in the soil organic matter (through microbial action), part is incorporated in the stable inorganic pool and part is available for uptake by the crop.

Available phosphorus is partitioned between immobilization and crop uptake, as a function of the ratio of total phosphorus 'requirement' of the organic matter (as defined above) and availability. This partitioning results in a situation, where with increasing quality (i.e. decreasing C/P-ratio) of the organic material, the proportion of the available phosphorus incorporated in the organic material is reduced. As for nitrogen, the quantitative relationship incorporated at the moment needs further reconsideration.

Crop water availability

Water availability to the crop is estimated on the basis of the ratio of rainfall and potential evapotranspiration during the growing season, taking into account the physical properties of the soil, derived from its texture class and soil depth. A major problem here is that the distribution of rainfall within each month is not known, which affects both the distribution between infiltration and runoff and the partitioning of infiltrated water among deep drainage, direct soil evaporation (both non-productive in terms of crop growth) and crop transpiration that contributes to dry matter production (cf. Van Keulen 1975).

The infiltration capacity of the soil is related to its sorptivity, which is a function of particle size distribution (Driessen 1986; Stroosnijder 1976), and is furthermore affected by soil and crop management (Stroosnijder and Koné 1982).

In the model, the 'base sorptivity' of the soil is calculated from the fraction clay, using a relation derived from Driessen (1986). Subsequently, actual sorptivity is calculated from this base value, taking into account surface roughness, as affected by the method of land preparation and the amount of crop residues in the field (Stroosnijder and Koné 1982). On the basis of this value, the annual fraction runoff is calculated, taking into account the slope of the field (Lal 1976).

From soil depth, and the water holding capacity per unit of soil depth, derived from its textural composition, soil water holding capacity is calculated. Seasonal drainage is derived from this value, by assuming that the sum of the monthly difference between precipitation and the sum of potential evapotranspiration and water holding capacity drains below the rooting zone. Seasonal soil evaporation is derived from the total amount of water stored in the root zone, taking into account the fraction of sand in the soil (higher sand contents reduce soil evaporation due to the formation of a 'self-mulching' layer), and soil cover, due to the crop and residues left on the soil.

The difference between total infiltration, and drainage and soil evaporation, is assumed to be available for transpiration by the crop. To derive dry matter production from this value, water use efficiency is introduced, defined as the ratio between the crop's *k*-value (Tanner and Sinclair 1983) and the average vapour pressure deficit during the growing season. Water-limited grain yield follows from total dry matter production through application of the harvest index.

Crop production

The nutrient balance module provides estimates of the availability of nitrogen and phosphorus for plant uptake. These availabilities are translated into crop dry matter production through application of the harvest index of the crop. Under conditions where either of the two elements is in short supply, the concentration in the tissue is diluted to crop-specific minimum concentrations in each of the organs (see e.g. Van Keulen 1986; Van Keulen and Van Heemst 1982), hence specific 'utilization efficiencies' (expressed here in kg economic product per kg nutrient taken up) can be applied. Water-limited yield follows from the water balance equations.

Actual grain yield is assumed to be the minimum of that determined by the availability of nitrogen and phosphorus on the one hand, and that of water on the other following 'Liebig's Law' (Von Liebig 1840). Straw and root yield are derived from the grain yield through application of the harvest index and shoot-root ratio.

The concentration of nitrogen and phosphorus in the crop products is derived from the yield and the uptake of both elements. The nutrient that is limiting is diluted to its minimum value, while the concentration of the other follows from uptake and yield.

Soil erosion

Soil erosion is calculated following the Universal Soil Loss Equation (USLE, Wischmeier and Smith 1978). That equation takes the form:

$$A = R * K * LS * C * P,$$

where A = calculated total soil loss; the unit of A is determined by the units applied for the various contributing factors, particularly R and K. In the current model, it is assumed that A represents average annual soil loss in metric tons per hectare. R = the rainfall and runoff factor, representing the erosive energy of the prevailing rainfall regime. It can be calculated from long-term rainfall records, that should preferable comprise detailed observations of rainfall duration and intensity. Its calculation is not incorporated in the current model, but it is derived from annual rainfall and a site-specific 'rainfall pattern' parameter (Wischmeier and Smith, op. cit.). K = the soil erodibility factor, defined originally by Wischmeier, as the soil loss per unit R, for a 'standard plot' (72.6 ft long under a uniform slope of 9%, continuously under clean-tilled fallow) of a specified soil type. In the current model an approximate empirical equation has been incorporated, derived from Williams (1975), that takes into account the particle size distribution of the soil and its organic matter content. LS = in fact a combination, that takes account of both slope length and slope steepness, to correct from the 'standard plot' to the actual field conditions. Again, originally they were defined in terms of ratios of soil loss of a plot with particular properties and the 'standard' plot. In this model an empirical relation has been incorporated that calculates the correction from slope length and slope steepness. C = crop cover and management factor, defined (originally) as the ratio of soil loss under specified soil and crop management conditions as compared to the clean fallow. Extensive tables are available that specify the value of this factor for various combinations of management and cover. It is incorporated in the current model as an external variable. It could be a suggestion to replace that by an extensive empirical equation that has been developed by Laflen *et al.* (1990) in the framework of the EPIC model. P = defined originally as the 'support practice factor', accounting for the effect of specific soil conservation measures, such as stripcropping, contouring or terracing. Again, values are available in tabulated form, and are introduced in the model as externals.

To calculate total soil loss, the year has been sub-divided into two seasons, the cropping season during which the soil is supposed to be covered, and the off-season during which crop residues are assumed to be present, the actual cover being a function of crop yield and management practices.

The soil erosion module estimates annual soil losses on the basis of soil and landscape characteristics, prevailing rainfall characteristics and soil and crop management practices. Effects of alternative practices, such as no-till, residue retainment, etc. can be explored in this way. In a recent paper (Risse *et al.* 1993) it has been shown that although soil loss predictions by the USLE on an annual basis may show considerable deviations from measured values, average long-term values were predicted with fair accuracy.

Nutrient loss, associated with the loss of soil through erosion is estimated using a function given in Knisel (1980). It relates nutrient loss to soil loss, nutrient content and an enrichment factor that accounts for the fact that the sediment usually has a higher nutrient content that the soil, because of a higher proportion of organic material and fine particles. The enrichment factor is defined as a function of total soil loss.

Crop and soil management

At the beginning of each season a selection is made from the options for soil and crop management (at the moment these options are introduced in the form of tables, in which the selected options are introduced as function of the year number in the sequence; it is the intention to develop an interface that allows the user to select the relevant options at the beginning of each year).

First it can be selected whether the current season will be a 'cropping' season or whether the soil will be fallowed. A selection is then made about the fate of last year's crop residues, which can be (partly) removed from the field, burned, left on the surface, or (partly) incorporated.

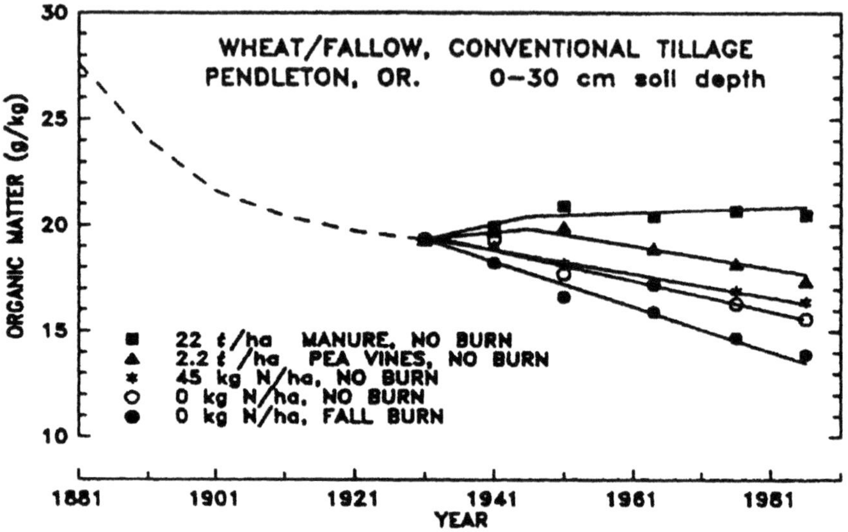

Fig. 1. Figure 1. Treatment effects on measured organic matter content in the 0–30 cm soil layer in the residue management experiment, Columbia Plateau Conservation Research Center, Pendleton, Oregon, 1931–1981.

The degree of land preparation and the implements used can be selected. This influences soil surface roughness, which affects both runoff and soil erosion.

The fertilizer regime selected considers the possibilities for application of animal manure, green manure, chemical (N and/or P) fertilizer and lime. Animal and green manure, for which the composition in terms of C, N and P content has to be specified, can be incorporated in the soil.

Model performance

One of the long-term experiments listed in a report of the Rockefeller Foundation (Steiner and Herdt 1993) is the 'Crop residue management experiment' at the Pendleton Experimental Station in Oregon, USA (22 °NL, 70.5 °WL, 138 m.a.s.l.) established in 1931. Results of this experiment have been published widely (Duff *et al.* 1994; Rasmussen and Parton 1994; Parton and Rasmussen 1994; Rasmussen and Smiley 1993; Collins *et al.* 1992; Rasmussen and Rohde 1989; Pikul and Allmaras 1986; Rasmussen *et al.* 1986; Douglas *et al.* 1984; Rasmussen *et al.* 1980; Oveson 1966), which makes it a useful set of data to test the performance of the model. Information about the behaviour of the system is given in figure 1 (treatment effects on organic matter dynamics) and figure 2 (treatment effects on grain yield). The site is characterized as semi-arid temperate. Climatic information, derived from an internal report (Smiley and Rasmussen 1993) is given in table 1.

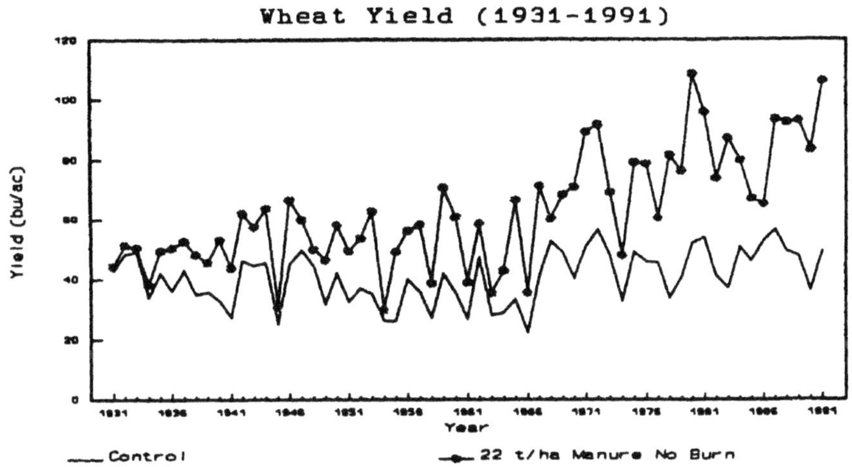

Fig. 2. Average grain yield for wheat in the control treatment and the manure treatment in the residue management experiment, Columbia Plateau Conservation Research Center, Pendleton, Oregon, 1931–1981.

Table 1. Meteorological information for Pendleton Oregon.

Month	Rainfall (mm)	Potential evapotranspiration (mm)	Temperature °C	Vapour pressure kPa
1	48	55	–0.3	0.46
2	38	55	3.0	0.55
3	43	86	7.2	0.62
4	38	128	9.7	0.75
5	36	177	13.4	1.27
6	32	223	17.1	1.47
7	9	292	21.2	1.89
8	12	254	20.4	1.47
9	19	172	16.0	1.36
10	33	98	10.2	1.00
11	54	75	4.4	0.82
12	52	60	1.2	0.63

The cropping system is wheat/fallow, with wheat being sown in October, and harvested in July, after which the plots are clean-fallowed till October of the following year. Details on the the various treatments, that comprised (i) burning of the wheat straw (either in fall or spring), (ii) incorporation of the wheat straw, (iii) incorporation with the addition of mineral N fertilizer, with (iiia) receiving a higher (90 vs. 45 kg ha^{-1}) dose of inorganic N from 1965

Table 2. Treatments in the crop residue management experiment, Pendleton, Oregon.

Code	Straw management	Organic addition kg ha $^{-1}$ crop^{-1}	Inorganic nitrogen kg ha $^{-1}$ crop^{-1}
fB-N_0	fall burned	0	0
nB-N_0	not burned	0	0
nB-N_{45}	not burned	0	34 (from 1967: 45)
nB-N_{90}	not burned	0	34 (from 1967: 90)
nB-PV	not burned	2240	0
nB-AM	not burned	22400	0

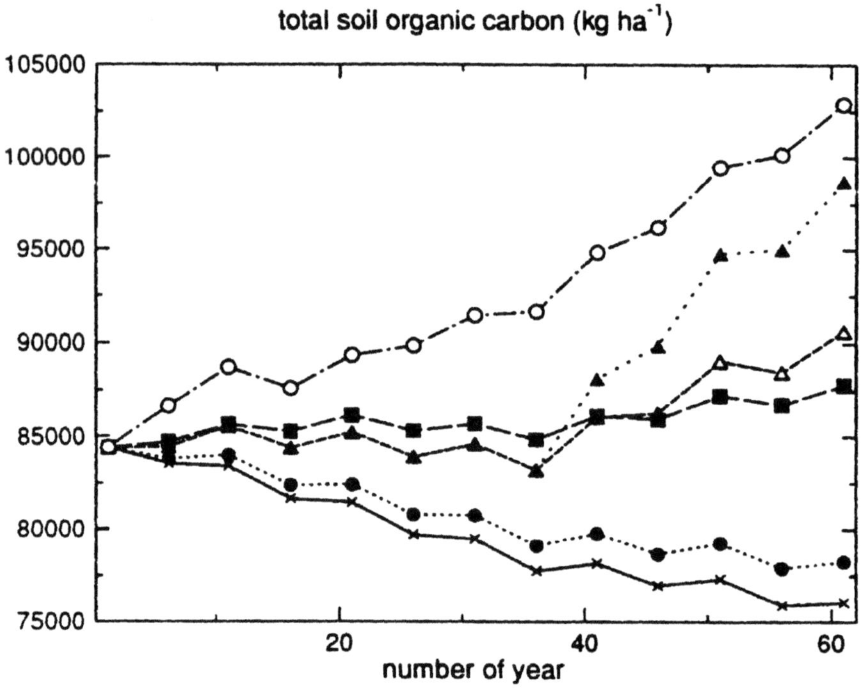

Fig. 3. Calculated organic carbon dynamics in the residue management experiment, Columbia Plateau Conservation Research Center, Pendleton, Oregon, 1931–1981.

onwards, (iv) incorporation with the addition of green manure (pea vine) and (v) incorporation with the addition of animal manure (table 2) are provided in Rasmussen and Parton (1994).

The wheat grown was changed in 1967 from a medium-tall variety (Rex M–1) to a semidwarf variety (cf. Nugaines until 1973, Hyslop till 1978 and subsequently Stephens).

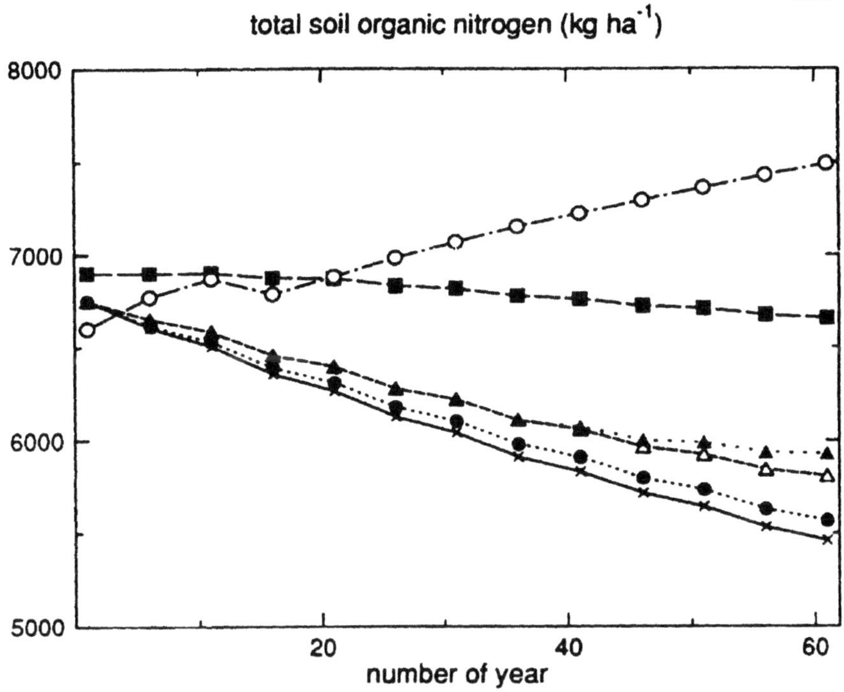

Fig. 4. Calculated organic nitrogen dynamics in the residue management experiment, Columbia Plateau Conservation Research Center, Pendleton, Oregon, 1931–1981.

Total depth of the root zone was set at 1.5 m. (Oveson 1966); initial amounts of organic components for the top 0.6 m. of the profile were derived from Rasmussen and Parton (1994), any organic material below that depth was disregarded in the analysis; initial soil pH was set at 6.8 (Oveson 1966), while it was assumed that no free $CaCO_3$ was present (Rasmussen and Parton 1994). It was assumed that during burning all N in the crop residues was lost (Rasmussen *et al.* 1986; Biederbeck *et al.* 1980), and 67% of its carbon (Rasmussen *et al.* 1980). The quantities of carbon and nitrogen incorporated with the green manure and animal manure were derived from data given by Rasmussen and Parton (1994).

The varietal characteristics included in the model are the harvest index (a caharacteristic that has changed drastically in the course of winter wheat breeding (cf. Austin *et al.* 1989, 1986) and the minimum concentrations of nitrogen and phosphorus in grain and straw (Van Keulen and Van Heemst 1982).

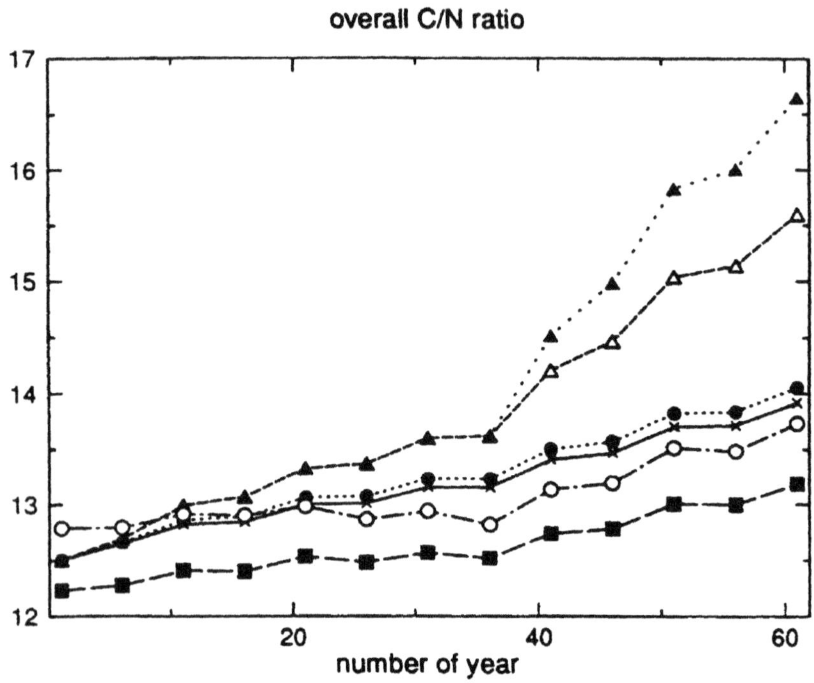

Fig. 5. Calculated dynamics of the overall carbon to nitrogen ratio of soil organic matter in the residue management experiment, Columbia Plateau Conservation Research Center, Pendleton, Oregon, 1931–1981.

Organic matter dynamics

The calculated dynamics of organic carbon in the soil (figure 3), show that the relative position of the treatments is in agreement with the measurements (figure 1), but that the calculated rate of carbon loss is too low, as only the burning treatment and that where straw is incorporated without any amendments, lead to net loss of soil carbon in the simulation, while in reality only the treatment with incorporation of animal manure showed a net gain. The average rates of change, both observed and calculated over the period 1931/1986, as given in table 3, are seemingly in disagreement with figure 1, but the latter refer to the 0–30 cm layer and table 3 to the 0–60 cm layer. In all treatments organic carbon in the 30–60 cm layer declined, irrespective of treatment (Rasmussen and Parton, op. cit.).

Comparison of the calculated and observed values suggests that the simulated rate of decomposition of organic matter is too low. However, with respect to the observed values it must be pointed out that the average rates over the 55-year period are strongly affected by the seemingly high rate during the last ten year period (table 3). Rasmussen and Parton (op. cit.) attribute

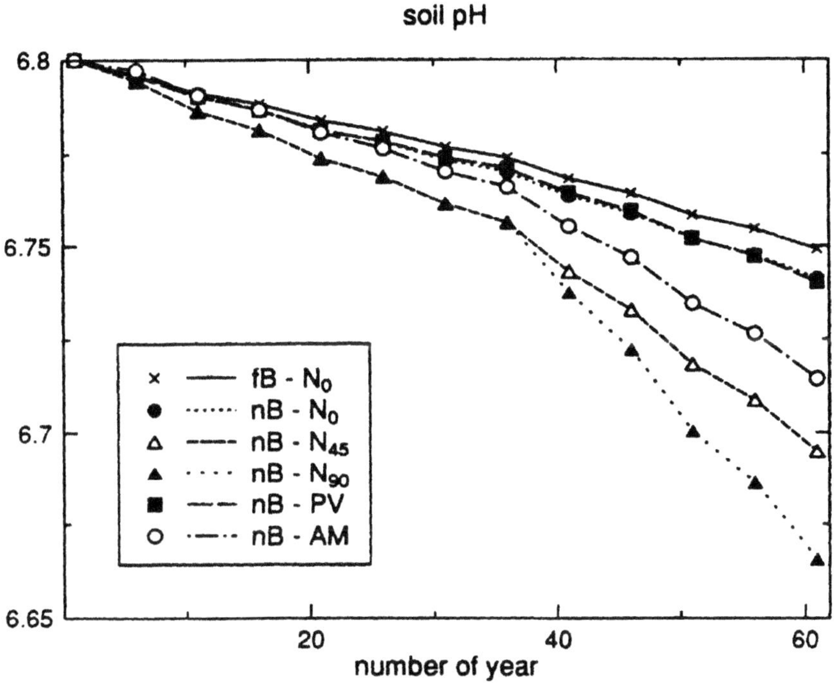

Fig. 6. Calculated time course of soil pH in the residue management experiment Columbia Plateau Conservation Research Center, Pendleton, Oregon, 1931–1981.

Table 3. Average measured and calculated rates of change in soil carbon and soil nitrogen (kg ha^{-1} yr^{-1}) over the periods 1931/86 and 1931/76.

Treatment	Soil carbon		Soil nitrogen	
	observed[a]	calculated	observed	calculated
fB-N_0	–425 (–347)[b]	–153	–24.5 (–23.3)	–22.5
nB-N_0	–356 (–253)	–124	–17.3 (–12.2)	–20.6
nB-N_{45}	–307 (–176)	76	–11.8 (–7.8)	–16.7
nB-N_{90}	–296 (–209)	193	–13.6 (–12.2)	–15.2
nB-PV	–252 (–103)	38	–8.2 (–5.5)	–3.5
nB-AM	–182 (–79)	291	–0.9 (5.5)	14.4

[a]Source: Rasmussen and Parton, 1993; initial values assumed identical, equal to average.
[b]Values in brackets refer to period 1931/76.

this phenomenon to sampling error and suggest that the rate of decline over the period 1976–1986 has essentially been similar to that in the preceding period. Nevertheless, the calculated rates are still too low.

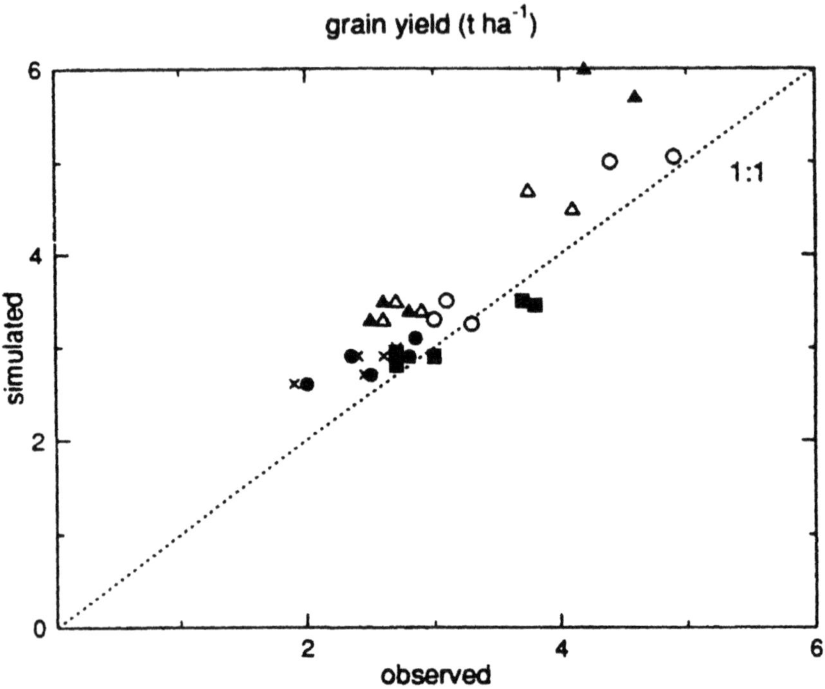

Fig. 7. Comparison of observed and simulated average nitrogen uptake for ten-year periods in the residue management experiment, Columbia Plateau Conservation Research Center, Pendleton, Oregon, 1931–1981.

In figure 4 the calculated dynamics of soil organic nitrogen are given, showing much closer agreement with the measured values than carbon dynamics, as also illustrated in table 3. The calculated values differ less than 15% from the measured ones, except for the farmyard manure treatment, where the difference is an order of magnitude. The result of this differential in simulated carbon and nitrogen dynamics is, that the C/N ratio that in the measurements is on average 11.2 at the end of the 55-year period, down from 12.7 at the start, with only slight differences among treatments (10.9 and 11.3), in the simulations varies between 12.5 and almost 16 (figure 5).

Soil acidity

The calculated dynamics of soil acidity (figure 6) indicate that in all treatments pH is decreasing too slowly (Rasmussen and Rohde 1989). It is not clear what the problem is in the description of soil acidity, but this needs further attention.

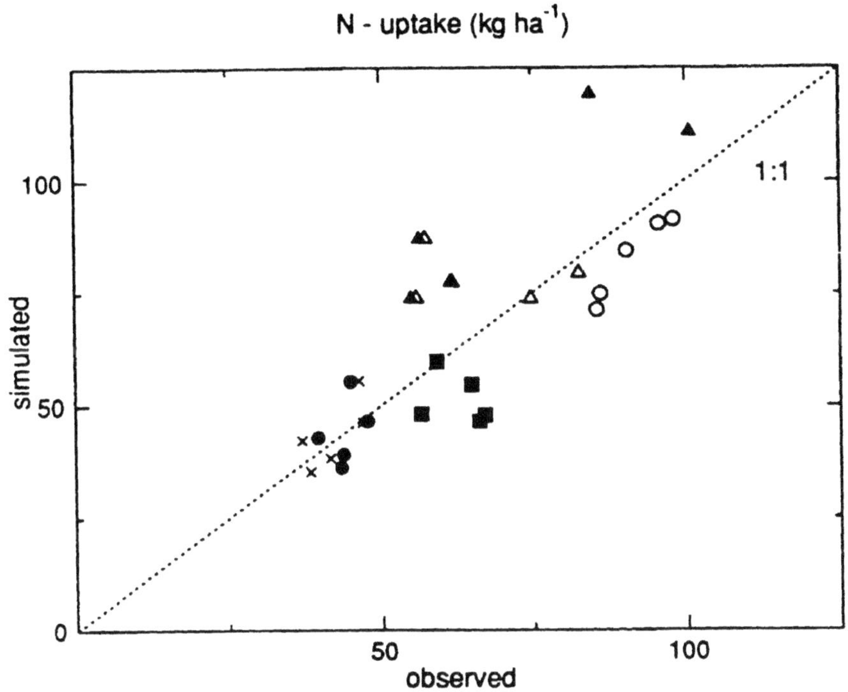

Fig. 8. Comparison of observed and simulated average grain yield for ten-year periods in the residue management experiment, Columbia Plateau Conservation Research Center, Pendleton, Oregon, 1931–1981.

Crop nitrogen uptake and grain yield

As average climatic data were used in the analysis, hence year-to-year variability cannot be taken into account, nitrogen uptake and grain yield have been compared as averages for approximately 10-year periods, following the analysis by Rasmussen and Parton (op. cit.) Nitrogen uptake is reproduced reasonably accurately (figure 7), albeit it is generally overestimated in the fertilizer treatments, while in the treatments with organic amendments it tends to be underestimated. The dynamics of nitrogen uptake are well-reperesented, with the introduction of the semidwarf variety generally leading to somewhat higher uptake values, that are associated with higher crop residue production. Simulated grain yields (figure 8) are generally higher than observed, which for the inorganic treatments is associated with the overestimated nitrogen uptake, but not for the treatments without amendments. The discrepancies for those treatments must thus originate in differences in nitrogen concentrations or in harvest index.

Conclusions

A modelling approach to the exploration of the dynamics of sustainability indicators seems promising, and may help in formulating criteria for different management strategies.

The performance of the model in terms of organic matter dynamics appears 'reasonable', and reproduces effects of soil and crop management in a recognizable form. The apparent discrepancy between organic nitrogen and organic carbon behaviour must be analysed in more detail, to establish its possible causes. It may well be that contrary to model predictions, additional losses of nitrogen through volatilization, denitrification or leaching have taken place.

Further testing of the model is necessary, however, before confidence can be placed in its predictive capacity.

Acronyms

CEC	Cation Exchange Capacity
USLE	Universal Soil Loss Equation

References

Anderson T, Domsch K H (1989) Ratios of microbial biomass carbon to total organic carbon in arable soils. Soil Biol. Biochem. 21:471–479.

Austin R B, Ford M A, Morgan C L (1989) Genetic improvement in the yield of winter wheat: a further evaluation. Journal of Agricultural Science, Camb. 112:295–301.

Austin R B, Bingham J, Blackwell R D, Evans L T, Ford M A, Morgan C L, Taylor M (1986) Genetic improvements in winter wheat yields since 1900 and associated physiological changes. Journal of Agricultural Science, Camb. 94:675–689.

Bache B W (1988) Measurements and mechanisms in acid soils. Comm. Soil Sci. Plant Anal. 19:775–792.

Biederbeck V O, Campbell C A, Bowren K E, Schitzner M, McIver R N (1980) Effect of burning cereal straw on soil properties and grain yields in Saskatchewan. Soil Sci. Soc. Am. J. 44:103–111.

Brams E A (1971) Continuous cultivation of West-African soils: organic matter diminution and effects of applied lime and phosphorus. Plant Soil 35:401–414.

Van Breemen N, Mulder J, Driscoll C T (1983) Acidification and alkalinization of soils. Plant Soil 75:283–308.

Van Cleemput O, Baert 1 (1984) Nitrite: a key compound in N loss processes under acid conditions? Plant Soil 76:233–241.

Coleman D C, Reid C P, Cole C V (1983) Biological strategies of nutrient cycling in soil systems. Adv. Ecol. Res. 13:1–55.

Collins H P, Rasmussen P E, Douglas C L (1992) Crop rotation and residue management effects on soil carbon and microbial dynamics. Soil Sci. Soc. Am. J. 56:783–788.

Dalal R C, Mayer R J (1986) Long-term trends in fertility of soils under continuous cultivation and cereal cropping in southern Queensland. IV. Loss of organic carbon from different density functions. Aust. Journal Soil Resources 24:293–300.

Douglas C L Jr, Allmaras R R, Roager N C (1984) Silicic acid and oxidizable carbon movement in a Walla Walla silt loam as related to long-term management. Soil Sci. Soc. Am. J. 48:156–162.

Driessen P M (1986) The water balance of the soil. Pages 76–116 in Van Keulen H, Wolf J (Eds.) Modelling of agricultural production: weather, soils and crops. Simulation Monographs, Pudoc, Wageningen, The Netherlands.

Duff B, Rasmussen P E, Smiley R W (1994) Wheat/Fallow systems in semi-arid regions in the pacific NW America. In Barnet V, Payne R, Steiner R (Eds.) Agricultural sustainability in economic, environmental and statistical terms. John Wiley and Sons, Chichester, UK (in press).

Helling C S, Chesters G, Corey R B (1964) Contribution of organic matter and clay to soil cation-exchange capacity as affected by the pH of the saturating solution. Soil Science Society Am. Proc. 28:517–520.

Helyar K R (1976) Nitrogen cycling and soil acidification. J. Aust. Inst. Agric. Sci. 42:217–221.

Helyar K R, Cregan P D, Godyn D L (1990) Soil acidity in New South Wales – current pH values and estimates of acidification rates. Aust. J. Soil Res. 28:523–537.

Janssen B H, Lathwell D J, Wolf J (1987) Modeling long-term crop response to fertilizer phosphorus. II. Comparison with field results. Agronomy Journal 79:452–458.

Jenkinson D S (1991) The Rothamsted long-term experiments. Are they still of use? Agronomy Journal 83:2–10.

Van Keulen H (1986) Crop yield and nutrient requirements. Pages 155–181 in Van Keulen H, Wolf J (Eds.) Modelling of agricultural production: weather soils and crops. Simulation Monographs, Pudoc, Wageningen, The Netherlands.

Van Keulen H (1975) Simulation of water use and herbage growth in arid regions. Simulation Monographs, Pudoc, Wageningen, The Netherlands.

Van Keulen H, Seligman N G (1987) Simulation of water use, nitrogen nutrition and growth of a spring wheat crop. Simulation Monographs, Pudoc, Wageningen, The Netherlands.

Van Keulen H, Van Heemst H D J (1982) Crop response to the supply of macronutrients. Agric. Res. Rep. Pudoc, Wageningen, The Netherlands.

Knisel W G (Ed.) (1980) CREAMS: A field-scale model for Chemicals, Runoff, and Erosion from Agricultural Management Systems. Conserv. Res. Rep. 26, US Dept. Agriculture, USA.

Kruseman G, Hengsdijk H, Ruben R (1993) Disentangling the concept of sustainability. Conceptual definitions, analytical framework and operational techniques in sustainable land use. DLV report no. 2, CABO-DLO, Wageningen, The Netherlands.

Ladd J N, Foster R C (1988) Role of soil microflora in nitrogen turnover. Pages 113–133 in Wilson J R (Ed.) Advances in nitrogen cycling in agricultural ecosystems. C.A.B.Intern., Wallingford, UK.

Ladd J N, Oades J M, Amato M (1981) Microbial biomass formed from ^{14}C, ^{15}N-labeled plant material decomposition in soils in the field. Soil Biol. Biochem. 13:119–126.

Laflen J M, Foster G R, Onstad C A (1990) Computation of universal soil loss equation R and C factors for simulating individual-storm soil loss. Pages 125–138 in Sharpley A N, Williams J R (Eds.) EPIC-Erosion/Productivity Impact Calculator: 1. Model description. US Dept. Agric. Techn. Bull. No. 1768.

Von Liebig J (1840) Organic chemistry in its application to agriculture and physiology. Playfair, London, UK.

Littleboy M, Silburn D M, Freebairn D M, Woodruff D R, Hammer G L (1989) PERFECT. A computer simulation model of Productivity Erosion Runoff Functions to Evaluate Conservation Techniques. Queensland Dept. Primary Industries, Brisbane, Australia.

Meerman F, Van de Ven G W J, Van Keulen H, De Ponti O M B (1992) Sustainable crop production and protection. Discussion paper, Ministry of Agriculture, Nature Management and Fisheries, Dept. OSL, The Hague, The Netherlands.

Mitchell C C, Westerman R L, Brown J R, Peck T R (1991) Overview of long-term agronomic research. Agron. J. 58:444–447.

Oveson M M (1966) Conservation of soil nitrogen in a wheat summer fallow farming practice. Agronomy Journal 58:444–447.

Parton W J, Rasmussen P E (1994) Long-term effects of crop management in wheat-fallow: II. CENTURY model. simulations. Soil Sci. Soc. Am. J. 58:530–536.

Parton W J, Sanford R L, Sanchez P A, Stewart J W B (1989) Modelling organic matter dynamics in tropical soils. Pages 153–171 in Coleman D C, Oades J M, Uehara G (Eds.) Dynamics of soil organic matter in tropical ecosystems. NifTAL Project, Dept. Agronomy and Soil Science, Coll. Trop. Agric. Human Res., University of Hawaii, USA.

Paul E A, Clark F E (1988) Soil microbiology and biochemistry. Academic Press, San Diego, USA.

Peech M (1965) Lime requirement. Agronomy 9:927–932.

Penning de Vries F W T (1982) Systems analysis and models of crop growth. Pages 9–19 in Penning de Vries F W T, Van Laar H H (Eds.) Simulation of plant growth and crop production. Simulation Monographs, Pudoc, Wageningen, The Netherlands.

Pichot J (1971) Etude de l'evolution du sol en presence de fumures organiques ou minerales. Cinq annees d'experimentation a la station de Boukoko, RCA. Agronomy Tropical 26:736–754.

Philip J R (1957) The theory of infiltration. 4. Sorptivity and algabraic infiltration. Soil Science 84:257–264.

Pikul J L, Allmaras R R (1986) Physical and chemical properties of a haploxeroll after fifty years of residue management. Soil Sci. Soc. Am. J. 50:214–219.

Rasmussen P E, Smiley R W (1994) Soil carbon and nitrogen change in long-term agricultural experiments at Pendleton. In Soil organic matter in the long-term plots of North-America. Lewis Publishers (in press).

Rasmussen P E, Parton W J (1994) Long-term effects of crop management in a wheat/fallow system: I. Inputs, crop yield and soil organic matter. Soil Sci. Soc. Am. J. 58:523–530.

Rasmussen P E, Rohde C R (1989) Soil acidification from ammonium-nitrogen fertilization in moldboard plow and stubble-mulch wheat-fallow tillage. Soil Sci. Soc. Am. J. 52:119–122.

Rasmussen P E, Rickman R W, Douglas C L Jr (1986) Air and soil temperatures during spring burning of standing wheat stubble. Agron. J. 78:261–263.

Rasmussen P E, Allmaras R R, Rohde C R, Roager N C Jr (1980) Crop residue influences on soil carbon and nitrogen in a wheat-fallow system. Soil Sci. Soc. Am. J. 44:596–600.

De Ridder N, Van Keulen H (1990) Some aspects of the role of organic matter in sustainable intensified arable farming systems in the West-African semi-arid-tropics (SAT). Fert. Res. 26:299–310.

Ridley A M, Helyar K R, Slattery W J (1990a) Soil acidification under subterranean clover (*Trifolium subterraneum* L.) pastures in north-eastern Victoria. Aust. J. exp. Agric. 30:195–201.

Ridley A M, Slattery W J, Helyar K R, Cowling A (1990b) The importance of the carbon cycle to acidification of a grazed annual pasture. Aust. J. exp. Agric. 30:529–537.

Risse L M, Nearing M A, Nicks A D, LaflenJ M (1993) Error assessment in the Universal Soil Loss Equation. Soil Sci. Soc. Am. J. 57: 825–833.

Seligman N G, Van Keulen H (1981) PAPRAN: A simulation model of annual pasture production limited by rainfall and nitrogen. Pages 192–222 in Frissel M J, Van Veen J A (Eds.) Simulation of nitrogen behaviour in soil-plant systems. Pudoc, Wageningen, The Netherlands.

Sharpley A N, Williams J R (Eds.) (1990) EPIC-Erosion/Productivity Impact Calculator: 1. Model description. US Dept. Agric. Techn. Bull. No. 1768.

Steiner R A, Heldt R W (Eds.) (1993) A global directory of long-term agronomic experiments (Vol. 1: Non-European experiments). The Rockefeller Foundation, 1133 Ave. of the Americas, New York, USA.

Stomph T J, Fresco L O, Van Keulen H (1994) Land use system evaluation: concepts and methodology. Agric. Syst. 44:243–255.

Stroosnijder L (1976) Infiltration and redistribution of water in soil. Agric. Res. Rep. 847, Pudoc, Wageningen, The Netherlands. (Dutch, with English summary).

Stroosnijder L, Koné D (1982) Le bilan d'eau du sol. Pages 133–165 in Penning de Vries F W T, Djitèye A M (Eds.) La productivité des pâturages sahéliens. Une étude des sols, des végétations et de l'exploitation de cette ressource naturelle. Pudoc, Wageningen, The Netherlands.

Tanner C B, Sinclair T R (1983) Efficient water use in crop production: Research or re-search? Pages 1–27 in Taylor H M, Jordan W R, Sinclair T R (Eds.) Limitations of efficient water use in crop production. American Society of Agronomy, ASA Monographs Inc., Madison, Wi., USA.

Van Veen J A, Ladd J N, FrisselM J (1984) Modelling C and N turnover through the microbial biomass in soil. Plant Soil 76:257–274.

Verberne E L J, Hassink J, De Willigen P, Groot J J R, Van Veen J A (1990) Modeling organic matter dynamics in different soils. Neth. J. Agric. Sci. 38:221–238.

Williams J R (1975) Sediment-yield prediction with universal soil loss equation using runoff energy factor. Pages 244–252 in Present and prospective technology for predicting sediment yield and sources. ARS-S-40, US Dept. Agriculture, USA.

Wischmeier W H, Smith D D (1978) Predicting rainfall erosion losses-a guide to conservation planning. US Dept. Agric. Agric. Handbook No. 537.

Wolf J, Van Keulen H (1989) Modeling long-term crop response to fertilizer and soil nitrogen. II. Comparison with field results. Plant Soil 120:23–38.

Wolf J, De Wit C T, Van Keulen H (1989) Modeling long-term crop response to fertilizer and soil nitrogen. I. Model description and application. Plant Soil 120:11–22.

Wolf J, De Wit C T, Janssen B H, Lathwell D J (1987) Modeling long-term crop response to fertilizer phosphorus. I. The model. Agron. J. 79:445–451.

Evaluating policies for sustainable land use: a sub-regional model with farm types in Costa Rica

R.A. SCHIPPER[1], H.G.P. JANSEN[2], J.J. STOORVOGEL[2] and D.M. JANSEN[2]

[1] *Agricultural Economist, Department of Development Economics, Wageningen Agricultural University, P.O. Box 8130, 6700 EW Wageningen, The Netherlands;* [2] *Respectively Economist and Coordinator; Soil Scientist and GIS Specialist; and Agronomist, Programa Zona Atlántica, Apartado 224, 7210 Guápiles, Costa Rica*

Key words: Costa Rica, farm types, land use analysis, linear programming, policy evaluation, scenarios, sub-regional models, sustainable agricultural development

Abstract. An interdisciplinary methodology for land use analysis at the sub-regional level is presented using data for the *Neguev* settlement in the perhumid tropical lowlands of Costa Rica. A linear programming (LP) model is employed for the maximisation of farm household income, given a flexible set of resource and sustainability related criteria. A large number of different land use systems with fixed input-output coefficients, developed on the basis of crop growth simulation and expert systems, are offered to the LP model for the optimisation of land use in terms of crop selection and technology choice. A geographical information system is used for the presentation of the results. The matrix of the LP model includes five sub-matrices, each encompassing a different farm type. Farm types are distinguished on the basis of their land-labour ratios and the availability of three different soil groups. Sustainability is explicitly taken into account through estimating environmental damage in terms of soil nutrient depletion using nutrient balances (for N, P and K) and through an index for biocide use. Different land use scenarios for the effects on land use of a number of policy interventions including changes in output, input and factor prices, capital availability, and regulatory measures, are analysed. For a number of these policy instruments, the results indicate trade-offs between sustainability and income objectives. The methodology, which is currently being upscaled towards the regional level, provides a valuable tool for *ex ante* assessment of different land use options and should become an integrated part of policy design and execution.

Introduction

Agricultural policies and economic incentives are important tools to achieve a more sustainable use of natural resources. Policies and incentives are defined with the objective of indirectly influencing land use decisions of individual producers. At the (sub-)regional and national level, the effects of alternative policies on land use can be analysed through land use evaluation models. This paper describes a methodology for such a model, developed at the Atlantic Zone Programme in Costa Rica, which is a cooperation between Wageningen Agricultural University (WAU), the Centre for Research and Education in Tropical Agriculture (CATIE), and the Costa Rican Ministry of Agriculture

J. Bouma et al. (eds.), Eco-regional approaches for sustainable land use and food production, 377–395.

Table 1. Land use in the Neguev: 1985, 1987, 1989 and 1991 (ha).

Major land use type/crop	1985[1]	1987[2]	1989[3]	1991[4]
Annuals	460	998	282	356
Perennials	238	364	335	383
Pastures	2346	1745	2519	2407
Forest and wasteland	1194	1073	1101	1090
Total	4236	4236	4236	4236
Maize	414	589	181	154
Cassava	30	118	76	187
Pineapple	–	90	13	18
Plantain	–	23	25	44
Palm heart	–	–	90	138

[1]IDA (1985): based on an inventory of all farms.

[2]Waaijenberg (1990): based on a random sample survey of 53 farms.

[3]Mücher (1992): based on interpretation of aerial photographs of six non-randomly selected sample areas with a total area of 1273 ha; the areas per land use type for the Neguev as a whole are obtained by weighing the sample data with the area of the three main soil types.

[4]Mücher (1992): based on 1989 land use, corrected for observed changes in 1991.

and Livestock (MAG). The model allows for the evaluation of the effect of different incentives and regulations on land use decisions. The model is currently operational for a small sub-region, i.e., the *Neguev* settlement in the Atlantic Zone of Costa Rica, comprising 5,340 ha (figure 1). Altitude is between 10 and 50 m above sea level in a region which climate is classified as very humid tropical, without dry months (Herrera and Gómez 1993). Average annual rainfall is 3,630 mm (1972–1988) with an air temperature of about 25°C (1976–1988; average between daily maximum and minimum temperatures). The settlement had its origin in an occupation by land squatters of the *hacienda* Neguev in September 1979. The IDA (*Instituto de Desarrollo Agropecuario*; Institute of Agrarian Development) parcelled the Neguev into 307 farms with a total area of 4,236 ha available for agriculture or forestry. Except for a few larger farms, farm size is rather uniform (Schipper 1993), with an average farm size of 13.8 ha. Pastures and forests are still the major land uses. As part of a structural adjustment programme, the government of Costa Rica changed its support price policies for basic food crops (including maize, rice and beans) in 1988, leading to some major changes in land use (table 1).

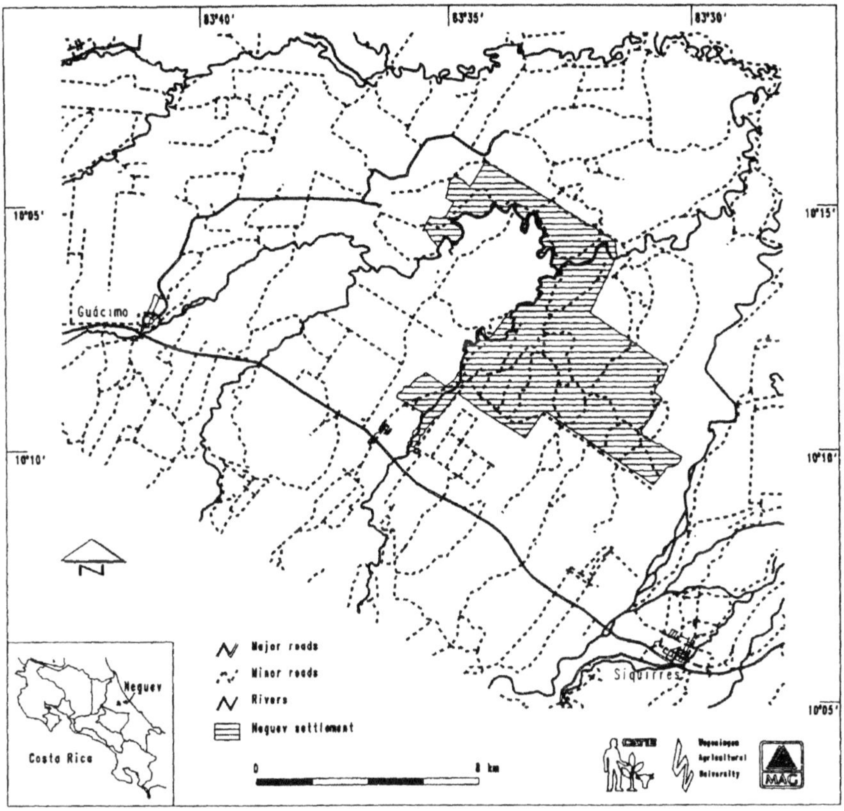

Fig. 1. Location of the Nequev settlement.

Trade-offs between different policy goals are studied through the use of scenarios in which the effects on land use are analysed for different (assumed) changes in the socio-economic environment, or policy instruments. Land use in this context refers to crop selection and technology choice. A commonly perceived trade-off between goals, particularly at the public level, and one which receives specific attention in this paper, is that between income generation and sustainability. To be useful in practical applications, the sustainability concept needs to be operationalised. Consequently, sustainability is measured in this paper in terms of biocide use and nutrient balances.

The present paper continues with a description of the methods employed in analysing land use in the case study area, including an approach to relevant sustainability aspects. Results of several scenarios affecting land use are subsequently presented. The paper ends with conclusions about the applied methodology.

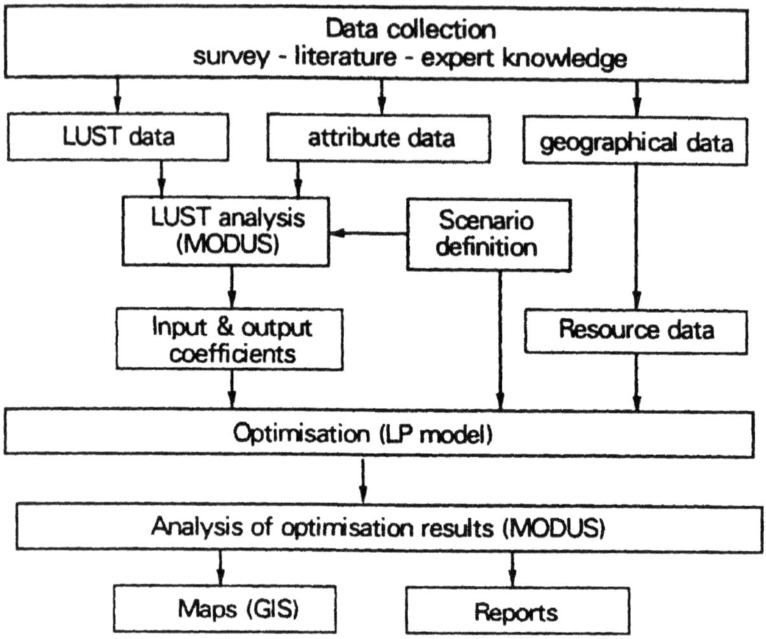

Fig. 2. The set-up og the USTED methodology.

Methods

The USTED methodology

To enable analysis of land use scenarios, various techniques are integrated (figure 2) in the methodology denominated USTED (*Uso Sostenible de Tierras En el Desarrollo*; Sustainable Land Use in Development). Options for land use are defined as Land Use Systems (combinations of a land unit and a land use type) with a specified Technology, called LUSTs. At present, the model includes six land units, based on three soil groups each either with or without a forest cover at present, and eight land use types, cassava, logged forest, maize, palm heart, pasture with cattle, pineapple, plantain and tree plantation. Each LUST describes a specific quantitative combination of physical inputs and outputs per hectare (Jansen and Schipper 1995). Attributes to these inputs and outputs (e.g. data on prices, chemical composition, and toxicity) are kept outside the LUST descriptions. The LUST data and attributes are stored in MODUS (MOdules for Data management in USted), as described in Stoorvogel *et al.* (1995). MODUS calculates input and output coefficients per ha for each LUST by combining the physical quantities of relevant inputs and outputs from the LUSTs with the required attributes.

Input and output coefficients include land use, labour requirements, costs of current inputs (sum of input quantities times prices), labour costs, production specified per product, soil nutrient depletion with regard to nitrogen (N), phosphorus (P) and potassium (K), and a biocide use index value. These coefficients are either averages per month (land and labour use) or per year (soil nutrient depletion of N, P and K, and the biocide use index value), or annuities of the net present value over the life-span of the land use systems (production, input costs and labour use), assuming constant prices over time. LUSTs are defined for both actual technologies, i.e. those currently in use by farmers, and alternative technologies, i.e. those designed by researchers and earmarked as promising. Thus, different crop and livestock related agricultural activities can be performed by different technologies which have different income generating capabilities as well as different sustainability implications. In this sense, LUSTs can be regarded as production functions with fixed input-output coefficients. While actual LUSTs are characterised mainly on the basis of farm survey data, alternative LUSTs are defined with the help of available literature, field experiments, crop growth simulation models, and expert knowledge.

The input and output coefficients of the LUSTs are offered to a sub-regional linear programming (LP) model (Stoorvogel *et al.* 1995; Schipper *et al.* 1995). Because of the relatively small size of the Neguev settlement, all prices are assumed exogenous, thus avoiding the issue of price decreases caused by supply increases, which is an important aggregation issue (Hazell and Norton 1986).

Policies influencing land use are defined by national or regional governments, but ultimate decisions regarding land use are made by individual producers. Consequently, even though the methodology focuses on the (sub-)regional level, individual farms are included as the ultimate decision makers. To deal with the relatively large number of farms in the Neguev, cluster analysis was used to distinguish five farm types, defined on the basis of size and soil distribution. Starting out from an original soil map distinguishing 21 soil series, the latter were aggregated into the following three groups, based on inherent fertility and drainage characteristics: (1) young, poorly drained volcanic soils of relatively high fertility (Entisols and Inceptisols), classified as fertile, poorly drained (SFP), (2) young alluvial, well drained volcanic soils of relatively high fertility (Inceptisols and Andisols), classified as fertile, well drained (SFW), and (3) old, well drained soils developed on fluvio-laharic sediments of relatively low fertility (Oxisols and Inceptisols), classified as infertile, well drained (SIW). Combined with an assumed constant average labour availability per farm household of 2.0 labour-years per year, each farm type is homogeneous in its land to labour ratio. In this way, the aggregation bias (Hazell and Norton 1986) is less than it would have been in case the Neguev would have been considered as one large farm. Table 2 summarises the results of the clustering procedure.

Table 2. Clustering of Neguev farms into five groups.

Farm type	Number of farms	Average area (ha)	Area with soil type (%) SFP	SFW	SIW	Land/labour ratio (ha/labour-year) total	SFP	SFW	SIW
1	33	15.7	60	12	28	7.9	4.7	1.0	2.3
2	4	32.1	12	10	78	16.2	1.9	1.7	12.7
3	46	13.5	7	52	41	6.4	0.5	3.5	2.8
4	35	14.1	3	91	6	7.1	0.2	6.5	0.4
5	189	13.1	6	6	88	6.6	0.4	0.4	5.8
total	307	13.8	13	23	64	7.0	0.9	1.6	4.6

The matrix of the LP model for the sub-region contains a specific sub-matrix for each of the farm types. Each sub-matrix encompasses the constraints for a specific farm type and the specific coefficients for the LUSTs. The model maximises ('objective function') the sum of the *economic surpluses* (calculated as the farm household income minus a valuation of on-farm household labour) of the different farm types (measured in *Colones* per year but, for the sake of convenience, converted into US dollars at the average 1991 exchange rate of 1 (US)\$ = ¢122), subject to resource constraints at both the farm and sub-regional levels.

A restricted sub-regional labour supply acts as an equilibrium condition for the model: the possibilities to work off-farm as hired labour on other farms inside the sub-region are limited by the aggregated demand for such labour (i.e. for every month the supply for off-farm labour is in equilibrium with the demand for labour in the *Neguev*). Off-farm work outside the sub-region on (banana) plantations is restricted to half of the available household labour in view of the structure of labour demand and of labour contracts offered by banana companies.

Farm household income is defined as the sum of the output value, income from off-farm work within the Neguev, and income from labour on plantations outside the settlement, minus the input costs and hired labour costs. Wages are set at \$0.82 per hour for hired labour, while work on other farms within the Neguev earns \$0.74 per hour, 10% less than the wage for hired labour because of assumed transaction costs. Work on a plantation outside the Neguev has an hourly wage of \$1.54.

To assess the valuation of on-farm household labour, a reservation wage of \$0.55 per hour is postulated. The rationale for a 'reservation' wage is that people are not willing to work on their own farms if they cannot earn a certain minimum return per hour worked. By imposing such a reservation wage, the LP model will not select an activity with a return to family labour lower than the reservation wage, unless forced to by other constraints. Alternatively, it can be said that a reservation wage reflects the preference for leisure of a household (Hazell and Norton 1986).

The methodology explicitly integrates agronomic, edaphologic, and socio-economic factors, and selects LUSTs to maximise the economic surplus for five different farm types given a certain set of resource and sustainability related criteria. The latter are flexible and in this sense the model can be used for assessing trade offs between income generation and sustainability. Scenario definitions are translated into changes in attribute data, technologies, parameters for the calculation of input-output coefficients, and resource constraints. Resource endowments with regard to land (per soil group) and household labour are specified per farm type and per month; the impact of the selected LUSTs on resources related to the sustainability parameters is appraised at both the farm as well as at the sub-regional level. The model outcomes are in terms of adjustments in land use and choice of technologies, relative to a base scenario. The structure of the methodology includes a Geographical Information System (GIS) which is used to store geo-referenced data and to visualise land use scenarios through the generation of maps.

Sustainability

To operationalise the concept of sustainability, the Pearce and Turner (1990) definition 'maximising the net benefits of economic development, subject to maintaining the services and quality of natural resources over time' was adopted. This implies accepting of the following rules: (1) utilise renewable resources at rates less than or equal to the natural rate at which they regenerate; (2) keep waste flows to the environment at or below the assimilative capacity of the environment; and (3) optimise the efficiency with which non-renewable resources are used, subject to substitutability between resources and technical progress. While the LP model optimises resource use in the sense of rule (3), soil nutrient depletion and biocide use are identified as the principal constraints on sustainability in the case of the Neguev, respectively corresponding to rules (1) and (2). Consequently, environmental damage is estimated through nutrient balances (for N, P and K) and a biocide index (Jansen *et al.* 1995). The latter is an indicator for the amount of active ingredients, their half life time and the toxicity. Nutrient balances are calculated by separate assessments of nutrient inputs (mineral and organic fertiliser, wet deposition, and N-fixation) and nutrient outputs (production, stover, denitrification, and leaching) with an adapted version of the NUTBAL nutrient balance model (Stoorvogel 1993).

The effects that policy induced changes have on producers, consumers and the environment can, in principle, be added up to arrive at an aggregated effect for the Neguev as a whole. For this purpose, a social welfare index can be constructed. Such an index would measure the sum of the farm household's economic surpluses (as reflected by the value of the objective function in the LP model), modified for effects on consumer surplus caused by price changes, and adjusted for environmental damage. Ignoring the world outside the Neguev, the impact of changes in output prices of certain products may be measured through changes in consumer surplus of Neguev settlement households. However, price change scenarios are calculated for palm heart only, and for this crop a zero consumption within the Neguev is assumed. Hence, the entire palm heart production is either consumed in other parts of Costa Rica or exported, thus having no effect on consumer surplus within the Neguev.

Environmental costs consist of losses of nutrients and costs associated with the use of biocides. Nutrient losses are valued by using actual nutrient prices prevailing in the settlement ($0.52/kg for N, $1.46/kg for P, and $0.42/kg for K) and approximations for fertiliser replacement efficiencies (40% for N, 95% for P, and 60% for K), which for P accounts for the high fixation in the soil. Even though, theoretically, a penalty can be imposed per unit increase of the biocide index to value environmental damage created by biocide use, such a penalty is considered too arbitrary given the currently available information. Consequently, environmental and human health damage caused by biocides are currently not yet part of the social welfare index. In this study, economic benefits (in terms of farm household economic surplus) and environmental costs (in terms of nutrient depletion) are summed on a one-to-one basis, thus neglecting distributional aspects between different farm types. It is, of course, perfectly possible to assign different, subjectively determined weights to impacts on different groups.

Analysis of policy instruments and their potential impact on sustainable land use

Scenarios run with the USTED methodology were evaluated in contrast to the base scenario. The base scenario reflects the maximisation of economic surplus with 1991 input and output prices. Each of the five farm types in the model can choose from 105 LUSTs. The discount rate for calculating annuities is set at 10%. No additional constraints on the sustainability parameters or the availability of capital are included.

The main activity on both fertile and infertile soils is palm heart (figure 3). Nearly 80% of the available area should be planted to palm heart. The optimal solution selects a zero fertilizer technology, yielding about 80% of the potential production of 10,000 palm hearts ha^{-1} year^{-1} on the SFW soils, three years after planting (5,000 plants ha^{-1}). On the less fertile SIW soils this technology yields about 5,000 units ha^{-1} year^{-1}. Given the relative prices, inputs and labour use, this appears to be an attractive technology and crop.

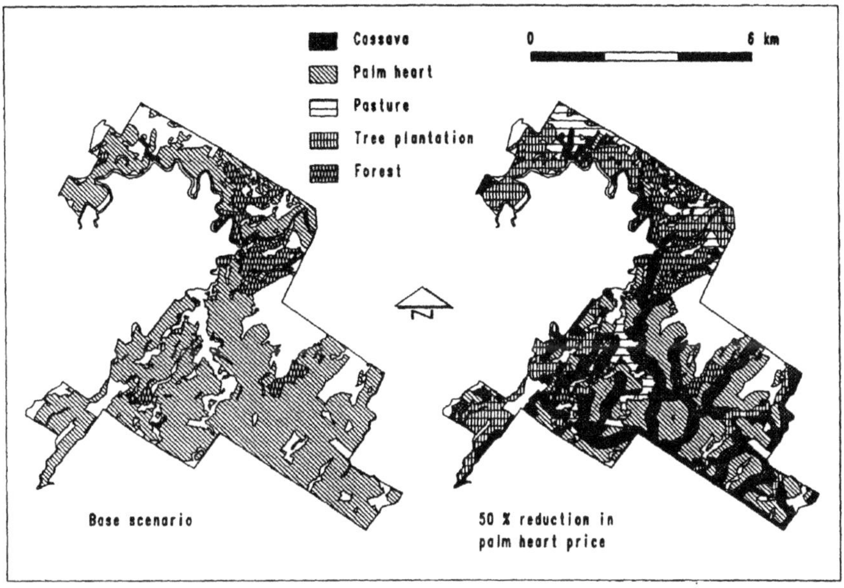

Fig. 3. Two alternative land use scenarios in the Nequev settlement.

Average farm household income is \$12,260 per year, while annual per farm values for economic surplus and nutrient balance came to \$11,150 and \$370, respectively. The value of the nutrient balance is 3.3% of the surplus, in other words, the environmental costs, as assessed by the value of the loss of soil nutrients, is small in comparison to the economic benefits. Negative nutrient balances were found for N (16.6 kg per ha) and K (12.6 kg per ha). A positive balance for P was obtained (2.4 kg per ha) due to high P applications and relatively low losses (no leaching of P occurs due to high P-fixation in the soil). These quantities amount to \$27 as costs per ha of nutrient depletion in the base scenario. As in the discussion of the other scenarios mostly percentage deviations of the base scenario are presented, it is useful to know that the average value of the biocide index value came to 30 per ha, while capital is used to the average extent of \$68 per ha.

The average values hide significant differences between farm types. For example, net annual income per farm varies from \$9,300 for farm type 1 to \$20,600 for farm type 2 (table 3). Nutrient losses and the biocide index value are, respectively, considerably less and much lower on farm type 1 than on the other farm types.

The economic performance of the different farm types is closely correlated with differences in available resources. For example, farm type 1 has an average farm area of 15.7 ha of which 60% is SFP soil, only 12% SFW soil and 28% SIW soil (table 2). SFP can hardly be used, while SIW is less suitable for agricultural production than SFW. Farm type 2, on the other hand,

Table 3. Basic economic data per farm type: base scenario ($\$10^3$ year^{-1}).

Farm type	1	2	3	4	5	average
Value of production[1]	7.2	22.4	12.5	17.3	10.5	11.4
Input costs[2]	0.7	1.5	1.0	1.0	0.9	0.9
Hired labour[3]	0	0.8	0	0.4	0	0.1
Gross margin[4]	6.5	20.1	11.5	15.9	9.6	10.4
Work on other farms[5]	0.3	0	0	0	0.1	0.1
Plantation work[6]	2.5	0.5	1.6	0.6	2.0	1.8
Farm household income[7]	9.3	20.6	13.0	16.5	11.6	12.3
Own labour[8]	0.7	2.2	1.2	1.8	1.0	1.1
Economic surplus[9]	8.6	18.4	11.8	15.7	10.6	11.2

[1]Value of production: physical output, valued at farm gate prices.

[2]Input costs: costs for current input goods (e.g. seeds, fertilizers, biocides) and capital services (e.g. use of machete, knapsack sprayer); in case of own capital goods, capital services include operation costs and depreciation per hour of use; in case of hired goods, capital services are expressed as a rental rate per hour.

[3] Hired labour: costs for hired labour.

[4]Gross margin = Value of production − Input costs − Hired labour.

[5]Work on other farms: remuneration for work on other farms within Neguev.

[6]Plantation work: remuneration for work on plantations outside Neguev.

[7]Farm household income = Gross margin + Work on other farms + Plantation work.

[8]Own labour: valuation of on-farm household labour at a reservation wage.

[9]Economic surplus = Farm household income − Own labour; it represents the returns to land, own capital and management of the farm, and to labour employed off-farm; calculated as a balance, it is an indicator of the postulated objective of the farm households. It coincides with the objective function of the LP model in all scenarios, except in the *valuation of soil nutrient depletion in the objective function* scenario, in which the objective function is the difference between the economic surplus and the value of the soil nutrient depletion.

is relatively large (32.1 ha) with 12% SFP, 10% SFW and 78% SIW. Because this farm type has more land, its farm performance is the best. Farm type 3 is small (13.5 ha) with 7% SFP, 52% SFW and 41% SIW. Thus, it is somewhat better endowed than an average farm, with corresponding performance. Farms of type 4 have an average area of 14.1 ha with 3% SFP, 91% SFW and 6% SIW. These farms are qualitatively best endowed. Their farm performance is also the best after that of the 'large' farm type 2. Finally, farms of type 5, the largest group, have an average area of 13.1 ha with 6% SFP, 6% SFW and 88% SIW. Its resources are less than average, while the same applies to its performance.

Table 4. Overview of scenario runs.

Scenario	Constraints	Data	Results
base	no restrictions on biocide use and nutrient depletion	actual and potential LUSTs	optimal land use under assumed base conditions
palm heart price	as in base scenario	50% reduction in the price of palm heart	relation between price of palm heart and land use
biocide price	as in base scenario	increase of biocide prices with 100%	relation between biocide price and biocide use
biocide regulation	quantitative restrictions on biocide index at Neguev level (percentages of average per ha value of base scenario)	as in base scenario	effects of restricting biocide use on land use
soil nutrient depletion	soil nutrient depletion at the level of land units within farm types	as in base scenario	effects of restricting soil nutrient depletion on land use
labour cost	as in base scenario	25% increase in labour costs	relation between labour costs and land use
capital availability	as in base scenario	quantitative restrictions on the availability of capital from $66.0 to $16.5 (in steps of $16.5).	relation between availability of capital and land use
valuation of soil nutrient depletion in the objective function	as in base scenario	incorporation of soil depletion in objective function	relation between environmental accounting and land use

Besides the base scenario, seven alternative scenarios were evaluated (table 4). Results of these alternative scenarios are presented as percentage deviations from the base run solution (table 5).

Palm heart price scenario

Whereas the area planted to palm heart has been steadily increasing since 1987, the actual palm heart area in 1991 (about 140 ha, table 1) was significantly less than the area calculated by the base scenario (nearly 2,700 ha). Therefore, the base scenario results can be interpreted as an indication that in the future more palm heart could be planted. Field observations in the Neguev

Table 5. Results of alternative scenarios (in % changes from base scenario).

Scenario	Objective function[1]	Gross margin	Farm household income	Nutrient balance per ha per year N	P	K	cost	Biocide index
Palm heart price –50%	–28	–38	–27	–16	–16	–16	–16	57
Biocide price +100%	–1	–1	–1	0	0	0	0	0
Biocide use 50% of base	–1	1	0	0	0	0	0	–50
Soil depletion	–3	–3	–2	–12	57	–14	–22	–4
Labour costs +25%	2	0	4	0	0	0	0	0
Available capital $66.0/ha	0	0	0	2	–27	2	6	1
Available capital $49.5/ha	–2	–3	–2	6	–99	4	20	–6
Available capital $33.0/ha	–6	–8	–5	2	–104	–4	15	–91
Available capital $16.5/ha	–25	–40	–25	–30	–125	–36	–19	–40
Valuation of soil nutrient depletion in the objective function	–3	0	0	0	0	0	0	0

[1]Objective function equals the economic surplus of the farm households, except in the last scenario in which the valuation of soil nutrient depletion is deducted from the economic surplus and thus taken into account for the optimisation of the LP model.

during 1993 and 1994 confirm that farms are still planting new palm heart. However, palm heart is a luxury product with relatively small domestic and international (mainly France and the USA) markets, crowded by competitors (especially Brazil and Colombia). The present production of palm heart in Costa Rica is supplied by about 2,000 ha, and an area of 2,700 ha with palm heart in the Neguev would double national production, with falling prices as a most likely consequence.

Consequently, alternative scenario runs were performed with reductions in the price of palm heart of 10 and 50%, respectively. A 10% price reduction hardly affects the model results. A 50% price reduction, however, causes a 60% decrease in palm heart area. Palm heart is partially replaced by cassava (figure 3), resulting in a 27% decrease (or $3,400) in farm household income (table 5). The selected cassava technologies use significantly more biocides than the palm heart LUSTs, resulting in a 57% jump in the average biocide index. However, aggregate nutrient losses decrease by 16%, in money terms about $4 per ha or $60 per farm. Therefore, aggregate losses (in terms of decreases in farm household income and reduced costs of soil depletion) per farm and for the Neguev settlement as a whole amount to $3,340 and $1,025,000, respectively.

Biocide price scenario

One of the most direct measures to reduce biocide use is an increase in biocide prices (e.g., through an environmental tax). For palm heart, which is the main crop in the base scenario, a number of alternative technologies are available in which the use of herbicides, the most commonly applied biocides in palm heart, is replaced by manual weeding. However, increases in the price of biocides as high as 100% had no significant impact on land use, with only marginal changes in income and sustainability indicators (table 5). The explanation for this result is that biocides represent only a relatively small part of total production costs for palm heart. In addition, even at a 100% increase in biocide price, substitution of herbicides by hand weeding is still relatively expensive.

Biocide regulation scenario

An alternative way to achieve reductions in biocides is to limit its use quantitatively, even though it is recognised that such a policy may meet practical implementation difficulties. In the model, such a restriction can be included in the set of constraints. A reduction in the biocide index value of 50% (distributed over the farm types in proportion to the area of each farm type) hardly changes farm household income. Sustainability is considerably improved since soil depletion is hardly changed in comparison to the base scenario (table 5). The rationale behind these results is that the land use pattern is very similar to the base scenario. Even though, in palm heart, herbicides are replaced by manual weeding, other inputs and outputs are maintained at the same level as in the base scenario.

Results of a further tightening of restrictions on the use of biocides are presented in table 6. Whereas the total area under palm heart remains constant, technology choice is pushed towards production techniques which use manual weeding instead of herbicides. As a result, hardly any reduction in farm household income occurs. Labour productivity, defined as the gross margin per labour day worked, decreases with increasing manual weeding. On the other hand, land productivity, defined as the difference of the value of production and the input costs per ha per year, slightly increases because of reduced herbicides costs. The shadow price of the biocide index increases with further tightening of biocide use.

Soil nutrient depletion scenario

In the base scenario, possible negative yield effects as a result of nutrient losses which build up over time are not taken into account. Consequently, in the current scenario, yearly permissible losses of N, P and K for each soil group within each farm type are used as values for the nutrient loss

Table 6. Influence of biocide use restrictions, relative to base scenario, on economic surplus, land and labour productivity, and areas with palm heart, both with and without herbicides.

Scenario	base[1]	50%[2]	40%	30%	20%	10%	5%	1%
economic surplus per farm ($\$10^3$ year^{-1})	11.15	11.05	11.03	11.01	10.99	10.97	10.95	10.94
shadow price of total biocide constraint[3] (\$ constraint unit^{-1})	0	0[4]	0.49	0.49	0.49	0.50	0.51	0.65
biocide use (index value ha^{-1} year^{-1})	30.0	14.8	12.0	9.0	6.0	3.0	1.5	0.3
land productivity[5] (\$ ha^{-1} year^{-1})	757	762	764	765	766	766	767	767
labour productivity[6] (\$ day^{-1})	5.08	4.91	4.88	4.84	4.81	4.78	4.76	4.75
palm heart with herbicides (ha)	2676	1177	893	596	299	29	3	0
palm heart without herbicides (ha)	0	1499	1783	2080	2377	2647	2673	2676

[1]Base scenario without any restriction on the use of biocides.
[2]Reduction of biocide use 50% in comparison to the base scenario.
[3]A shadow price of a constraint indicates the decrease in the objective function as a result of tightening the constraint with one unit. In this case the shadow price refers to the total biocide constraint rather than to the constraints per farm type.
[4]The overall biocide constraint is not binding, and neither is the constraint for farm type 1; however, the constraints for farm types 2 to 5 are binding. The constraints per farm type have a weighted average shadow price of \$0.43. The weights are the respective right hand side values of the constraints.
[5]Land productivity is defined as the difference of the value of production and the input costs per ha per year.
[6]Labour productivity is defined as the gross margin per labour day worked.

Table 7. Total amount of nutrients (kg ha^{-1}) and permissible yearly losses (kg ha^{-1} year^{-1}) in the top 20 cm of the three soil types. (Source: Jansen and Schipper 1995)

Soil type	N		P		K	
	total	loss	total	loss	total	loss
SFP	3696	37	539	27	446	22
SFW	4831	48	299	15	355	18
SIW	3610	36	278	14	297	15

constraints in the LP model. It is assumed that these critical values remain constant over a ten-year period and that yields remain constant as long as depletion does not exceed the critical values. For each nutrient in each soil group a separate assessment is made on the basis of nutrient and soil specific factors (table 7).

For part of the palm heart area, the restriction on the depletion of soil nutrients results in a substitution of the zero fertiliser technology with a technology that uses 50% of the amount of fertiliser needed to reach the highest possible yield (Jansen and Schipper 1995). In this way, the depletion caused by the zero fertiliser technology is compensated, while the overall area planted to palm heart hardly changes. The most critical nutrient is K for which the critical loss per ha per year limit is reached on the well drained soils for all farm types. Sustainability is substantially improved with 22 and 4% reductions in soil depletion and the biocide index, respectively (table 5). However, this occurs at the cost of a 2% reduction of the farm household income in comparison to the base scenario.

Labour costs scenario

A priori, the effects of a change in the price of labour on farm household income can be expected to be ambiguous. Even though labour constitutes a significant part of total production costs for most LUSTs, it is both a cost (own household and hired labour) and a benefit (off-farm work on other farms within the Neguev and on plantations). By construction, the total amount of hired labour equals the amount of off-farm work on other farms and, since its wage is 10% higher than the wage for off-farm work on other farms, hired labour implies a net cost to the model. Since the (reservation) wage in the base scenario for own household labour is lower than the wage for plantation work, an equal percentage increase in all wages decreases the attractiveness of on-farm activities relatively to working on plantations. A 25% increase in the price of labour does not affect the gross margin earned from crop cultivation. However, a substantial increase in off-farm work results in an increase in farm household income (table 5).

Capital availability scenario

Capital is often one of the most limiting farm resources and frequently the focus of policy interventions. The base scenario involved a capital use of $68 per ha. The impact on land use of lower capital availability, i.e., $66.0, $49.5, $33.0, and $16.5, was evaluated. The effects of increased capital scarcity are twofold (table 5). First, relatively expensive (i.e. input-intensive) technologies are replaced by technologies requiring less inputs, followed by a similar change of land use types. For example, even though palm heart remains an attractive crop as reflected in a virtually constant area (at least with a capital availability of at least $33.0 per ha), herbicide intensive technologies are increasingly replaced by technologies that involve manual weeding operations. The dramatic decrease in the biocide index (91%) when moving from a capital availability of $49.5 to $33.0 is mainly due to a switch from palm heart technologies that use herbicides to technologies that involve hand weeding.

Then, a capital availability of $16.5 per ha results in a severe decrease in palm heart area, as it becomes impossible to finance the necessary investments. As a consequence, palm heart is partially replaced by cassava. The cassava technologies selected under a severely restricted capital regime use significantly more biocides and fertiliser than the palm heart technologies thus replaced, inducing a considerable rise in the average biocide index. However, compared to the base scenario, the biocide index is still 40% lower. With regard to soil nutrient depletion, increasing capital scarcity results in increasing aggregate nutrient losses at first, followed by a decrease compared to the base scenario (table 5).

The second effect of rising capital scarcity is that decreases in income from crop cultivation are compensated in part by working more off-farm. As a result, farm household income decreases less than the gross margin from crop cultivation.

Valuation of soil depletion in objective function scenario

This scenario differs from all others in that the costs of changes in nutrient balances are directly accounted for in the objective function of the model. Thus, nutrients are valued at their replacement costs (Ehui and Spencer 1993; WRI 1991) which receive weights equal to changes in economic surplus of the farm households. However, the model outcomes were virtually identical to that of the base scenario, with the exception of the objective function being $370 per year per farm lower, which is exactly the value of the soil nutrient losses. In effect, the incorporation of the valuation of the nutrient balance in the objective function did not alter the outcome of the optimisation process with regard to land use and income generation. It is shown elsewhere (Stoorvogel 1994) that, even with four times the original nutrient depletion penalties, no significant changes in land use take place. These results are due to the small value of nutrient losses relative to farm household economic surplus.

Conclusions

This paper describes an interdisciplinary methodology (called USTED) for land use analysis. The methodology integrates linear programming (LP) techniques for optimisation and a module for the storage of data on combinations of land use types and soils with specific technologies (called LUSTs), linked to a geographical information system. Each LUST has fixed technical coefficients, estimated on the basis of farm surveys, literature searches, expert knowledge and crop growth simulation models.

The methodology has thus far been developed for the sub-regional level, and the LP model is currently operational for the Neguev settlement in the

Atlantic Zone of Costa Rica. Because the LP model contains sub-matrices for a number of farm types within the sub-region, it takes account of different resource availabilities at the farm level. At the same time, the different farm types are not optimised in isolation, since each type has to take into account sub-regional labour supply and demand. Besides the focus on maximising (aggregate) economic surplus, the methodology explicitly takes into account sustainability-related factors which are operationalised through nutrient balances and a biocide use index. As such, the methodology can be used to provide the policy maker with information regarding trade-offs between different goals. This is illustrated through the analysis of the effects on land use of changes in the socio-economic environment, such as decreases of the palm heart price and changes in wages. Furthermore, the effects of policy instruments were studied, in particular price instruments like a doubling of biocide prices, and regulatory measures, as reductions in the capital availability (credit) and limits on biocide use and nutrient depletion.

Research to refine and extend the methodology currently continues along various lines. First, since land use policy is typically formulated at the regional or national level, work is now focusing on upscaling the methodology towards the regional level (i.e., the Atlantic Zone of Costa Rica). Among other things, this requires information on price and cost differences at the sub-regional level, as well as estimates for own and cross-price elasticities to relax the assumption of perfectly elastic demand curves. In addition, a larger number of land use types may be included in the model. For example, in the Neguev model, cattle is connected to pasture at fixed stocking rates with no supplementary feeding from other land use types. However, this assumption is no longer realistic for more intensive management systems with improved pastures, legumes, supplementary feeding, etc. Thus, there is a need to define animal production systems with specific technology levels (APSTs), which use products (pasture, cobs, bananas, leguminous leaves) from LUSTs as inputs. While these inputs provide the necessary calories, proteins and dry matter to the APSTs, products of APSTs (e.g. dung) can be used by LUSTs as inputs.

Second, while the methodology performs well for the Neguev, its development and associated data collection have taken several years. Consequently, efforts are ongoing which aim at a low-input data validation of USTED for the Guanacaste region of Costa Rica.

Third, adequate incorporation of aspects of sustainability in the model warrants continuing attention. Soil nutrient depletion is estimated as a flow variable for each LUST, comparing the depletion with estimates of the stock of nutrients in the soil, while assuming a period in which the depletion does not affect the land use type in question, thus indicating the limits for resource use. However, the effects over time of decreasing soil fertility on yields are not taken into account. Similarly, the real costs associated with the use of biocides in terms of environmental degradation and human health damage are not yet part of the methodology.

Acronyms

APST	Animal Production System with a specified Technology
CATIE	Centro Agronómico Tropical de Investigación y Enseñanza (Center for Research and Education in Tropical Agriculture)
IDA	Instituto de Desarrollo Agropecuario (Institute of Agrarian Development)
GIS	Geographical Information System
MODUS	MOdules for Data management in USted
LP	Linear Programming
LUST	Land Use System with a specified Technology
USTED	Uso Sostenible de Tierras En el Desarrollo (Sustainable Land Use in Development)
WAU	Wageningen Agricultural University

References

Ehui S K, Spencer D S C (1993) Measuring the sustainability and economic viability of tropical farming systems: a model from sub-Saharan Africa. Agricultural Economics 9:279–96.

Hazell P B R, Norton R D (1986) Mathematical programming for economic analysis in agriculture. Macmillan Publishing Company, New York, USA.

Herrera W, Gómez L D (1993) Mapa de unidades bióticas de Costa Rica. Instituto Geográfico de Costa Rica, San José, Costa Rica.

IDA (1985) Condición de la parcela. Instituto de Desarrollo Agropecuario, San José, Costa Rica (unpublished).

Jansen D M, Schipper R A (1995) A static descriptive approach to quantify land use systems. Netherlands Journal of Agricultural Science (submitted).

Jansen D M, Stoorvogel J J, Schipper R A (1995) Using sustainability indicators in agricultural land use analysis: an example from Costa Rica. Netherlands Journal of Agricultural Science (submitted).

Mücher C A (1992) A study on the spatial distribution of land use in the settlement Neguev. Centro Agronómico Tropical de Investigación y Enseñanza, Turrialba, Costa Rica. (Atlantic Zone Programme, Phase 2, Report No. 9).

Pearce D W, Turner R K (1990) Economics of natural resources and the environment. Harvester Wheatsheaf, New York, USA.

Schipper R A (1993) Una caracterización de fincas en Neguev, Río Jiménez y Cocori; Zona Atlántica, Costa Rica: informe en base a una encuesta dentro del marco de estudios de base. Centro Agronómico Tropical de Investigación y Enseñanza, Turrialba, Costa Rica. (Atlantic Zone Programme, Phase 2, Report No. 49).

Schipper R A, Jansen D M, Stoorvogel J J (1995) Sub-regional linear programming models in land use analysis: a case study of the *Neguev* settlement. Netherlands Journal of Agricultural Science (submitted).

Stoorvogel J J (1993) Optimizing land use distribution to minimize nutrient depletion: a case study for the Atlantic Zone of Costa Rica. Geoderma 60:277–292.

Stoorvogel J J (1994) Systems integration to evaluate alternative land use scenarios. Paper presented at the ICASA symposium 'Role of agronomic models in interdisciplinary research', annual ASA meetings, 13–18 November 1994, Seattle, USA.

Stoorvogel J J, Schipper R A, Jansen D M (1995) USTED: a methodology for quantitative analysis of land use scenarios. Netherlands Journal of Agricultural Science (submitted).

Waaijenberg H (1990) Sistemas de producción. In De Oñoro M T. El Neguev: interacción de campesinos y estado en el aprovechamiento de los recursos naturales. Centro Agronómico Tropical de Investigación y Enseñanza, Turrialba, Costa Rica. (Serie Técnica, Informe Técnico No. 162).

WRI (World Resources Institute) (1991) Accounts overdue: natural resource depreciation in Costa Rica. Tropical Science Center, San José, Costa Rica; World Resources Institute, Washington, DC, USA.

Response to section F

M.S. DICKO, N.G. TRAORÉ and K. SISSOKO
Mali

To develop strategies for improved food production without concomitant degradation of the basic natural resources, especially soils, is a challenge for all countries, but in particular for those experiencing spurts in population. For many observers and members of the Technical Advisory Committee (TAC) of the Consultative Group on International Agricultural Research (CGIAR), such strategies should have an eco-regional spread, and in order to sustain productions a balance should be maintained between research on crop improvement and management of natural resources.

The three papers in section F demonstrate the importance of modeling for the comprehension and exploration of the complexity of agricultural production systems. Models allow, among other things, a quantitative system description and *ex ante* evaluation of impact of interventions and environmental variations productivity. Assessment of risks associated to profits can also be made.

Model performance is clearly presented in the papers by Thornton *et al.*, Van Keulen, and Schipper *et al.* They used a simulation model of the whole farm, a soil nutrient model, and a linear programming model for optimal use of lands, respectively. The results in terms of areas occupied by arable lands, nutritive value of soils, crop production, or monetary reserves may help programming of efficient research and planning of appropriate development actions.

In addition, Thornton *et al.* examined the role of their model at the enterprise and the farm level analysis and discussed how to achieve aggregation between these levels and the regional level. However, they did not examine the analysis at the village level, which is an interesting step in the aggregation process between farm and region level. Also, they did not describe any relationship between their crop simulation model and other models such as animal production and linear programming models. Such comparisons could, for instance, show how to utilize their model for establishing better crop and livestock integration mechanisms.

Van Keulen also did not refer to relationships between models, while in their conclusions, Schipper *et al.* observed connections between the model they used and simulation models that allow evaluation of technical coefficients.

J. Bouma et al. (eds.), Eco-regional approaches for sustainable land use and food production, 397–399.

Thornton *et al.* defined an enterprise as an activity to achieve a given production, and a farm as a collection of enterprises. The importance of farm and enterprise data in modeling was described by Schipper *et al.* who pointed out that land use policy is determined more by farmers than by regional decision makers.

When one talks about land use and food production in a region, one mentions crop and livestock production, which are generally both carried out at the farm level. However, in some cases, like in animal transhuman systems, the land involved may extend to one or two eco-regions. In addition, one should not forget other types of food, such as hunting and gathered products from quite large bushes in some developing countries. Unfortunately, the papers presented here were biased towards crops. Most of the models presented made few or no allusion to others foods. In sub-Sahara Africa relevant eco-regional approaches for sustainable land use and food productions require taking into account all the above mentioned production types. Consequently, enterprise and farm data are not everywhere available and are not always sufficient for developing regional food production strategies. Furthermore, one should point out difficulties in extrapolating enterprise and farm data to a region. How to integrate data from different farms, each with different resources and objectives in order to derive a satisfactory and profitable policy for all, is an important question which has not yet received and adequate answer.

The following comments will focus on some factors which should not be forgotten when elaborating models for sub-Sahara Africa situations. Already, quite a number of physical and economical factors can be transcribed and introduced in models. As far as socio-cultural factors are concerned, few of them are yet considered. However, they are important for the purpose of this meeting, particularly when developing approaches in sub-Sahara Africa. For instance, regarding land tenure, sustainable soil fertility is of lesser importance to a farmer who rents the land being cultivated. Similar, degradation of pastures generally occurs in communal grazing systems. Difficulties in the quantification and transcription of socio-cultural factors, which are somehow subjective, might probably explain the present marginal cause of this type of information. According to Thornton *et al.*, another reason is our inability to understand the decision making process by farmers.

Female labour constitutes an important input in agriculture and has a major impact on land use, tapping forests, and the improvement of soil fertility. In southern Mali, crops such as vandzou, groundnut and sesamc, specifically cultivated by women, are introduced in crop rotation after cotton, maize and sorghum. When one investigates agriculture in the sub-Sahara Africa, there is certainly a need to explore scenarios for incorporating female contributions to agriculture.

In harmony with the authors of the papers, I agree that eco-regional approaches for sustainable land use and food production cannot by-pass enterprise and farm data. However, more information on foods that are produced beyond farm boundaries might be needed for developing consolidated strategies for regions. Since models constitute tools for analyzing and exploring production systems, there is a need for improving their performance by taking into account all factors influencing land use and food productions. Thus, planning of top-down projects should be avoided.

Lastly, a tool is only useful if one can use it. Efficient utilisation of models in developing counties requires training of people able to elaborate, manipulate, and adapt models to the specific conditions of their environment. Presently, this can and should be stimulated through international cooperation on model development and utilisation.

Response to secion F

J. BAIDU-FORSON, M.V.K. SIVAKUMAR and J. BROUWER
ICRISAT Sahelian Center, B.P. 12404, Niamey, Niger

Introduction

Judicious reviews and discussions of the papers require clear understanding of the key issues that the authors should address: approaches for an eco-region based on enterprise/farm considerations; sustainable land use and food production.

An eco-region encompasses a broad contiguous biophysical environment that is not necessarily confined within the boundaries of one country. Admittedly, from this description, research with an eco-regional focus will have to deal with many stakeholders who differ in socio-economic characteristics; enterprise/farm variations; and in some cases, policy variations that affect outcomes or strategies. There is also the disquieting reality of unreliability in production trends unless records are available for 10 to 20 years (Monteith 1990). Yet, we simply cannot depend on costly long-term experiments to determine sustainable land use patterns and food production. In addition, results of such experiments may have limited validity where significant spatial and temporal diversities occur within contiguous and seemingly similar biophysical environments. This makes it hazardous to attempt to propose a 'blanket recommendation' for an eco-region.

We have to reconcile the challenges of biophysical and socio-economic variations with donor expectations of short to medium term research results that contribute to widespread improvement in the welfare of farm populations. Modelling offers a reasonable short-term approach. An important option in this regard is to use results from previous experiments or characterization surveys to model concepts that can be widely applied at targeted decision points within an eco-region. A realistic requirement of this approach is to incorporate enterprise/farm analysis and realities in conception, development and validation of models. This will minimize the divergences which underlie the observation of Thornton *et al.* that "researchers' and farmers' mindset rarely meet".

We now define sustainability requirements that are quite appealing and form the basis of assessment of the approaches proposed by Thornton *et al.*, Van Keulen, and by Schipper *et al.* Sustainability should reconcile two imperatives: maintenance of year-to-year benefits and non-deterioration of a

J. Bouma et al. (eds.), Eco-regional approaches for sustainable land use and food production, 401–408.

system (Monteith 1990). While the first imperative requires the enhancement of production efficiency and profitability, the second suggests that trends in viability of resources are central to the assessment of sustainability. Attainment of these two goals need not necessarily be incompatible.

Response to paper by Thornton *et al.*

The authors first discuss conceptual issues relevant to their paper and then illustrate the potential use of simulation models in two case studies: providing management recommendations on input use to stabilize returns at enterprise-level; and highlighting the role of price and weather risk in analysis of management options at the farm level. The authors conclude with four specific areas where major progress will be needed prior to extending analysis to eco-regional level.

We now briefly comment on some of the conceptual issues highlighted by Thornton *et al.* The key hypothesis in designing an agricultural technology is that it responds to the needs and preferences of an identifiable target to provide improvements over an existing situation. If this hypothesis is true then we need not worry too much about predicting adoption, if a conducive policy and institutional environment exists to support adoption once the technology is right. Adoption prospects become a problem when the needs and preferences of the target group are not properly accounted for in conception, and development of technologies or relevant supporting policies and institutions are not in place.

Thornton *et al.* dramatize the problem of searching for optimality to bring home the difficulties in dealing with human preferences in a real world environment. It is essential to identify the full range of attributes which members of a target group aim at optimizing. However, an important prerequisite is to have a clear definition, from the target group's perspectives, of the problem to be solved and the goal. It is equally important to identify priorities assigned to the various attributes. Admittedly, priorities attached to attributes – and hence the coefficients that reflect them – are likely to change over time and we are a bit far away from correctly predicting the magnitude of such changes. However, on the basis of priorities at a point in time, it should be possible to narrow down the range of attributes that need to be optimized. Otherwise we run the risk of seeking to optimize an infinite range of attributes and possibly their interactive effects. When we finally end up with more than one objective function, goal programming (Hillier and Lieberman 1980) and multi-criteria decision making techniques (Cohon *et al.* 1979; Romero *et al.* 1987; Berbel 1993; Maino *et al.* 1993) offer opportunities to handle multiple objectives at farm level. However, there is the important question of determining appropriate relative weights to attribute to the different objectives. Also, the use of representative situation(s) may minimize the problem of many players. Obviously, this requires identification of homogenous groups (Byerlee *et al.* 1980; Harrington and Tripp 1984; Williams 1994) using appropriate classifying variable(s), within the enterprise/farm type or eco-region.

The key argument of the case study on maize production in Malawi is that satisfaction of potential food demands depends on improved adaptive crop management set in the context of an enabling institutional and policy environment. It is hard to disagree with this viewpoint for rainfed agriculture in semi-arid regions of sub-Saharan Africa.

The authors hypothesize that "validated crop models offer real potential in the efficient derivation and provision of management recommendations appropriate to a location and a season type". To the extent that the model properly characterizes the location(s) and season-type, it may be hard to argue to the contrary assuming the goal is the same as that of the target population. It is a truism that farmers re-evaluate production decisions either in making choices or in sequencing activities on the basis of environmental – biophysical, socioeconomic, institutional and policy – conditions. However, the date of onset of rains is not sufficient. Weather-based enterprise-level response farming model could benefit from reflecting the influence of durations and probability of dry spells in choice of strategy. This is because of the need to match crop phenology with dry spell lengths, particularly to meet water requirements at sensitive stages of growth (Sivakumar 1992). Moreover, the lengths of dry spells within a season influence decisions to weed, apply a second dose of fertilizer or change variety, even after the onset of rains.

Some limitations identified, e.g. effects of disease/pests on productivity, resource constraints, etc., show the need for integrated weather-crop-pest/disease model(s), coupled with discrete stochastic sequential programming (Cocks 1968; Rae 1971; Kaiser *et al.* 1993) to reflect production strategy decisions at different temporal or spatial stages. However, the appropriate level for such analysis will be the farm or higher scale unless the objective(s) and constraints can be clearly specified at the enterprise-level, and there is no interdependence of enterprises.

Although Thornton *et al.* emphasize "improved adaptive crop management" to achieve increased production, there are important eco-regional implications of weather-based response farming for breeding strategy. In particular, there is the need to focus on provision of an array of cultivars with different maturity cycles, tolerance to dry spells or physiological flexibility to adjust to difficult conditions that could be targeted to categorized decision points within an eco-region. This is particularly relevant for rainfed agriculture characterized by high variability in length of growing season and dry spells. However, the successful development of an appropriate breeding strategy would require the exploitation of data on season and dry spell lengths and probabilities to determine a limited set of cultivars that should be targeted to specific decision points.

The case study for the hillside farm is an interesting, well-constructed whole-farm analysis. However, some of the weaknesses of the simulation model components identified have serious consequences for the validity of recommendations if the aim is to promote sustainable production. The emphasis was on searching for cropping system options for farmers who wish to replace coffee on hillsides of the marginal coffee-growing area in the Cauca River watershed of the lower Andes. farmers do derive a high but variable income from coffee. Also, coffee provides continuous ground cover to protect 'steep andic soils from erosion'. Since replacement of coffee with annual crops exposes soils to erosion, the benefits and costs associated with soil/nutrient and chemical transfer should be internalized in the modelling framework if we are to reflect concerns about sustainable production. This analysis may need to be done on a watershed scale. It is hoped that this wille be attempted in further model refinements because it is likely that the cropping systems evaluated differ in their environmental effects.

A minor point is the clarification of the representativeness of the farmer chosen for analysis. This is important for judging the appropriateness of results for making broad recommendations. Are the sensitivity analyses of preference rankings to simulated scenarios of changes in discount rate and risk sufficient? What about the sensitivity analyses of resource endowments?

The concluding comments raise thought-provoking issues, some of which are hard to disagree with. However, it is doubtful that we are able to "understand and predict many biophysical processes" with a high degree of certainty. Some may claim that considerably more modelling advances have occurred in the biophysical sciences than in the socio-economics area, particularly in modelling human behavior. However, rainfall distribution and amounts cannot as yet be predicted well enough, particularly in eco-regions such as the Sahel where climatic variability is high. Happily, model refinements are still going on. In the case of farmer decision making, the need for more work on analytical tools cannot be disputed. However, do we want to clamor for an ability to predict temporal changes in the coefficients of augments of preference functions of individuals? We may be straying onto very slippery grounds if we clamor for perfect prediction of human behavior. Otherwise, existing mathematical programming methodologies provide some avenues for modelling choice strategies reasonably well. Nonetheless, reliance on biophysical models – no matter how perfect they seem – will be of limited value unless we incorporate socioeconomic information, however imperfect. Appropriately, Thornton *et al.* suggested four specific areas where 'collaborative' progress is needed. These should provide points of reflection during the general discussions and for modelers.

Response to paper by Van Keulen

In recognition of the problematic nature of long-term experimental approach to judging sustainability of cropping systems and environmental complexities, Van Keulen focussed on modelling processes that determine production. The key argument is the need for a modelling approach whose data requirements can be satisfied and thereby overcome deficiencies of earlier modelling efforts: data requirements that can hardly be satisfied; problem of validation of variables that cannot be adequately measured; unbalanced descriptions due to partial knowledge of many processes. In furtherance of this objective, Van Keulen outlined "some of the main principles" guiding the formulation of a preliminary version of such a model. The main areas explored are: environmental factors; soil organic matter; soil acidity; soil erosion; crop nutrient availability; crop water availability and crop and soil management.

The systematic approach, adopted by the author, to harnessing knowledge into modelled concepts and the emphasis on the search for a model whose data requirements can be satisfied are quite commendable and appropriate to obtaining a model that could have wide applicability in an eco-region. From a sustainability perspective, assessment of temporal variability of soil organic matter and nutrient pools, under alternative management strategies, may provide a rational basis for assessment of appropriate management strategies. We have few comments that hopefully could enrich model developments subsequent to the preliminary one presented in the paper.

Firstly, we have a query related to the author's definition of location: "location is defined by the first and last month of the growing season". Does this definition imply that the dynamics of soil organic matter and nutrients during the off-seasons are irrelevant? Research conducted at the ICRISAT Sahelian Center showed the incorporation of organic matter by termites, significant continued conversion of nutrients in moist subsoil during the dry season and the leaching of nutrients – including organic carbon, nitrogen and phosphorus (Brouwer and Powell 1994). For adaptation of the model to semi-arid situations, the inclusion of the effects of termite activity on nutrient dynamics is very important. Furthermore, nutrient input through dry dust depositions – originating from eroded areas – are quite important, particularly in the drier regions of the semi-arid tropics. Ideally, a comprehensive physical-chemical-physiological crop growth model needs to be coupled with models of population dynamics of pests and diseases. What may appear promising from only water and nutrient use point of view may appear less attractive when dynamics of pests and diseases are taken into account.

Secondly, one wonders whether the use of monthly climatic data can effectively capture variability in the factors examined. This is because the use of monthly data may be questionable in the semi-arid tropics where timely sowing with first 'good' rains at locations with short growing season holds the key to good yields. For example, in the West African Semi-Arid Tropics, a delay in sowing for 20 days can result in two-fold difference in millet yields. Also, rainfall distribution in time has a much greater impact on crop growth and yield rather than the total amount of rain *per se*. Therefore, can one disregard intra-seasonal dynamics in the semi-arid tropics?

Thirdly, Van Keulen has not included multiple cropping patterns – intercropping, relay cropping and sequential cropping. Residual moisture in the soil profile after the first crop influences the yields of the second crop.

Fourthly, can the average monthly temperatures used by the author adequately simulate the effects of temperature range, maxima and minima on different physical and physiological processes?

Fifthly, the assumption that soil surface is covered during the cropping season, ignores the soil loss computation implications of only partial coverage which is commonly observed in most areas of the semi-arid tropics.

Finally, since this exploratory study of dynamics of sustainability indicators "may help in formulating criteria for different management strategies", it stimulates other interests as well. In particular, it is hoped that a model obtained from this approach could provide guidance to the formulation of criteria for advising when and what changes in strategies are needed to maintain sustainable production. Also, for the measurement of total factor productivities of cropping systems it is increasingly advised that particular attention be given to valuation of natural resource stocks and flows (Ehui and Spencer 1993). This requires accounting for implicit costs in terms of forgone productivity when stocks of resources are decreased – declines in Ca and Mg leading to increased acidity for example – and implicit benefits when stocks of resources are increased (eg. through nitrogen fixation). Hopefully, the modelling effort will later link the dynamics of sustainability indicators to crop productivity and facilitate the computation of such implicit gains or losses.

Response to paper by Schipper *et al.*

Schipper *et al.* analyze land use patterns in a sub-region based on segmented representative farms. Admittedly, the segmentation of farms and incorporation of resource availabilities relevant to representative farms could contribute to minimizing the first aggregation problem: "a region seen as a whole farm produces more than the sum of farms of that region because of more favorable ratios of resource availabilities". However, side-stepping the problem of identification of variables that change from being exogenous at farm level to endogenous at regional or sectoral level is not very helpful. This is borne out by the model results which predict palm heart production levels that would lead to price declines whereas the authors assume that the region is "small" for output prices to be affected by supply from the sub-region. Also, it is not clear how the assumption of farm household objective function resolves the third aggregation issue: "decisions about land use are being made at different levels, at each level there might be different objectives, furthermore, each level is not fully informed about decisions at the other levels". This is especially the case if policymakers at national or regional levels are not well informed of farm-level situation or pursue social objectives which may not necessarily lead to privately optimal solutions.

Explicit consideration of risk is needed when either the decision-maker is risk averse or production is dynamic (Antle 1983). This requires knowledge of the properties of price and output distributions faced by farmers. It is hoped that this aspect will be incorporated in future model refinements.

'Logged forest' is one of the land use types considered in the model for the Neguev sub-region. However, the authors do not explain how the forest is managed to ensure that its exploitation would not be relevant to sustainability of production in the Atlantic Zone. Is the "tree plantation" a re-afforestation requirement in the model to ensure replacement of "logged forest"? If yes, how was the rate of logging, with replacement, determined so as to avoid detrimental exploitation. This needs to be clearly spelt out. Otherwise, the depletion of forest cover has implications for sustainable agricultural production because the forest contributes to the maintenance of soil quality and reduction of soil erosion. In many cases when forests are cleared, the physical and chemical properties of soils undergo significant changes that could lead to nutrient loss, accelerated erosion and declining yields. The relevance of forest logging for sustainable production provided a motivation for Ehui and Hertel (1989), using a two-sector – agriculture vs forestry – optimal control model, to analyze optimal steady state forest stock.

To conclude, the paper reveals some problems to be resolved in building sub-regional level land use and planning models. However, land use planning for an eco-region is likely to be quite rare, particularly if parts of several countries are within the eco-regional boundary. Even within the confines of a single country, centralized planning seems to be on the wane. However, sectoral/regional/national land use analysis may provide opportunities for examining policies needed to modify human behavior and achieve socially desirable management of resources and guarantee sustainable production.

References

Antle J M (1983) Incorporating risk in production analysis. American Journal of Agricultural Economics 65(5):1099–1106.

Berbel J (1993) Risk programming in agricultural systems: a multiple criteria analysis. Agricultural Systems 41:275–288.

Brouwer J, Powell J M (1994) Soil aspects of nutrient cycling in a manure application experiment in Niger. Proceedings of the International Workshop on the Role of Livestock in Nutrient Cycling in Mixed farming Systems of Sub-Saharan Africa, ILCA, Addis Abeba, Ethiopia, November 22–26, 1993 (in press).

Byerlee D, Collinson M, Perrin R, Winklemann D, Biggs S, Moscardi E, Martinez J, Harrington L, Benjamin A (1980) Planning Technologies Appropriate to Farmers: Concepts and Procedures. CIMMYT, El Batan, Mexico.

Cocks K D (1968) Discrete stochastic programming. Management Science 15:72–79.

Cohon J L, Church R L, Sheer D P (1979) Generating multiobjective trade-offs: an algorithm for bicriterion problems. Water Resources Research 15(5):1001–1010.

Ehui S K, Hertel T W (1989) Deforestation and agricultural productivity in the Côte D'Ivoire. American Journal of Agricultural Economics 71 9(3):703–711.

Ehui S K, Spencer D S C (1993) Measuring the sustainability and economic viability of tropical farming systems: a model from sub-Saharan Africa. Agricultural Economics 9:279–296.

Harrington L, Tripp R (1984) Recommendation domains: A Framework for On-farm Research. Economics Program Working Paper 02/84, CIMMYT, El Batan, Mexico.

Hillier F S, Lieberman G J (1980) Introduction to Operations Research. Third edition. Holden-Day Inc., Oakland, CA, USA.

Kaiser H M, Riha S J, Wilks D S, Rossiter D G, Sampath R (1993) A farm-level analysis of economic and agronomic impacts of gradual climate warming. American Journal of Agricultural Economics 75(2):387–398.

Maino M, Berdegué J, Rivas T (1993) Multiple objective programming; an application for analysis and evaluation of peasant economy of the VIIIth region of Chile. Agricultural Systems 41:387–397.

Monteith J L (1990) Can sustainability be quantified? Indian Journal of Dryland Agricultural Research & Development 5(1&2):1–15.

Rae A N (1971) Stochastic programming, utility and sequential decision problems in farm management. American Journal of Agricultural Economics 53:448–460.

Romero C, Amador F, Barco A (1987) Multiple objectives in agricultural planning: a compromise programming application. American Journal of Agricultural Economics 69:78–86.

Sivakumar M V K (1992) Empirical analysis of dry spells for agricultural applications in West Africa. Journal of Climate 5(5):532–539.

Williams T O (1994) Identifying target groups for livestock improvement research: the classification of sedentary livestock producers in western Niger. Agricultural Systems 46:227–237.

Response to section F

S. FERNANDEZ-RIVERA[1], P. HIERNAUX[1], R. VON KAUFMANN[2], M.D. TURNER[1] and T.O. WILLIAMS[1] *

[1] *ICRISAT/ILCA Sahelian Center, B.P. 12404, Niamey, Niger;* [2] *International Livestock Centre for Africa (ILCA), P.O. Box 5689, Addis Ababa, Ethiopia*

Introduction

The three presentations in section F represent a useful collection of papers on different aspects of eco-regional research. Their relevance lies in the fact that they all touch on the core idea behind the concept of eco-regional research, i.e. the integration of resource management with productivity concerns. The three papers use modelling as a research technique. One of them (Van Keulen's) presents a model of biological processes to investigate long-term effects of soil and crop management strategies. The other two papers present approaches to study crop production strategies at the enterprise and farm levels (Thornton *et al.*) and to identify production activities that maximize profit at the sub-regional level (Schipper *et al.*).

For reviewers coming from a livestock research institute, a major disappointment is that none of the papers treats in any detail livestock production and the interactions between livestock, crop and soil that are known to have significant effects on soil fertility, agricultural productivity and farm income.

Response to paper by Thornton *et al.*

This paper provides a broad and interesting overview of the scope and limits of modelling in an eco-regional context. They use simulation models to identify management options that could improve agricultural income and long-term viability at the enterprise and farm levels. In the process they make two important points. On the one hand, they identify gaps in our knowledge, particularly with respect to methods for measuring farmers' decision-making processes, the factors that affect them, and modelling of the socio-economic environment. On the other hand, they show how a careful manipulation of

* Reviewers' names appear in alphabetical order. Responsibility for comments is equally shared.

J. Bouma et al. (eds.), Eco-regional approaches for sustainable land use and food production, 409–413.

what we know about biophysical processes can be used to improve farm decision-making, even if only in a second-best sense.

An initial observation is that we would have liked a clearer presentation of the whole-farm model structure. This would have facilitated our understanding of the novelty and relevance of the functions included in the model.

Unfortunately, they do not fully address the issues of sustainability at the enterprise and farm levels. Although they incorporate biophysical components in their enterprise and household models, they do not address the sustainability of the production systems they model. This shortcoming is glaring in the case of their model of annual cropping on the slopes of the Cauca River watershed in the lower Andes. The major development question for such a location would be the persistence of the production patterns given the obvious environmental hazards. To say that increased tomato growing would provide an economic benefit only tells a part of the story.

Given some of the unanswered questions posed in the paper, particularly with respect to competition between enterprises, risk, and long-term productivity, it appears that there could be a useful role for the joint utilization of simulation and mathematical programming models in addressing these unresolved issues.

The assertion that animal models are less developed than crop models may be true, but this does not justify ignoring livestock. In fact, the family farm was considering converting most of the land into pasture to produce milk and meat. The list of deficiencies in present modelling approaches provided does have broader relevance, and we agree that there is a strong need to incorporate different social and spatial scales, especially for exploring resource management and sustainability questions.

Response to paper by Van Keulen

This paper proposes a new soil-crop simulation model focussed on the dynamics of organic matter, nitrogen and phosphorus in the soil and their interaction with C_3 cereal crops (wheat, barley and rye). The model is based on the basic biological processes that sustain crop production. Validation with extensive experimental data showed that the model predicts reasonably well the long-term influences of management strategies on soil and crop productivity. It is hoped that, after further validation, this new model can become a valuable tool, especially if it can be adapted to C_4 cereals as well as legume and tuber crops.

The new model is proposed as an alternative to existing models which have extensive data requirements, are difficult to validate, and do not describe appropriately the underlying processes. While all these criticisms are plausible, these reviewers find no clear evidence of the author's success in overcoming these disadvantages. The data inputs required in the described model are not much different from those needed in the CENTURY or EPIC models. The need for long-term experimental data for model validation is also recognized as a condition for any future predictive application of the new model. Furthermore, there appears to be an important gap that act against a well-balanced description of the processes studied – for instance, the influence of run-off/run-on on water balance and soil erosion is only predicted in relative terms using empirical adjusting factors, in spite of their major role in the soil-crop model.

We would have liked to see in the paper the mathematical functions that substantiate the model. Not only would this have more clearly presented the underlying assumptions but it would have also facilitated the use or adaptation of the model by potential users.

For resource management decision-making, we are uneasy with the level of aggregation at which the model functions. Without detracting from the potential and useful applications of the model at the level of crop fields, we would have welcomed an effort to integrate the processes at a higher level of organization (farm, watershed, or landscape). Notwithstanding the difficulties of such a task, it is imperative to accept the fact that it is at these levels that most production units in the developing world operate. This is particularly true in mixed crop-livestock systems, where animals have a large impact on the dynamics of nutrients. Under these conditions, modelling efforts should consider dynamics of nutrients with not only a temporal but also a spatial perspective. We hope progress can be made in this direction in the future.

The application of the proposed model to crop-livestock systems will likely need to consider the impact of animals on the processes that drive the model. For instance, in such systems the rate of decomposition of organic matter would depend not only on the soil and environmental conditions considered in the model, but also on factors such as trampling, vegetation removal and chemical nature of livestock manure.

The author's approach to exploring the long-term influence of soil and crop management strategies is commendable. The paper clearly demonstrates the potential of modelling as a research tool and as an aid in exploring options for the development of sustainable crop production systems.

Response to paper by Schipper *et al.*

This paper employs a linear programming (LP) model, as part of a set of tools, to optimize land use at a sub-regional level. They argue that the building blocks for regional land-use analysis should be farms, with data on objectives, constraints, opportunities and sustainable land-use practices at the farm level being essential inputs into the analysis. Two sustainability indicators – soil nutrient depletion and level of biocide application – are estimated and several land use scenarios are analyzed to determine whether the income of farms in the study region would increase through improved and sustainable land use.

An initial observation is that 53 square kilometres (of which only 42 km^2 was included in the study) appears to be a small area for even a 'sub-regional' study. The authors fail to mention if this area is representative of a larger zone in Costa Rica or other countries in Central America. A discussion of the representativeness of this site will help determine whether or not the model results can be extrapolated to other areas.

There is insufficient discussion of the methodological problems associated with attempting to move from farm-level to regional-level analysis. For example, cluster analysis have been used to minimize the aggregation bias that arises from collecting data from farms that differ in several respects. This approach is commendable as it is much better than *ad-hoc* methods of grouping farms. However, it is not clear that tests were conducted to establish the stability and validity of the identified clusters. The need for these tests becomes apparent when one considers the wide variation, from one survey to another, in areas planted to annuals and perennials in table 1. Also, it is likely that the identified farm clusters would have been less uniform (palm heart oriented) if demographic and social criteria were considered in addition to the two main variables – land size and soil type – used for the derivation of the clusters.

There is also insufficient discussion of the origin of the productivity coefficients assumed in the LP model, as well as their validation under different real situations. Similarly, the predictions of the model appear to be taken as valid, without further tests or validation. Without such tests, it becomes difficult to ascertain how realistic are the results.

Issues can be raised concerning the adequacy and empirical basis of the sustainability indicators used in the analysis. Were these parameters actually measured in the field? One suspects that nutrient depletion rates were estimated simply from production/harvest index calculations. Such calculations have often not been found to accurately predict changes in nutrient availability, especially in situations where livestock are important.

The usefulness for policy analysis of a sub-regional linear programming model in contrast to a regional, computable general-equilibrium model can also be questioned.

The authors recognize some of the aforementioned shortcomings, but some of them are indeed serious. For instance, by paying insufficient attention to the livestock component of the system they are neglecting about 60% of the land, since approximately this fraction is used for pastures. The inclusion of livestock in the model could change dramatically the nutrient balance and the spatial distribution of key nutrients.

In future revisions of the model, one scenario that would be worth exploring would be to determine the consequences for production, farm income, sustainable land use, and rural income distribution of better access to improved technologies on the part of the poorer farms typified by farm types 1 and 5 in the paper.

It would also be useful if the linear programming components are more closely integrated with the geographical information system. This would allow spatial relationships between different land-use and farm types to be incorporated into the analyses of labour use, synergism of different production systems (e.g. crop-livestock) and sustainability. If the geo-referenced data is detailed enough to show distribution of not only soil but farm types, one could potentially do a more sophisticated spatial analysis of production which would provide a better sense of sustainability. Spatial pattern of soil depletion is as important as the total amount of depletion over the whole area.

Concluding comments

These papers indicate that eco-regional research needs to be broad enough to cover analysis at the micro, meso and macro levels. Realism dictates that a solid foundation is needed at the lower levels before issues at the macro level are addressed. The challenge is to be able to assess accurately the amount of resources and division of labour between research partners needed to add to or refine the available knowledge at the lower levels, without necessarily compromising work at the macro level. Most importantly, these papers highlight the importance of well-focused interdisciplinary work in eco-regional activities.

To demonstrate the potential for including crop-livestock interactions in regional analysis, an example of regional planning that will take livestock in to account is provided in another paper by Kaufmann *et al.* (1994)

References

Von Kaufmann R, Fischer G, Mohamed Saleem M, Van Velthuizen H (1994) Assessing the potential of forage legumes for intensification of crop-livestock production systems in West Africa. ILCA, Ethiopia, IIASA, Austria, FAO, Rome.

SECTION G

Case studies of eco-regional approaches at the crop level

Using systems approaches to design and evaluate ideotypes for specific environments

M.J. KROPFF[1,2,3]*, A.J. HAVERKORT[2], P.K. AGGARWAL[1] and P.L. KOOMAN[2,3]

[1] *The International Rice Research Institute, P.O. Box 933, 1099 Manila, Philippines;* [2] *DLO-Research Institute for Agrobiology and Soil Fertility (AB-DLO), P.O. Box 14, 6700 AA Wageningen, The Netherlands;* [3] *Department of Theoretical Production Ecology, Wageningen Agricultural University, P.O. Box 430, 6700 AK Wageningen, The Netherlands*

Key words: breeding, G×E, ideotype, crop model, plant type design, rice

Abstract. An eco-regional approach is needed to find ways to increase agricultural production to meet the growing demand for food in large parts of the world, while sustaining the environment. The selection of germplasm is a key element in such an approach. Selection of varieties for particular environments is mostly based on the intuition of the breeder. With the increasing knowledge of processes that determine crop production and the availability of systems approaches, the opportunity emerged to design crop ideotypes and improve the efficiency of germplasm evaluation programs.

A systems framework to target plant types for different production environments is presented. Specific examples are given of on-going research to design and evaluate the suitability plant types for different environments for major crops such as rice and potato.

Introduction

The objective of any breeding program is the development of improved breeding lines/varieties for a specific product with a defined quality for target environments. For most crops this breeding process involves a time investment of 10–15 years (Hunt 1993), although the process is shorter for tropical crops like rice, where several crop cycles can be completed within a year. Some kind of conceptual design has been used since the onset of breeding activities through the identification of parents with complementary traits which combined were expected to improve existing cultivars. Donald (1962) identified the importance of basic knowledge on the processes driving the production of dry matter, and economic yield formation. This led to a design driven breeding approach (Donald 1968). He argued that the chance of finding a strongly improved plant type in conventional breeding without a clear design would be small. In a recent review, however, Hunt (1993) concluded that only a few breeding programs have adopted the concept of a formal ideotype design as a major breeding activity.

* Present address: 2 and 3

J. Bouma et al. (eds.), Eco-regional approaches for sustainable land use and food production, 417–435.

An example of an improved plant type design that had a major impact on agricultural production is the short stature design for cereals that was developed in the 1960s. The introduction of semi-dwarf varieties like the rice variety IR8 was successful because fertilizers came available at that time to which semi-dwarf varieties responded much better than traditional varieties. This resulted in the so-called green revolution which resulted in a tripling of rice yields in Asia (IRRI 1993). The yield increases were obtained by a new plant type and an adjustment of the agronomic environment (fertilizers, irrigation and pesticides) to optimize genotype × environment interactions. This can be characterized as a seed based technology with a focus on wide adaptation.

Although the wide adaptation concept worked for some systems like the homogeneous irrigated rice systems in the tropics and subtropics and wheat in South Asia, it has been recognized that strong genotype × environment × management interactions in most agricultural systems make it necessary to use genotypes that are adapted to the specific agro-environments. Besides that, the issue of designing ideotypes has to be seen in the context of required developments of the agricultural production systems in the future. There is an increasing awareness that food production must strongly be increased in the coming decades to meet the demand created by growing populations (IRRI 1989). This production increase has to be achieved, on less land, with less labor, less water, and less pesticides and they must be sustainable through conserving scarce natural resources. To meet the challenges to increase food production, ways have to be found to improve the productivity and profitability of agricultural production systems. That can only be done by using a systems approach that helps to optimize local agricultural production systems. That will involve improved crop management systems, but also new crop types. For rice, for example, new varieties with a higher yield potential will be needed early in the next century because yields will quickly reach the potential yield of current varieties (Penning de Vries 1993).

Crop ideotype designs should then be based on sound physiological understanding of the system and its interaction with edaphic, climatic, biotic and management factors. Such an understanding should facilitate the prediction of the effects of different combinations of physiological/morphological traits in different environments. With the advancement of ecophysiological systems modelling of crops in the past decades, the potential to integrate physiological/morphological process level knowledge to design new plant types has emerged. Several examples of such an approach were presented at the international symposium on systems approaches in agricultural development in 1991 (Penning de Vries *et al.* 1993).

This paper reviews the opportunities of using systems approaches to design plant types targeted for specific environments. First a systems framework for targeting plant types for different production environments is discussed. Then an approach is presented to design ideotypes for potential yields, resource limited yields and yields expected under yield reducing (pests) conditions, followed by suggestions for a more efficient breeding program in which systems approaches are used in the different stages of the breeding process. Specific examples are given for rice and potato.

A systems framework for targeting crop ideotypes designs to different environments

To understand the differences in the performance of genotypes in the wide ranges of environments in which crops are grown, it is necessary to use a systems framework in which the role of the different factors that determine yield are distinguished. First it has to be recognized that single plants grow differently from plants grown in a crop situation at a specific density (Kropff *et al.* 1994a). Because one is only interested in crop production per unit of cropped area and not per plant, it would be better to use the term crop ideotype instead of plant ideotype.

A simple widely used generic model for crop growth and production can be used as a framework for designing crop ideotypes. The yield of a crop (Y, g m^{-2}) is determined by the amount of photosynthetically active solar radiation it intercepts (R_i, J m^{-2}), its conversion efficiency into dry matter (E, g J^{-1}), and the proportion of total dry matter in the harvested parts (H_i,-) and its dry matter content (D_{mc}):

$$Y = R_i E H_i / D_{mc}.$$

The value of each parameter in this simple model of crop growth is determined by a number of underlying processes. R_i depends on the time course of the leaf area index and light interception characteristics of the canopy, E depends on photosynthesis and respiration rates, H_i depends on dry matter partitioning processes and the dry matter content (for potato) depends on the soil moisture content and the temperature during growth.

Several production situations have to be distinguished in which different environmental factors determine the parameter values in the simple model. In the potential production situation, where water and nutrients are available in ample supply and the crop is free of pests and weeds, only the growth determining factors radiation, temperature and crop physiological characteristics play a role. Plant processes that govern potential yield are related to light interception, light use efficiency, growth duration and dry matter distribution. Potential yields may vary considerably across environments. For example, the potential yield of rice varies between 6 t ha^{-1} in the tropical wet season to 15 t ha^{-1} in temperate environments (Kropff *et al.* 1994d). Crop ideotypes for increasing potential yields have to focus on those physiological traits that increase the light interception during the season, the efficiency of light use or the harvest index.

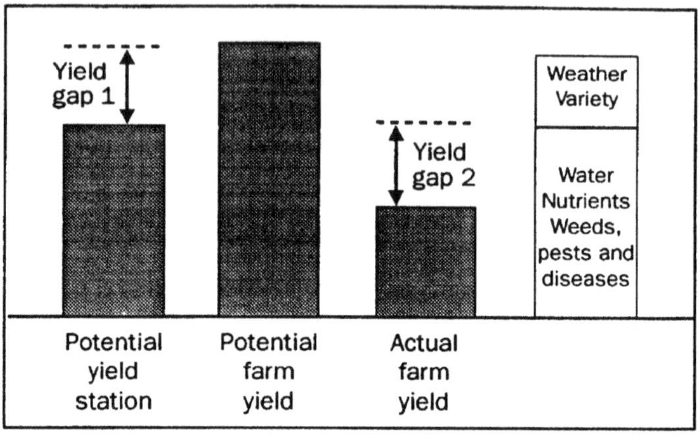

Fig. 1. Schematic presentation of the different yield gaps: yield gap 1 indicates the difference in potential yield between environments (or between station and a site) which is the result of differences in radiation and temperature regime after sowing and the varietal characteristics. The second yield gap indicates the difference between the potential yield in a given environment and the actual yield. This yield gap can be split in several components.

In most production situations, actual yields are much lower than potential yields because of insufficient availability of growth limiting resources to the crop (water and/or nutrients, mainly nitrogen) and growth reduction by insects, diseases and weeds. In these situations, the system is much more complex. In addition to the processes involved in the potential production situation, processes related to the efficiency of water uptake and water-use efficiency, the efficiency of uptake of nutrients and the nutrient-use efficiency in the different physiological processes and crop-pest interactions may affect the R_i (through leaf canopy development) or E. In these production situations, yields range from yield levels close to the potential yield to extremely low yields in heavily exploited areas where no fertilizers are used like in many of the upland rice ecosystems. Ecophysiological models for the effect of water stress and nitrogen limited yields have been developed for several crops with varying results (e.g. Penning de Vries *et al.* 1993).

Because of the different processes that are involved in the production situations, new crop type designs have to be developed for a specific production situation. That should be based on thorough understanding of the factors that are responsible for the so-called yield gap between actual and potential yields in that particular environment. The way in which yield gaps are defined in the framework that is used here is illustrated in figure 1. In situations where there is a considerable yield gap, it might be more efficient to first identify the major constraints and management strategies to improve the resource-use efficiency for the limiting factors or to reduce pest effects before designing new crop ideotypes, so that the breeding effort, which is a long term investment, can be focused on the major problems that cannot be solved by improved management.

Designing improved crop types using systems approaches

Increasing potential yields

The potential yield of a crop is defined only by varietal characteristics and the seasonal pattern of the environmental variables temperature, daylength and radiation. The potential yield of a specific cultivar differs among environments, in different years and between seasons in the same year. In a particular environment only cultivar characteristics determine the potential yield.

Rice

Rice breeding programs of National Research Systems in rice growing countries and the International Rice Research Institute (IRRI) collaborate intensively to develop improved rice varieties for specific environments. Through the international network for the exchange of germplasm, the best varieties from many countries are evaluated in 95 countries in ecosystem oriented nurseries and so-called stress oriented nurseries. In this program there is a need for systems approaches to target genotypes to specific environments and to reduce the number of sites and years for evaluation. The timing of phenological events is important in rainfed lowland systems to ensure flowering occurs well before the end of the rainy season, and in temperate environments, where the total crop duration has to match the length of the growing season as determined by the annual temperature cycle. Work has been initiated recently to model phenological events to help targeting varieties to those specific environments.

However, most rice is grown in tropical and subtropical environments, where the duration has to be minimized without sacrificing potential yield to allow the cultivation of several crops per year. Since the 1960s breeders were successful in this aspect: varieties have been released with a similar potential yield as IR8 despite differences in growth duration (IR72 has a growth duration of about 110 days vs. 125 days for IR8) (Kropff *et al.* 1993ab; Cassman *et al.* 1993). Therefore, the key issue for rice is to increase the yield potential in rice with a minimum increase in the total duration in tropical environments.

IRRI scientists proposed modifications to the present high-yielding crop types that would enhance the potential yield for rice in irrigated direct-seeded conditions considerably (IRRI 1989). The proposed characteristics for the designed new crop type came from different perspectives, mainly based on physiological analyses and include: a reduced tillering capacity, no unproductive tillers, over 200 grains per panicle, sturdy stems, harvest index of 0.6,

13–15 t ha^{-1} yield potential in the tropical dry season. Breeding activities started in 1989 with the identification of donors for the various traits, most of which were so-called javanicas from Indonesia (Peng *et al.* 1994).

The physiological characteristics needed to support greater yield potential were evaluated in more detail using models at different level of complexity (Penning de Vries 1991; Dingkuhn *et al.* 1991; Kropff *et al.* 1994ab). In a simple model analysis, Kropff *et al.* (1994a) concluded that rice yields beyond 10 t ha^{-1} in the tropics should either result from (i) increased accumulation and allocation of stem reserves, (ii) from a prolonged grain filling period (R_i), or (iii) from an increased growth rate (E) during grain filling. They argued that an increase of stem reserves allocation from 2 to 2.5 t dry matter ha^{-1} would be the maximum without prolonging the vegetative period. Such a prolongation is not agronomically desirable. Besides that, a sturdy stem and a high LAI (leaf area index) are needed to support a larger grain yield. For a grain yield of 15 t ha^{-1} at 14% moisture content, 40 days of effective grain filling would be needed at the same growth rate as the current varieties during 25 days (Penning de Vries 1991; Kropff *et al.* 1994a,b). Thus, a higher yield potential requires a longer duration, unless the vegetative phase is shortened. Therefore it is needed to determine how much the vegetative phase can be shortened without losing yield potential.

To achieve such yields, an increased sink size is also required. A yield of 15 t ha^{-1} requires 60,000 spikelets m^{-2}, which is 13% greater than the total number of filled plus unfilled spikelets produced by a recently tested hybrid, and 33% greater than for the inbred variety IR72 (Kropff *et al.* 1994b). Thus the fundamental parameters are: increased sink size (i.e. spikelets m^{-2}), a longer period of effective grain filling, and a longer duration of green leaf area and active canopy photosynthesis to match the increase in grain fill duration. These parameters are similar for an increased yield potential in any rice growing environment.

In a more detailed analysis, we used the ecophysiological model ORYZA1, which was developed and evaluated using the data sets from field experiments conducted in 1991–1993 at the IRRI experimental farm which were designed to restore the yield potential in rice (Kropff *et al.* 1994c). This model is based on the Wageningen models SUCROS (Spitters *et al.* 1989) and MACROS (Penning de Vries *et al.* 1989). The model simulates growth of a rice crop on basis of the light distribution within the canopy, single leaf photosynthesis as a function of leaf N concentration and light intensity and respiration. The accumulated dry matter is partitioned among the various plant organs, which is a function of phenological development. The model simulates leaf area index development and accounts for sink limitation.

Total potential above-ground biomass (ranging from 11.5–17.8 t ha^{-1} in wet and dry seasons, respectively) as well as panicle dry matter (ranging from 5.7–9.8 t ha^{-1}) was simulated accurately by the model (table 1). The parameters used in these simulations were derived from the highest yielding

Table 1. Observed and simulated crop characteristics for field experiments on yield potential in rice in different environments (standard error between brackets). (Source: Kropff *et al.* 1994c)

	IRRI WS	Japan	IRRI DS	Yanco
Year	1991	1987	1992	1992
Variety	IR72	Nipponbare	IR72	YRL39
Biomass obs. (t ha^{-1})	11.5 (0.6)	12.8 (0.5)	17.8 (0.6)	28.9 (1.1)
Biomass sim. (t ha^{-1})	11.5	13.2	15.4	27.3
Yield obs. (t ha^{-1}, panicle dry wt)	5.7 (0.2)	7.8 (0.5)	9.8 (0.2)	14.7 (0.6)
Yield sim.(t ha^{-1}, panicle dry wt)	7.26	7.41	9.49	14.3
Maximum LAI obs. (m^2 m^{-2})	4.3 (0.4)	5.3 (0.7)	6.0 (0.5)	14.0 (1.0)
Maximum LAI sim. (m^2 m^{-2})	4.9	8.9	6.4	12.8
Crop duration (d)	109	153	110	159
Grain filling duration sim. (d)	29	32	30	39
Mean temperature				
sowing-flowering (_C)	28.1	23.3	26.0	21.3
flowering-maturity (_C)	27.7	25.6	27.9	22.9
Av. daily incoming radiation				
sowing-flowering (MJ m^{-2})	15.2	16.3	17.4	22.9
flowering-maturity	18.9	13.5	21.9	22.6
Absorbed total radiation				
sowing-flowering (MJ m^{-2})	189	195	278	413
flowering-maturity	215	301	371	1747
Light use efficiency (g MJ^{-1})	2.8	2.6	2.7	2.5

dry season experiment at IRRI (development rates, partitioning coefficients, leaf-N content, relative growth rate of the leaf area). The ORYZA1 model was also used to predict yield potential for a temperate environment. In Yanco Australia, yields of close to 15 t ha^{-1} have been observed (Williams and Lewin, personal communication). Model predictions for these independent data were very close to that level (table 1). Model analysis showed that the differences in yield potential between for example the IRRI wet season (6 t ha^{-1}) and Yanco (15 t ha^{-1}) were the result of a low temperature in Yanco causing a long duration of grain filling and a high radiation level in Yanco which increased the R_i.

These simulations confirm that a genotype with a 15 t ha^{-1} yield potential in the tropics must have a grain filling duration in tropical environments that is comparable to that of current varieties grown at higher latitudes where temperatures are cooler as in Yanco. For several crop species, increased 'stay green' (i.e. longer green leaf duration during grain filling) indeed has been a major achievement of breeders and agronomists in the past decades (Evans

1990). New physiological insight is needed in the grain filling process to understand why the grain filling process stops in current varieties, while there are many unfilled spikelets as observed in several experiments (Kropff *et al.* 1994b). It was concluded that the proposed characteristics of the IRRI new plant type (Peng *et al.* 1994), such as more spikelets per panicle and a larger flag leaf will not by definition lead to the required longer grain filling duration with a sustained high growth rate of the crop. Currently, prototypes of the new crop type is being evaluated by IRRI and many of the desired characteristics, like a large thick flag leaf, more spikelets m^{-2} and less tillers per plant, have successfully been introduced in the lines, although higher yields have not been observed yet (Peng *et al.* 1994).

To analyse the need for different crop types for different environments, Aggarwal *et al.* (1995) determined the importance of various traits for the dry season and the wet season using ORYZA1. Crop characteristics of IR72 (Kropff *et al.* 1994) and weather data of Los Banos for 1992 were used. The critical model parameter values were varied between +50% and –50% with respect to those of IR72 to simulate the effect of a change in leaf area development, sink size, leaf N distribution in the canopy, light extinction coefficient during vegetative and reproductive development stages, leaf N content, maintenance respiration of leaves, fraction of stem reserves, dry matter partitioning, and crop development rates during juvenile phase and grain filling period. It was assumed that 50% change in model inputs will represent the extremes of variation available for most crop parameters. The results indicated that the relative importance of different traits may change with season and a 50% change in individual traits can provide a maximum of 19% increase in yield potential. The magnitude of response to input change may however be dependent on cropping year, location and level of management.

To further examine the opportunities for increasing rice yields substantially in different seasons, an additional analysis was done by using 'hypothetical' plant types with different combinations of traits. These crop types simulated various combinations of source capacity, sink size and grain filling duration (table 2) (Aggarwal *et al.* 1995). The probability distributions of yield for the Los Banos, Philippines situation are presented in figure 2. Increased source availability (crop type 1) always increased yield. Average increase was between 1.0 and 1.4 t ha^{-1} irrespective of the season. Increase in grain filling duration alone (crop type 2) resulted in increased grain yield, the magnitude of increase in yield relative to the control was less in the dry season than in the wet season. In 50% of the years, increase in the wet season was more than 0.8 t ha^{-1}. In the dry season, the increase was between 0 and 0.6 t ha^{-1} only. In many seasons, grain weight had already reached its maximum permissible value and thus there was a sink limitation.

Increase in spikelet number (crop type 3) resulted in hardly any change in yield in the dry season and in the wet season. When the number of spikelets increased simultaneously with LAI and leaf N content (crop type 4), 50%

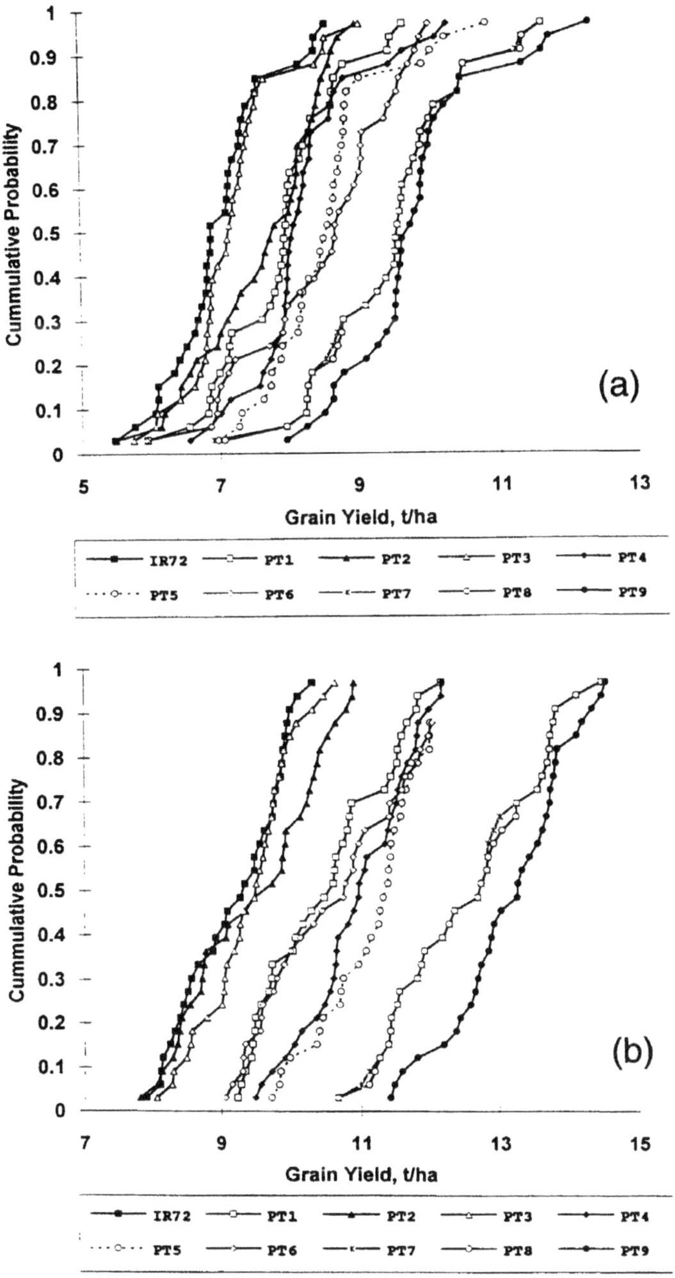

Fig. 2. Simulated grain yield probabilities of various plant types (see table 2) in Los Banos, Philippines, for the dry season (A) and the wet season (B), 1992.

Table 2. Characteristics of 9 hypothetical crop types in comparison to the check variety IR72. Traits: SLA (specific leaf area); NFLV (leaf N content); SPGF (spikelet formation factor), WGRMX (maximum 1000 grain weight) and DVRR (development rate during grain filling).

Crop type	Percent change in input parameter				
	SLA	NFLV	SPGF	WGRMX	DVRR
IR 72	0	0	0	0	0
1	+20	+20	0	0	0
2	0	0	0	0	–20
3	0	0	+20	0	0
4	+20	+20	+20	0	0
5	0	0	+20	+20	–20
6	+20	+20	0	0	–20
7	+20	+20	+20	0	–20
8	+20	+20	0	+20	–20
9	+20	+20	+20	+20	–20

yields exceeded 10.9 t ha^{-1} in the dry season and 8.1 t ha^{-1} in the wet season. Increased sink capacity due to more spikelets and higher 1000 grain weight together with longer grain filling duration (crop type 5) resulted in substantial yield increases in both seasons. However, only crop types 7, 8 and 9 in which the source, sink and grainfilling duration are increased, substantial yield gains of 4 t ha^{-1} were simulated.

Potato

Breeding programmes, of centres such as the International Potato Centre (CIP), but also of private companies in Canada, France and The Netherlands focus on the breeding of potatoes for specific environmental conditions. The most important aspect is to match the growth cycle of the cultivar to the length of the growing season as defined by the temperature. Kooman and Haverkort (1995) developed a simple potato simulation model that can be used for that purpose. The 'LINTUL-POTATO' model overcomes the limitations of other models that only perform well for either tropical conditions or temperate conditions (Haverkort 1990; Haverkort and Harris 1987).

The model 'LINTUL-POTATO' simulates potential total dry matter production and yield on basis of the light use efficiency of intercepted light by a potato crop (Kooman and Haverkort 1995). To predict growth and development over areas with widely varying temperature and daylength regimes, the growth cycle is divided into four phases each starting and ending by a characteristic stage of development. For each phase the key processes are different (figure 3). Phase 0 starts at planting and ends at emergence. The main process is the rate of emergence as a function of temperature. Phase 1 is from emergence to tuber initiation. The main process is foliar expansion

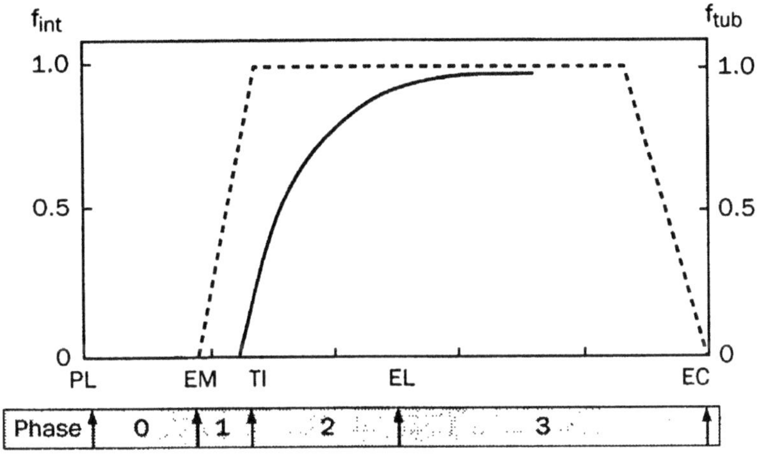

Fig. 3. Schematic representation of the fraction of solar radiation intercepted by the crop (F_{int}) and the fraction of daily produced assimilates partitioned to the tubers (fr_{tub}) from planting (PL) to emergence (EM), the start of tuber growth (TI), the moment that fr_{tub} (EL) reaches the value of 0.9 until the end of crop growth (EC).

which is a function of temperature. After this, leaf expansion is calculated from leaf dry weight increases. Daylength and temperature both determine the duration of this phase and thus the amount of leaves produced. Short days increase the development rate until tuber initiation. Phase 2 is from tuber initiation until the end of leaf growth (defined as the moment when 90% of the assimilates which are daily produced, are partitioned to the tubers). Temperature determines the duration of this phase. Phase 3 is from the end of leaf growth until the end of crop growth. In Phase 3 all assimilates are partitioned to the tubers. Leaf senescence is a function of temperature.

The model simulated adequately differences between genotypes within the same environment and differences between environments with the same genotype (Kooman, in prep.).

Model analyses showed that the key to the optimization of potential yield in potato is the proper timing of the moment of tuber initiation. If that moment is too late for a particular environment, too much foliar tissue is formed and the tuber growth period can be too short. If tuber initiation is too early, not enough foliar tissue is formed to sustain growth until the end of the growing season. LINTUL-POTATO can be used to determine the cultivar characteristics with respect to temperature and daylength response that give the highest yield in a particular environment. This means that for each environment the optimal moment of tuber initiation can be calculated and only genotypes with such desired earliness have to be evaluated in this environment.

Nitrogen: a key yield limiting factor
Nitrogen is one of the key nutrients that limit crop growth and production in many production situations. Its main effect is on leaf expansion (and thus R_i), but also the light use efficiency (E) of a crop is affected by N. Although there is variability in the N-use efficiency of crops, it appears that optimization of N management has potentially more impact on yield than breeding for increased N-use efficiency. In farmers fields, the efficiency of N management may vary considerably as was shown for intensive rice systems in the Philippines (Cassman *et al.* 1993). Because N fertilizers are relatively inexpensive, it is not clear if crop type designs have to be developed for N limited conditions. In the unfavorable environments, where cash flow often does not allow the purchase of inputs, several factors can be limiting and an identification of the most limiting factors has to be conducted first.

The importance of critically evaluating management practices, before starting to develop crop ideotype designs, was illustrated in recent studies at IRRI. A decline in rice productivity had been observed in high yielding intensive rice cropping systems (Flinn *et al.* 1982, Cassman *et al.* 1994). On IRRI's experimental farm, rice yielded 10 t ha^{-1} in the dry season and 6 t ha^{-1} in the wet season in the late sixties with the first semi-dwarf variety IR8, but dry season yields in the 1980s generally did not exceed 6–7 t ha^{-1} in the dry season and 4–5 t ha^{-1} in the wet season at recommended fertilizer rates. It was not known if this was related to genetic factors in recent varieties, because the early semi-dwarf varieties were yielding even lower in recent experiments because of the lack of resistance to diseases. In recent studies where systems analysis and experimentation were combined, yields of about 6 t ha^{-1} in the wet season and 9–10 t ha^{-1} in the dry season were achieved, indicating that the genetic potential remained the same, despite differences in growth duration (Kropff *et al.* 1993b; Cassman *et al.* 1993).

Varieties that strongly differ in duration have different N requirements. For example, late potato cultivars will build up a higher leaf area index, which requires more N. In the development of crop ideotypes the nitrogen requirements (and other nutrients as well if not abundantly available) have to be identified. For example, a 15 t ha^{-1} rice crop will require about 300 kg of N uptake, whereas a 10 t ha^{-1} crop only requires 200 kg N ha^{-1} (Penning de Vries 1991; Kropff *et al.* 1994a).

Water stress
Water is one of the most limiting factors for growth and quality of rainfed crops. The effect of water stress on the crop depends on the timing and duration of the drought spell. Drought stress affects many processes like leaf area development, the light use efficiency (reduction of photosynthesis when stomata close) and phenological development. These effects may be divided

Table 3. Simulated results of a SUCROS potato model.

Treatment	Irrigated		Early drought		Late Drought	
cultivar	Late	Early	Late	Early	Late	Early
Total biomass (t ha^{-1})	28.9	16.4	13.6	5.5	13.3	9.1
Tuber dry matter (t ha^{-1})	22.4	13.2	10.2	3.7	7.4	6.0
Harvest index	0.78	0.81	0.75	0.67	0.55	0.65
Highest LAI value	7.4	3.8	4.8	2.5	7.4	3.8
Transpiration (mm)	471	283	239	105	238	166
Transp. efficiency (g kg^{-1})	6.33	5.99	5.92	5.52	5.95	5.85

into short-term effects and long term effects. Short-term effects like stomatal closure or leaf rolling are reversible effects. The longer-term effects, like reduced leaf area development, phenological development are most important to consider here.

Quantitative understanding of the impact of drought on crop yield on basis of the effects on the different processes through crop models can be instrumental in targeting genotypes to different environments. A key characteristic in drought prone environments is the duration of the growing cycle of a variety. Depending on the probability of the occurrence of drought in different periods of the growing season, early or late varieties may have an advantage.

For potato in The Netherlands, Haverkort and Goudriaan (1994) found that yield losses due to drought were greatest with relatively early cultivars. This was mainly due to a reduction of the amount of intercepted radiation because of earlier senescence. The light-use efficiency was affected less than intercepted radiation. The tolerance of the late cultivars of drought resulted from its abundant formation of foliage associated with its lateness. Long-term effects of drought on crop growth in different potato varieties were simulated by Haverkort and Goudriaan (1994). They used 30 years average weather data and simulated potato growth assuming a sandy loam soil. When subjected to drought, leaf senescence was advanced by 30 days in the late cultivar and by about 20 days in the early cultivar which prevented recovery of the early cultivar in the second half of the growing season. The simulation results are shown in table 3. An early dry spell had a stronger impact on yield of an early cultivar, whereas a late dry spell affected the yield of the late cultivars more strongly. The late drought strongly reduced both tuber yield and harvest index as hardly any dry matter was partitioned to the tubers.

It may be concluded that the length of the growth cycle of a cultivar is the major factor determining tolerance of short transient or of late prolonged dry periods. Late cultivars initiate their tubers late and develop a lush canopy that is able to loose much leaf area before interception of solar radiation is affected. An early cultivar may be more suited to (Mediterranean) conditions in which drought gradually increases towards the end of the growing season. Increasing dry matter partitioning to the tubers leads to an escape from drought.

However, it is important to determine the trade-offs between traits that are useful in different production situations. Breeding for stress tolerance may sometimes inadvertently lead to a lower productivity under optimal conditions. For instance, cultivars with a low harvest index which produce abundant foliage are more tolerant of drought but have lower yields under optimum conditions than genotypes that produce less foliage (Haverkort and Schapendonk 1994).

A detailed simulation study for the performance of several crops (sorghum, maize and kenaf) in the semi arid tropics in Australia was presented by Muchow and Carberry (1993). They showed the great advantage of using models for the evaluation of crop types in extremely variable environments with respect to drought. They simulated yield probabilities for different crop types in different environments using many years of weather data. Their results indicated that the choice of crop type depends on the attitude to risk. For example, an earlier maturing variety improved yields in lower yielding years, and increased yield stability, but the variety performed less in higher yielding years. This analysis shows that well validated models can help fine tune variety design and selection based on long term yield assessments, whereas conventional agronomic evaluation would require enormous investments in time and money.

Crop type designs and pest interactions

Besides adaptation to the climatic conditions, resistance or tolerance to pests and diseases is important. The disease or pest multiplies less rapidly when the genotype is more resistant, whereas tolerance indicates the ability of the variety to grow while infected. Screening for specific pest resistance is relatively easy to conduct experimentally by infecting a wide range of genotypes in a plot. However, tolerance involves quantitative traits and may vary by location. Systems approaches may help to quantify physiological damage mechanisms and possibilities to compensate. Once the mechanism is known, traits can be identified for crop type designs.

An example is the compensation of early insect damage in rice. Systems approaches have helped to identify the role of the tillering capacity in compensating the effects of leaf removal by insects (Elings and Rubia, pers. comm.). The new rice crop type design, mentioned earlier in this paper, has a reduced tillering capacity to reduce the contribution of non productive tillers to increase yield potential. In view of designing crop ideosystems that meet objectives at the eco-regional level as stated in the introduction, it is essential to quantify the trade offs between the reduction of non productive tillers and

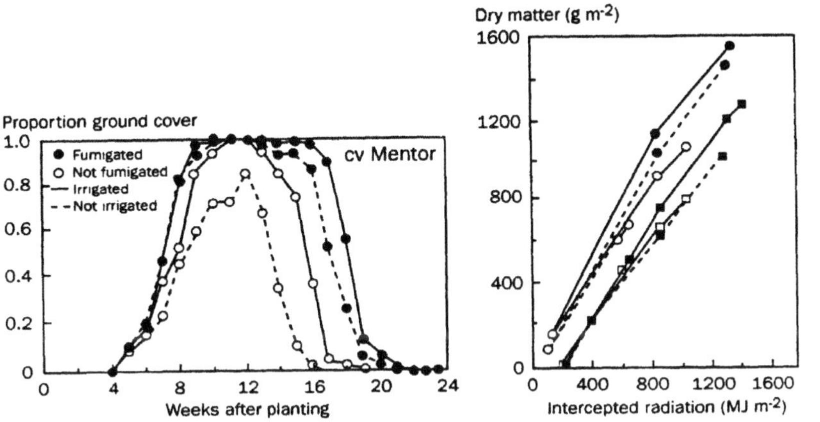

Fig. 4. Proportion of ground cover measured by infrared reflectance of potato crops subjected to drought and potato cyst nematodes. Irrigated, unirrigated, nematodes controlled (fumigated), nematodes not controlled (after Haverkort and Schapendonk 1994).

Table 4. Relative values of yield components in irrigated and unfumigated plots (mean data 1989 and 1990, values of irrigated and fumigated plots = 100). Parameters: (crop yield Y, g m^{-2}), the amount of photosynthetically active solar radiation intercepted (R_i, J m^{-2}), the conversion efficiency into dry matter (E, g J^{-1}), and the proportion of total dry matter in the harvested parts (H_i,-) and its dry matter content (D_{mc}) (Haverkort *et al.* 1992).

Cultivar	Yield	= R_i	× E	× H_i	/ D_{mc}
Darwina	48	61	90	94	105
Desiree	52	71	86	93	102
Elles	73	86	95	102	107
Mentor	49	57	92	100	101

the compensation capacity to minimize pest damage after early leaf removal. Integrated experimental and modelling studies can help reveal these trade offs.

The effect of nematodes on potatoes was studied in detail by Haverkort *et al.* (1991, 1992). Figure 4 demonstrates the strong impact of nematodes and water stress on leaf area development (measured as infrared reflectance), resulting in a strong reduction of R_i. Infra-red reflectance and hence ground cover were reduced more strongly by nematodes than by drought. Table 4 shows the values of intercepted radiation, radiation-use efficiency, harvest index and dry matter content of the unfumigated plots relative to those of the fumigated controls. Apparently cyst nematodes and drought act similarly on the components of yield formation: yields are mainly reduced because crops intercept less solar radiation. Other factors reducing fresh tuber yields are lower light-use efficiencies whereas reduced harvest indices and increased dry matter contents also play a role. The cultivar least tolerant of nematode damage (Darwina) also was least tolerant of drought whereas cv. Elles was most tolerant of both stress factors.

These results indicate that varieties that quickly form a closed canopy and senesce slowly will be more tolerant to nematode and drought effects. This means that a larger specific leaf area (the formation of thinner leaves) a higher leaf water content (leading to larger leaves) and more and longer stems (allowing a better spread of the leaves over the soil surface with less overlap) are the major factors contributing to light interception with a given amount of dry matter partitioned to the crop tops.

Another example is the design of genotypes that have a higher competitive ability versus weeds. An ecophysiological simulation model for interplant competition was used to identify traits that determine the competitive ability of a crop. The most important traits were: rapid early leaf area – tiller – and height development, and more horizontally oriented leaves in early growth stages (vertical ones later on because of yield potential) (Kropff and Van Laar 1993). In experiments, rice varieties that differed in these traits were evaluated with respect to their competitive ability versus a standard purple colored variety. The variety with all required traits, Mahsuri, reduced the growth of the purple variety so much, that all purple rice plants had died before the final harvest (Kropff, unpublished results).

The studies mentioned show that quantitative ecophysiological understanding of the damage mechanisms and plant traits involved, can help designing crop ideotypes with a minimum pest damage which will help to reduce the use of pesticides to control pests. However, for insects and diseases, breeding for resistance will have to be continued, although it is known by now that the use of resistant varieties quickly leads to the development of new pest populations to which the crop lacks resistance. Approaches as the ones mentioned are complementary.

Conclusions

The examples given in this paper illustrate the potential of using systems approaches to increase our quantitative understanding of agricultural production systems and how such understanding can be used in the designing of crop ideosystems for target environments. This includes the variety and its management. A simple crop model based on intercepted radiation and the light use efficiency can be used as a first step in the analysis of experimental data, followed by more mechanistic analyses using detailed ecophysiological models. Once the mechanisms are identified, the models can be used to predict the performance of genotypes for different environments. However, it is an essential requirement to identify the yield constraints in the target environments through yield gap analysis. Models can be helpful as a guideline.

With respect to breeding for increased potential yield, the approach used has to be suitable for the specific crops in their environments. In temperate environments, for most crops the duration has to be optimized in such a way that the crop uses the full growing season as determined by temperature. In potatoes the optimization of yield can be easily defined by using the model to identify the optimum duration to tuber initiation so that a leaf canopy is produced that can sustain a long tuber filling period. In rice the flowering date has to be optimized to match the season. This has been done by breeders, who developed varieties in temperate environments that require a low temperature sum to flowering. In the tropics the story is different, as more crops can be grown per year. Therefore the maximum yield with a relatively short duration has to be achieved for both rice and potatoes.

The paper clearly illustrates that a simple seed based technology as used in the in the past (improve the genotype, homogenize the environments through external inputs thereby minimizing G×E) may not be the most suitable approach for sustainable eco-regional development. For that purpose, detailed G×E studies are needed in a wide range of environments, where experimentation and modelling are combined. Therefore, teamwork of breeders, physiologists, agronomists and modellers is an essential requirement in the process of designing ideosytems. Such an effort is underway in the Systems Analysis and Simulation in Rice Production project in which IRRI collaborates with National Agricultural Research Institutes in the region (Ten Berge *et al.* 1994) with the germplasm evaluation projects of IRRI and the NARS. The focus of this effort is to target crop types for specific environments based on understanding of the genotype × environment interactions.

Acronyms

CIP	Centro Internacional de la Papa (International Potato Center)
G×E	Genotype by Environment
IRRI	International Rice Research Institute
LAI	Leaf Area Index
LINTUL	Light INTerception and Utilization model
MACROS	Modules for an Annual CROp Simulator
NARS	National Agricultural Research System
SUCROS	Simple and Universal CROp growth Simulator

References

Aggarwal P K, Kropff M J, Matthews R B, McLaren G (1995) Using simulation models to relate physiological traits to the performance of rice genotypes. Proceedings of the

IRRI-ICRISAT-Univ. of Queensland meeting on genotype × environment interactions (in press).
Ten Berge H F M, Kropff M J, Wopereis M C S (Eds.) (1994) The SARP project Phase III (1992–1995) Overview, Goals, Plans. SARP Research Proceedings, IRRI, Los Banos, Philippines, and AB-DLO, Wageningen, The Netherlands.
Cassman K G, Kropff M J, Gaunt J, Peng S (1993) Nitrogen use efficiency of rice reconsidered: What are the key constraints? Plant and Soil 155/156:359–362.
Cassman K G, De Datta S K, Olk D C, Alcantara J M, Samson M I, Descalsota J P, Dizon M A (1994) Yield decline and the nitrogen economy of long-term experiments on continuous, irrigated rice systems in the tropics. Advances in Soil Science (in press).
Dingkuhn M, Penning de Vries F W T, De Datta S K, Van Laar H H (1991) Concept for a new plant type for direct seeded flooded tropical rice. Pages 17–39 in Direct seeded flooded rice in the tropics. Selected papers from the International Rice Research Conference, August 27–31, 1990, Seoul, Korea. IRRI, Los Banos, Philippines.
Donald C M (1962) In search of yield. J. Aust. Inst. Agric. Sci. 28:171–178.
Donald C M (1968) The breeding of crop ideotypes. Euphytica 17:385–403.
Evans L T (1990) Assimilation, allocation, explanation, extrapolation. Pages 77–87 in Rabbinge R, Goudriaan J, Van Keulen H, Penning de Vries F W T, Van Laar H H (Eds.) Theoretical Production Ecology: reflections and prospects. Simulation Monographs 34, Pudoc, Wageningen, The Netherlands.
Flinn J C, De Datta S K, Labadan, E (1982) An analysis of long-term rice yields in a wetland soil. Field Crops Research 5:201–216.
Haverkort A J (1990) Ecology of potato cropping systems in relation to latitude and altitude. Agricultural Systems 32:251–272.
Haverkort A J, Goudriaan J (1994) Perspectives of improved tolerance of drought in crops. Aspects of Applied Biology 38:79–92.
Haverkort A J, Harris P M (1987) A model for potato growth and yield under tropical highland conditions Agricultural an Forest Meteorology 39:271–282.
Haverkort A J, Schapendonk A H C M (1994) Crop reactions to environmental stress factors. Pages 339–346 in Struik P C, Vreedenberg W J, Renkema J A, Parlevliet J E (Eds.) Plant production on the threshold of a new century. Kluwer Academic Publishers, Dordrecht, The Netherlands.
Haverkort A J, Boerma M, Velema R, Van de Waart R, Van der Waart M (1992) The influence of drought and cyst nematodes on potato growth. 4. Effects on crop growth under field conditions of four cultivars differing in tolerance. Netherlands Journal of Plant Pathology 98:179–191.
Haverkort A J, Uenk D, Veroude H, Van de Waart M (1991) Relationships between ground cover, intercepted solar radiation, leaf area index and infrared reflectance of potato crops. Potato Research 34:113–121.
Hunt L A (1993) Designing improved plant types: a breeder's viewpoint. Pages 3–17 in Penning de Vries F W T, Teng P S, Metselaar K (Eds.) Systems Approaches for agricultural development. Kluwer Academic Publishers, The Netherlands.
IRRI (1989) IRRI towards 2000 and beyond. International Rice Research Institute, Los Banos, Philippines.
IRRI (1993) The IRRI rice almanac, The International Rice Research Institute, Los Banos, Philippines.
Kooman P L (1995) Yielding abilities of potato crops influenced by temperature and daylength. PhD thesis, Wageningen Agricultural University (in press).
Kooman P L, Haverkort A J (1995) Modeling development and growth of the potato crop influenced by daylength and temperature: LINTUL-POTATO. In Haverkort A J, MacKerron D K L (Eds.) Ecology and modeling of potato under conditions limiting growth. Kluwer Academic Publishers, Dordrecht, The Netherlands (in press).
Kropff M J, Van Laar H H (Eds.) (1993) Modeling crop-weed interactions. CAB, Wallingford and IRRI, Los Banos, Philippines.

Kropff M J, Cassman K G, Penning de Vries F W T, Van Laar H H (1993a) Increasing the yield plateau and the impact of global climate change. J. Agric. Meteorol. 48:795–798.
Kropff M J, Cassman K G, Van Laar H H, Peng S (1993b) Nitrogen and yield potential of irrigated rice. Plant and Soil 155/156:391–394;
Kropff M J, Cassman K G, Van Laar H H (1994a) Quantitative understanding of the irrigated rice ecosystems and yield potential. Pages 97–114 in Virmani S S (Ed.) Hybrid rice technology: new developments and future prospects. IRRI, Los Banos, Philippines.
Kropff M J, Cassman K G, Peng S, Matthews R B, Setter T L (1994b) Quantitative undestanding of yield potential. Pages 21–38 in Cassman K G (Ed.) Breaking the yield barrier. IRRI, Los Banos, Philippines.
Kropff M J, Van Laar H H, Matthews R B (Eds. (1994c) ORYZA1: An ecophysiological model for irrigated rice production. SARP Research Proceedings, AB-DLO, Wageningen, The Netherlands, and IRRI, Los Banos, Philippines.
Kropff M J, Williams R L, Horie T, Angus J F, Singh U, Centeno H G, Cassman K G (1994d) Predicting the yield potential in different environments. In Yanco Symposium Research Proceedings (in press).
Muchow R C, Carberry P S (1993) Designing plant types for the semiarid tropics: agronomists' viewpoint. Pages 37–61 in Penning de Vries F W T, Teng P S, Metselaar K (Eds.) Systems approaches for agricultural development. Kluwer Academic Publishers, The Netherlands.
Peng S, Khush G S, Cassman K G (1994) Evolution of the new plant ideotype for increased yield potential. Pages 5–20 in Cassman K G (Ed.) Breaking the yield barrier. IRRI, Los Banos, Philippines.
Penning de Vries F W T (1991) Improving yields: designing and testing VHYVs. Pages 13–19 in Penning de Vries F W T, Kropff M J, Teng P S, Kirk G J D (Eds.) Systems Simulation at IRRI. IRRI Research Paper Series, November 1991, Number 151, Los Banos, Philippines.
Penning de Vries F W T (1993) Rice production and global climate change. Pages 175–189 in Penning de Vries F W T, Teng P S, Metselaar K (Eds.) Systems approaches for agricultural development. Kluwer Academic Publishers, The Netherlands.
Penning de Vries F W T, Jansen D M, Ten Berge H F M, Bakema A (1989) Simulation of ecophysiological processes of growth of several annual crops. Simulation Monographs 29. Pudoc, Wageningen, The Netherlands, and IRRI, Los Banos, Philippines.
Penning de Vries F W T, Teng P S, Metselaar K (Eds.) (1993) Systems Approaches for agricultural development. Kluwer Academic Publishers, Dordrecht/Boston/London, and IRRI, Los Banos, Philippines.
Spitters C J T, Van Keulen H, Van Kraalingen D W G (1989) A simple and universal crop growth simulator: SUCROS87. Pages 147–181 in Rabbinge R, Ward S A, Van Laar H H (Eds.) Simulation and systems management in crop protection. Simulation Monographs, Pudoc, Wageningen, The Netherlands.

An eco-regional perspective of crop protection problems

J.C. ZADOKS[1], P.K. ANDERSON[1,2] and S. SAVARY[3]

[1] *Department of Phytopathology, Wageningen Agricultural University, P.O. Box 8025, 6700 EE Wageningen, The Netherlands;* [2] *Programa Epidemiología de Virus, Universidad Nacional Agraria, Apartado OR–8, Managua, Nicaragua;* [3] *ORSTOM-IRRI Shuttle Project on Rice Pest Characterization, 911, Ave. Agropolis, BP 5045, 34032 Montpellier, France, and P.O. Box 933, 1099 Manila, Philippines*

Key words: disease, dispersal, endemic, epidemic, exodemic, green bridge effect, perennation, pest

Abstract. Botanical epidemiology deals with outbreaks of pests, using the word 'pest' in a broad sense. These outbreaks, which may be once-only or recurrent, have specific characteristics in time and space. Temporal and spatial parameters of epidemics depend on intrinsic and extrinsic characteristics of the pest organism involved. Among the intrinsic characteristics are perennation and dispersal abilities. Among the extrinsic characteristics are cropping patterns and 'green bridge' effects. There exists a geography of crops and pests, and of their combinations in 'pathosystems', and there may be a geographic component in the management of such pathosystems, at different scales of time and space. A coarse-grained geographical pattern may contain fine-grained patterns, as exemplified by 'recommendation domains'. Such patterns, at different levels of resolution, warrant a fresh, holistic approach to eco-regional crop protection problems.

Introduction

In medicine, attention is focussed on the individual patient. The preoccupation with the individual may lead to forget that many diseases, whether contagious or not, have a typical pattern in time and space when a population of individuals is considered. Analysis of such patterns is the task of medical epidemiology (Macdonald 1957).

In agriculture, the individual plant is seldom a matter of concern. Attention focuses on the plot, greenhouse, field or plantation. The individual grower is definitely aware of patterns in time and space within his own field, but his notion about patterns in time and space in his community, valley, region, country or continent may be vague. Analysis of these patterns is the task of botanical epidemiology.

Discerning patterns in time and space is an art. Documenting the existence of patterns and pattern variation, and analyzing the underlying causes of pattern variation is the science of epidemiology. Botanical epidemiology has gone far in this art and science (Gregory 1973; Gregory and Monteith 1967; Pedgley 1982; Zadoks 1961, 1967, 1973), and has come to the state that

J. Bouma et al. (eds.), Eco-regional approaches for sustainable land use and food production, 437–452.

Table 1. Magnitudes an epidemics. Modified after Heesterbeek and Zadoks (1987).

Epidemic order	Sub-division	Temporal dimension	Spatia dimension
0-order		within season	meters
1st order		within season	kilometers
2nd order	2A	between seasons	kilometers
2nd order	2B	over years	continental

predictions can be made (Zadoks and Van den Bosch 1994). This is amazing, because of the great variety in crops, climates and cultural methods.

Eco-regional approaches are the bread and butter of field-oriented crop protectionists, because most pests move, at least within limited regions, and they are often fine-tuned to specific regions. 'Pest' is used here in a generic sense to indicate all yield-reducing organisms, animals from large to minute, fungi, bacteria and viruses. Pest insects damage plants by boring, gnawing or sucking. 'Vectors' are organisms, often insects, which transmit pest organisms from plant to plant, primarily viruses, more rarely fungi and bacteria. Sometimes, pests can be controlled by means of 'beneficials', among which are insects, fungi, bacteria and viruses.

Description of temporal and spatial patterns in botanical epidemiology

Magnitude of epidemics in time and space. A discussion of patterns in time and space requires appropriate definitions. Botanical epidemiology distinguishes epidemics at three orders of magnitude. These orders represent three processes through which pests conquer space. These processes may or may not share common mechanisms. A zero order epidemic is localized, a matter of meters, and develops within one growing season. A first order epidemic is a regional affair, to be measured in kilometers, and develops within one growing season. A second order epidemic has continental dimensions and develops over a sequence of years (table 1; Heesterbeek and Zadoks 1987).

The terminology of zero, first and second order epidemics was developed for once-only phenomena. Although very disruptive from an economic and social point of view, once-only outbreaks are the exception rather than the rule. The ususal pattern consists of repetitive outbreaks of insect pests, vectors and diseases. Thus, we distinguish endemics and exodemics.

Endemics. Most pest insects, vectors and pathogens, once established in an area, become endemic. They may survive within the area even when there

Table 2. Strategies and patterns of pest (insect, vector, pathogen) survival (perennation), with examples.

Frequency	Survival stategy	
	Endemic	Exodemic
Annual	EnA	E×A
Whitefly	transmit- ted geminiviruses Latin America	Yellow stripe rust China
Incidental	EnI	E×I
	Yellow stripe rust Netherlands	Desert locust Afrasian desert belt

is little or no crop available, and flare up when crops appear and conditions become favourable. The patterns depend on the survival and perennation strategy of the insect, the vector or the pathogen.

Viruses, which need a vector, form a special case. Considering survival systems for viruses, Harrison (1981) coined the concepts CULPAD (CULtivated Plant ADapted) and WILPAD (WILd Plant ADapted) viruses, that is, viruses that survive primarily in cultivated crops and viruses that survive, or perennate, in non-cultivated weeds or wild plants. Similarly, vectors with a wide host range survive on plant remnants everywhere, in all fields, at high spatial density whereas others with a narrow host range survive at low spatial density and have to reconquer space to cause zero and first order epidemics.

Exodemics. The exodemic is the epidemic caused by a migrant pest. Such a pest is endemic in restricted areas, where it usually causes little harm. It migrates to a target area where the damage can be catastrophic.

Two strategies. We discern two major survival strategies which, combined with two frequency types, lead to four patterns (table 2). Survival may be endemic, scattered within the region even though thinly spread, or exodemic, perennating outside the region and re-invading. The frequency of outbreaks may be regular or incidental, and in either case the intensity of the epidemics may vary considerably from case to case.

The EnA strategy is represented by whitefly (*Bemisia tabaci*)-transmitted geminivirus outbreaks in various crops. This whitefly has over 500 recorded host plants (Greathead, 1986). *B. tabaci* vectors bean golden mosaic virus (BGMV), a devastating disease in tropical Latin America. It is estimated that over 2.5 million ha are currently affected by BGMV, and that at least an additional million ha per year cannot be planted because of the risk of total loss, mainly during the dry season (Anderson and Morales 1994).

Table 3. Examples of long-range exodemics with annual occurrence (E×A).

Plant host	rice	wheat	wheat	peanut
Organism	*Nilaparvata translugens*	Aphids	*Puccinia striiformis*	*Puccinia arachidis*
	Brown plant hopper		Yellow rust fungus	Peanut rust fungus
Role	vector	vector	disease	disease
Disease transmitted	grassy stunt, rice ragged stunt	barley yellow dwarf	yellow rust	peanut rust
Source area	Philippines	US (S plains)	China (W mountains)	Ivory Coast (forest area)
Target areas	China, Japan, Korea	US (NC plains)	China (E plains) Japan	Ivory Coast Korhogo area Burkina Faso
Distance	2000 km	2000 km	4000 km	1000 km
Reference	Rosenberg and Magor (1983)	Wallin and Loonan (1971)	Zeng (1991)	Savary *et al.* (1988)

The EnI strategy is exhibited by yellow stripe rust (*Puccinia striiformis*) of wheat in The Netherlands, where the pathogen oversummers and overwinters annually within the country at various intensities, but where the disease flares up about once in ten years (Daamen *et al.* 1992; Zadoks 1961).

The E×A strategy is shown in China by yellow stripe rust, which overwinters in the Gangsu mountains and spreads annually over China (Zeng 1991; Yang and Zeng 1992) and into Japan (Hogg *et al.* 1969) by means of wind-borne spores, over a distance of 4,000 km (table 3).

The E×I strategy is exemplified by the desert locust (*Locusta migratoria*), which normally leads a hidden life in the bush, but once in twenty years flares up and spreads over a region stretching from Morocco to India, well over 8,000 km.

Scaling the outbreaks. The problem is to define the scale of such phenomena in units of length. On a small scale we will see incidental zero order epidemics, followed by a first order epidemic. The small scale can be seen as an isolated valley, an oasis in the desert, an irrigation project, a distinct agro-ecological area such as the Dutch sea polders. There may be a hierarchical nesting of such units (figure 1; Map 51 for *P. striiformis* in Zadoks and Rijsdijk 1984), so that for different purposes we need a different key parameter for size.

The maximum flight range of airborne pathogens and insects is of a completely different order, and the size parameter runs up to 6,000 km (Hogg *et al.* 1969). Most of the long-distance outbreaks could be characterized as first order epidemics, although an outbreak of the desert locust or of a new physiological race of yellow stripe rust may last a few years and thus may assume a second order character.

Intrinsic and extrinsic contributing to epidemic patterns

Perennation strategies primarily depend on the intrinsic characteristics of the insect pest, vector or pathogen involved. Intrinsic characteristics refer to the existence of, for example, pupation in insects and perennation structures in fungi. They are survival strategies that have evolved over centuries. These characteristics are usually well known and stable, and thus lead to good predictability of perennation. Outbreaks may be predicted if their weather dependence is known.

However, there are also extrinsic characteristics, mainly determined by cropping practices such as crop deployment, which affect survival and outbreaks. Cropping practices are determined by decisions of humans, individual farmers and authorities at various organizational levels. Such decisions have an effect on the dynamics of pest insects, vectors and pathogens. Again, the size parameters of these effects may show a hierarchical pattern, ranging from small scale decisions affecting a single field (range 0.01–1 km) to decisions at national or even a regional level (range 100–1,000 km). A decision taken at a local scale may well affect crops at 10 to 1,000 km away.

Examples. (i) The decisions of peasant farmers in the south of Ivory Coast to sow plots of peanuts measuring ca. 100 m^2 affect the large-scale peanut crops up to one thousand km further north in Ivory Coast, Mali and Burkina Fasso because of the peanut rust (*Puccinia arachidis*) (figure 2; Savary *et al.* 1988). (ii) In Brazil, soybean is a successful reproductive host for the whitefly, *Bemisia tabaci*. The whitefly population, reproducing on soybeans, increased dramatically and subsequently migrated into other crops, including *Phaseolus* beans, causing an epidemic of bean golden mosaic virus that resulted in up to 85% yield loss in Brazilian bean production (Costa 1975; Costa and Cupertino 1976). The decision to step up soybean production created an epidemic in the bean crops. (iii) Poorly managed irrigation schemes provide permanent refuges for pests and diseases, as exemplified by the golden snail (*Pomacea*

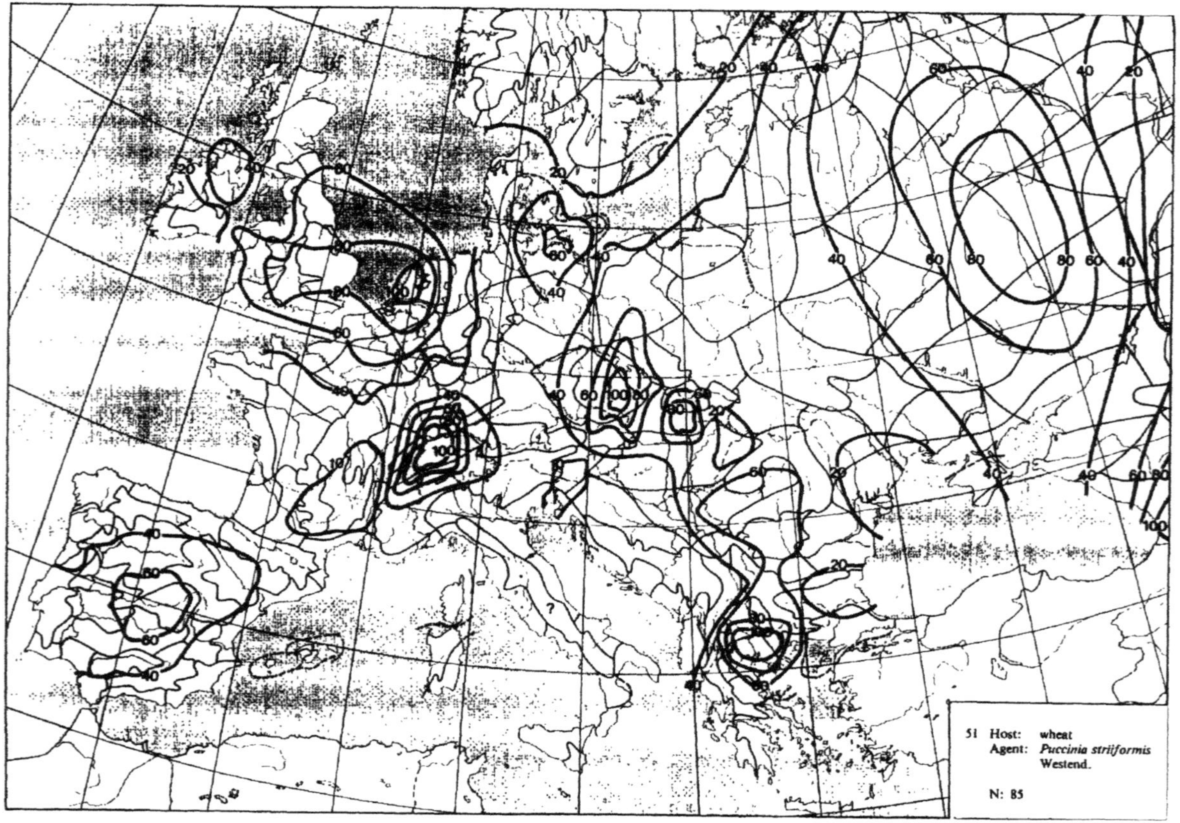

Fig. 1. Hierarchical nesting of epidemics in an eco-regional context. Represented is the wheat-yellow stripe rust pathosystem in Europe. Isolines connect points with equal damage averaged over many (ca. 20) years. Units attached to siolines represent average yield loss in per mil. (Source: Zadoks and Rijsdijk 1984, map 51)

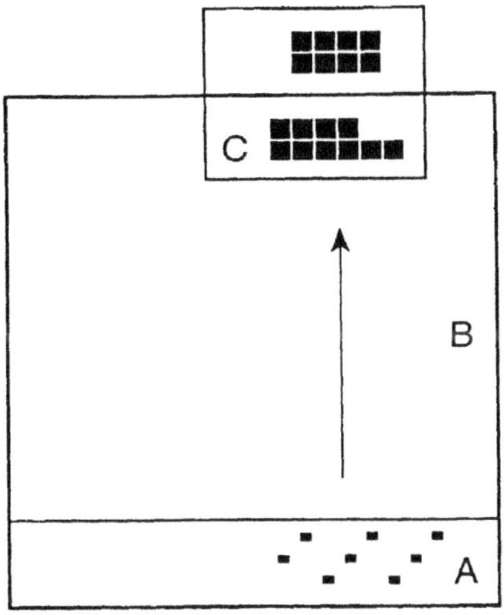

Fig. 2. Schematic representation of the annual migration of *Puccinia arachidis* on peanut, pattern E×A, in Ivory Coast. (A) Forest zone with few, small, scattered fields functioning as perennation sources. (B) Zone with little on no peanut fields. (C) Zone with intensive, seasonal cultivation at both sides of the border. ↑ Direction of prevalent wind.

caniculata) in the Philippines, where tail-end farmers are the victims of decisions on irrigation systems taken without their participation (Ketelaar 1993).

The green bridge. These examples belong to the broad category of 'green bridge' effects. In most climates, agriculture is subject to seasonality, where a cropping season alternates with a crop-free period due to winter temperatures or seasonal drought. The breaks between cropping seasons considerably reduce endemic populations of pest insects, vectors and pathogens, especially those which depend for survival on living, green crop plants, such as leafhoppers and some biotrophic fungi. With a metaphor we say that the green bridge (Zadoks 1984) from one crop season to the next is narrow and long. Only few individuals are able to pass the bridge to provide continuity in time for their species.

There are several subtle and few gross methods to make the bridge shorter and wider, so that more pest individuals can pass it. In north-west European agriculture the methods for winter wheat were subtle indeed. Split nitrogen applications retarded the ripening of the crop, whereas sowing dates were advanced one month. The green bridge was shortened by 4 to 6 weeks and powdery mildew (*Erysiphe graminis*), yellow stripe rust and aphids tripped over happily. In addition, Dutch farmers undersowed the wheat with grass, omitted summer tillage of the soil, and ploughed the grass under in the late fall, to improve the soil texture. Self-sown wheat abounded, was not controlled, and broadened the green bridge by a factor of one thousand to one million.

In the tropics, double and triple cropping of rice resulted in similar green bridges leading to problems with rats (Oka, pers. comm.) and brown planthoppers (Oka 1979), tungro and soil-borne diseases. In the tropics, irrigation systems are notorious for their green-bridge effect, especially at the tail end.

A classic example comes from Nicaragua. In the early 1980s serious shortages in maize, and an agricultural policy emphasizing national food security, resulted in a 'Contingency Plan' to install irrigation systems in large tracts of the Pacific Coastal Plain and to plant irrigated maize during the traditional fallow period from January to April. The first irrigated plantings were made in 1984. By 1986 an epidemic caused by leafhopper (*Dalbulus maidis*)-transmitted corn stunting pathogens (virus, mycoplasma, spiroplasma) resulted in the loss of nearly 30,000 tonnes of maize, the equivalent of the per capita intake of 13% of the Nicaraguan people. By 1987, the losses were up to 60% over a 12,000 ha area. Paradoxically, the attempt to increase food production created an epidemic that resulted in the necessity to increase maize imports (Anderson 1991).

Another social aspect is the rapid change in rural labour availability. In newly industrializing countries, such as Indonesia, specifically the island of Java, and China the men migrate to the cities where job opportunity seems high, leaving the women, the elderly and the children to tend the crops. We expect a detrimental effect on crop protection, with unpredictable eco-regional consequences.

Analysis of spatial and temporal patterns in time and space

Pathosystems. We use the term pathosystem (Robinson 1976) and take it here in a wide sense, comprising the target crop and the harmful agent, its eventual vector, alternate hosts, refuge and beneficials. As each pathosystem is different, even if the target crop is the same, the geography of pathosystems can be symbolised by Venn diagrams (Weltzien 1978).

A hierarchy of sizes for a single pathosystem is shown in figure 3. Such a hierarchy may occur in an area with valleys, separated by high ridges. Typical valley patterns may occur (Zogg 1949), as the senior author observed in Switzerland with yellow stripe rust of wheat, where one valley had a severe epidemic and the other valley but a light infection even though wheat cultivar, altitude and environmental conditions were the same.

The situation of a multiple pathosystem (one target crop, several harmful agents) is shown in figure 4. Subdivisions may be needed, because cultural

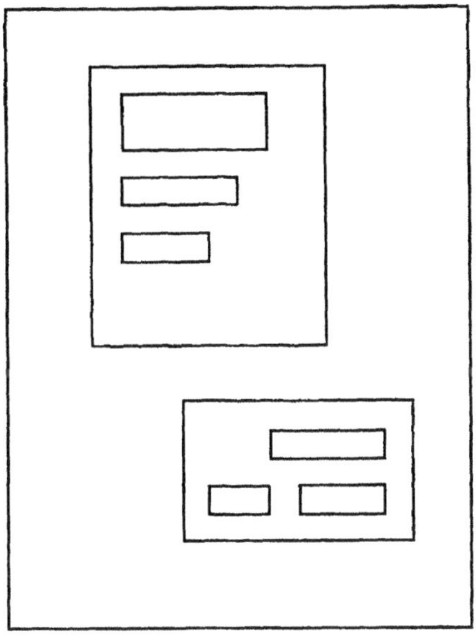

Fig. 3. A hierarchy of pathosystems rendered by means of Venn diagrams. Hierarchical nesting within a single pathosystem may occur when e.g. two larger areas are separated by a high mountain range and when each area consists of valleys separated by ridges.

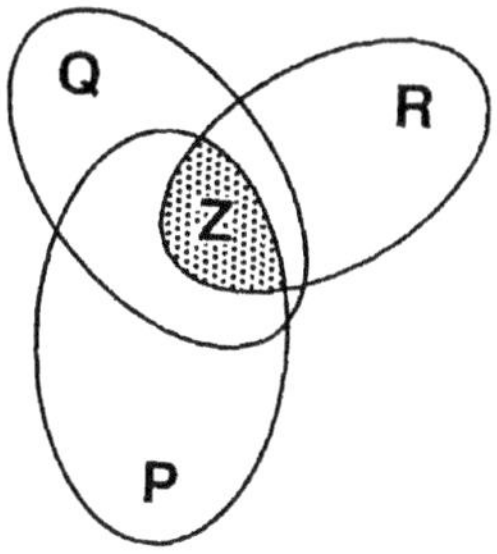

Fig. 4. A multiple pathosystem rendered by means of Venn diagrams. The multiple pathosystem of crop C comprises pathosystems P, Q and R, which may mutually affect each other. The crop area is the unison of the pathosystem areas. The high risk area Z is the intersection.

conditions may vary within a region. For example, the green bridge effect in winter wheat and its various diseases was caused in the Dutch sea polders by undersowing grass, in England by very early sowing of the crop.

In the tropics, well tended and watered home gardens may be an ideal refuge for whiteflies and various gemini viruses in some and not in other areas. Herewith, the social element is introduced into our geographic considerations. Cultivar preference and practices of intercropping and relay cropping may vary over short distances (100 km) and so enrich the patterns. Whereas some cropping zones are enclosed by others, devoid of beneficials, other cropping

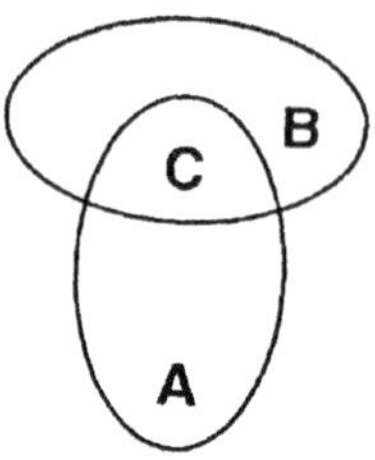

Fig. 5. Agricultural areas with and without biological control by beneficials coming from surrounding non-managed areas. The cultivated area A is neigboured by a wild area B. The intersection C is the sub-area of A which is positively affected by beneficials from B.

zones might benefit from relatively undisturbed bush area which produces beneficials (figure 5). Beneficials, such as e.g. coccinellids, have their own migration patterns which are often poorly known.

Geographic information systems. In the USA, effective control of black stem rust (*Puccinia graminis*) on wheat was obtained after barberry eradication, at great costs. After a prolonged struggle, scientists finally convinced authorities and public to take action. Convincing power was build up by carefully mapping the spread of disease (Hogg *et al.* 1969; Roelfs 1982), an early use of geographic information. Today, it is easy to introduce known data into a Geographic Information System (GIS), which already contains the geography of a region, its climatology, and a gross outline of its agriculture.

The problem is to identify the relevant characteristics of agricultural practices and of social habits for any pathosystem (Liebhold *et al.* 1993), and to describe the migration and outbreak patterns. Pioneer work was done by Weltzien (1972), who called his work 'geophytopathology'. He was able to predict new outbreaks by matching the physical requirements of a pathogen with the climatic characteristics of a region. An example is his statement that California would be a good environment for sugar beet mildew (*Erysiphe betae*), a European pathogen (Drandarevski 1969). A first order epidemic appeared in California in 1974 (Ruppel *et al.* 1975). Quarantine services have made data sheets of many harmful agents, which can be matched to climatic data anywhere in the world. A typical example is the study on soybean rust by Yang *et al.* (1991).

Correspondence analysis in epidemiology. New survey methods, supported by modern statistics, help in rapid fact finding (Savary 1987; Savary *et al.* 1993). These techniques can easily incorporate characteristics of cropping practices, pests, and various social factors. A survey is a conventional but also a very modern tool for research at a high hierarchical level. A survey can relate reductionist laboratory research to the requirements of policy makers by means of 'ground truth' (Habtu 1994; Savary 1986; Schouten 1991;

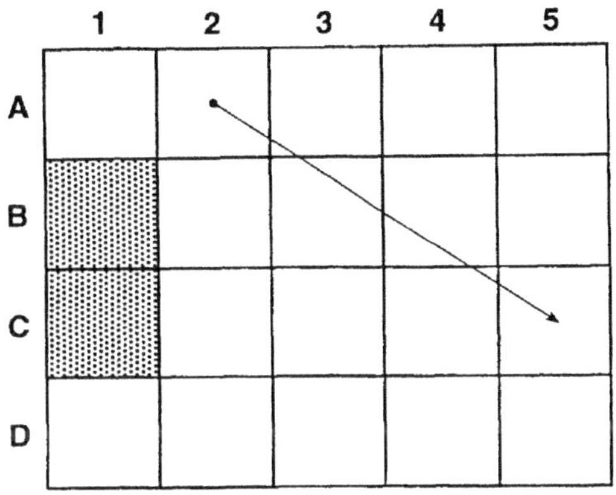

Fig. 6. The eco-region A1 ... D5 is divided into $4 * 5 = 20$ units. Units B1 and C1 contain the perennation sources of a pathogen. For any two units there exists, at any time, a transition probability $P_{ij,kl}(t)$ describing the probability of infection of one unit by the other, e.g. as indicated by the arrow $P_{A2,5C}(t)$. P depends on intrinsic characteristics of the pathogen and on extrinsic characteristics such as crop and weather. Selfinfection and perennation are given by $P_{ij,ij}$.

Zadoks 1961). In modern, policy oriented crop protection research, a survey must be the basis for any eco-regional approach.

It may be possible to cut the geographic area (determined by physical of political borders) into small units, each with their own weather characteristics. Then, transition probabilities may be established for each pair of units (figure 6). If so, general patterns of large scale migration can be described and simulated (Frantzen 1994; Zeng 1992). On-line prediction of outbreaks becomes feasible. The predictions could be of use to policiy makers, for logistic purposes (pest control, harvest operations, selling, purchasing or shipping of commodities) and for the establishment of research policies.

Even if an area is apparently homogeneous (e.g. in climatological terms), it may, upon detailed inspection, turn out to be a patchwork of agriculturally and, sometimes, socially different units, which need different agrotechnical approaches. Thus, two neighbouring fields may belong to different 'recommendation domains' (Singh *et al.* 1994) whereas two fields a hundred kilometers apart may belong to the same recommendation domain (figure 7). Here, the eco-regional approach will experience a dilemma. Acceptance of the patchwork idea makes sweeping statements at an eco-regional level almost meaningless, while rejection of the patchwork idea makes most decisions at the eco-regional level ineffective.

Notions such as 'resolution', the technically feasible resolution and the administratively desirable resolution, must come to the aid. To appreciate the eco-regional approach we may need a coarse-grained picture, while to successfully implement an eco-regional approach, a finer-grained understanding may be necessary. Social variation imposes additional dimensions to the pic-

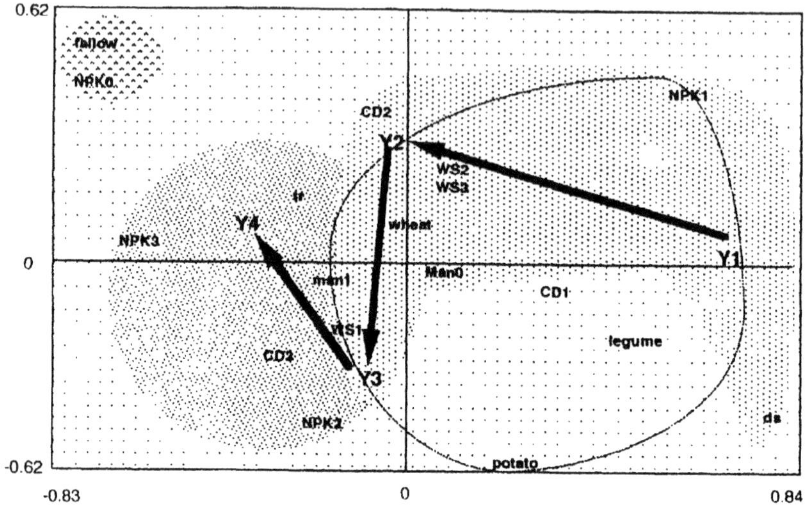

domain	cropping practices	pest constraints
A	- High fertilizer input - Some manure - previous crop: wheat predominent - transplanted rice - medium-high yield	- deadhearts - whiteheads - army worms - rice bugs - sheath blight - sheath rot - neck blast
B	- low fertilizer input - some manure - diverse previous crops - water stress often high - transplanted rice predominent - short cycle - low yield	- glume discoloration - (leaf blast)
C	- low fertilizer input - no manure - previous crop: wheat only - medium-low water stress - transplanted or direct-seeded rice - long cycle varieties	- weeds above the rice canopy - glume discoloration
D	- no fertilizer - fallow	- deadhearts - whiteheads - sheath rot

Fig. 7. Patterns of cropping practices and corresponding yield levels and pest constraints in Uttar Pradesh, India. After Singh *et al.* (1994). The figure is the result of ordination applied to a large data set on cropping practices using a non-parametric multivariate analysis of categorized data. Abscissa and ordinate represent two orthogonal axes explaining most of the inertia of the data set. Figures along axes are chi-square distances. The arrows represent a path or trajectory from low (Y1) to high (Y4) yields. Yield levels Y1 and Y2 are associated with domain C, Y3 and especially Y4 are associated with domain A. Low fertilizer use (NPK1) and rotation with legumes are associated with low yields (Y1). Rotation with wheat is associated with intermediate yields. High yield levels (Y1) are associated with high fertilizer levels (NPK3).

ture. Crop protectionists have dug deeply in the laboratory, but in these field issues they have just begun to scratch the surface.

Research and policy requirements

Policy. Intensification of the world's agriculture is a necessity, but the limits to intensification posed by pests and diseases are clear and predictable in principle. These limits have to be respected in a context of sustainable land use and food production. The knowledge and experience that already exist can aid in planning the intensification, if the expert knowledge is utilized. Policy makers must include crop protection specialists in the planning process.

For example, where continuities in time and space threaten sustainability, discontinuities should be built into our agro-ecosystems, or be re-introduced (Zadoks and Schein 1979). In some areas of Indonesia, an annual break in rice growing is mandatory. The National Commission for Whitefly Management of the Ministry of Agriculture (SEA) of the Dominican Republic recommended crop-free periods during which whitefly host crops could not be planted. The recommendations were legalized and enforcement has resulted in lower yield losses from the whitefly transmitted BGMV in the Azua region (Alvarez *et al.* 1993; Morales 1994). Genetic discontinuities were recommended in northern India to interrupt the spread of brown leaf rust (*P. recondita*) of wheat (Nagarajan and Singh 1990).

Organizational alignment (Useem *et al.* 1992) of all actors in crop protection within a certain area is essential, because otherwise social problems may disturb the patterns.

Research. Researchers must scale up their research focus. On a small scale we will see incidental zero order epidemics, followed by a first order epidemic. The small scale can be seen as an isolated valley, an oasis in the desert, or a distinct agro-ecological area such as the Dutch sea polders. There may be hierarchical nesting of such units, so that for different purposes we need different scales. The individual producer will be more concerned with zero and first order epidemics, while the regional research institute will be called on to consider first and second order epidemics.

We feel that a quantitative, predictive, policy-oriented eco-regional approach to crop protection, while still in its infancy, is feasible. A large amount of inspiring data is available, that could be utilized in the desired eco-regional approach (Aylor 1986). Recommendations on an eco-regional basis were given for the management of peanut rust in Ivory Coast (Savary *et al.* 1988) and for bean rust (*Uromyces appendiculatus*) in Ethiopia (Habtu, 1994). Data handling, storage and retrieval techniques are available. Advanced simulation models describe yield and crop loss if disease dynamics are known (Bastiaan 1993). Elementary mathematics exist to describe and predict outbreaks

(Campbell and Madden 1990; Zadoks and Schein 1979). A new methodology to interpret survey data has been developed (Savary *et al.* 1994; Singh *et al.* 1994) which allows for the (rapid) characterization of pest constraints over large areas, and the construction of hypotheses for management oriented research.

There is no obstacle to a fresh, holistic approach to crop protection problems, aimed at understanding large-scale phenomena in a policy-oriented manner, but relevant at farm level.

Acronyms

BGMV	Bean Golden Mosaic Virus
GIS	Geographic Information System
CULPAD	CULtivated Plant ADapted
WILPAD	WILd Plant ADapted

References

Alvarez P, Alfonseca L, Abud A, Villar A, Rowland R, Marcano E, Borbon J C, Garrido L (1993) Las moscas blancas en la Republica Dominicana. Pages 34–37 in Hilje L, Arboleda O (Eds.) Las moscas blancas en America Central y el Caribe. Turrialba, CATIE, Costa Rica.

Anderson P K (1991) Epidemiology of insect-transmitted plant pathogens. Ph.D Thesis. Harvard, USA. 304 pp.

Anderson P K, Morales F J (1994) The emergence of new plant diseases: The case of insect-transmitted plant viruses. Annals of the New York Academy of Sciences 740 (in press).

Aylor D E (1986) A framework for examining inter-regional aerial transport of fungal spores. Agriculture and Forest Meteorology 38:263–288.

Bastiaan L (1993) Understanding yield reduction in rice due to leaf blast. Ph.D Thesis, Wageningen Agricultural University, Wageningen, The Netherlands. 127 pp.

Campbell C L, Madden L V (1990) Introduction to plant disease epidemiology. John Wiley & Sons, New York, USA. 532 pp.

Costa A S (1975) Increase in the population density of *Bemisia tabaci*, a threat of widespread virus infection of legume crops in Brazil. In Bird J, Maramorosh K (Eds.) Tropical diseases of legumes. Academic Press, New York, USA.

Costa A S, Cupertino F P (1976) Avaliacao das perdas na producao do fejoeira causada pelo virus do mosaico dourado. Fitopatologia Brasileira 1:18–25.

Daamen R A, Stubs R W, Stol W (1992) Surveys of cereal diseases and pests in the Netherlands. 4. Occurrence of powdery mildew and rusts in winter wheat. Netherlands Journal of Plant Pathology 98:301–312.

Drandarevski C (1969) Untersuchungen über den echten Mehltau *Erysiphe betae* (Vanha) Weltzien. III. Geophytopathologische Untersuchungen. Phytopathologische Zeitschrift 65:201–218.

Frantzen P A M J (1994) Studies on the weed pathosystem *Cirsium arvense - Puccinia punctiformis*. Ph.D Thesis, Wageningen Agricultural University, Wageningen, The Netherlands. 91 pp.

Greathead A H (1986) Host plants. Pages 17–25 in Cock M J W (Ed.) *Bemisia tabaci* - A literature survey. CAB International, Berks., UK. 121 pp.

Gregory P H (1973) Microbiology of the atmosphere. 2nd Ed. Aylsbury, Leonard Hill. 377 pp.
Gregory P H, Monteith J L (Eds.) (1967) Airborne microbes. Cambridge University Press, Cambridge, UK. 385 pp.
Habtu A (1994) Epidemiology of bean rust in Ethiopia. Ph.D Thesis. Wageningen Agricultural University, Wageningen, The netherlands. 171 pp.
Harrison B D (1981) Plant virus ecology: ingredients, interactions and environmental influences. Annals of Applied Biology 99:195–209.
Heesterbeek J A P, Zadoks J C (1987) Modelling pandemics of quarantine pests and diseases; problem and perspectives. Crop Protection 6:211–221.
Hogg W H, Hounam C E, Mallik A K, Zadoks J C (1969) Meteorological factors affecting the epidemiology of wheat rusts. World Meteorological Organization, Technical Note No. 99, Geneva, Switzerland. 143 pp.
Ketelaar J W (1993) Strategies for solving the Philippine snail problem. A system perspective. Wageningen, The Netherlands. 100 pp.
Liebhold A M, Rossi R E, Kemp W P (1993) Geostatistics and geographic information systems in applied insect ecology. Annual Review of Entomology 38:303–327.
Morales F J (Ed.) (1994) Bean golden mosaic virus research advances. Centro Internacional de Agricultura Tropical, Cali, Colombia. 193 pp.
Nagarajan S, Singh D V (1990) Long-distance dispersion of rust pathogens. Annual Review of Phytopathology 28:139–153.
Oka Ida Nyoman (1979) Cultural control of the brown plant hopper. Pages 357–369 in Brown planthopper: threat to rice production in Asia. IRRI, Los Baños, Philippines. 369 pp.
Pedgley D E (1982) Windborne pests and diseases: Metorology of airborne organisms. John Wiley & Sons, Chichester, UK. 240 pp.
Robinson R A (1976) Plant pathosystems. Springer, Berlin, Germany. 184 pp. Rosenberg L J, Magor J I (1983) A technique for examining the long-distance spread of plant virus disease transmitted by the brown plant hopper *Nilaparvata lugens* (Homoptera: Delphacidae), and other wind-borne insect vectors. Pages 229–238 in Plumb R T, Thresh J M (Eds.) Plant virus epidemiology. Blackwell, Oxford, UK. 377 pp.
Ruppel E G, Hills F J, Mumford D L (1975) Epidemiological observations on the sugarbeet powdery mildew in Western USA. Plant Disease Reporter 59:293–286.
Savary S (1986) Etudes épidémiologiques sur la rouille de l'arachide en Côte d'Ivoire. Ph.D Thesis. Wageningen Agricultural University, Wageningen, The Netherlands. 154 pp.
Savary S (1987) Enquête sur les maladies fongiques de l'arachide (*Arachis hypogaea*) en Côte-d'Ivoire. II. Epidémiologie de la rouille de l'arachide (*Puccinia arachidis*). Netherlands Journal of Plant Pathology 93:215–231.
Savary S, Bosc J P, Noirot M, Zadoks J C (1988) Peanut rust in West Africa: A new component in a multiple pathosystem. Plant Disease 72:1001–1009.
Savary S, Fabellar N, Tiongco E R, Teng P S (1993) A characterziation of rice tungro epidemics in The Philippines from historical survey data. Plant Disease 77:376–382.
Savary S, Elazegui F A, Moody K, Litsinger J A L, Teng P S (1994) Characterization of rice cropping practices and multiple pest systems in the Philippines. Agricultural Systems 46:385–408.
Schouten H J (1991) Studies on fire blight. Ph.D Thesis. Wageningen Agricultural University, Wageningen, The Netherlands. 143 pp.
Singh H M, Srivastava R K, Singh R K, Savary S (1994) Illustrating the recommendation domain concept in integrated pest management: an Indian case study. International Rice Research Notes 19:28–30.
Useem M, Setti L, Pincus J (1992) The science of Javanese management: organizational alignment in an Indonesion development program. Public Aministration and Development 12:447–471.
Yang X B, Dowler W M, Royer M H (1991) Assessing the risk and potential impact of an exotic plant disease. Plant Disease 75:976–982.

Yang X B, Zeng S M (1992) Detecting patterns of wheat stripe rust in time and space. Phytopathology 82:571–576.

Wallin J R, Loonan D V (1971) Low-level jet winds, aphid vectors, local weather and barley yellow dwarf virus outbreaks. Phytopathology 61:1068–1070.

Weltzien H C (1972) Geophytopathology. Annual Review of Phytopathology 10:277–298.

Weltzien H C (1978) Geophytopathology. Pages 203–222 in Horsfall J G, Cowling E B (Eds.) Plant diseases, an advanced treatise. Vol. III. Academic Press, New York, USA.

Zadoks J C (1961) Yellow rust on wheat, studies in epidemiology and physiologic specialization. Tijdschrift over Plantenziekten (Netherlands Journal of Plant Pathology) 67:69–256.

Zadoks J C (1967) International dispersal of fungi. Netherlands Journal of Plant Pathology 73, Suppl. 1:61–80.

Zadoks J C (1973) Long-range transmission of phytopathogens. Pages 392–396 in Hers J F Ph, Winkler K C (Eds.) Airborne transmission and airborne infection. Oosthoek, Utrecht, The Netherlands. 300 pp.

Zadoks J C (1984) Disease and pest shifts in modern wheat cultivation. Pages 215–225 in Gallagher E J (Ed.) Cereal production. Butterworths, London, UK. 354 pp.

Zadoks J C, Rijsdijk F H (1984) Agro-ecological atlas of cereal growing in Europe. III. Atlas of cereal diseases and pests in Europe. Pudoc, Wageningen, The Netherlands. 169 pp.

Zadoks J C (Ed.) (1993) Modern crop protection: developments and perspectives. Wageningen Pers, Wageningen, The Netherlands. 309 pp.

Zadoks J C, Schein R D (1979) Epidemiology and plant disease management. Oxford University Press, New York, USA. 427 pp.

Zeng S M (1991) PANCRIN, a prototype model of the pandemic cultivar-race interaction of yellow rust on wheat in China. Plant Pathology 40:287–295.

Zogg H (1949) Untersuchungen über die Epidemiologie des Maisrostes *Puccinia sorghi* Schw. Phytopathologische Zeitschrift 15:143–192.

Site specific management on field level: high and low tech approaches

J. BOUMA[1], J. BROUWER[2], A. VERHAGEN[1] and H.W.G. BOOLTINK[1]

[1] *Department of Soil Science and Geology, Wageningen Agricultural University, P.O. Box 37, 6700 AB Wageningen, The Netherlands;* [2] *ICRISAT Sahelian Center, BP 12404, Niamey, Niger*

Key words: global positioning systems, modelling, nitrate leaching, potato growth, pedotranfer functions, spatial variability, soil fertility

Abstract. Spatial variability of soil conditions and potato growth were studied in a 6 ha farmers field in a Dutch polder. Potato yields, measured in 65 small plots varied between 30 and 45 tons ha^{-1}, while yields of commercially attractive large potatoes varied between 3 and 15 tons ha^{-1}. An experiment in Niger indicated a major effect of spatial variability on growth of millet, with yields from different plots within a 0.3 ha field varying by as much as a factor 3.6, for the same treatment. Such differences are economically significant in both areas. A high-tech system for site-specific management is discussed for Dutch conditions including site specific sampling for soil fertility and use of dynamic simulation modeling to characterize soil water regimes and nutrient fluxes, e.g. of nitrate. A low-tech system for Niger includes field sampling and site specific interpretation of data obtained. In both cases, uniform fertilization rates based on one mixed sample for the entire field, are bound to result in local over- and underfertilization, implying inefficient use of natural resources. Modeling can be used to balance production and environmental aspects of soil fertilization, as was demonstrated for the Dutch study. Data needs of the WAVE model, used for simulation of yields and nitrate fluxes, are discussed including distinction of only four 'functional layers' for the 6 ha field, which define all variability in basic hydraulic characteristics. Fine-tuning of management practices, including fertilization, taking into account natural variability patterns appears to be an attractive and practical procedure to increase the use-efficiency of natural resources. The Niger example illustrates use of possible low-tech procedures involving field experimentation and improved advisory practices.

Introduction

Sometimes a technological breakthrough has unexpected side-effects. The development of global positioning systems (GPS), initially in secret for the military but later openly aimed at a large group of prospective buyers, has by now resulted in the availability of relatively cheap gadgets allowing accurate determinations of locations on the earth surface at any time. When applied on harvesters which are also equipped with sensors for continuous yield monitoring ('yield monitoring on-the-go'), some interesting results are being obtained in the USA and western Europe (e.g. Cahn *et al.* 1994). Differences

J. Bouma et al. (eds.), Eco-regional approaches for sustainable land use and food production, 453–473.

in crop production within agricultural fields, which are the management units for a farmer, turned out to be much higher than anticipated, varying by a factor two to four. These results are new. Farmers would know, of course, that differences occur. However, such impressions were always hard to quantify because documentation by making a series of small harvests within a field, is obviously not feasible from a management point of view. Yields are therefore always expressed in terms of e.g. tons ha^{-1}, by dividing total yields by the total area of the farm being covered by a particular crop, ignoring local differences. Application of GPS and yield-sensors does, however, allow expression of such differences.

Differences of yields within fields are very obvious in many developing countries, even without access to GPS and automatic yield monitoring equipment. In this paper attention will be paid,next to a farming system in the Netherlands, to a farming system in the Sahel in Africa, where millet is the main crop. Studies made in sandy soils near ICRISAT Sahelian Centre in Sadore, Niger, have indicated major differences in yield within a few meters distance which appear to be related to differences in water infiltration and to local activities of termites (Brouwer *et al.* 1993).

What are the implications of knowing yield differences within fields? The first challenge is to find the reasons why these differences occur and the second challenge is to then develop management procedures which can reduce or make use of these differences. The overall expectation would be that reducing differences would be economically and ecologically attractive for the farmer but, particularly, for society at large because natural resources are used more efficiently. Clearly, for Sahelian conditions the economic aspect has most emphasis.

Even though such expectations would appear not to be unreasonable, specific research is clearly needed to prove the point. Classical agronomic research is poorly equipped to investigate such issues as attention has traditionally been focused on measuring yields in plots which are implicitly assumed to be internally homogeneous. Differences in crop response among different plots are next attributed to differences in treatment and are not related to spatial variability patterns within the plots.

Yield differences within fields can be due to many reasons, such as differences in actual soil fertility or to erratic application rates of fertilizers or biocides; occurrence of compacted layers; low and wet spots or high and dry spots; local occurrence of pests and diseases and many other reasons. Once causes have been established, site-specific management procedures have to be devised which allow local rectification of differences. The high-tech approach is to develop new technology, where, again, GPS plays a central role (e.g. Robert *et al.* 1994). However, a low-tech approach is equally feasible. In a developing country such as Niger, farmers don't have the luxury of restricting fertilizer use: they do not have money to buy it in the first place! Rather, they can place available cow- or sheep manure or weeds and crop residues at certain spots only. In fact, farmers are known to do this already.

High-tech research on site specific management, as discussed above, has been in progress in several countries with a clear focus on soil fertility. The traditional manner to collect soil fertility samples is to obtain a mixed sample from a field and to derive a corresponding fertilizer recommendation by using standard tables relating fertilization rate to yield. Obviously, this procedure will overestimate rates for some areas and underestimate them for others. Several studies have been reported where many seperate samples were taken and where fertilization rates were determined for each point (Franzen and Peck 1994). Information technology, such as Geographic Information Systems, and computer-guided application devices were used to achieve what has been called: 'soil or site specific management'. As discussed above, other factors than soil fertility could well be the cause of yield differences within fields. Recently, therefore, the term 'Precision Agriculture' has been coined to cover all factors of location-specific management,including technologies and software.

This paper will address the question which soil related research is necessary to allow execution of either high- or low-tech Precision Agriculture, and how research should proceed. Emphasis in this paper on field-level research may at first sight appear odd, considering the general focus of this book on eco-regional approaches. However, activities in an (eco-)region cannot be discussed without considering actual, but also potential conditions at the farm- and field level which have to be integrated towards higher aggregation levels. In fact, there is an increasing awareness that a productive and environmentally sound agriculture anywhere in the world can only be realized by the immediate stakeholders, the farmers, who operate at farm level.

Soil research for precision agriculture

The following elements may be distinguished when defining soil research for precision agriculture to be focused on specific fields of a given farm. In principle, these elements apply for both high- and low-tech systems:

1. Establish a soil database, which contains relevant soil characteristics.
2. Monitor crop growth and physical and chemical conditions during one or more growing seasons. Use data in expert systems or for calibration and validation of simulation models for crop growth and solute fluxes.
3. Define threshold values for yields and chemical fluxes from an economic and environmental point of view and determine when they would be exceeded under different well defined forms of management and variable weather conditions. Obviously, options for different forms of management vary widely among developed and developing countries.

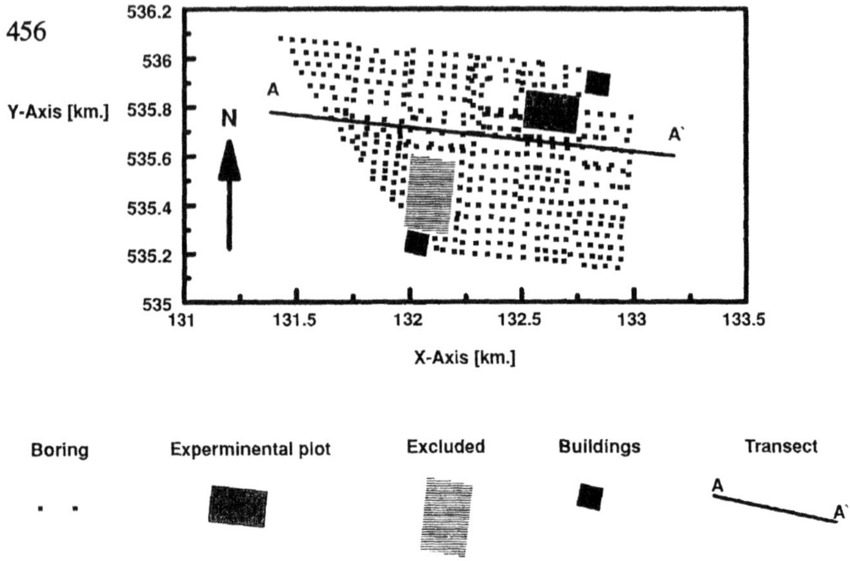

Fig. 1. Layout of the experimental farm 'Prof.Dr J.M. van Bemmelenhoeve', Wieringermeer, The Netherlands.

4. Use expert systems or modeling techniques to define management schemes that avoid exceedance of threshold values while maintaining yields at economic levels.
5. Implement the schemes by developing and using site-specific technology, which can be high- or low-tech.
6. Develop operational decision support systems to be used in practice. These can be computer driven but they can also consist of simple, easily accessible advisory schemes.

The above elements will now be discussed in more detail, providing examples from literature and from ongoing case studies in The Netherlands for an experimental plot at the experimental farm 'Van Bemmelenhoeve' in Wieringermeer (figure 1), and in Niger for a farmers field near ICRISAT-Sahelian Centre.

Soil database

The Netherlands

The field to be characterized measured 300 m by 200 m. An exploratory soil survey was made and soils were classified as *Typic Udifluvents* (Soil Survey Staff 1975). A geostatistical analysis indicated that the observed spatial variability could be optimally characterized by using a 50 m by 50 m grid,

Table 1. Description of the functional layers and corresponding Van Genuchten parameters for the hydraulic functions.

Layer	ρ [kg dm^{-3}]	sd	org. matter %	clay %	K_{est} [cm day^{-1}]	θ_{sat}	sd	θ_{res}	α	n	l
F1	1.48	0.039	0–2	0–4	183	0.40	0.02	0	0.03096	2	2.2842
F2	1.21	0.097	0–2	4–11	128	0.53	0.03	0.05	0.02000	1.5	0.5
F3	1.08	0.231	0–2	11–23	7	0.57	0.05	0.05	0.18650	1.2676	0.4205
F4	1.30	0.045	0–3	4–23	335	0.52	0.02	0.02	0.24986	1.2154	1.0791

which resulted in 65 detailed point observations of soil characteristics such as texture and organic matter content. Each observation was geo-referenced. In contrast to a traditional soil survey, attention was focused here on 'functional' soil horizons and not on the traditional genetic horizons. Functional horizons consist of combinations of genetic horizons with identical behaviour. Here, emphasis is on hydraulic properties (Finke 1993). In this field, soils were strongly layered and distinction of functional layers was based on descriptions of soil texture, as observed during augering, and a prelimainary classification that was finalized after measurement of hydraulic conductivity and moisture retention, using modern techniques (e.g. Finke 1993). Four functional horizons were distinguished, as summarized in table 1, which also lists the Van Genuchten coefficients for the measured hydraulic characteristics. Average curves based on measured hydraulic conductivity and moisture retention data of each of the four layers are shown in figure 2. Use of functional horizons is attractive because the number of vertically successive horizons (layers) to be distinguished at each point can be represented by only a few, and sometimes only one, functional horizon, rather than the relatively high number of pedological horizons which are distinguished according to pedological criteria. A representative cross section through the field of study, in which each depth is characterized by a functional layer is shown in figure 3. Basic hydraulic soil data can be used for simulation modeling as will be discussed later.

Geo-referenced soil data, as collected here, are quite different from data derived from conventional soil maps, even highly detailed ones. Soil maps define mapping units in which a particular soil type is supposed to occur, while attention in this study is on defining point data in terms of characteristics which are relevant for modelling. Expressions for areas of land are obtained by interpolation of point data (e.g. Finke 1993).

Niger

ICRISAT Sahelian Center (ISC) is located 40 km south of Niamey, Niger. Annual rainfall has averaged 562 mm over the last 80 years. The rainy season is from May to September. The sandy soils are deep, red *Psammentic*

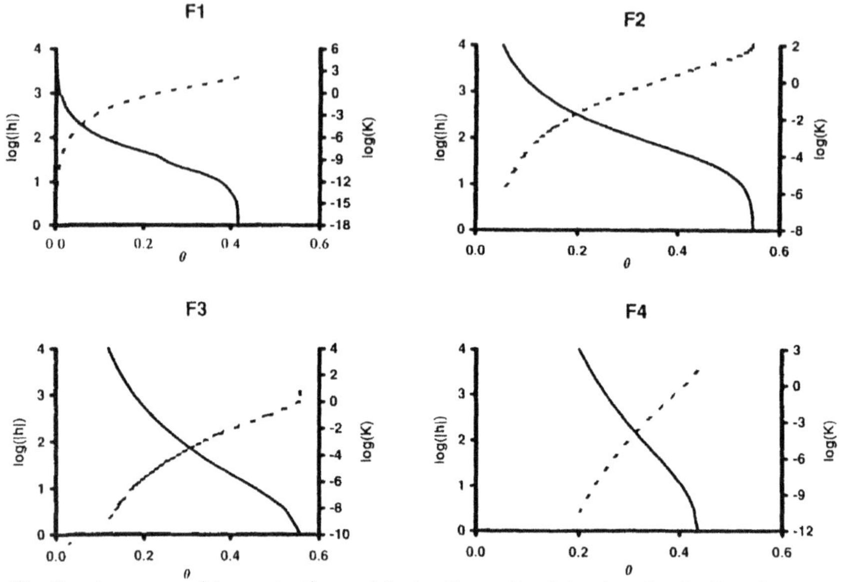

Fig. 2. Average moisture retention and hydraulic conductivity data for the four functional layers. Averages are based on five replicate measurements.

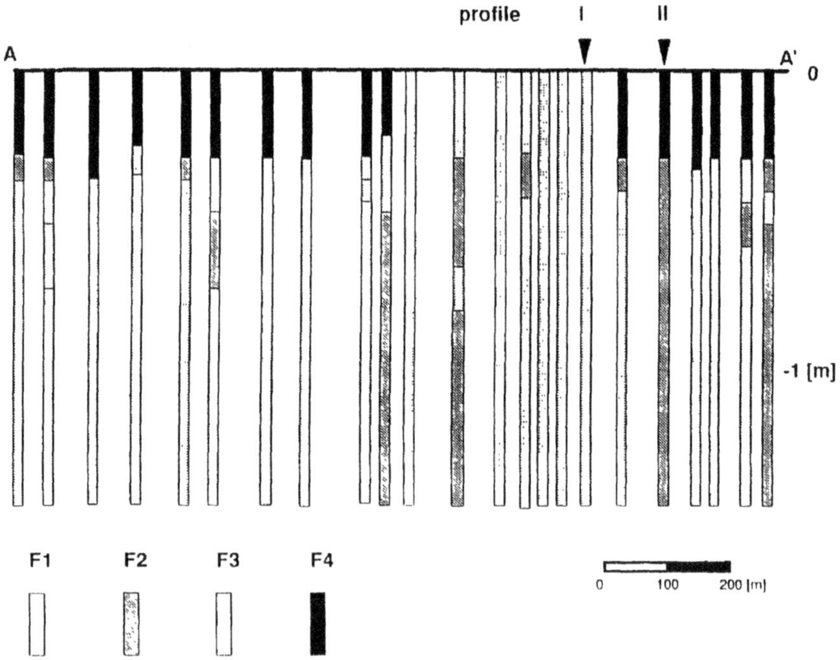

Fig. 3. Cross section of soil profiles along the line A–A′ in figure 1. Each profile is composed of a number of functional layers.

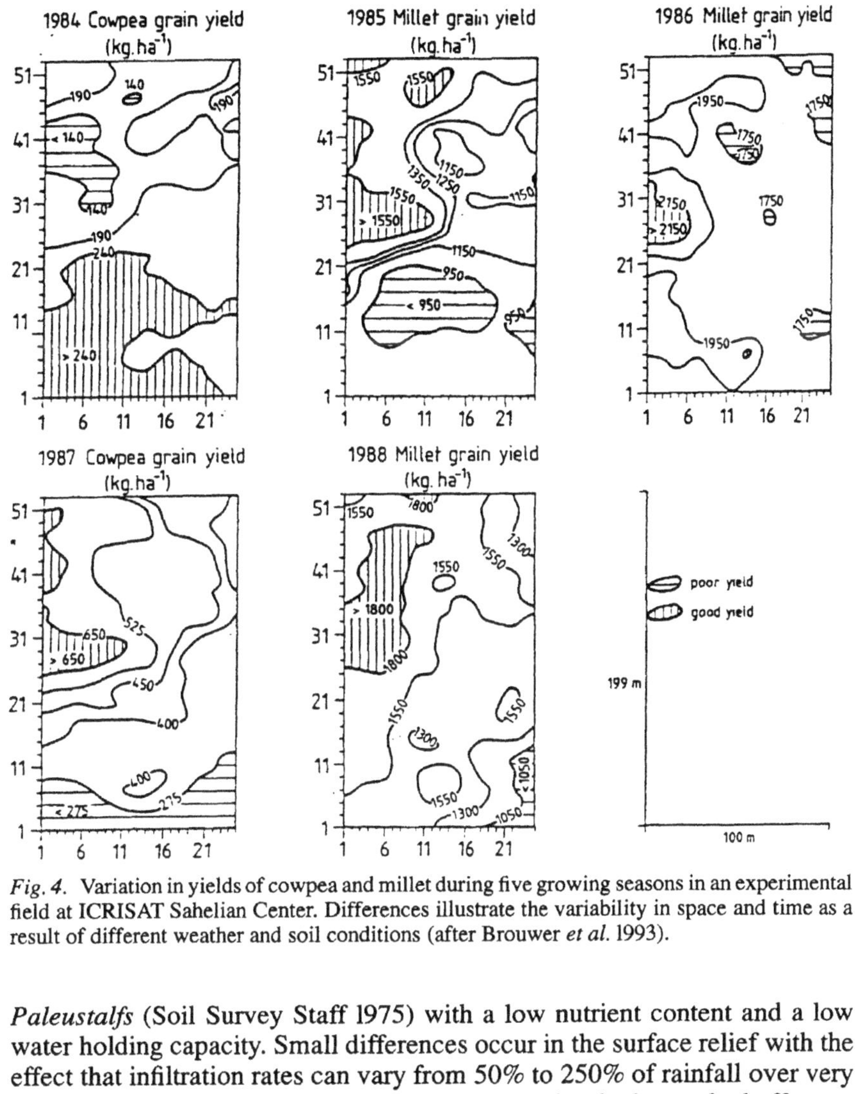

Fig. 4. Variation in yields of cowpea and millet during five growing seasons in an experimental field at ICRISAT Sahelian Center. Differences illustrate the variability in space and time as a result of different weather and soil conditions (after Brouwer *et al.* 1993).

Paleustalfs (Soil Survey Staff 1975) with a low nutrient content and a low water holding capacity. Small differences occur in the surface relief with the effect that infiltration rates can vary from 50% to 250% of rainfall over very short distances (Gaze *et al.* 1995). This variation has had a marked effect on soil properties. The higher spots are less leached than the lower spots. In moist, average, years growth can be best on the more fertile high spots and much lower on leached low spots, while in dry years little growth may occur on the high spots while better growth could still be possible on the relatively moist, low spots. Spatial variability patterns of yields vary considerably, therefore, not only within one growing season but also, and particularly, among different growing seasons (Brouwer *et al.* 1993; see figure 4 which shows only different yields).

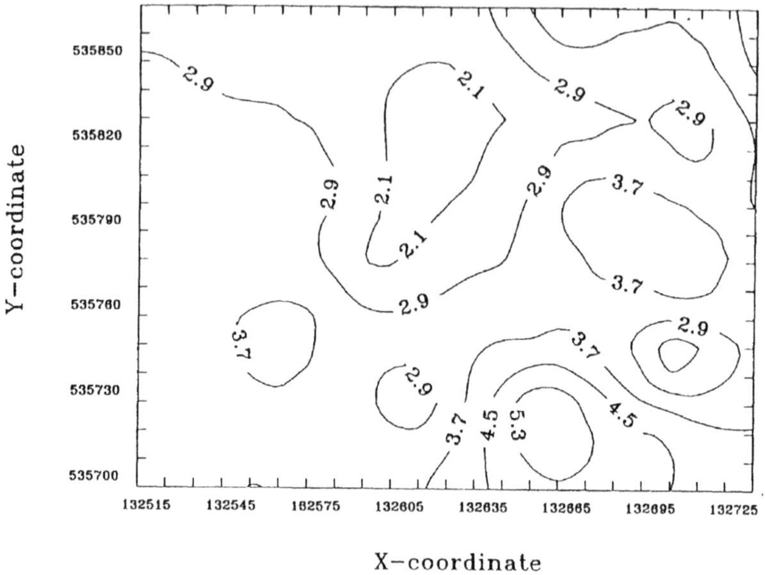

Fig. 5. Total N (100 kg ha^{-1}) at June 8, for a depth of 0–60 cm below surface.

Crop growth and soil conditions: expert systems and models

The Netherlands

Advisory schemes for soil fertilization have been used in The Netherlands quite succesfully for more than 60 years. They relate actual soil fertility to fertilization rate and expected yield and are based on field experimental data. Of course, such relations are complex and depend on many factors, which are not all expressed in the schemes. The factors include, for instance, effects of different weather conditions in different years and soil hydrology as well as effects of different soil types. In general, the advisory schemes take into account soil differences but only in broad terms, such as clay soils versus sand soils. As coarse as the schemes may be, they proved to be sufficiently reliable to allow estimates between actual soil fertility, fertilization rates and expected yields, although the latter are often not specified. The schemes are exclusively focused on crop production. No attention is paid to possibly adverse environmental side effects such as soil and water pollution.

Advisory schemes are suitable to explore expected effects of site specific management. In the field being investigated, 30 soil fertility samples were taken. Also a composite sample was made. Each sample was interpreted in terms of fertilization rate and expected yield. Results are summarized in figure 5, which shows total N as measured on June 8. The map was obtained by interpolation, using the kriging technique, of 30 point data. Using computer

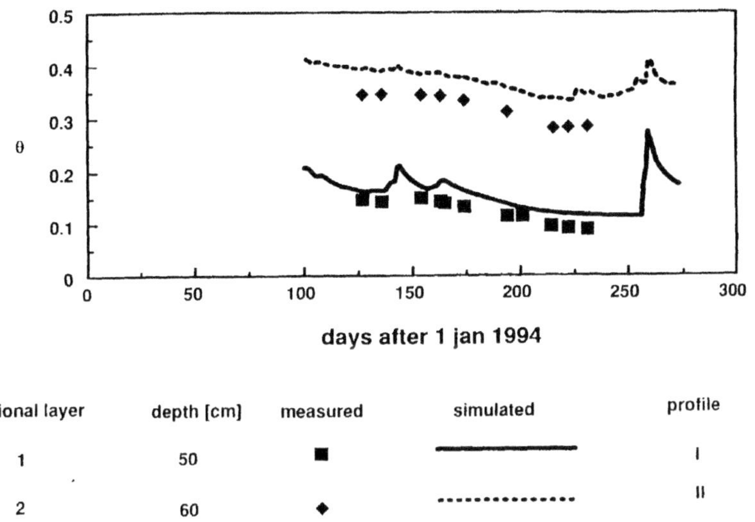

Fig. 6. Simulated versus measured moisture content for a sandy (I) and loamy (II) soil profile (for locations see figure 4).

simulation models for crop growth and solute fluxes represents a modern and more detailed method to express relations between crop growth and soil conditions (e.g. Teng and Penning de Vries 1992). Models can be used to express both crop growth and environmental effects. Different types of models can be distinguished. In our work, use is made of the WAVE model for simulating water movement, nitrogen transformations and crop growth. This model was developed in the context of an international program financed by the European Union. This program includes calibration and validation procedures for different experimental sites in UK, Belgium and The Netherlands. The model was run for the observation points in the study field. Calibration and validation was, so far, based on measured and calculated water contents in the soil during the growing season of 1994. A representative example of calculated water contents using the independantly calibrated and validated model is shown in figure 6 for both a relatively sandy and a clayey spot. The two locations are indicated on figure 3.

In this study potato yields were measured by harvesting 65 small plots (figure 1). These values were interpolated and results are shown in figure 7a. Yields varied between 30 and 45 tons ha^{-1}, which represents a very large range. More important than total yield is the yield of potatoes larger than 50 mm. These potatoes are well marketable for chips. Differences among yields of potatoes larger than 50 mm were pronounced, and ranged from 3 to 15 tons ha^{-1} (figure 7b). The smallest values are obtained in areas with the lowest yields. For example, total yields of 30 tons ha^{-1} correspond with a yield of large potatoes of 3 tons ha^{-1} (10%), while total yields of 45 tons ha^{-1}

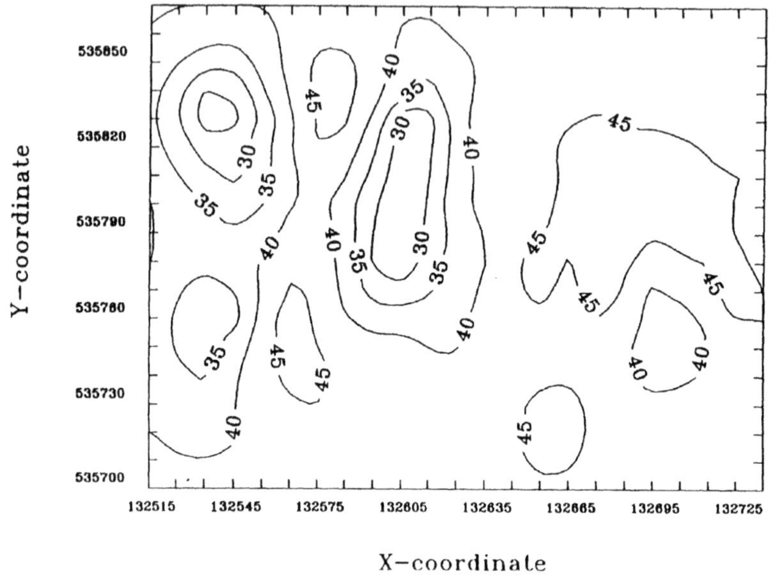

Fig. 7a. Total potato yield (ton ha^{-1}) in 1994.

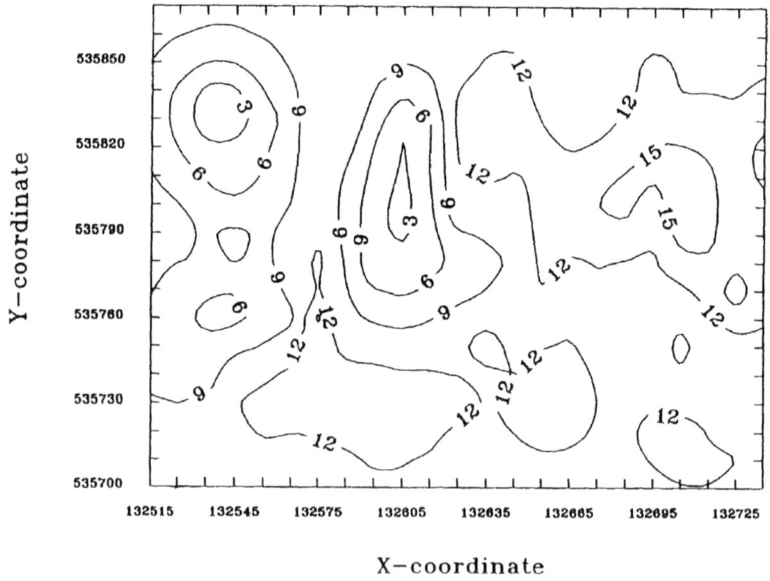

Fig. 7b. Yield (ton ha^{-1}) for tubers with diameter larger than 50 mm.

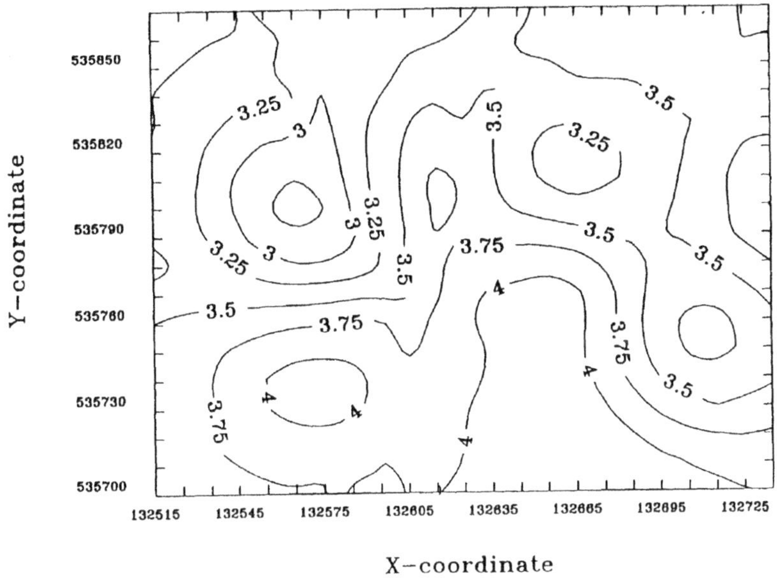

Fig. 7c. Leaf area indexes (LAI) measured on August 3 1994. Patterns were obtained by interpolating 80 point measurements.

correspond with yields of large potatoes of 12–15 tons ha^{-1} (30%). Here, yields were measured. They can also be simulated. However, measurements are generally to be preferred, if feasible, in view of uncertainties associated with modeling. In this context we also show figure 7c showing measured leaf area indexes, obtained with a (hand-held) crop scan apparatus. In general terms, patterns of leaf area indexes measured on August 3 correspond reasonably well with the yield patterns measured in September. In all these examples, point data were interpolated to areas of land by using geostatistical techniques (Finke 1993).

The WAVE model was used to calculate water and nitrate fluxes for the growing season of 1994. Some selected results are shown in figures 8a and 8b, using again two contrasting point locations that were illustrated in figures 1 and 3. The figures show nitrate profiles as a function of time, and demonstrate clearly that penetration of nitrates proceeds quicker and to a relatively greater depth in the more sandy soil. The model was used here for real time conditions, allowing comparisons with measured data. Of course, use of simulation models is particularly attractive when making runs for weather and soil conditions in different years.

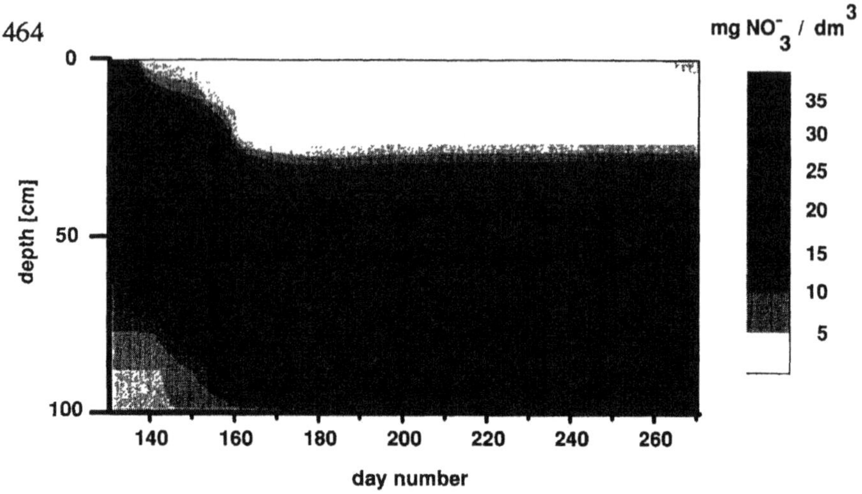

Fig. 8a. Simulated nitrate profile during the growing season in a sandy soil (I).

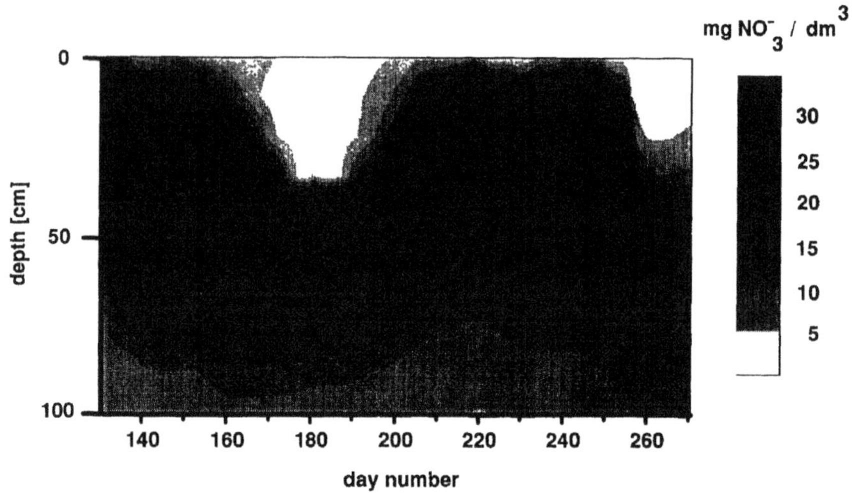

Fig. 8b. Simulated nitrate profile during the growing season in a loamy soil (II). High nitrate contents in surface soil between days 220 and 250 are a result of upward unsaturated flow, which did not occur in the more sandy soil (I).

Niger

Observations were made on a large manure application experiment. The treatments, for which detailed measurements and calculations were made, consisted of four repetitions of a factorial combination of application of cattle or sheep manure, the equivalent of 1 or 3 nights production. Plots measured 5 × 5 m. For the application of manure, cattle and sheep grazed natural fallow pastures 10 hours daily. Six cattle or 23 sheep camped each night in one of two 4x4m corrals in the center of one of the plots, untill all plots had been camped on the designated number of nights. Manure was hand-gathered daily, weighed and sampled for N and P, then returned to the plots. In 1990 millet was sown in late May and harvested in September. Within treatments, yields per plot varied by as much as a factor 3.6. When cattle were present one night, grain yields varied widely between 356–1288 kg ha^{-1} and the above-ground dry matter between 1783–5717 kg ha^{-1}. There was no clear difference between yields on plots on relatively low and 'wet' locations and those on relatively high and 'dry' locations. This may have been partly due to the fact that the year 1990 was neither dry nor wet. However, yields of plots that had received three nights of either cattle or sheep manure were much more uniform than those that had only received one night's manure (table 2). This points to the crop growth variability being at least in part related tot the availability of nutrients in the soil. Adding a lot of manure reduced crop growth variation in 1990. At the same time, high yielding plots with only one night's manure had yields that were similar to those of plots with three nights of manure, while low yielding plots with only one night of manure were much poorer. This points to some of the 3-night plots having received too much manure and some of the one night plots having received too little in 1990.

A nutrient budget was made based on soil samples taken in May, July, August and October 1990 and May 1991 (Brouwer and Powell 1994). Some selected results are presented in table 2. Even though the selection of 'wet' and 'dry' spots was based on a qualitative analysis, a distinct difference was obtained for those locations. Even when considering crop uptake, the N content of the soil increased after manuring, due in part to accelerated breakdown of coarse organic matter already present in the soil at the start of the experiment and possibly to increased microbial and termite activity. However, the net balances for the wettest plot in each treatment averaged only 45 kg N ha^{-1} while those for the other plots averaged 315 kg N ha^{-1}. The difference is attributed to leaching of nitrates in the 'wet' spots. The picture for P is comparable. There is a large net loss of P in the wettest plot of each of the four treatments averaging −157 kg P ha^{-1}. For the other plots the average loss is −44 kg P ha^{-1}. The net losses are probably due to leaching of (organic) phosphates in part already present before the start of the experiment, desorbed by the organic matter of the manure and possibly by the increase in the pH due to the sheep and cattle manure. The experiments indicated very large differences among yields within one treatment while nutrient budgets were also rather different within small distances. With little manure available, it is obviously unattractive to allow substantial leaching. In contrast to the Dutch study, concern does not focus on environmental pollution of groundwater but on reduction of potential crop growth.

Table 2. Yields, N and P balances (in kg/ha for the top 2.0 m of soil), May 1990–May 1991, using average background values for May, 1990. Data from ILCA–ICRISAT–Wageningen Agricultural University manure application experiment at ICRISAT Sahelian Center, Niamey, Niger (after Brouwer and Powell 1994).

plot	spec.	nights	wetness	yield in kg ha^{-1}			ratio highest: lowest yield			START	IN	EX	END	Net.	START	IN	EX	END	Net
				grain	stover	d.m.	grain	stover	d.m.	SOIL	MANU	CROP	SOIL	Bal.	SOIL	MANU	CROP	SOIL	Bal.
										tot. N	N	N	N	N	tot.P	P	P	tot.P	P
79	C	1	3	356	1428	1783	2486	87	20	2730	177	2008	5.4	5.3	841	–167			
30	C	1	2–3	1288	4429	5717	3.6	3.1	3.2	2486	105	77	2968	454	2008	6.5 11.6	1846	–157	
2	C	1	2	814	3970	4785	2486	37	70	2713	260	2008	2.3	6.0	1763	–241			
9	C	3	3	974	3914	4888	2486	203	50	2628	–11	2008	12.7	10.9	1943	–67			
83	C	3	2	751	3814	4566	1.4	1.3	1.3	2486	289	*	3076	251	2008	17.7	*	1991	–43
69	C	3	2	709	3078	3787	2486	260	46	3029	329	2008	16.0	11.6	2257	244			
56	S	1	2	1154	3763	4917	1.6	1.7	1.7	2486	93	51	2842	314	2008	6.0	7.9	1720	–108
44	S	1	1	717	2181	2898	2486	112	29	3034	464	2008	7.0	4.2	1993	–18			
33	S	3	3	875	3955	4830	2486	425	66	2644	–200	2008	26.2	7.9	1918	–108			
88	S	3	2	626	3966	4593	1.4	1.1	1.1	2486	184	43	2859	236	2008	11.8	12.3	1857	–151
13	S	3	1	802	3480	4282	2486	349	47	2998	210	2008	21.8	5.8	2082	58			

SPEC.: C = cattle, S = sheep

WETNESS: 1 = low, 2 = average, 3 = high (estimation of infiltration of rainfall)

START SOIL = storage in soil at start of experiment, May 1990 (kg ha^{-1})

IN MANU = nutrient import in manure and urine (kg ha^{-1})

EX CROP = nutrient export in millet above-ground dry matter (straw, grain and rest of heads) (kg ha^{-1})

END SOIL = storage in soil 12 months into the experiment, May 1991 (kg ha^{-1})

Net. Bal. = (END SOIL) – (START SOIL) – (IN MANU) + (EX CROP) (kg ha^{-1})

Exceedance of threshold values of selected indicators

Introduction

The overall objective of 'Precision Agriculture' is to manipulate nutrient (and biocide) fluxes in such a way that conditions for crop growth and development are maximized while unfavorable environmental side effects are minimized. In developed countries this discussion is governed by an abundance of cheap fertilizer.

To judge both the level of crop growth and environmental side effects we need indicators and their threshold values as a reference. Indicators are defined as environmental statistics that measure or reflect environmental status or change in condition, while threshold values represent levels of environmental indicators beyond which a system undergoes significant change: points at which stimuli provoke significant response (FAO 1993).

The Netherlands

For our case study, soil water content and nitrate fluxes in the soil are two obvious indicator values. Both were shown in figures 6 and 8, which demonstrate, in principle, the importance of using simulation models when dealing with the operationalization of the indicator and threshold value concept. Without the models, there simply would not be enough time nor funding to generate the large number of data needed to test and implement the concept. Implementation gives rise to a number of questions: (i) which indicators relate to crop production and which to environmental aspects; (ii) what are the threshold values for these indicators, and (iii) how feasible is it to have not only common indicators but common threshold values as well for the soil functions being considered. Both indicators, mentioned above, are important for crop growth and environmental conditions. First, then, the water content of the soil. The dry months of July and August have reduced crop growth considerably. The WAVE model calculates potential versus actual transpiration (figure 9, for two arbitratrily chosen points in the study field). The data presented in figure 9 can also be used to show how much water should have been applied at what time to allow potential transpiration to occur. The nitrogen content is more difficult to characterize. Crop N demand is a function of the crop growth stage and the growth rate; downward fluxes of nitrate are a function of the amount of N in the rootzone and the water flux, which is directly associated with the water content. Simulated downward nitrate fluxes, which can be derived from data shown in figure 8, can be interpreted directly in terms of exceedance of a critical threshold value for leaching by drawing a horizontal line at the threshold level and by counting the number of days that the concentration of nitrate in the percolating soil water was higher than a selected critical threshold of, say, 50 mg liter^{-1}. At the same time, N needs of the crop can be determined for

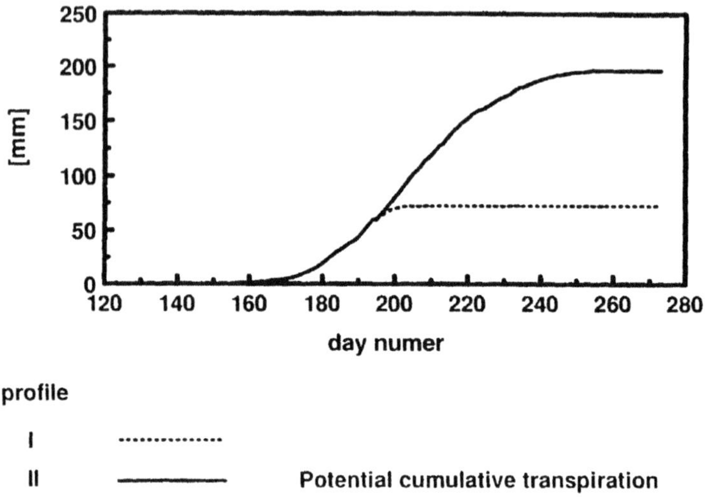

Fig. 9. Simulated cumulative transpiration for a potatoe crop at the two locations I and II.

any period in the growing season when aiming for nutrient unlimited yield. This determination represents a more sophisticated approach to the problem of determining relations between nutrient status, fertilization and yield as discussed earlier. Clearly, in defining an optimal system, requirements defined for production have to be balanced against requirements for environmental quality.

Niger

In this case study emphasis is only on nutrient availability for crop growth. Since chemical fertilizer is too expensive for the average farmer, placement of organic fertilizer or crop residue forms the only option to maintain soil fertility. Indicators would be the soil-water and the soil N- and P content. Threshold values could be derived from traditional agronomic experiments, but this study shows that nutrient budgets differ significantly as a function of the location in the field. A given fertilization application in a low, 'wet' spot will be less effective than one given in a high, 'dry' spot. To complicate matters even more, the story may well be different in wet and dry years. There is a need therefore to take into account the effects of the site specific microrelief when planning agronomic experiments. This requires a new paradigm in fertility research.

Balancing yields and solute fluxes

The Netherlands

The example discussed in the previous section demonstrates the need to balance requirements for production on the one hand and for environmental conditions, on the other. The analysis for the indicator 'water content' resulted in a definition of periods during which water could be applied to achieve potential production. However, this could result in unacceptably high downward flows of water and nitrates, considering accepted threshold values for N-contents of the soil water. So before a recommendation about irrigation can be made, the model should be used to check whether such fluxes would indeed be probable. If so, it may not be possible to add all the water that would be needed to achieve maximum production from a hydrological point of view. Consideration of the other indicator 'nitrogen content' leads to a similar analysis. Adding more nitrogen at particular times during the growing season on the basis of crop requirements, may lead to unacceptable nitrate losses,certainly when additional water would be applied to combat drought stress.

When analysing this balance between yield and solute fluxes, the scenario approach can be used in which a series of variants are run by the model. One can be chosen as being the best compromise.

So far, only real weather conditions in the growing season of 1994 were considered. Conditions are bound to be different in different years and models can, again, be used to characterize such conditions. This will be done as the study is being continued.

Niger

As stated above, there is no question here when choosing between production and environmental effects. Production should have the primary focus. Considering the usually marginal rates of fertilization, substantial leaching of N and P is unlikely even though fluxes can be significant from the plant-growth point of view (Brouwer and Powell 1994).

Implementation by site specific technology

Once point or plot data are obtained with procedures described in the previous sections, we have to consider spatial differences. As stated above, these can be derived from point data by interpolation techniques (see figures 5 and 7). Next, technical means have to be defined to use those differences with the purpose to obtain a better production system from a balanced ecological point of view. No new original research results are presented in this section, but work elsewhere is briefly reviewed. Interesting high-tech developments occur in the USA, UK and Germany in terms of development of site specific technology, including GPS guided fertilization machinery (e.g. Robert *et al.* 1994; Murphy *et al.* 1994). Machines need to be fed with proper spatial data and this still presents a major challenge to research. Automatic monitoring equipment is devised to sense 'in situ' N-contents, which are transmitted to a fertilizer spreader which is adjusted automatically 'on-the-go'. Sensors are placed on sprinkling guns to spray weeds only when they are sensed by equipment passing over a field. Indeed, only the sky is the limit.

However, low-tech approaches are equally valid. By dividing a Sahel field in subareas requiring different treatments, the treatment itself can de implemented by hand. Farmers have been seen putting weeds and crop residues on eroded spots. More research is needed to obtain response curves for the different subareas to be distinguished.

Operational decision support systems

So far, precision agriculture appears to be mainly focusing on defining site specific fertilization. Modern machinery has been developed to allow spatial differentiation of fertilization. To define site specific rates, most often classical assessment schemes are used which are based on field experiments and which relate actual nutrient status to advised fertilization rate and expected yield. Clearly, simulation modeling can help to fine tune such rather crude prediction systems. Still, models need to be validated and have high data demands which cannot always be satisfied. The questions needs to be raised how models can play a role within operational decision support systems, which need to be *pro-active* rather than *re-active*. A farmer is primarily interested to know what he should do for the coming weeks and months, not in what happened last year. At this time, it is unclear which type of data are needed in an operational decision support system. Likely, the accuracy of medium -term weather predictions could be a deciding factor in determining the degree of detail of other agronomic and soil data needed in the DSS.

In the Sahel as well, weather conditions are crucial because of different reactions of the soil and the crops to wet and dry years. In the Sahel, site specific management is at the same time traditional and still in its infancy. It is traditional, because farmers know very well that not all parts of their fields are the same and that different parts can be managed in different ways. But site specific management is also in its infancy because farmers and researchers are often not aware what happens below the surface. Farmers often do not realize that part of the rain actually percolates down to the groundwater. It is, then, quite a revelation that water going down may take with it nutrients that are crucial for plant growth. Even though manuring rates vary strongly over short distances, farmers make little or no attempt to redistribute manure after it has been deposited on their field by livestock. Perhaps they assume that when manure is put on the ground it will benefit the crop unless it is carried away by wind or runoff. An educational effort is therefore necessary.

Conclusions

1. Experimental field work in a marine clayey soil in The Netherlands has demonstrated occurrence of significant differences in potato yields within a farmer's field of 6 ha. Total yields varied between 30 and 45 tons ha^{-1}, while yields of commercially attractive large potatoes varied between 3 and 15 tons ha^{-1}. In Niger, millet dry matter production within a 1 ha field varied between 1.8 and 5.7 tons ha^{-1} for the same treatment. In both cases the differences encountered are not at all unusual for field conditions and quite significant from an economic point of view.

2. Differences in yield can be due to many factors that were explored in two (as yet incomplete, ongoing) case studies by field measurements. Computer simulation techniques were used succesfully in the Dutch study to demonstrate the importance of the water supply capacity in governing (simulated) yield. In addition, simulation allowed an estimate of nitrate fluxes in the soil which govern environmental side effects of the production system. Simulation techniques are important as exploratory tools in finding an acceptable balance between production and environmental pollution. In Niger, field sampling taking into account differences within fields was effective even without additional modeling efforts.

3. Thirty fertility samples, taken within the 6 ha Dutch field, showed a large variation. So did samples taken from a randomized-block experiment in Niger. Recommended fertilization rates, based on seperate point data are bound to differ significantly from those based on one mixed sample for the entire field. The latter procedure clearly results in local over- and under fertilization, which represents inefficient use of natural resources in both cases.

4. Consideration of the important effects of spatial variation requires a different approach to soil fertility studies, rather than using classical statistical techniques. Soil within a plot cannot a priori be considered to be homogeneous. Rather, agronomic experiments should be made that include consideration of spatial variation. This is true everywhere. Thus, a new paradigm for agronomic field experimentation has to be developed. This paradigm is highly relevant as well for eco-regional studies, where effects of farm-level activities are integrated.

Acronyms

ICRISAT	International Crops Research Institute for the Semi-Arid Tropics
ISC	ICRISAT Sahelian Centre
PGS	Global Positioning System

Acknowledgements

The reported Dutch work was partly funded by the European Union (AIR 3 program: 94–1204) and by the Wageningen Agricultural University, Wageningen, The Netherlands. The Niger work at ICRISAT Sahelian Centre is partly financed by the Directorate General for International Cooperation (DGIS), Ministry of Foreign Affairs, The Netherlands. The manure application experiment was set up and managed by Dr M. Powell of the International Livestock Centre for Africa (ILCA). Diafarou Amadou assisted with the collection of soil samples, which were analysed under the supervision of Ilyassou Oumarou.

References

Brouwer J, Fussel L K, Hermann L (1993) Soil and crop growth variability in the West-African semi-arid tropics: a possible risk-reducing factor for subsistence farmers. Agriculture, Ecosystems and Environment 45:229–238.

Brouwer J, Powell J M (1994) Soil aspects of nutrient cycling in a manure application experiment in Niger. Proc. Intern. Workshop on the role of livestock in nutrient cycling in mixed farming systems in sub-Saharan Africa. International Livestock Centre for Africa (ILCA), Addis Ababa, Ethiopia. November 22–26, 1993.

Cahn M D, Hummel J W, Goering C E (1994) Mapping maize and soybean yields using GPS and a grain flow sensor. In Robert P C, Rust R H, Larson W E (Eds.) Site specific management for agricultural systems. Spec. Publ. SSSA Madison, Wis, USA (in press).

FAO (1993) FESLM: An international framework for evaluating sustainable land management. World Resources Report 73. FAO, Rome, Italy.

Finke P A (1993) Field scale variability of soil structure and its impact on crop growth and nitrate leaching in the analysis of fertilizing scenarios. Geoderma 60:89–107.

Franzen D W, Peck T R (1994) Sampling for site-specific fertilizer application. In Robert P C, Rust R H, Larson W E (Eds.) Site specific management for agricultural systems. Spec. Publ. SSSA Madison, Wis, USA (in press).

Gaze S R, Brouwer J, Simmonds L P, Bouma J (1995) Measurement of surface redistribution of rainfall and its effect on the water balance of a millet field in SW Niger. J. Hydrology (Special Issue HAPEX Sahel).

Murphy D P, Haneklaus S, Schnug E (1994) Innovative soil sampling and analysis procedures for the local resource management of agricultural soils. Pages 613–631 in Proc. 15th World Congress of Soil Science, Vol. 6a.

Robert P C, Seeley M W, Anderson J L, Cheng H H, Lamb J A, Moncrief J F, Rehm G E, Rosen C J, Schmitt M A (1994) The Minnesota precision farming initiative. In Robert P C, Rust R H, Larson W E (Eds.) Site specific management for agricutural systems. Spec. Publ. SSSA Madison, Wis. USA (in press).

Soil Survey Staff (1975) Soil Taxonomy: A basic system of soil classification for making and interpreting soil surveys. USDA_SCS Agric. Handbook 436. US Gov't printing Office. Washington DC, USA.

Teng P S, Penning de Vries F W T (1992) Systems approaches for agricultural development. Agricultural Systems 40: 1–3.

Response to section G

R.A. QUIROZ

Consortium for the Sustainable Development of the Andean Eco-Regio (CONDESAN, CIP/IDRC). Plaza Espana esq. Mendez Arcos 710, P.O. Box 5783, La Paz, Bolivia

Introduction

Three papers were presented at the crop level. All three case studies agreed on the main goal; increasing productivity. Authors described complementary approaches for increasing crop productivity: implementation of precision agriculture, minimization of adverse effect through appropriate management and development of crop ideotypes for specific target environments. To visualise the link among the three approaches in relation to the main goal, a simplified scheme is proposed (figure 1).

Response to paper by Bouma *et al.*

Spatial variability is an intrinsic characteristic of most experimental units. Even highly controlled environments, such as growth chambers, show spatial variability (Chong-soon and Rawlings 1982), so dynamic systems such as soil are expected to vary not only on a spatial scale but on a time scale as well. The recent technological development described by the authors, will be useful for maximizing productivity through precision agriculture.

Measuring actual variability in soil properties is complex and expensive, especially in developing countries, even when conventional soil mapping techniques are used. In traditional farming systems, farmer's expertise can be utilized for mapping soils according to fertility. In a recent study in Bolivia (unpublished) it was demonstrated that indigenous soil classification did not vary from a physical-chemical one. Plot selection for different crops was quite accurate. For instance, the plot selected for potato had a 2.6-fold yield increase compared to that obtained on non-selected plots.

The introduction of inappropriate technologies can be deleterious for the sustainability of food production in fragile ecosystems. The introduction of tractors in the southern 'Altiplano' of Bolivia, without a proper training for plowing sandy soil, has induced wind erosion of soils. Quinoa production in soils cultivated with the use of tractors instead of animal traction, showed a

J. Bouma et al. (eds.), Eco-regional approaches for sustainable land use and food production, 475–479.

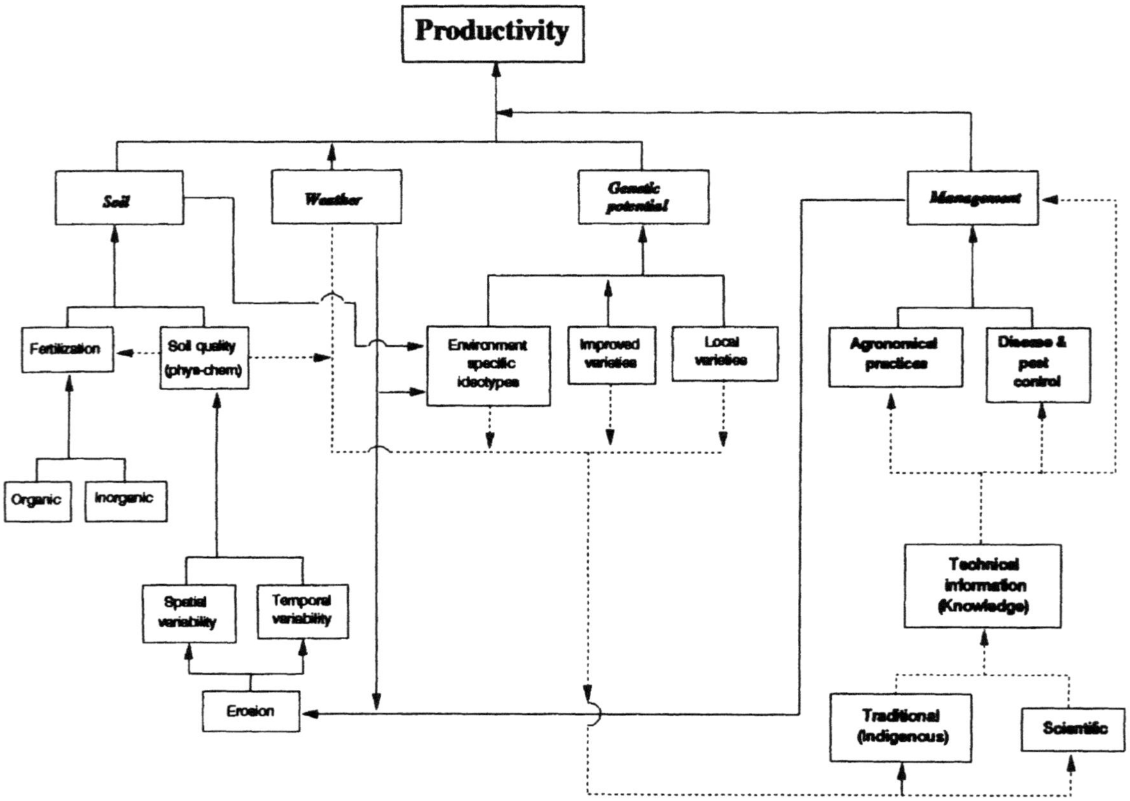

Fig. 1. Schematic representation of the main factors affecting crop productivity and their inter-relationship.

16-fold decrease in yield after 20 years of cultivation (4 years inter-annual cultivation and 6 years fallow).

The application of precision agriculture in the Andean eco-region requires a detailed knowledge of frost risk. Zonation of these risks for the Altiplano is being conducted by ABTEMA (Asociación de Teledetección para el Medio Ambiente; member of CONDESAN). Combining fertility patterns, frost risk knowledge and other site-specific limitations such as salinization risk with indigenous knowledge, may prove advantageous. Precision agriculture could be used to substantially improve crop production in the Andes, with a minimum damage to the environment.

As pointed out by the authors, pollution due to the application of large quantities of chemicals is generally of little concern in developing countries. However, other pollutants such as silt deposition in dams due to inappropriate soil conservation practices on hillsides, are very common and must be considered in eco-regional approaches.

Response to paper by Kropff *et al.*

Developing high-producing crop ideotypes for specific target environments looks interesting and promising. The approach presented in this paper, using quantitative ecophysiology, is appealing. Simulation models are useful to integrate available knowledge and to point out where information is lacking. These are important aspects to be considered in allocating research funds. However, in this paper, little attention is given to product quality, which is an important characteristic for consumption.

The prioritization of crops and environments where ideotypes will have economic impact in a short time must be considered. The approach presented in this paper should be followed in high-producing environments and for crops of which the actual production level is close to its potential. Crops with yield levels that are way below their potential, demand different eco-regional approaches. In these environments, the determination of bottle-necks and the suggestion of solutions as to how to cope with them has a higher priority. In this respect, simulation models and expert systems have been used in the Andean eco-region with promising results.

Response to paper by Zadoks *et al.*

Effective crop protection is mandatory when productivity is expected to be increased. As the authors point out, prediction of perennation and pest outbreaks, and the establishment of appropriate policies for pest control are key issues for achieving effective crop protection. Now that prediction of outbreaks seems possible (using elementary mathematics), it may be a problem to communicate such information to farmers, researchers, extension agents and policy makers.

Discussion

As shown in figure 1, management decisions play a key role in the fate of crop productivity. Farmers' decisions are based on available knowledge, either indigenous or scientific. However, in most developing countries, the information flow is inappropriate and farmers seldom receive the necessary information at the right time. This is due to several reasons. Probably the most important reason is the lack of appropriate extension services. The number of extension agents is generally small, and it is common that recent graduates 'do the job'. Moreover, in many cases, extension agents despise indigenous knowledge that could be relevant for solving many of the limitations encountered in a specific area.

The flow of scientific as well as indigenous knowledge must be improved to achieve better management. Eco-regional approaches should be useful for the exchange of results and experiences. Multiple site tests and validation of technologies are advantageous to solve common problems across a particular region. The knowledge acquired should be processed and delivered to the potential users in a friendly form. Expert systems are a good means to communicate both scientific and indigenous know-how. An example of an expert system for potato production in the Andean eco-region is described by Arze (1992).

One of the advantages of eco-regional approaches is the integration of different institutions, financial sources, interests, expertise, etc. Most of the basic research should be done by the IARCs. NARS and NGO's may participate in multilocational tests and develop technologies for conditions not encountered in other areas. Simulation models and expert systems should be developed by IARCs and specialized universities. Professionals in NARS and NGOs should be trained in how to adapt the models to their needs and how to use the results. Sharing costs and benefits of developing appropriate technologies for a particular eco-region, will benefit farmers as well as the participating institutions and countries.

Improving world's agriculture requires analyzing and solving the main production constraints. In most areas, farmers rely on mixed farming systems for their subsistence. It is therefore desirable to look at other components of the production systems than crops alone. Animals play an important role in the survival of many peasant families across the world, and little attention has been paid to them in this eco-regional symposium. In addition, market prices for peasant agricultural products should be considered for improving the productivity of the agricultural production systems within a eco-region.

Acronyms

CONDESAN	Consorcio para el Desarollo Sostenible de la Ecorregion Andina (Consortium for the Sustainable Development of the Andean Eco-region)
IARC	International Agricultural Research Center
NARS	National Agricultural Research System
NGO	Non-Governmental Organization

References

Arze J (1992) Desarrollo de modelos para la transferencia de agrotecnología en el altiplano peruano. Pages 47–66 in Argüelles L, Estrada R D (Eds.) Perspectiva de la Investigación Agropecuaria para el Altiplano. INIAA-PISA. Lima, Peru.

Chong-soon L, Rawlings J O (1982) Design of experiments in growth chambers – Uniformity trials in the North Carolina State University phytotron. Crop Science 22:551–558.

Response to section G

I.O. AKOBUNDU*

Moist Savanna Program, International Institute of Tropical Agriculture (IITA), PMB 5320, Ibadan, Nigeria

Introduction

In the past decades when funds were readily available and demands for them were few, fragmented research and development projects and unintentional duplications of efforts were common in many developing countries. Dwindling resources and concern for environmental degradation have led to demands for consolidation and re-orientation of research to address more productive and sustainable resource management and food production strategies. Current discussions on eco-regional approaches to research and development are attempts to ensure that problems that have eco-regional dimensions can be addressed in a manner that will ensure consolidation of efforts; and that partnerships will involve all relevant interest groups in a region. There is recognition that implementation of partnership principles is necessary to achieve common goals with the limited resources that are available for agricultural research and development. Unequal distribution of resources and access to inputs are among the problems that have militated against establishment of partnerships in the past. The on-going eco-regional discussions should make partnership principles the central theme of all eco-regional initiatives. The fact that there is heterogeneity in biophysical and socio-economic factors affecting agricultural activities, requires that any eco-regional initiative to sustainable food production that is to be undertaken should permit careful targeting of improved crop varieties to environments where the crops have the best chance of adaptation and of making efficient use of production resources. The eco-regional approach to technology development must take this heterogeneity into account.

Response to papers by Kropff *et al.*, Zadoks *et al.*, and Bouma *et al.*

Three papers have addressed crop level issues concerning eco-regional approach. The paper by Kropff *et al.* discussed a systems framework for targeting plant

* I.O. Akobundu is the current leader of the Moist Savanna Program (MSP). The author has drawn from comments made by the IITA Scientists B.T. Kang, M. Tamo and F.M. Quin. It includes the current view of IITA/MSP on the subject.

J. Bouma et al. (eds.), Eco-regional approaches for sustainable land use and food production, 481–486.

types for different production environments. Specific examples of on-going research to design and evaluate rice and potato types for different environments were given. We agree with the authors on the potential benefits of using systems approach to design and evaluate genotypes for specific environments. However, we feel that more emphasis should be placed on genotype by environment (G×E) management approaches to crop improvement research. A G×E approach has the advantage of bringing together a multi-disciplinary team that can attempt to fit crop types into cropping systems in which the genotypes will be needed.

The paper by Zadoks *et al.* discussed basic concepts of epidemiology in relation to eco-regionality. We agree with the authors on the dilemma between eco-regionality and the patchwork idea. The fact is that insect pest occurrence and severity does not fit neatly into the defined boundaries of eco-regions. Vegetation plays an important role as an alternate host for insects both in propagation and perpetuation of the pests. Insect pest incident is likely to follow a gradient determined by vegetation from the humid to the arid zones. Some of the other points that require attention in eco-regional approach to plant health management include the following: (i) different resistance levels are needed in different zones depending on the insect pest complex and on the infestation levels, and (ii) some biological control agents cannot be confined to specific eco-regions as has been demonstrated by the parasitoid of the cassava mealybug.

The paper by Bouma *et al.* highlighted the effects of spatial variability on crop yield normally seen in farmers' fields. It draws attention to the need for site specific interpretation of data. Dobbermann and George (1994) recently reviewed this subject and stressed its effect on crop performance. The approach proposed by the authors for site specific fertilization (precision agriculture) has been used for years in northern Sumatra, Indonesia, for wrapper tobacco leaf. The technique is used for metering fertilizer based on micro-site variations. What precision agriculture does at the micro level can, with proper site description, be done with reasonable level of success at the meso and macro levels using benchmark research approaches. If composite soil samples are based on uniformity studies, the minimum sample number can be determined statistically and reasonably accurate soil fertility status of a field can be determined. As regards the work reported from Niger, the problem of low yields in low spots in wet years due to water-logging has been minimized in most of the moist areas of the semi-arid tropics by use of tied ridges. We feel that variability in soil fertility status of arable fields can be managed in the context of an eco-regional approach by using the benchmark approach discussed later in this paper.

Eco-regional approach at crop level: IITA's views

Eco-regional approaches at the crop level include germplasm selection for both endowed environments and those at risk, protection of crops from pests and management of the resource base in a manner that will minimize irreversible degradation. IITA's approach to sustainable crop production in an agro-ecological zone is to first understand the types and levels of interaction that exist between selected genotypes and the environment in which the crop will express itself. This genotype by environment interaction (G*x*E) encompasses the physical environmental factors, the biological factors and the socio-economic factors that affect phenotypic expressions of a given genotype. The scope of this analysis includes multilocational testing, site selection, germplasm characteristics, biotechnology-assisted selection, end-user perspective, and adoption studies. This genotype is then tested for suitability in the cropping systems in which it will be used by farmers.

Component technologies developed as a result of crop improvement research need to be fitted into specific cropping and farming systems within defined environmental limits. To achieve this goal, the developed technologies should accommodate a wide range of biophysical and socio-economic conditions of the farmers. Benchmark research constitutes a strategy within an eco-regional approach to disaggregate the heterogeneity of biophysical and socio-economic environments in the process of technology development.

Benchmark research strategy

A benchmark location is an area within an agro-ecological zone. In the Moist Savanna Zone (MSZ) of West and Central Africa, a benchmark location is an area within an MSZ subzone, such as the Coastal Savanna (CS), Derived Savanna (DS), Southern Guinea Savanna (SGS) and Northern Guinea Savanna (NGS). An entire subzone covering thousands of hectares could be designated a benchmark location. Within each benchmark location there could be several benchmark sites representing specific combinations of subzonal biophysical and socio-economic characteristics. Benchmark sites are field laboratories or stations (village farm or farms; or several farms in a cluster of villages) where prototype technologies resulting from both laboratory and on-station research are tested in close collaboration with NARS and farmers. These sites provide a focal point for collaboration with NARS and non-NARS partners. Investigations carried out at these benchmark sites range from strategic research through applied research to farmer-managed field testing of technologies. In benchmark sites, available technologies are tested for their performances and rapid feedback. Testing is done to the stage where adoption can be estimated, or if farmers' reactions are negative, farmers' own criteria for evaluating technology will be noted for possible incorporation into the package. Recommendations are made to NARS through a steering committee mechanism. In the IITA's benchmark research strategy, results from in-depth studies within benchmark sites and locations can be extrapolated to larger environments that are represented in the subzone or in a given agro-ecological zone.

Criteria for benchmark site selection

Biophysical and socio-economic factors that affect technology adoption and food production within an agro-ecological zone are taken into consideration in the selection of benchmark locations and sites. The main sources of heterogeneity within a given zone or subzone are used to locate target areas. For example, socio-economic factors such as access to market and land use intensity are important determinants for inclusion in a framework to distinguish population-driven systems from market-driven systems. Major type of soils and pests may be used in the process of site selection if their heterogeneity can be highlighted. The general principle is that selection of criteria should be followed by a priority setting exercise that is based on research needs and realistically achievable goals. Following prioritization, sites should be identified based on where interventions can have a comparative advantage. A meaningful eco-regional approach to crop level study should therefore be based on careful identification of variability and their spatial distribution within a given agro-ecological zone. This information will be useful in extrapolating results obtained in a benchmark site to other sites within the zone.

Site selection process: example from Nigeria

The first benchmark sites in the MSP were selected in 1993 in the most northern part of the NGS of Nigeria, using available information from detailed characterization studies. Socio-economic criteria were applied after the choice of the sub-zone. Once the distinction was made between population-driven and market-driven areas, and the different needs for appropriate technologies were known, it was decided to focus first on a population-driven area. The rationale for this prioritization was that farmers in population-driven areas rely mainly on natural resources to develop sustainable production systems, have poor to fair market access, little or no use of improved technologies, and little or no use of external inputs. Farmers in market-driven areas have many more technology options that are often based on external inputs.

In selecting a second benchmark site in the NGS, the MSP included another selection criterion, namely the presence of inland valleys cultivable in the rainy and dry seasons. A Landsat satellite image complemented by ground-truthing located inland valleys of high potential utilization during the dry season. The suitability class of the valleys chosen coincided with the land use intensity noted during an earlier socio-economic characterization that had identified the same sites as falling within a market-driven area. This example of benchmark site selection in NGS in Nigeria demonstrates the type of flexibility we consider necessary in both the definitions of criteria for and the process of selecting benchmark sites.

Operational strategies

IITA scientists are applying this benchmark strategy in their eco-regional approach to suitable food production. At the crop level, we are using the benchmark research approach to carry out resource management research at the process level in soil and vegetation management in both the humid and subhumid zones. For example, in the MSZ, two of the research themes that we are targeting are vegetation management and soil management. In vegetation management research, we focus on the integration of legumes into farming systems of subzones of the MSZ. We have already developed a decision-support system called LEXSYS (Legume Expert System) as part of the tools for integrating legumes into farming systems of the MSZ.

Our research and that of others working in the savannas has identified the important role that plant residues play in soil fertility maintenance and weed suppression. Therefore, we have undertaken systems level studies on fallow management as part of sustainable crop production in the savannas. We have had a long history of research on soil organic matter maintenance and biological nitrogen fixation. As we expand our data base at benchmark sites on determinants of sustainability, technology development and testing, a logical next step will be to extrapolate benchmark-tested technologies to similar environments in other sites within a given eco-region, using geographical information tools that are available to us.

Conclusions

Eco-regional approaches at the crop level should reflect the eco-regional concept with the same objective of achieving sustainable improvement in agricultural productivity. Activities at the crop level should employ unique site selection criteria that will make extrapolation of results possible with a good degree of success. The success of benchmark approaches depends on several factors such as the representativity of the selected sites for major farming areas, selection criteria, how representative the selected factors and sites are in relation to the larger environment of the given agro-ecosystem. Both socio-economic and biophysical factors need to be included in characterization so as to be properly monitored in micro-level characterization at the benchmark level. Involvement of NARS partners at the priority setting stage is necessary for full implementation of the benchmark approach.

Acronyms

CS	Coastal Savanna
DS	Derived Savanna
IITA	International Institute for Tropical Agriculture
MSP	Moist Savanna Program
MSZ	Moist Savanna Zone
NARS	National Agricultural Research System
NGS	Northern Guinea Savanna
SGS	Southern Guinea Savanna

References

Dobermann A, George T (1994) Field-scale soil fertility variability in acid tropical soils. Pages 610–627 in Transactions of the 15th World Congress of Soil Science. Vol. 5a. Symposium ID–11. Soil acidity and its ammelioration in the tropics.

Discussions

General discussions

Four discussion groups met on four consecutive days, resulting in 16 different dicussion sessions, each session having different chairpersons and rapporteurs. The discussions focussed on four different aspects: (i) socio-economic issues, (ii) agro-ecological issues, (iii) institutional issues, and (iv) identification of clients of eco-regional research. The 16 individual reports were summarized by M. Dingkuhn and P.G. Pardey.

What is an 'eco-region'?

The term 'eco-region' was discussed at length despite previous efforts by speakers to define it. While participants in this symposium came to no consensus on what constitutes an 'eco-regional approach', there seemed fairly widespread agreement that the eco-region concept as currently conceived by TAC (1993) is found wanting in some important respects. In its present form TAC's eco-regions comprise an overlay of selected agro-climatic zones adjusted to coincide with spatial (national or regional) boundaries. According to Duiker's and Goldworthy's (1994) interpretation, the term evolved from the need for greater emphasis in international agricultural research on natural resource management while establishing spatial boundaries for institutional mandates and activities. The resulting eco-regions are broad aggregates of geographical space that bear little or no spatial correspondence to many of the production and policy problems that constrain agricultural production. Thus their use as a research planning and -evaluation device is severely compromised.

A, perhaps, cynical view could be that the CG's eco-regions represent little more than 'window dressing' designed to appease a shift in donor preferences in a post-UNCED world and thereby make it possible to capture some of the 'green' funding that may otherwise bypass agriculture. A more charitable view is that an emerging – but still poorly quantified and understood – awareness of the environmental consequences of agriculture, coupled with the pressure to further intensify agricultural production was the primary impetus behind these developments. These concerns over the state of the natural resources that sustain agricultural production coincided with accelerating improvements in the availability of georeferenced data and the computer

J. Bouma et al. (eds.), Eco-regional approaches for sustainable land use and food production, 489–494.

hardware and software required to cost-effectively process and analyse it. In reality all these factors played a part in bringing eco-regional approaches to the fore within the CG.

Agro-ecological and socio-economic aggregation

Several issues were raised repeatedly at this symposium. The first and fundamental point is how to meaningfully merge biophysical phenomena with economic behaviour. The objects of ecology-oriented agricultural research are generally complex systems, and eco-regionality in the sense it was understood in this symposium forces researchers to consider biophysical, socio-economic, and policy-related system properties simultaneously as they target 'their' system. A major sticking point is that agricultural technologies, and the research that gives rise to them, are often best conceived in an agro-ecological domain, whereas economic behaviour and policy interventions have impact that can span agro-ecological zones and operate on different scales. Thus the economist's notion of a market, which forms the basis for much policy modelling and analysis, rarely fits neatly into an agro-ecological domain that makes sense from a biophysical point of view.

The common way of dealing with this problem is to aggregate markets or agro-ecological zones until a tractable correspondence has been achieved. But the process of aggregation raises many of its own conceptual as well as practical difficulties. For biophysical models the criteria used and data available to classify agro-ecological domains varies with the scale being studied and the production processes being modeled. And the model details themselves are sensitive to scale; modelling plant growth and performance at the level of an individual plant is different from modelling the same processes for a crop or plant community.

Analagous issues of aggregation and scale apply to economic models. Many of these issues revolve around *ceteris paribus* assumptions, i.e. what variables are appropriately treated as exogenous to the model. While it may make sense to treat input and output prices as constants when modelling farm household behaviour, this may not be so when modelling regional or national markets. Similarly, the multi-attribute objective functions specified for household models may not be, and often are not, appropriate for modelling more aggregate economic behaviour. So simply scaling up the modelling strategies used for studying biophysical and economic phenomena at the micro level (or even the results coming from such analyses) to address issues at a macro level is usually not a meaningful option. The modelling strategy employed must be sensitive to the scale of analysis in addition to the purpose of the model, data availability, and so on.

The purpose of eco-regional approaches

'Eco-region' thus being a pragmatic term with somewhat unclear scientific substance, the term's appropriateness for guiding research was controversial, particularly as its creation seemed to be partially motivated by the more plentiful resources and public support for ecological than for productivity-oriented research. Despite this opportunistic element, however, the final consensus was that eco-regional (or at least agro-ecological) approaches are appropriate and timely, and are bound to play a continuing and indeed probably growing role in planning, funding, and executing research within national and international agencies, because they provide a strong incentive to integrate (i) divergent research goals (productivity vs. improved management of natural resources), (ii) research disciplines; and (iii) institutions (IARCs, NARS, policy-makers, development agencies). The desire to engage in a more transparent and replicable research priority setting process than was hitherto the case was a primary motivation for making eco-regional approaches more explicit in the CGIAR. While a worthy goal in principle, it is clear there are some serious difficulties with the current CG eco-regional approaches in practice. But real progress is possible. With limited resources, the obvious suggestion is to target the effort to where the likely payoff is highest.

Related to the questions of scale and modelling strategy are issues concerning the purpose of and clients for eco-regional analyses. At one level there is a desire to understand the causal nature of fundamental biophysical phenomena (say, down to the level of cell chemistry, or regarding plant morphology or plant interaction effects, for example) or the economic constraints and technological options constraining farm household behaviour. But if the intent is to study more aggregate behaviour to allow policy choices to be made, an altogether different level of detail, modelling methods and timeframe, and presentation of results is often warranted.

While introducing an eco-regional approach into studies of biophysical and economic behaviour can be useful, it is not universally so. A good deal of the basic sciences that underpin biophysical models are derived from site specific effects while many policy questions are successfully tackled in an economic framework that does not explicitly take an eco-regional perspective. Nonetheless, those production practices that are strongly conditioned by aspects of climate, soil, terrain, and so forth are the best candidates to be studied by an eco-regional approach. One obvious area of concern is the environmental conseqences of agricultural production. We are still struggling to understand the precise nature, extent, and (agricultural) causes of agricultural degradation processes and it is here that taking explicit account of the agro-ecological basis of agricultural production can play an important role. Precisely how to define and deploy eco-regional approaches to deal meaningfully with these issues is a question that is yet to be resolved but one that is worth continuing to tackle (Wood and Pardey 1995).

Implementation of eco-regional approaches

Workshop participants were concerned with a number of constraints to the implementation of eco-regional agricultural research. They can be grouped into (a) scientific-technical concerns, (b) institutional concerns, and (c) problems in project conceptualization.

(a) Major scientific-technical challenges

New, integrated research approaches require new tools. Systems research tools (e.g. simulation models) are supposed to link system properties and components with each other, thereby integrating research disciplines. The still embryonic role of systems research, however, is indicated by the frequent lack of functioning interfaces to field research (the most important source of detailed information) and potential clients (e.g. policy-makers). In many cases, available information is too scarce, poorly structured or costly to permit meaningful aggregation or disaggregation. The following major challenges were noted:

- the development of long-term scenarios requires long-term data sets from the field. In many cases, however, funding modi permit only short-term, quick-return activities;
- lack of data and data infrastructure, e.g. for spatial analyses. Neither research organizations nor the institutions specialized on data pooling frequently are prepared for the quantity and degree of availability of data required in eco-regional systems research;
- technical problems in modelling the spatial dimension of system dynamics, particularly for diverse systems having strong horizontal interactions (e.g. hydrology of watersheds);
- our inability to understand and predict seemingly erratic behavior of some system components, e.g. some highly mobile pests;
- the problem of quantifying some parameters, e.g. the value of non-marketable resources;
- the problem of how to measure the impact of eco-regional research, particularly where not productivity but instead, the prevention of degradation is the main objective.

(b) Major institutional challenges

The inter- and intra-institutional implications of eco-regional research approaches have led to the creation of numerous regional networks (frequently serving spatial 'horizontal' coverage) and consortia (which provide an additional 'vertical' coverage across complementary types of institutions). The objective is in all cases the mobilization of inter-institutional complementarities (e.g. the Task Forces of WARDA and NARSs in West-Africa) and sometimes, the systematic creation of new complementarities through institution-building (e.g. the SARP-project in Asia). The following major challenges were identified in the institutional context:

- sharing of knowledge (data, experience) and research tools is frequently a limiting factor in cross-institutional research. This is not always due to poor relationships among administrations, but in many cases also to inter-institutional gaps in terms of resources and training. Frequently, appropriate mechanisms for the sharing of databases are not in place;
- institutions have different mandates. For example, NARS have a national mandate. For the eco-regional (cross-national) research involving NARS scientists, particular mechanisms and resources must be developed;
- inter-institutional collaboration must not be limited to the research sector. A direct interaction with the ultimate clients (e.g. policy makers or extension systems) is sometimes warranted at the planning and diffusion stages, but not necessarily in the process of research. Consortia, for example, can provide for such links;
- inter-institutional teamwork has high operational and administrative cost.

(c) Critical elements in project conceptualization

There was a general feeling that the conceptualization and planning phases of eco-regional research projects need more attention. Since eco-regional research is of applied nature, goals and objectives must explicitly drive the technical and institutional approach. The following principles were emphasized:

- the demand side for research must not only be respected, but must drive project design. For example, biophysical studies should be relevant to their respective socio-economic context;
- systems research activities should be appropriately classified as exploratory (mainly providing feedback to research), analytical-descriptive (to provide structured knowledge to users beyond the research context), or predictive (aid to decision-making at whatever level the study aims). Projects may serve more than one of these purposes, however. But the presentation and transfer of research outputs is usually more effective if the activity's nature is clearly defined at the start;
- there is a common tendency to define system boundaries by convenience (availability of data, tools or disciplinary expertise) rather than objectives;
- there generally is an optimum scale for the study of any specific problem. However, results should regularly be checked against the next higher and

lower scales. It was even recommended to systematically test all results of systems research at the smallest socio-economic level, the farm or household, because only at this level biophysical and economic phenomena interact directly.

A major conclusion was that methodology development on eco-regional research should take place in close contact with potential clients (to ensure output applicability) and the source of detailed information (field/farm level; to ensure output validity). For this, as some participants felt, the CGIAR-centers do not (yet) have the appropriate mechanisms in place. Full implementation of eco-regional research requires mechanisms for (i) more inter-institutional and inter-disciplinary cohesion in the context of collaboration, and (ii) appropriate incentives and recognition for scientists who get involved. So serious thought needs to be given on how best to proceed.

References

Duiker S W, Goldsworthy P R (1994) A summary of selected characteristics of some current and planned eco-regional initiatives of the CGIAR. Discussion paper, International Service for National Agricultural Research (ISNAR), The Hague, The Netherlands.

TAC (1993) The eco-regional approach to research in the CGIAR. Report of the TAC/Center Directors Working Group, March 1993. Technical Advisory Committee, CGIAR, Rome, Italy.

Wood S, Pardey P G (1994) Methods for and Limitations of Agro-ecological Analysis. Paper presented to the IFPRI Eco-regional/2020 Vision Workshop, Airlie House Conference Center, Virginia, USA, November 1994 (revised March 1995).

Closing panel discussion

The closing panel consisted of Dr. Rudy Rabbinge (chairman), Dr. Michael Dingkuhn (representing agronomists' viewpoints), and Dr. Phil Pardey (representing economists' viewpoints). The conclusions were summarized by F.W.T. Penning de Vries and A. Kuyvenhoven.

Conclusions

1. 'Eco-regional approaches', as presented in this meeting, are a new tool. However, they are as yet predominantly used for traditional subjects (arable cropping) and still neglect livestock, fisheries, and in particular the environment.
   An eco-regional emphasis should succeed the traditional commodity orientation of the CGIAR-institutes.
2. Eco-regional research should be broad-based, multidisciplinary and integrative, but also specific and targeted to regional issues and needs of the population. The administrative, physical and socio-economic boundaries of a region all need to be specified, as they are related to the objective of the research.
   Such eco-regional approaches require (i) adapted research organisations, (ii) peer recognition, and (iii) explicit attention to scientific quality.
   Untargeted eco-regional approaches may yield top heavy, unwieldy instruments, and may go the way of Farming Systems Research. An eco-regional approach should start at its own level of complexity. Starting on top of traditional commodity research is unbalanced.
3. The nature of eco-regional study should be determined from the start: is it explorative, predictive or descriptive? The nature of the objective and the type of study should be expressed in the methods employed, the disciplines involved and the data required.
4. There is a need for eco-regional tools for decision makers. Methodological development should take place in close contact with end users, including governments, input suppliers, banks, and other policy making institutions. It was noted that this meeting lacked the direct input of 'stakeholders'.
   Use of these tools, including simulation, geographic information systems, multiple goal linear programming, and policy analysis should be promoted by CGIAR-institutes.

J. Bouma et al. (eds.), Eco-regional approaches for sustainable land use and food production, 495–496.

Increasingly, regions are being given economic and institutional tools for their own rural development; we should provide operational tools to assist them in developing options and decision making. A case should be made to design socially acceptable and cost-effective policy measures to achieve the desired infrastructure. Development of such tools requires tripartite co-operation (NARCs, IARCs and AROs).

5. 'Eco-regional tools' presented at this meeting need improvement with respect to spatial and temporal heterogeneity, animal production, natural vegetation, the environment, and to institutional issues. Stronger ties between physical and socio-economic sciences are required.
   Technical issues (handbooks, documentation, software quality control, commercialisation), matters of substance (see (5)) and of efficiency (tools for decision makers, not vice versa) of ecoregional tools were briefly discussed. Advanced institutes are urged to take a lead in solving these. Shortage of regional data (physical, socio-economic) was generally acknowledged. This leads to requests for more data acquisition, but also to invitations to scientists to extract the maximum information out of existing data. A major task for CG-institutes and NARCs was identified here.
6. The accessibility of tools for eco-regional approaches to NARCs and CGIAR-institutes should be improved by courses and interactive training. The CGIAR should promote this; groups like those present in 'Wageningen', ICASA, are invited to be involved.
7. Processes underlying 'sustainability' and 'rural development' are not scale neutral. Eco-regional research should address them at the appropriate level.

Acronyms

AEZ	Agro-Ecological Zone
AB-DLO	DLO-Instituut voor Agrobiologisch en Bodemvruchtbaarheidsonderzoek (DLO-Research Institute for Agrobiology and Soil Fertility) (former CABO-DLO)
CAP	Common Agricultural Policy
CATIE	Centro Agronómico Tropical de Investigación y Enseñanza (Center for Research and Education in Tropical Agriculture)
CGIAR	Consultative Group on International Agricultural Research
CIAT	Centro Internacional de Agricultura Tropical (International Center for Tropical Agriculture)
CIFOR	Center for International Forestry Research
CIMMYT	Centro Internacional de Mejoramiento de Maiz y Trigo (International Center for the Improvement of Maize and Wheat)
CIP	Centro Internacional de la Papa (International Potato Center)
CIRAD	Centre de Coopération Internationale en Recherche Agronomigue pour le Développement (Centre for International Cooperation and Agricultural Development Research)
CONDESAN	Consorcio para el Desarollo Sostenible de la Ecorregion Andina (Consortium for the Sustainable Development of the Andean Eco-Region)
DGIS	Directoraat Generaal voor Internationale Samenwerking (Directorate General for International Cooperation)
DLO	Dienst Landbouwkundig Onderzoek (Agricultural Research Department)
DLV	Duurzaam Landgebruik en Voedselvoorziening (Sustainable Land Use and Food Security)
DSS	Decision Support System
EAP	Eco-regional Action Plan
EC	European Community
ERA	Eco-Regional Approach
EU	European Union
FAO	Food and Agriculture Organization of the United Nations
FSA	Farming Systems Analysis
FSR	Farming Systems Research
FSR&D	Farming Systems Research and Development
GATT	General Agreement on Tariffs and Trade
GIS	Geographic Information System
GPS	Global Positioning System
$G \times E$	Genotype by Environment
IARC	International Agricultural Research Center
IBSNAT	International Benchmark Systems Network for Agrotechnology Transfer
ICARDA	International Center for Agricultural Research in the Dry Areas
ICASA	International Consortium for Agricultural Systems Applications
ICLARM	International Center for Living Aquatic Resources Management
ICRAF	International Centre for Research in Agroforestry
ICRISAT	International Crops Research Institute for the Semi-Arid Tropics

J. Bouma et al. (eds.), Eco-regional approaches for sustainable land use and food production, 497–498.

IER	Institut dÉconomie Rurale (Rural Economy Institute)
IFPRI	International Food Policy Research Institute
IIMI	International Irrigation Management Institute
IITA	International Institute of Tropical Agriculture
ILCA	International Livestock Centre for Africa
ILRAD	International Laboratory for Research on Animal Diseases
ILRI	International Livestock Research Institute
IMGP	Interactive Multiple Goal Programming
IMGLP	Interactive Multiple Goal Linear Programming
IPM	Integrated Pest Management
INIBAP	International Network for the Improvement of Banana and Plantain
IPGRI	International Plant Genetic Resources Institute
IRM	Integrated Resource Management
IRRI	International Rice Research Institute
ISC	ICRISAT Sahelian Centre
ISNAR	International Service for National Agricultural Research
ISRIC	International Soil Reference and Information Centre
KIT	Koninklijk Instituut voor de Tropen (Royal Tropical Instituut)
LEFSA	Land Evaluation and Farming Systems Analysis
LEISA	Low External Input Sustainable Agriculture
LEU	Land Evaluation Unit
LNV	Ministerie van Landbouw, Natuur en Visserij (Netherlands Ministry of Agriculture, Nature Management and Fisheries)
LP	Linear Programming
LUST	Land Use Systems with a specified Technology
MGLP	Multiple Goal Linear Programming
MPU	Manpower Unit
NARC	National Agricultural Research Center
NARS	National Agricultural Research System
NGO	Non-Governmental Organization
NPK	Nitrogen-Phosphorous-Potassium
NRM	Natural Resource Management
OR	Operational Research
R&D	Research & Development
SARP	Simulation and Systems Analysis for Rice Production
TAC	Technical Advisory Committee (of the CGIAR)
TLU	Tropical Livestock Unit
UN	United Nations
UNCED	United Nations Conference on the Environment and Development
WARDA	West Africa Rice Development Association
WAU	Wageningen Agricultural University
WRR	Wetenschappelijke Raad voor het Regeringsbeleid (Netherlands Scientific Council for Government Policy)

List of participants

Aggarwal, P.K. Water Technology Center, Indian Agricultural Research Institute, (IARI-WTC). 110–012, New Delhi, India. Tel: (+91) 11 5788682. Fax: (+91) 11 5752006.

Akobundu, I.O. International Institute of Tropical Agriculture (IITA). Oyo Road, PMB 5320, Ibadan, Nigeria. Tel: (+234) 22 400300. Fax: (+874) 1772276.

Atta-Krah, K. International Centre for Research in Agroforestry (ICRAF). United Nations Avenue, P.O. Box 30677, Nairobi, Kenya. Tel: (+254) 2 521450. Fax: (+254) 2 520023.

Baidu-Forson, J. International Crops Research Institute for the Semi-Arid Tropics (ICRISAT), ILCA/ICRISAT Sahelian Center. BP 12404, Niamey, Niger. Tel: (+227) 722529. Fax: (+227) 734329.

Becker, B. Faculty of Agricultural and Environmental Sciences, University of Kassel. P.O. Box 1252, 37213 Witzenhausen, Germany. Tel: (+49) 5542 981280. Fax: (+49) 5542 981309.

Van den Berg, R.D. Directorate General for International Cooperation (DGIS), Netherlands Ministry of Foreign Affairs. P.O. Box 20061, 2500 EB The Hague. The Netherlands. Tel: (+31) 70 3486486. Fax: (+31) 70 3484848.

Ten Berge, H.F.M. DLO-Research Institute for Agrobiology and Soil Fertility (AB-DLO). Centre the Born, P.O. Box 14, 6700 AA Wageningen, The Netherlands. Tel: (+31) 8370 75700. Fax: (+31) 8370 23110.

Bhatia, R. International Irrigation Management Institute (IIMI). P.O. Box 2075, Colombo, Sri Lanka. Tel: (+94) 1 867404. Fax: (+94) 1 866854.

Bouma, J. Department of Soil Science and Geology, Wageningen Agricultural University (WAU). P.O. Box 37, 6700 AA Wageningen, The Netherlands. Tel: (+31) 8370 84410. Fax: (+31) 8370 82419.

Bouman, B.A.M. DLO-Research Institute for Agrobiology and Soil Fertility (AB-DLO). Centre the Born, P.O. Box 14, 6700 AA Wageningen, The Netherlands. Tel: (+31) 8370 75700. Fax: (+31) 8370 23110.

Breman, H. DLO-Research Institute for Agrobiology and Soil Fertility (AB-DLO). Centre the Born, P.O. Box 14, 6700 AA, Wageningen, The Netherlands. Tel: (+31) 8370 75700. Fax: (+31) 8370 23110.

Dicko, M.S. BP 9032, Bamako, Mali. Tel: (+223) 231594. Fax: (+223) 231932.

J. Bouma et al. (eds.), Eco-regional approaches for sustainable land use and food production, 499–502.

Dingkuhn, M. West Africa Rice Development Association (WARDA). 01 BP 2551, Bouaké, Ivoty Coast. Tel: (+225) 634514. Fax: (+225) 634714.

Eyzaguirre, P.B. International Plant Genetic Resources Institute (IPGRI). c/o Food and Agriculture Organization of the U.N. Via delle Sette Chiese 142, 00145 Rome, Italy. Tel: (+39) 6 518921. Fax: (+39) 6 5750309.

Fresco, L.O. Department of Agronomy, Wageningen Agricultural University (WAU). P.O. Box 341, 6700 AH Wageningen, The Netherlands. Tel: (+31) 8370 83040. Fax: (+31) 8370 84575.

Gillison, A. Center for International Foresty Research (CIFOR). P.O. Box 6596, JKPWB Jakarta 10065, Indonesia. Tel: (+62) 251 343652. Fax: (+62) 252 326433.

Goldsworthy, P.R. International Service for National Agricultural Research (ISNAR). P.O. Box 93375, 2509 AJ The Hague. The Netherlands. Tel: (+31) 70 3496100. Fax: (+31) 381 9677.

Haverkort, A.J. DLO-Research Institute for Agrobiology and Soil Fertility (AB-DLO). Centre the Born, P.O. Box 14, 6700 AA, Wageningen, The Netherlands. Tel: (+31) 8370 75700. Fax: (+31) 8370 23110.

Hobbs, P. Centro Internacional de Mejoramiento de Maiz y Trigo (CIMMYT), office Nepal. P.O. Box 5186, Kathmandu, Nepal. Tel: (+977) 1 417791. Fax: (+977) 1 414184.

Hodgkin, T. International Plant Genetic Resources Institute (IPGRI). c/o Food and Agriculture Organization of the U.N. Via delle Sette Chiese 142, 00145 Rome, Italy. Tel: (+39) 6 518921. Fax: (+39) 6 5750309.

Jansen, H.G.P. Programa Zona Atlántica. Apartado 224, 7210 Quápiles, Costa Rica. Tel: (+506) 7106595. Fax: (+506) 7102327.

Janssen, W.G. International Service for National Agricultural Research (ISNAR). P.O. Box 93375, 2509 AJ, The Hague. The Netherlands. Tel: (+31) 70 3496100. Fax: (+31) 381 9677.

Johansen, C. International Crops Research Institute for the Semi-Arid Tropics (ICRISAT). Patancheru P.O., Andra Pradesh 502 324, India. Tel: (+91) 40 596161. Fax: (+91) 40 241239.

Jones, M. International Center for Agricultural Research in the Dry Areas (ICARDA). P.O. Box 5466, Aleppo, Syria. Tel: (+963) 21 225012. Fax: (+963) 21 225105.

Jones, P. Centro Internacional de Agricultura Tropical (CIAT), Apartado Aereo 6713, Cali, Columbia. Tel: (+57) 2 6675050. Fax: (+57) 2 6647243.

Von Kaufmann, R. International Livestock Centre for Africa (ILCA). P.O. Box 5689, Addis Ababa, Ethiopia. Tel: (+251) 1 613215. Fax: (+251) 1 611892.

Van Keulen, H. DLO-Research Institute for Agrobiology and Soil Fertility (AB-DLO). Centre the Born, P.O. Box 14, 6700 AA, Wageningen, The Netherlands. Tel: (+31) 8370 75700. Fax: (+31) 8370 23110.

Knapp, E.B. Centro Internacional de Agricultura Tropical (CIAT), Apartado Aereo 6713, Cali, Columbia. Tel: (+57) 2 6675050. Fax: (+57) 2 6647243.

Kuyvenhoven, A. Department of Development Economics, Wageningen Agricultural University (WAU). De Leeuwenborch, P.O.Box 8130, 6700 EW Wageningen. The Netherlands. Tel: (+31) 8370 84360. Fax: (+31) 8370 84037.

Van Latesteijn, H.C. Netherlands Scientific Council for Government Policy (WRR). P.O. Box 20004, 2500 EA The Hague. The Netherlands. Tel: (+31) 70 3564472. Fax: (+31) 70 3562695.

Luyten, J.C. DLO-Research Institute for Agrobiology and Soil Fertility (AB-DLO). Centre the Born, P.O. Box 14, 6700 AA, Wageningen, The Netherlands. Tel: (+31) 8370 75700. Fax: (+31) 8370 23110.

Manichon, H. Centre de Coopération Internationale en Recherche Agronomigue pour le Développement (CIRAD). 2477 avenue du Val de Montferrand, BP 5035, 34032 Montpellier, Cédex 1, France. Tel: (+33) 67615779. Fax (+33) 67615512.

Mateo, N. International Network for the Improvement of Banana and Plantain (INIBAP). Parc Scientifique Agropolis, 34397 Montpellier Cédex 5, France. Tel: (+33) 67611302. Fax: (+33) 67610334.

Norton, R.D. Proyecto PROMESA, El Salvador/Panama. Miami Express #N0004, P.O. Box 52–7948, Miami, FL 33152–7948, USA.

Pardey, P.G. International Food Policy Research Institute (IFPRI). 1200 17th street N.W. Washington, DC 20036–3006, USA. Tel: (+1) 202 8625600. Fax: (+1) 202 4674439.

Penning de Vries, F.W.T. DLO-Research Institute for Agrobiology and Soil Fertility (AB-DLO). Centre the Born, P.O. Box 14, 6700 AA, Wageningen, The Netherlands. Tel: (+31) 8370 75700. Fax: (+31) 8370 23110.

Pinstrup-Andersen, P. International Food Policy Research Institute (IFPRI). 1200 17th street N.W. Washington, DC 20036–3006, USA. Tel: (+1) 202 8625600. Fax: (+1) 202 4674439.

Prein, M. International Center for Living Aquatic Resources Management (ICLARM). MC P.O. Box 2631, Makati Central Post Office, 0718 Makati, Metro Manila, Philippines. Tel: (+63) 2 8180466. Fax: (+63) 2 8163183.

Pullin, R. International Center for Living Aquatic Resources Management (ICLARM). MC P.O. Box 2631, Makati Central Post Office, 0718 Makati, Metro Manila, Philippines. Tel: (+63) 2 8180466. Fax: (+63) 2 8163183.

Quiroz, R.A. Highland Farming Systems Project, IBTA-IDRC. Plaza Espana esq. Mendez Arcos 710, P.O. Box 5783, La Paz, Bolivia. Tel: (+591) 2 358680. Fax: (+591) 2 358680.

Rabbinge, R. Department of Theoretical Production Ecology, Wageningen Agricultural University (WAU). P.O. Box 430, 6700 AK Wageningen, The Netherlands. Tel: (+31) 8370 82141. Fax: (+31) 8370 84892.

Schipper, R.A. Department of Development Economics, Wageningen Agricultural University (WAU). De Leeuwenborch, P.O. Box 8130, 6700 EW, Wageningen. The Netherlands. Tel: (+31) 8370 84360. Fax: (+31) 8370 84037.

Teng, P.S. Division of Plant Pathology, International Rice Research Institute (IRRI). P.O. Box 933, 1099 Manila, Philippines. Tel: (+63) 2 8181926. Fax: (+63) 2 8182087.

Thornton, P.K. Research and Development Division, International Fertilizer Development Center (IFDC). P.O. Box 2040, Muscle Shoals, Alabama 35662, USA. Tel: (+1) 205 3816600. Fax: (+1) 205 3817408.

Tims, W. Centre for World Food Studies (SOW-VU). De Boeleaan 1105, 1081 HV, Amsterdam, The Netherlands. Tel: (+31) 20 5484622. Fax: (+31) 20 6425525.

Weise, S.F. International Institute of Tropical Agriculture (IITA), office Cameroon. Humid Forest Station. B.P. 2067, Yaounde (Messa), Cameroon. Tel: (+237) 237434. Fax: (+237) 237437.

Zadoks, J.C. Department of Phytopathology, Wageningen Agricultural University (WAU). P.O. Box 8025, 6700 EE, Wageningen. The Netherlands. Tel: (+31) 8370 83410. Fax: (+31) 8370 83412.

Zandstra, H.G. Centro Internacional de la Papa (CIP), Apartado 5969, Lima, Peru. Tel: (+51) 14 366920. Fax: (+51) 14 351570.

Subject index

Systems Approaches for Sustainable Agricultural Development

1. Th. Alberda, H. van Keulen, N.G. Seligman and C.T. de Wit (eds.): *Food from Dry Lands.* An Integrated Approach to Planning of Agricultural Development. 1992 ISBN 0-7923-1877-3
2. F.W.T. Penning de Vries, P.S. Teng and K. Metselaar (eds.): *Systems Approaches for Agricultural Development.* Proceedings of the International Symposium (Bangkok, Thailand, December 1991). 1993 ISBN 0-7923-1880-3; Pb 0-7923-1881-1
3. P. Goldsworthy and F.W.T. Penning de Vries (eds.): *Opportunities, Use, and Transfer of Systems Research Methods in Agriculture to Developing Countries.* Proceedings of a International Workshop (The Hague, November 1993). 1994 ISBN 0-7923-3205-9
4. J. Bouma, A. Kuyvenhoven, B.A.M. Bouman, J.C. Luyten and H.G. Zandstra (eds.): *Eco-regional Approaches for Sustainable Land Use and Food Production.* Proceedings of a Symposium (The Hague, December 1994). 1995 ISBN 0-7923-3608-9

KLUWER ACADEMIC PUBLISHERS – DORDRECHT / BOSTON / LONDON

GPSR Compliance

The European Union's (EU) General Product Safety Regulation (GPSR) is a set of rules that requires consumer products to be safe and our obligations to ensure this.

If you have any concerns about our products, you can contact us on ProductSafety@springernature.com

In case Publisher is established outside the EU, the EU authorized representative is:

Springer Nature Customer Service Center GmbH
Europaplatz 3
69115 Heidelberg, Germany

Batch number: 10816188

Printed by Printforce, the Netherlands